Ado Jorio, Riichiro Saito,
Gene Dresselhaus and
Mildred S. Dresselhaus

Raman Spectroscopy
in Graphene Related Systems

The Authors

Ado Jorio *is a Professor in the Physics Dept. of the Federal University of Minas Gerais, Brazil, where he also earned his PhD, in 1999. His Post-doctoral research was done at MIT, USA, where his collaboration with the Dresselhaus group and with Professor Saito started. He has authored and co-authored several book chapters and books on carbon science and has been active in science policy in Latin America.*

Mildred Dresselhaus *received her Ph.D. at the University of Chicago in 1958 and started research on carbon science in 1960 with Gene Dresselhaus while working at the MIT Lincoln Laboratory. She has been an MIT professor since 1967, and started working with Riichiro Saito in 1991 and with Ado Jorio since 2000.*

Riichiro Saito *received PhD degree from the University of Tokyo in 1985. After being a Research Associate at the University of Tokyo in 1985 and an Associate Professor at the University of Electro-Communication in Tokyo in 1990, he became Professor at Tohoku University in Sendai since 2003.*

Gene F. Dresselhaus *received his PhD degree from the University on California, Berkeley under the supervision of Charles Kittel. He has actively worked on a variety of problems in condensed matter physics. He has taught courses in condensed matter physics at the University of Chicago and at Cornell University. He currently holds a Research appointment at MIT and jointly leads a research group at the MIT Center for Materials Science and Engineering which studies graphite intercalation compounds, fullerenes, graphene, and carbon nanotubes. He has Co-authored or Co-edited six books on Carbon Science.*

Ado Jorio, Riichiro Saito, Gene Dresselhaus, and Mildred S. Dresselhaus

Raman Spectroscopy in Graphene Related Systems

WILEY-VCH

WILEY-VCH Verlag GmbH & Co. KGaA

The Authors

Prof. Ado Jorio
Departamento de Física
Universidade Federal de Minas Gerais
Av. Antonio Carlos, 6627, CP 702
30.123-970 Belo Horizonte, MG
Brazil

Prof. Riichiro Saito
Tohoku University
Dept. of Physics
6-3 Aoba, Aramaki, Aoba-ku
Sendai 980-8578
Japan

Prof. Mildred S. Dresselhaus
Dr. Gene Dresselhaus
MIT
Room 13-3005
77 Massachusetts Ave.
Cambridge, MA 02139-4307
USA

Library of Congress Card No.: applied for

British Library Cataloguing-in-Publication Data:
A catalogue record for this book is available from the British Library.

Bibliographic information published by the Deutsche Nationalbibliothek
The Deutsche Nationalbibliothek lists this publication in the Deutsche Nationalbibliografie; detailed bibliographic data are available on the Internet at http://dnb.d-nb.de.

Typesetting le-tex publishing services GmbH, Leipzig
Printing and Binding Fabulous Printers Pte Ltd, Singapore
Cover Design Formgeber, Eppelheim

Printed in Singapore
Printed on acid-free paper

ISBN 978-3-527-40811-5

A. J. and R. S. dedicate this book to the 80th birthday of Professor Gene Dresselhaus (born Nov. 7, 1929) and Professor Mildred S. Dresselhaus (born Nov. 11, 1930).

Raman Spectroscopy in Graphene Related Systems. Ado Jorio, Riichiro Saito, Gene Dresselhaus, and Mildred S. Dresselhaus

ISBN: 978-3-527-40811-5

Contents

Raman Spectroscopy in Graphene Related Systems. Ado Jorio, Riichiro Saito, Gene Dresselhaus, and Mildred S. Dresselhaus

ISBN: 978-3-527-40811-5

Preface

Raman spectroscopy is the inelastic scattering of light by matter. Being highly sensitive to the physical and chemical properties of materials, as well as to environmental effects that change these properties, Raman spectroscopy is now evolving into one of the most important tools for nanoscience and nanotechnology. In contrast to usual microscopy-related techniques, the advantages of using light for nanoscience relate both to experimental and fundamental aspects. Experimentally, the techniques are widely available, relatively simple to perform, possible to carry out at room temperature and under ambient pressure, and require relatively simple or no special sample preparation. Fundamentally, optical techniques (normally using infrared and visible wavelengths) are nondestructive and noninvasive because they use the photon, a massless and chargeless particle, as a probe.

For understanding Raman spectroscopy, a combination of experiments and theory is important because some concepts of basic solid state physics are needed for explaining the behavior of the Raman spectra as a function of many experimental parameters, such as light polarization, the energy of the photon, temperature, pressure and changes in the environment. In this book, starting from some known example of physics and chemistry, we will explain how to use the basic concepts of molecular and solid state physics, together with optics to understand Raman scattering. Graphene, nanographite and carbon nanotubes (sp^2 carbons) are selected as the materials to be studied, due to their importance to nanoscience and nanotechnology, and because the Raman technique has been extremely successful in advancing our knowledge about these nanomaterials. It is possible to observe Raman scattering from one single sheet of sp^2-hybridized carbon atoms, the two-dimensional (2D) graphene sheet, as well as from a narrow strip of a graphene sheet rolled-up into a 1 nm diameter cylinder to form the one-dimensional (1D) single-wall carbon nanotube. These observations are possible simply by shining light on the nanostructure focused through a commonly available microscope. This book therefore focuses on the basic concepts of both Raman spectroscopy and sp^2 carbon nanomaterials, together with their interaction. The similarities and differences in the Raman spectra for different sp^2carbon nanomaterials, such as graphene and carbon nanotubes, provide a deep understanding of the Raman scattering capabilities that are emphasized in this book.

Raman Spectroscopy in Graphene Related Systems. Ado Jorio, Riichiro Saito,
Gene Dresselhaus, and Mildred S. Dresselhaus

ISBN: 978-3-527-40811-5

There is a general feeling that Raman spectroscopy is too complicated for a non-specialist. Often, common users of Raman spectroscopy as a characterization tool for their samples only touch the surface of the capabilities of the Raman technique. This book is aimed to be sufficiently pedagogic and also detailed to help the general nanoscience and nanotechnology user of Raman spectroscopy to better utilize their instrumentation to yield more detailed information about their nanostructures than before. Our challenge was writing a book that would build from the most basic concept, the Schrödinger equation for the hydrogen atom, going up to the highest level use and application of Raman spectroscopy to study nanocarbons in general.

The book was initially structured for use in a course for graduate students in the Federal University of Minas Gerais (UFMG), Brazil, and it is organized in two parts. The first part gives the basic concepts of Raman spectroscopy and nanocarbons, addressing why we choose nanocarbons as prototype materials for writing this Raman book. The text is suitable for physicists, chemists, material scientists, and engineers, building a link between their languages, a link that is necessary for the future development of nanoscience. The second part gives a detailed treatment of the Raman spectroscopy of nanocarbons, addressing both fundamental material science and the use of Raman spectroscopy towards material applications. Again nanostructured sp^2-hybridized carbon materials are model systems, both due to the common interest that physicists, chemists, material scientists, and engineers have in these systems and because these systems are pertinent to the length scales where these fields converge. By giving more details, the second part gives examples of the large amount of physics one can learn from studying nanocarbons.

Even though the Raman effect was first observed in the early 1920s, we believe this book is the starting point for lots of new scientific perspectives that the "nano" generation is making possible. We hope the reader will be interested in Raman spectroscopy and will accept the challenges that many researchers are now trying to solve in applying this technique to study nanostructures. Problem sets are included at the end of each chapter, designed to provide a better understanding of the concepts presented in this book and to reinforce the learning process. We appreciate if the readers are willing to solve our problems and send the solutions to the authors to post on the web. The answers by the readers and students using this book can be posted on the following web page: http://flex.phys.tohoku.ac.jp/book10/index.html.

Finally, we strongly acknowledge all students and collaborators who have contributed to the development of this book.

September, 2010

Ado Jorio, Belo Horizonte, MG, Brazil
Riichiro Saito, Sendai, Japan
Gene Dresselhaus and *Mildred S. Dresselhaus*,
Cambridge, MA, USA

Part One Materials Science and Raman Spectroscopy Background

1
The sp^2 Nanocarbons: Prototypes for Nanoscience and Nanotechnology

This chapter presents the reasons why we focus on nanostructured carbon materials as a model materials system for studying Raman spectroscopy and its applications to condensed matter, materials physics and other related science fields. In short, the answer for "why carbon" and "why nano" is the combination of simplicity and richness [1, 2], making possible an unprecedented and accurate exploitation of both the basic fundamentals that link the broad field of condensed matter and materials physics to the applications of Raman spectroscopy, which provides a highly sensitive and versatile probe of the nano-world.

1.1
Definition of sp^2 Nanocarbon Systems

The concept of sp^2 hybridization, where hybridization means the mixing of valence electronic states, is presented here. Carbon has six electrons, two are in $1s$ states, and four are valence electrons, occupying the $2s$ and $2p$ orbitals. The $1s$ orbitals at around $E = -285$ eV are occupied by two electrons and the $1s$ electrons are called core electrons. These core electrons are strongly bound to the nucleus and do not participate in atomic bonding. Thus, they have a small influence on the physical properties of carbon-based materials, and mostly serve as sources for dielectric screening of the outer shell electrons. The second shell $n = 2$ is more flexible. The energy difference between the $2s$ and $2p$ orbitals is less than the energy gain through C–C binding. For this reason, when carbon atoms bind to each other, their $2s$ and $2p$ orbitals can mix with one another in sp^n ($n = 1, 2, 3$) hybridized orbitals. To form the diamond structure, the orbitals for one $2s$ and three $2p$ electrons mix, forming four sp^3 orbitals, binding each carbon atom to four carbon neighbors at the vertices of a regular tetrahedron. In contrast, in the sp^2 configuration, the $2s$ and two $2p$ orbitals mix to form three in-plane covalent bonds (see Figure 1.1). Here, each carbon atom has three nearest neighbors, forming the hexagonal planar network of graphene. Finally, the sp hybridization, mixing the orbitals of only one $2s$ and one $2p$ electron is also possible, and gives rise to linear chains of carbon atoms, the basis for polyene, the filling of the core of certain nanotubes [3], and providing a step in the coalescence of adjacent nanotubes [4].

Raman Spectroscopy in Graphene Related Systems. Ado Jorio, Riichiro Saito, Gene Dresselhaus, and Mildred S. Dresselhaus

ISBN: 978-3-527-40811-5

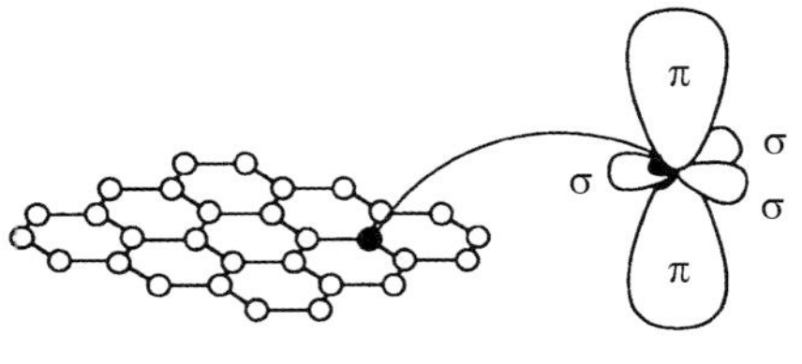

Figure 1.1 The carbon atomic σ and π orbitals in the sp^2 honeycomb lattice [5].

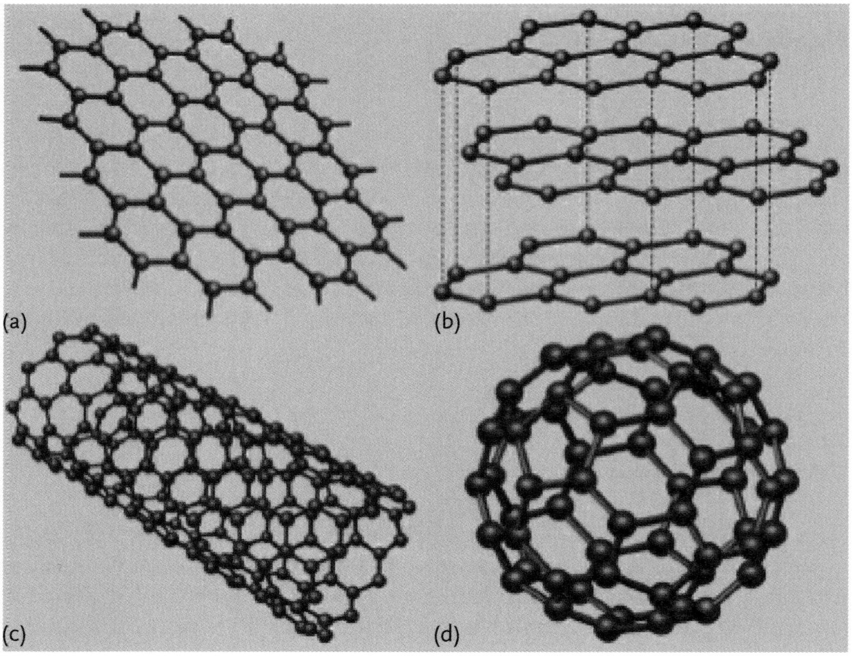

Figure 1.2 Examples of sp^2 carbon materials, including (a) single-layer graphene, (b) triple-layer graphene, (c) a single-wall carbon nanotube, and (d) a C_{60} fullerene, which includes 12 pentagons and 20 hexagons in its structure [7, 8].

Having defined the sp^2 hybridization, we now define nanocarbons. The nanocarbons discussed in this book are structures with sizes between the molecular and the macroscopic. The Technical Committee (TC-229) for nanotechnologies standardization of the International Organization for Standardization (ISO) defines nanotechnology as "the application of scientific knowledge to control and utilize matter at the nanoscale, where size-related properties and phenomena can emerge (the nanoscale is the size range from approximately 1 nm to 100 nm)."

The ideal concept of sp^2 nanocarbons starts with the single graphene sheet (see Figure 1.2a), the planar honeycomb lattice of sp^2 hybridized carbon atoms, which is denoted by 1-LG. Although this system can be large (ideally infinite) in the plane, it is only one atom thick, thus representing a two-dimensional sp^2 nanocarbon.

By stacking two graphene sheets, a so-called bilayer graphene (2-LG) is obtained. Three sheets gives three-layer graphene (3-LG), as shown in Figure 1.2b, and many graphene layers on top of each other yield graphite. A narrow strip of graphene (below 100 nm wide) is called a graphene nanoribbon. Rolling-up this narrow strip of graphene in a seamless way into a cylinder forms what is called a single-wall carbon nanotube (SWNT, see Figure 1.2c). Conceptually nanoribbons and nanotubes can be infinitely long, thus representing one-dimensional systems. Add one-, two-layer concentric cylinders and we get double-, triple-wall carbon nanotubes. Many rolled-up cylinders would make a multi-wall carbon nanotube (MWNT). A piece of graphite with small lateral dimensions (a few hundred nanometers and smaller) is called nanographite, which represents a zero-dimensional system. Finally, the "buckyball" (or fullerene) is among the smallest sp^2–sp^3-like nanocarbon structure (see Figure 1.2d, the most common C_{60} fullerene) having revolutionized the field of molecular structures. The fullerenes have special properties and can be considered as another class of materials, which are discussed in detail in [6]. As we see, this very flexible sp^2 carbon system gives rise to many different materials with different interesting physics-chemistry related properties that can be studied in depth. And besides its scientific richness, these sp^2 nanocarbons also play a very important role in applications, as discussed in Section 1.2.

1.2 Short Survey from Discovery to Applications

The ideal concept of the different sp^2 nanocarbons starting from graphene, as described above, is didactic, but historically these materials came to human knowledge in the opposite order. Three-dimensional (3D) graphite is one of the longest-known forms of pure carbon, being found on the surface of the earth as a mineral, and formed by graphene planes arranged in an ABAB Bernal stacking sequence [2].[1] Of all materials, graphite has the highest melting point (4200 K), the highest thermal conductivity (3000 W/mK), and a high room temperature electron mobility (30 000 cm^2/Vs) [9]. Synthetic 3D graphite was made for the first time in 1960 by Arthur Moore [10–15] and was called highly oriented pyrolytic graphite (HOPG). Graphite and its related carbon fibers [16–18] have been used commercially for decades [19]. Their applications range from use as conductive fillers and mechanical structural reinforcements in composites (e. g., in the aerospace industry) to their use as electrode materials exploiting their resiliency (e. g., in lithium ion battery applications) (see Table 1.1) [19, 20].

In 1985 a unique discovery in another sp^2 carbon system took place: the observation of the C_{60} fullerene molecule [21], the first isolated carbon nanosystem. The fullerenes stimulated and motivated a large scientific community from the time of their discovery up to the end of the century [6], but fullerene-based applications

1) ABAB Bernal stacking is the stacking order of graphene layers as shown in Figure 1.2b. One type of carbon atom (A) aligns in the direction perpendicular to the graphene layer, while the other type (B) aligns in every other layer.

Table 1.1 Some of the main applications of traditional graphite-based materials including carbon fibers [19].

Traditional graphite materials	Commercial applications
Graphite and graphite-based products	Materials-processing applications such as furnaces/crucibles, large electrodes in metallurgical processes, electrical and electronic devices such as electric brushes, membrane switches, variable resistors, etc., electrochemical applications for electrode materials in primary and secondary batteries, separators for fuel cells, nuclear fission reactors, bearings and seals (mechanical) and dispersions such as inks. (Estimated market in 2008: 13 billion USD)
Carbon-fiber-based products	Carbon-fiber composites Aerospace (70%), sporting goods (18%), industrial equipment (7%), marine (2%), miscellaneous (3%) (Total market in 2008: $\sim$ 1 billion USD)
Carbon-carbon composites	High-temperature structural materials, Aerospace applications, such as missile nose tips, re-entry heat shields, etc., Brake-disc applications (lightweight, high thermal conductivity, stability), Rotating shafts, pistons, bearings (low coefficient of friction), Biomedical implants such as bone plates (biocompatibility) (Estimated market in 2008: 202 million USD).

remain sparse to date. Carbon nanotubes arrived on the scene following the footsteps of the emergence of the C_{60} fullerene molecule, and they have evolved into one of the most intensively studied materials, now being held responsible for co-triggering the nanotechnology revolution.

The big rush into carbon nanotube science started immediately after the observation of multi-wall carbon nanotubes (MWNTs) on the cathode of a carbon arc system used to produce fullerenes [27], even though they were identified in the core structure of vapor grown carbon fibers as very small carbon fibers in the 1970s [28–30] and even earlier in the 1950s in the Russian literature [23] (see Figure 1.3). However, single-wall carbon nanotubes (SWNTs), the most widely studied carbon nanostructure, were first synthesized intentionally in 1993 [25, 26]. The interest in

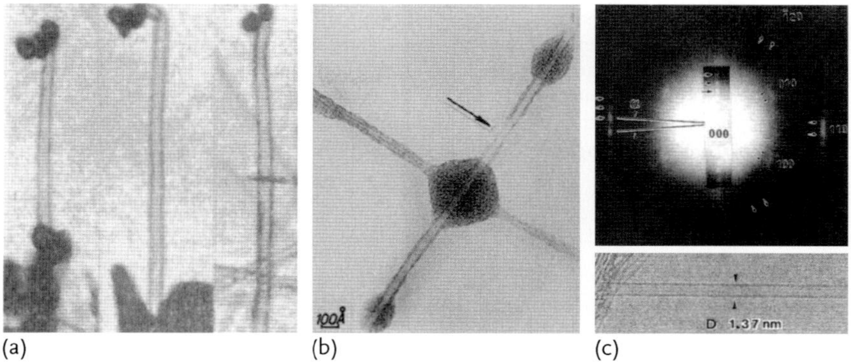

Figure 1.3 The transmission electron microscopy images of carbon nanotubes [22]. The early reported observations (a) in 1952 [23] and (b) in 1976 [24]. In (c) the observation of single-walled carbon nanotubes that launched the field in 1993 [25] together with [26].

the fundamental properties of carbon nanotubes and in their exploitation through a wide range of applications is due to their unique structural, chemical, mechanical, thermal, optical, optoelectronic and electronic properties [20, 31, 32]. The growth of a single SWNT at a specific location and pointing in a given direction, and the growth of a huge amount of millimeter-long tubes with nearly 100% purity have been achieved [33]. Substantial success with the separation of nanotubes by their (n, m) structural indices, metallicity (semiconducting and metallic) and by length has been achieved by different methods, as summarized in [33], and advances have been made with doping nanotubes for the modification of their properties, as summarized in [34]. Studies on nanotube mechanical properties [35, 36], optical properties [37–43], magnetic properties [44], optoelectronics [45, 46], transport properties [47] and electrochemistry [48, 49] have exploded, revealing many rich and complex fundamental excitonic and other collective phenomena. Quantum transport phenomena, including quantum information, spintronics and superconducting effects have also been explored [47]. After a decade and a half of intense activity in carbon nanotube research, more and more attention is now focusing on the practical applications of the many unique and special properties of carbon nanotubes (see Table 1.2) [19]. All these advanced topics in the synthesis, structure, properties and applications of carbon nanotubes have been collected in [20].

In the meantime, the study of nanographite was under development as an important model for nano-sized π-electron systems [50]. Its widespread study was launched by the discovery by Novoselov *et al.* [51] of a simple method using Scotch tape to transfer a single atomic layer of sp^2 carbon called graphene (1-LG) from the c-face of graphite to a substrate suitable for the measurement of the electrical and optical properties of monolayer graphene [52]. While the interest in monolayer graphene preparation goes back to the pioneering theoretical work of Wallace in 1947 [53], the Novoselov finding in 2004 led to a renewed interest in what was before considered to be a prototypical system highly valued for theoretical calcu-

Table 1.2 Applications of nanotubes grouped as present (existing), near-term (to appear in the market within ten years) and long-term (beyond a ten-year horizon), and as categories belonging to large-volume (requiring large amounts of material) and limited-volume (small volume and utilizing the organized nanotube structure) applications [19].

	Large-volume applications	**Limited-volume applications (mostly based on engineered nanotube structures)**
Present	Battery electrode additives (MWNT) Composites (sporting goods, MWNT) Composites (electrostatic shielding applications, MWNT)	Scanning probe tips (MWNT) Specialized medical appliances (catheters)
Near-term (less than ten years)	Battery and supercapacitor electrodes Multi-functional composites (3D, electrostatic damping) Fuel-cell electrodes (catalyst support) Transparent conducting films Field emission displays/lighting CNT-based inks for printing	Single-tip electron guns Multi-tip array X-ray sources Probe array test systems CNT brush contacts CNT sensor devices Electromechanical memory device Thermal-management systems
Long-term (beyond ten years)	Power transmission cables Structural composites (aerospace and automobile, etc.)	Nanoelectronics (FET, interconnects), flexible electronics CNT-based biosensors CNT in photovoltaic devices CNT filtration/separation membranes, drug-delivery

lations for sp^2 carbons, thereby providing a basis for establishing the structure of graphite, fullerenes, carbon nanotubes and other sp^2 nanocarbons. Surprisingly, this very basic graphene system, which had been studied by researchers over a period of many decades, suddenly appeared with many novel physical properties that were not previously imagined [7, 52]. In one or two years, the rush on graphene science began.

Besides outstanding mechanical and thermal properties (breaking strength $\sim$ 40 N/m, Young's modulus $\sim$ 1.0 TPa, room temperature thermal conductivity $\sim$ 5000 W m^{-1} K^{-1} [54]), the scientific interest in graphene was stimulated by the widespread report of the relativistic (massless) properties of the conduction electrons (and holes) in a single graphene layer less than 1 nm thick, which is responsible for the unusual electrical transport properties in this system (see Figure 1.4) with the state-of-the-art mobility for suspended graphene reaching $\mu =$ 200 000 cm^2/Vs [55, 56]. Other unusual properties have been predicted and demonstrated experimentally, such as the minimum conductivity and the half-integer quantum Hall effect [57], Klein tunneling [58–64], negative refractive index and Veselago lensing [62], anomalous Andreev reflection at metal-superconductor junc-

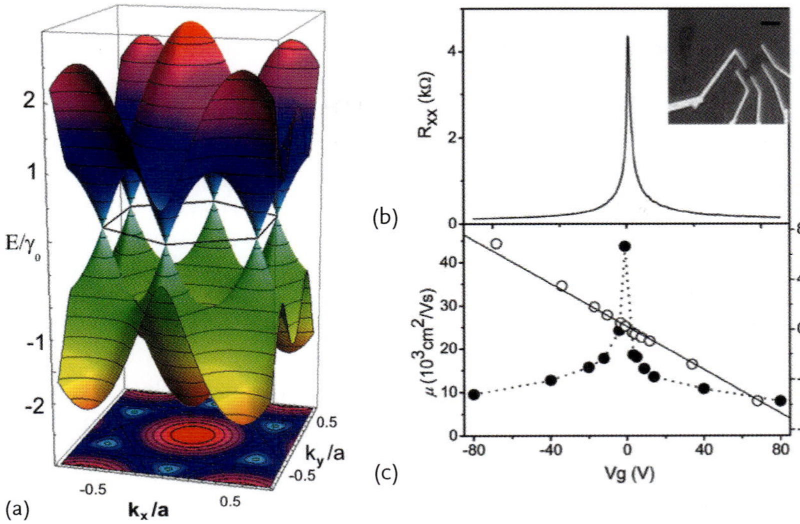

Figure 1.4 (a) Electronic structure of graphene. The valence and conduction bands touch each other at six points, each called the "Dirac point". Near these Dirac points, the electron energy (E) depends linearly on the electron wave vector ($E \propto k$), giving rise to the Dirac cones, similar to massless particles, like in light cones ($E = cp$, where c is the speed of light). Parts (b) and (c) show transport experiments in a single-layer graphene field effect transistor device. (b) Gate voltage V_g-dependent in-plane resistance R_{xx} showing a finite value at the Dirac point. The resistivity ρ_{xx} can be calculated from the resistance R_{xx} using the geometry of the device. The inset is an image of a graphene device sitting on a Si:SiO_2 substrate. The Si is the bottom gate; five top electrodes formed via e-beam lithography are shown. The scale bar is 5 μm. (c) Mobility μ (dotted curve) and carrier density n_S (solid line) as a function of V_g (for holes $V_g < 0$ and for electrons $V_g > 0$). The mobility vs. V_g diverges but to a finite value at the Dirac point due to the finite resistivity. Adapted from [5].

tions [58, 63–66], anisotropies under antidot lattices [67] or periodic potentials [68], and a metal–insulator transition [69]. Applications as a filler for composite materials, supercapacitors, batteries, interconnects and field emitters have been exploited, although it is still too early to say whether graphene will be able to compete with carbon nanotubes and other materials in the applications world [70].

Finally, graphene can be patterned using high-resolution lithography [71] for the fabrication of nanocircuits with graphene-nanoribbon interconnects. Many groups are now making devices using graphene and also graphene nanoribbons, which have a long length and a small width, and where the ribbon edges play an important role in determining their electronic structure and in exhibiting unusual spin polarization properties [72]. While lithographic techniques have limited resolution for the fabrication of small ribbons (< 20 nm wide), chemical [73] and synthetic [74] methods have been employed successfully, including the unzipping of SWNTs as a route to produce carbon nanoribbons [75, 76].

1.3
Why sp^2 Nanocarbons Are Prototypes for Nanoscience and Nanotechnology

The integrated circuit represents the first human example of nanotechnology, and gave birth to the information age. Together with the nonstop shrinking of electronic circuits, the rapid development of molecular biology and the evolution of chemistry from atoms and molecules into large complexes, such as proteins and quantum dots, have together with other developments launched nanotechnology. It is not possible to clearly envisage the future or the impact of nanotechnology, or even the limit for the potential of nanomaterials, but clearly serious fundamental challenges can already be identified:

- To construct nanoscale building blocks precisely and reproducibly;
- To discover and to control the rules for assembling these nano-objects into complex systems;
- To predict and to probe the emergent properties of these assembled systems.

Emergent properties refer to the complex properties of ensembles of components which exhibit much simpler interactions with their nearest neighbors. These challenges are not only technological, but also conceptual: how to treat a system that is too big to be solved by present day first-principles calculations, and yet too small for using statistical methods? Although these challenges punctuate nanoscience and nanotechnology, the success here will represent a revolution in larger-scale scientific challenges in the fields of emergent phenomena and information technology. Answers to questions like "how do complex phenomena emerge from simple ingredients?" and "how will the information technology revolution be extended?" will probably come from using nanoscience in meeting the challenges of nanotechnology [77].

It is exactly in this context that nanocarbon is expected to play a very important role. On one hand, nature shows that it is possible to manipulate matter and energy the way integrated circuits manipulate electrons, by assembling complex self-replicating carbon-based structures that are able to sustain life. On the other hand, carbon is the upstairs neighbor to silicon in the periodic table, with carbon having more flexible bonding and having unique physical, chemical and biological properties. Nevertheless carbon nanoscience holds promise for a revolution in electronics at some point in the future. Three important factors make sp^2 carbon materials special for facing the nano-challenges listed in the previous paragraph: First is the unusually strong covalent sp^2 bonding between neighboring atoms; second is the extended π-electron clouds coming from the p_z orbitals; and third is the simplicity of the sp^2 carbon system. We briefly elaborate on these three factors in the following paragraphs.

In the sp^2 configuration, the $2s$, p_x and p_y orbitals mix to form three covalent bonds, 120° from each other in the xy plane (see Figure 1.1). Each carbon atom has three neighbors, forming a hexagonal (honeycomb) network. These sp^2 in-plane bonds are the strongest bonds in nature, comparable to the sp^3 bonds in diamond, with a measured Young's modulus on the order of 1.0 TPa [54, 78, 79]. This strength is advantageous for sp^2 carbons as a prototype material for the development of nanoscience and nanotechnology, since different interesting nanostructures (sheets, ribbons, tubes, horns, fullerenes, etc.) are stable and strong enough for exposure to many different types of characterization and processing steps.

The p_z electrons that remain perpendicular to the hexagonal network (see Figure 1.1) form delocalized π electron states which collectively form valence and conduction energy bands. For this reason, sp^2 carbons, which include graphene, graphite, carbon nanotubes, fullerenes and other carbonaceous materials, are also called π-electron materials. The delocalized electronic states in monolayer graphene are highly unusual, because they behave like relativistic Dirac fermions, that is, these states exhibit a massless-like linear energy-momentum relation (like a photon, see Figure 1.4a), and are responsible for unique transport (both thermal and electronic) properties at sufficiently small energy and momentum values [5, 7, 52]. This unusual electronic structure is also responsible for unique optical phenomena, which will be discussed in depth in this book for the case of Raman spectroscopy.

These two physical properties accompany a very important aspect of sp^2 carbons, which is the simplicity of a system formed by only one type of atom in a periodic hexagonal structure. Therefore, different from most materials, sp^2 nanocarbons allow us to easily access their special properties using both experimental and theoretical approaches. Being able to model the structure is crucial for the development of our methodologies and knowledge.

1.4 Raman Spectroscopy Applied to sp^2 Nanocarbons

Raman spectroscopy has historically played an important role in the study and characterization of graphitic materials [16, 80], being widely used in the last four decades to characterize pyrolytic graphite, carbon fibers [16], glassy carbon, pitch-based graphitic foams [81, 82], nanographite ribbons [83], fullerenes [6], carbon nanotubes [31, 80], and graphene [84, 85]. For sp^2 nanocarbons, Raman spectroscopy can give information about crystallite size, clustering of the sp^2 phase, the presence of sp^3 hybridization and chemical impurities, mass density, optical energy gap, elastic constants, doping, defects and other crystal disorder, edge structure, strain, number of graphene layers, nanotube diameter, nanotube chirality and metallic vs. semiconductor behavior, as discussed in this book.

Figure 1.5 shows the Raman spectra from different crystalline and disordered sp^2 carbon nanostructures. The first spectrum shown is that for monolayer graphene, the building block of many sp^2 nanocarbons. What is evident from Figure 1.5 is that every different sp^2 carbon in this figure shows a distinct Raman spectrum, which can be used to understand the different properties that accompany each of these different sp^2 carbon structures. For example, 3D highly oriented pyrolytic graphite (labeled HOPG in the figure) shows a distinctly different spectrum from that for monolayer graphene (1-LG) in Figure 1.5, which in turn is shown to be distinct from the Raman spectra characteristic of the various few layer-graphene materials, for example 2-LG and 3-LG [86].

Figure 1.5 also shows the Raman spectrum for single wall carbon nanotubes (SWNTs). Here we see a variety of features such as the radial breathing mode (RBM) or the splitting of the G-band into G^+ and G^--bands that distinguish a SWNT from any other sp^2 carbon nanostructure. Carbon nanotubes are unique materials in many ways, one being their ability to exhibit transport properties that are either metallic (where their valence band and conduction band touch each other at the $K(K')$ points in the respective graphene Brillouin zone) or semiconducting (where a band gap typically of several hundred meV separates their valence and conduction bands). Nanotubes are also unique in that their Raman spectra differ according to whether the nanotube is semiconducting (as shown in Figure 1.5) or metallic (not shown).

The introduction of disorder breaks the crystal symmetry of graphene and activates certain vibrational modes that would otherwise be silent, such as the D-band and the D'-band features and their combination $D + D'$ mode, shown in the spectrum labeled damaged graphene in Figure 1.5. The different types of defects do in fact show their own characteristic Raman spectra, as illustrated in Figure 1.5 by comparing the spectra labeled damaged graphene and SWNH (denoting single-wall carbon nanohorns, another nanostructured form of sp^2 carbon which may include pentagons with a small content of sp^3 bonding [87]). However the topic of distinguishing between the Raman spectra of one and another type of defective graphene remains an area to be explored in detail in the future. When the disorder is so dominant that only near neighbor structural correlations are present (labeled

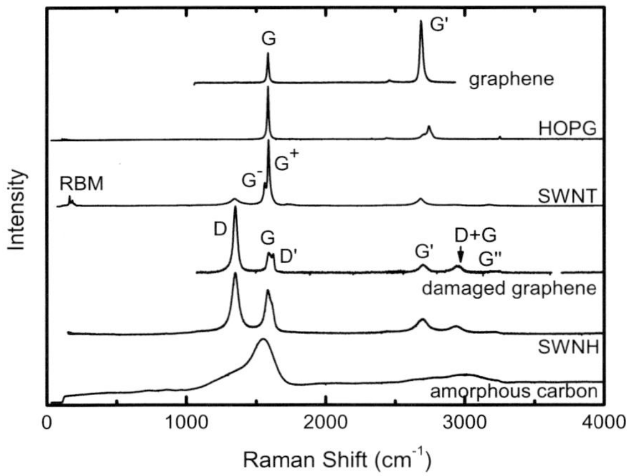

Figure 1.5 Raman spectra from several sp^2 nanocarbons. From top to bottom: crystalline monolayer graphene, highly oriented pyrolytic graphite (HOPG), a single-wall carbon nanotube (SWNT) bundle sample, damaged graphene, single-wall carbon nanohorns (SWNH) and hydrogenated amorphous carbon. The most intense Raman peaks are labeled in a few of the spectra [85].

amorphous carbon in the figure), broad first-order and second-order features are seen, with both sp^2 and sp^3 bonding present. Some hydrogen uptake can also occur for such materials to satisfy their dangling bonds [88].

The extremely exciting and rapid development of Raman spectroscopy in sp^2 carbon materials has promoted many advances occurring in this field: graphite is already well-established and commercialized. Carbon nanotubes are by now also mature, after having had an exciting and fast moving research agenda for nearly 20 years. In fact, carbon nanotubes are now ready to make a transition from science to applications, that is at a critical juncture where the laboratory demonstrations of applications need to get translated into product lines. Graphene is younger, but is now attracting many researchers to address the exciting new science hidden in this prototype nanostructure. While the study of the fundamental properties of graphite was essential for understanding the properties of new nanostructured sp^2 carbon forms, further developments of the field are showing how these younger sp^2 carbon nanostructures are revealing many new and unexpected physical phenomena. It is fascinating that Raman spectroscopy has, from the beginning, provided a tool for understanding sp^2 carbon systems. Even after almost a century since the first observation of Raman spectra in carbon-based systems by Sir C. V. Raman himself [89, 90], the Raman spectra from sp^2 carbon materials still puzzle chemists, physicists and material scientists, and these materials offer a challenging system where the worlds of chemistry and physics feed each other.

Problems

[1-1] The carbon–carbon distance of graphene (see Figure 1.1) is 1.42 Å. How much area is occupied by a single carbon atom in the graphene plane?

[1-2] The interlayer distance in multi-layer graphene or graphite (see Figure 1.2b) is 3.35 Å. How much volume is occupied by a single carbon atom in graphite? From this information, estimate the density of graphite in g/cm^3. Compare your estimate with the literature value of $2.25\ g/cm^3$.

[1-3] Figure 1.2b shows the AB stacking of graphene layers in forming graphite. Explain how two graphene layers are stacked in the AB stacking sequence.

[1-4] There are several ways to stack graphene layers. When we put a third layer on the two AB-stacked graphene layers, there are two possible ways of doing this stacking, which we call ABAB and ABC stacking. Show a graphic picture of both ABAB and ABC stacking and explain your answer in words, including the relation between the location of carbon atoms in each relevant plane.

[1-5] C_{60} molecules form face-centered cubic (fcc) structures. The density of the C_{60} crystal is $1.72\ g/cm^3$. From this value, estimate the C_{60}–C_{60} distance and the fcc lattice constants.

[1-6] Each carbon atom in a C_{60} molecule has one pentagonal and two hexagonal rings. Calculate the dihedral angles (a) between the two hexagonal rings and (b) between the hexagonal ring and the pentagonal ring.

[1-7] Diamond crystallizes in a cubic diamond structure with four (sp^3) chemical bonds. All bond angles for any pair of chemical bonds are identical. Calculate the bond angle between two chemical bonds by using an analytical solution and also give the numerical value in degrees. The C–C distance in diamond is 1.544 Å. Estimate the cube edge length and density in g/cm^3.

[1-8] In spectroscopy, a wave vector is defined by $1/\lambda$ (where λ is the wavelength) while in solid state physics, the definition of a wave vector is $2\pi/\lambda$. Show that a 1 eV photon corresponds to 8065 cm^{-1} (wavenumbers). In Raman spectroscopy, the difference between the wave vectors for the incident and scattered light is called the Raman shift whose units are generally given in cm^{-1}.

[1-9] Raman spectroscopy involves the inelastic light scattering process. Part of the energy of the incident light is lost or gained, respectively, in materials in which some elemental excitation such as an atomic vibration (phonon) absorbs or releases the energy from or to the light. We call these two Raman processes Stokes and anti-Stokes processes, respectively. When light with the wavelength 632.8 nm is incident on the sample and loses energy by creating a phonon with an energy of 0.2 eV, what is the scattered wavelength? Also give the scattered wavelength for the anti-Stokes Raman signal.

[1-10] Consider the optical electric field of the incident light with an angular frequency $\omega_0 = 2\pi\nu_0$ and amplitude E_0,

$$E = E_0 \cos \omega_0 t \ .$$

Then the dipole moment P of a diatomic molecule is proportional to E such that $P = \alpha E$, in which α is called the polarizability. When the molecule is vibrating with a frequency ω, then α is also vibrating with the frequency ω,

$$\alpha = \alpha_0 + \alpha_1 \cos \omega t \ .$$

When substituting α into the formula $P = \alpha E$, show that there are three different frequencies for the scattered light (or P), ω_0 (elastic, Rayleigh scattering) and $\omega_0 \pm \omega$ (inelastic, Stokes ($-$) and anti-Stokes ($+$) Raman scattering).

[1-11] Let us consider a resonance effect. Here we consider a particle with a mass m which is connected to a system by a spring with spring constant K. When we apply an oscillatory force $f \exp(i\omega t)$, the equation of motion for the amplitude u of the vibration is

$$m\ddot{u} + Ku = f \exp(i\omega t) \ .$$

Solve this differential equation and plot u after a sufficiently long time as a function of ω. Show that a singularity occurs when $\omega = \omega_0 \equiv \sqrt{K/m}$ and discuss the significance of this singularity.

[1-12] In a more realistic model than the previous model, we can consider the friction term $\gamma \dot{u}$ in the vibration

$$m\ddot{u} + \gamma \dot{u} + Ku = f \exp(i\omega t) .$$

Plot u as a function of ω for this case. Show that now a singularity no longer occurs. How does γ appear in the plot of $\omega(u)$? Consider the limits of weak damping and strong damping and find what determines the transition between these limits.

2
Electrons in sp^2 Nanocarbons

Usually Raman spectra only involve phonons explicitly, being independent of the laser energy used to excite the Raman spectra and the electronic transitions in the material (to the extent that the electron–phonon interaction is weak). Furthermore, the usual Raman scattering signal is weak. However, the scattering efficiency gets much larger and the Raman signal much stronger when the laser energy matches the energy between optically allowed electronic transitions in the material. This intensity enhancement process is called resonance Raman scattering (RRS) [91]. Under the RRS regime, the resonance Raman intensity is further enhanced by the large density of electronic states (DOS) available for the optical transitions. This large density of states is especially important for one-dimensional systems, which have singularities in their density of states at the energy onset of an allowed optical transitions.

This chapter has the goal of reviewing the important concepts needed for understanding the Raman spectroscopy of sp^2 nanocarbons, making a link between molecular and solid state science. Due to the peculiar π-electron structure (delocalized p_z orbitals, as discussed in Sections 1.3 and 2.2.2), the Raman spectroscopic response in sp^2 nanocarbons depends strongly on their electronic structure due to the ubiquitous resonance processes that dominate their inelastic scattering of light. For this reason, it is important to review the electronic properties of these systems.

We start by reviewing the basic concepts relevant to the electronic energy levels of isolated molecules and what happens when these molecules are assembled in the solid state. In Section 2.1 we present the one-electron system for the hydrogen atom and then move to more and more complex systems, discussing the formation of molecular orbitals and finally building the transition to solid state systems in Section 2.1.5, and to sp^2 nanocarbon systems in particular (Sections 2.2 and 2.3). Here both the molecular orbital theory (bonding and antibonding states) and the valence bonding theory (hybridization) are introduced and, while the discussion of the intermixing may not be fully rigorous, it is useful for gaining an understanding of sp^2 carbon systems. In Section 2.2.1 we present the crystal structure of graphene, which is followed by the tight-binding model for the π-band electronic structure for monolayer graphene in Section 2.2.2. The π-bands extend over an energy range that goes from the Fermi point up to the ultra-violet, and the π-bands are thus responsible for all transport and optical phenomena. In Section 2.2.3 the σ-bands

Raman Spectroscopy in Graphene Related Systems. Ado Jorio, Riichiro Saito, Gene Dresselhaus, and Mildred S. Dresselhaus

ISBN: 978-3-527-40811-5

are reviewed to yield the electronic structure for graphene which contains both π and σ-bands. In flat graphene the σ-bands are not important for optical phenomena. However, when curvature is present, like in the case of carbon nanotubes, σ-π hybridization can occur, with consequences on the optical response. The remaining sections of this chapter extend the picture to few-layer graphene and then to many-layer graphene in Section 2.2.4 and to quantum confinement phenomena occurring in nanoribbons (Section 2.2.5). The effect of quantum confinement on the electronic structure of nanotubes is next discussed in Section 2.3. The structure of carbon nanotubes is introduced in Section 2.3.1 followed by a discussion of the zone-folding procedure (Section 2.3.2) and the density of electronic states (Section 2.3.3), which is important to understand Raman spectroscopy in these materials, as discussed in Section 2.3.4. This chapter ends with a short discussion in Section 2.4 of the physics beyond the simple tight-binding and zone-folding approximations. This final section comes here just as a brief introduction to concepts that will be developed in later chapters.

2.1 Basic Concepts: from the Electronic Levels in Atoms and Molecules to Solids

Before discussing the electronic properties of the crystalline sp^2 systems, we remind the reader about the basic concepts used to describe the electronic levels of a mono-atomic system, the hydrogen atom in Section 2.1.1, and we then move to molecular systems like the H_2 in Section 2.1.2, to NO in Section 2.1.3, and C_2H_2 in Section 2.1.4, and finally leading into the electronic structure of a linear chain of atoms in a periodic lattice in Section 2.1.5. With this procedure, we hope that the reader will feel comfortable when looking at the electron wavefunctions for graphene, carbon nanotubes and other sp^2 carbon systems.

2.1.1 The One-Electron System and the Schrödinger Equation

We start by reviewing the most basic system, that of the hydrogen atom, with one electron of charge $-e$ and mass m orbiting about a nucleus with mass M. The Schrödinger equation for the hydrogen atom [92] is written as:

$$\left[-\frac{\hbar^2}{2\mu}\Delta + V(\boldsymbol{r})\right]\Psi(\boldsymbol{r}) = E\Psi(\boldsymbol{r}) , \tag{2.1}$$

where μ is the reduced mass given by

$$\frac{1}{\mu} = \frac{1}{m} + \frac{1}{M}, \quad \text{or} \quad \mu = \frac{M}{m+M}m , \tag{2.2}$$

and the reduced mass is shown in Figure 2.1.

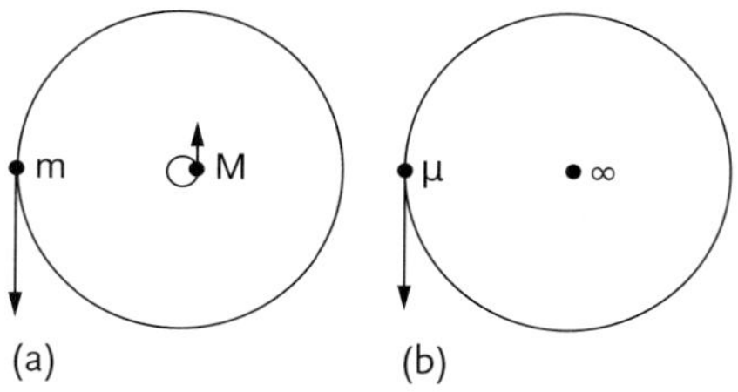

Figure 2.1 Schematic definition of a system with masses m and M in motion (a), and the corresponding reduced mass model system (b), in which a particle with a reduced mass μ is moving around the center of mass, indicated by ∞.

The Coulomb potential $V(\boldsymbol{r})$ for the hydrogen atom has a spherical symmetry which is represented by $r = \sqrt{x^2 + y^2 + z^2}$ and

$$V(\boldsymbol{r}) = -\frac{Ze^2}{4\pi\epsilon_0 r}, \tag{2.3}$$

in which Z is the charge on the nucleus ($Z = 1$ for hydrogen), and ϵ_0 is the dielectric constant of vacuum. Since the Hamiltonian has a spherical symmetry around the center, the wavefunction Ψ of Eq. (2.1) can be written as:

$$\Psi(\boldsymbol{r}) = R(r)\Theta(\theta)\Phi(\phi), \tag{2.4}$$

so that Eq. (2.4) can be decomposed into three partial differential equations for $R(r)$, $\Theta(\theta)$ and $\Phi(\phi)$. Since there is no term depending on θ and ϕ in $V(r)$, the solution of Eq. (2.4) simply replaces $\Theta(\theta)$ and $\Phi(\phi)$ by the solution for free space given by the spherical harmonics $Y_\ell^m(\theta,\phi)$. As for the radial part of the wave function $R(r)$, we can solve this by considering Laguerre polynomials. Although we do not go into detail for this solution, the end result for the energy eigenvalues is [92]

$$E_n = -\frac{Z^2}{(4\pi\epsilon_0)^2} \cdot \frac{\mu e^4}{2\hbar} \cdot \frac{1}{n^2} \quad (n = 1, 2, 3, \ldots), \tag{2.5}$$

and $R(r)$ for a given n is expressed by

$$R_{n\ell}(\boldsymbol{r}) = \exp\left(-\frac{Zr}{na_0}\right)\left(\frac{Zr}{a_0}\right)^\ell G_{n\ell}\left(\frac{Zr}{a_0}\right), \tag{2.6}$$

where $G_{n\ell}$ denotes the Laguerre polynomials depending on the variable Zr/a_0 (where a_0 is the Bohr radius, $a_0 = \hbar^2/me^2$). The eigenvalues are characterized by four quantum numbers: the principal quantum number n, the angular momentum quantum number ℓ, the z component of the angular momentum m_ℓ, and the spin of the electron m_s, which are not explicitly written in Eq. (2.5). These quantum numbers assume the following values:

$$n = 1, 2, 3, \ldots \tag{2.7}$$

$$\ell = 0, 1, 2, \ldots, n-1 . \tag{2.8}$$

$$m_\ell = -\ell, -\ell+1, \ldots, \ell-1, \ell . \tag{2.9}$$

$$m_s = 1/2, -1/2 . \tag{2.10}$$

The common designations for the atomic orbitals $s, p, d, \ldots$ correspond to $\ell = 0, 1, 2, \ldots$, respectively.

For the case of carbon $Z = 6$, we predominantly consider the principal quantum numbers $n = 1, 2$ where the $n = 1$ orbital ($1s$) is fully occupied with one spin up and one spin down core electron, and $n = 2$ is half occupied with four electrons with orbitals ($2s, 2p_x, 2p_y$ and $2p_z$) having energies comparable to that of the hydrogen atom. In the lowest energy state, these $n = 2$ electrons occupy the hybridized graphene $sp^2 + p_z$ orbitals, while the four electrons in diamond occupy a symmetric sp^3 hybridized orbital, which is higher in energy at room temperature and under ambient pressure (see also Section 1.1).

2.1.2
The Schrödinger Equation for the Hydrogen Molecule

Now we recall what happens to the electrons when the two H atoms are combined into the H_2 molecule. In the two-electron system of a hydrogen molecule, the Schrödinger equation can be written in matrix form resulting in the solution of a secular equation, written generically as:

$$|\langle \Psi_i | H | \Psi_j \rangle - E \langle \Psi_i | \Psi_j \rangle| = 0 , \tag{2.11}$$

where $\langle \Psi_i | H | \Psi_j \rangle$ and $\langle \Psi_i | \Psi_j \rangle$ denote, respectively, the Hamiltonian and overlap matrices for basis functions.[1] Here we consider the hydrogen molecule H_2 and Ψ_i is taken as the hydrogen $1s$ atomic orbital for each H atom. If we adopt the approximation that Ψ_1 and Ψ_2 are orthogonal to each other, then $\langle \Psi_1 | \Psi_2 \rangle = 0$ and the Schrödinger Equation (2.11) yields

$$\begin{cases} E_{1s} \Psi_1 + V_0 \Psi_2 = E \Psi_1 \\ V_0 \Psi_1 + E_{1s} \Psi_2 = E \Psi_2 , \end{cases} \tag{2.12}$$

where E_{1s} is the energy of an unperturbed H atom, and the Hamiltonian matrix element is $V_0 \equiv \langle \Psi_1 | H | \Psi_2 \rangle < 0$. In evaluating the Hamiltonian matrix elements, we should also consider the Coulomb interaction between the two electrons in the Hamiltonian.[2] Here we simply consider that the Coulomb interaction is included

1) Basis functions are atomic orbitals or molecular orbitals that are used by variational principles to obtain the energy by $E = \langle \Psi_i | H | \Psi_j \rangle / \langle \Psi_i | \Psi_j \rangle$.

2) If we use the Hartree–Fock approximation for the Coulomb interaction [93], the interaction further consists of a direct Coulomb term and an exchange term. The exchange term corrects for the overestimation of the direct Coulomb interaction term and arises from the fact that two electrons with the same spin cannot be at the same location in accordance with the Pauli exclusion principle.

in both V_0 and E_{1s}. In matrix form Eq. (2.12) can be written as:

$$\begin{pmatrix} E_{1s} & V_0 \\ V_0 & E_{1s} \end{pmatrix} \begin{pmatrix} \Psi_1 \\ \Psi_2 \end{pmatrix} = E \begin{pmatrix} \Psi_1 \\ \Psi_2 \end{pmatrix} . \tag{2.13}$$

Equation (2.13) can be diagonalized by solving the secular equation[3] (Eq. (2.11)),

$$(E - E_{1s})^2 - V_0^2 = 0 , \tag{2.14}$$

which gives $E = E_{1s} \pm V_0$. The diagonalization can be done by a unitary transformation of the Hamiltonian matrix H given by $U^\dagger H U$, where the unitary matrix U is given by[4]

$$U = (1/\sqrt{2}) \begin{pmatrix} 1 & 1 \\ 1 & -1 \end{pmatrix} . \tag{2.15}$$

The resulting symmetrized eigenvectors will be two molecular orbitals formed by a linear combination of atomic orbitals (LCAOs), given by the symmetric (S) and antisymmetric (AS) combinations [94]

$$\Psi_S = (1/\sqrt{2})(\Psi_1 + \Psi_2) \tag{2.16}$$

$$\Psi_{AS} = (1/\sqrt{2})(\Psi_1 - \Psi_2) . \tag{2.17}$$

The spatial dependence of the electronic wavefunctions for the hydrogen molecule is shown in Figure 2.2, where the symmetric combination $\Psi_S = (1/\sqrt{2})(\Psi_1 + \Psi_2)$ has the lower energy ($E_S = E_{1s} + V_0$), in which V_0 has a negative value, resulting in an enhancement in the probability for finding an electron at the center between the two H atoms. For this reason, this state is usually called the *bonding state*, describing the ground state for the H_2 molecule by occupying two electrons (one spin up, one spin down). The antisymmetric combination $\Psi_{AS} = (1/\sqrt{2})(\Psi_1 - \Psi_2)$, with energy $E_{AS} = E_{1s} - V_0$, is named the *antibonding state* with a node in the wave function at the center between the two H atoms, as shown in Figure 2.2.

2.1.3 Many-Electron Systems: the NO Molecule

In this section we show how the molecular electronic complexity increases when the number of electrons in a diatomic molecule increases. Figure 2.3 shows the schematics of the electronic levels for the heterogeneous diatomic NO molecule [94]. The $1s^2$ levels (core electrons, not shown) lie much lower in energy. These electrons are tightly bound to their respective atoms and do not contribute

3) The secular equation is given by the determinant of the matrix of Eq. (2.13) when it becomes zero. If the determinant is not zero, we get the inverse matrix and multiplying the inverse matrix by Eq. (2.13), we get the meaningless solution of ${}^t(\Psi_1, \Psi_2) = {}^t(0, 0)$.

4) $U^\dagger$ is the transpose and complex conjugate of U. In the case of a unitary matrix, $U^\dagger = U^{-1}$.

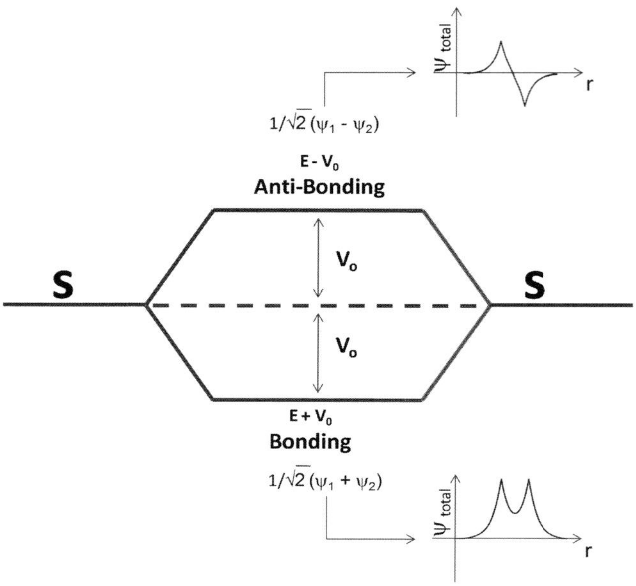

Figure 2.2 Bonding and antibonding molecular levels of the H_2 molecule. The energy separation between the bonding and antibonding orbitals for this symmetric diatomic molecule is given by $2V_0$. The wavefunctions $\Psi(r)$ for the bonding and antibonding states are also shown.

to molecular bonding and molecular properties. The $2s$ electrons form bonding and antibonding states, which are fully occupied by four electrons, similar to the discussion in Section 2.1.2. Next, considering the bonding for the p electrons, the lowest energy is for the $2p_z$ orbitals if we take the z-axis to be along the NO bond direction. The diatomic potential of the NO molecule will break the atomic

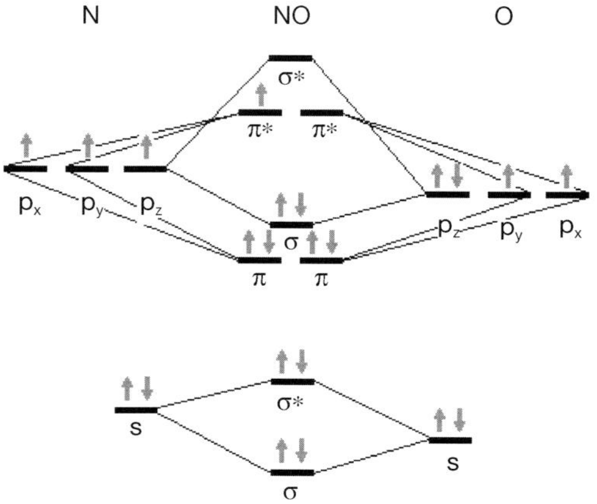

Figure 2.3 Energy levels for the heterogeneous diatomic NO molecule, showing the bonding and antibonding states filled with spin up and spin down electrons (gray).

degeneracy between the $2p_z$ and $2p_{x,y}$ orbitals, while the NO bonding will mix the N and O levels which have the same angular momentum L around the z axis ($L = 0$ for $2s$ and $2p_z$, $L = 1$ for $2p_x$ and $2p_y$), forming bonding and antibonding molecular orbitals,[5] as shown in Figure 2.3 [94]. The 11 electrons pertaining to both the N and O atoms in the $n = 2$ atomic shell (N: $2s^2, 2p^3$ and O: $2s^2, 2p^4$) will fill the five lowest energy levels as shown in Figure 2.3 (accounting for both spin up and spin down states) plus one extra electron in the highest π^* antibonding state. The *highest occupied molecular orbital* and the *lowest unoccupied molecular orbital* are often called the HOMO and LUMO levels, respectively. In the case of NO, π^* is called the singly occupied MO (SOMO). Considering a NO^+ ionized molecule, the highest π^* antibonding state would be empty and become the LUMO level. The p_z-based σ level would represent the HOMO level.

2.1.4
Hybridization: the Acetylene C_2H_2 Molecule

Now we address the problem of hybridization, whereby atomic orbitals mix with each other within an atom to form a chemical bond in a specific direction. Considering Figure 2.3, imagine that the $2s$ level from one atom is closer in energy to the $2p$ level of the other atom. This indeed happens in the CO molecule. Otherwise, imagine that the bonding interaction is strong enough to mix the 2s and the $2p_z$ orbitals which have the same symmetry. This happens in some cases where the minimization of energy for the molecular bonding requires an elongation of the electronic wavefunctions to the other atoms. Such an elongation can be represented by the hybridization (mixing) of different atomic orbitals from the same atom, as occurs in the case of acetylene C_2H_2 [31]. Considering the bonding along the x direction, the p_x electrons from the two carbon atoms will be involved in the strongest interatomic bonding. This bonding is called σ bonding, resulting in an elongation of the electronic wavefunctions, as shown in Figure 2.4, where we see a mixing of $|2s\rangle$ and $|2p_z\rangle$ orbitals (s–p hybridization).

The two linear combinations of atomic orbitals (LCAOs) for the H_2C_2 acetylene molecule will be the

$$|sp_a\rangle = \frac{1}{\sqrt{2}}(|2s\rangle + |2p_x\rangle) \tag{2.18}$$

and the

$$|sp_b\rangle = \frac{1}{\sqrt{2}}(|2s\rangle - |2p_x\rangle) \tag{2.19}$$

orbitals, which are elongated along the +x and $-x$ directions, respectively (Figure 2.4), where $|sp_a\rangle$ and $|sp_b\rangle$ are hybridized orbitals for the left and right atoms, respectively. Furthermore the $|sp_a\rangle$ and $|sp_b\rangle$ form the symmetric and antisymmetric combinations $|sp_a\rangle \pm |sp_b\rangle$, respectively, and they are usually named σ and

5) We must consider hybridization of $2s$ and $2p_z$ in the case of NO. This is the reason why σ of $2p_z$ lies higher in energy than π of $2p_x$ and $2p_y$.

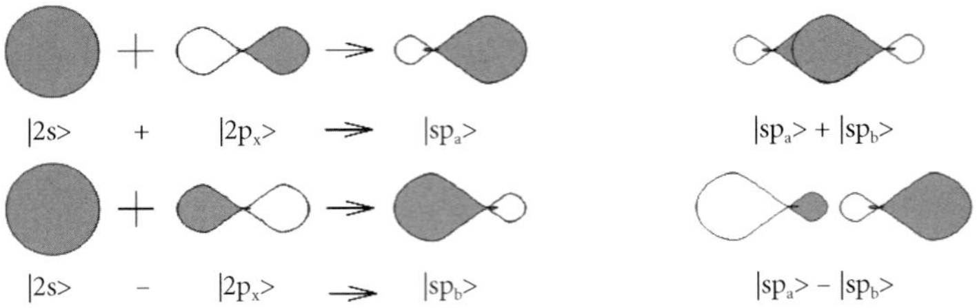

Figure 2.4 Schematics for the sp hybridization and bonding formation [31]. The shading represents the positive value of the wave function $|sp_a\rangle = |2s\rangle + |2p_x\rangle$, where the bond is elongated in the positive direction of x, while the state $|sp_b\rangle = |2s\rangle - |2p_x\rangle$ is elongated in the negative direction. The symmetric ($|sp_a\rangle + |sp_b\rangle$) and antisymmetric ($|sp_a\rangle - |sp_b\rangle$) combinations constitute "bonding" and "antibonding" σ states, respectively.

σ^* states. The σ state is responsible for the strong covalent bonding between two carbon atoms, while the σ^* state is an unoccupied state. The two remaining p_y and p_z electronic states, which are perpendicular to the bonding x direction, form the (weak) so-called *π bonds*, giving rise to the symmetric and antisymmetric orbital combinations that occur in the acetylene molecule HC≡CH [31]. While the symmetric and antisymmetric combinations retain a similarity with the concept of bonding and antibonding orbitals, this connection is not fully correct. In sp^2 carbon systems, hybridization occurs by mixing $2s$, $2p_x$ and $2p_y$ orbitals which make three hybridized orbitals which are elongated to the three nearest neighbor atoms (sp^2 hybridization). The sp^2 hybridized orbitals form three σ (bonding) and three σ^* (antibonding) orbitals. The remaining $2p_z$ form π and π^* orbitals. In the case of sp^2 carbon, π and π^* orbitals correspond to the HOMO and LUMO, respectively. All these concepts are broadly used in the description of the sp^2 nanocarbons.

2.1.5
Basic Concepts for the Electronic Structure of Crystals

Next, we consider the electronic structure of a crystalline solid and attempt to make a connection to the simple concepts used in molecular electronics. The Schrödinger equation for an electron in a crystal is written as [95]:

$$\left[-\frac{\hbar^2}{2m}\nabla^2 + V(\boldsymbol{r})\right]\Psi = E\Psi\,, \tag{2.20}$$

where $V(\boldsymbol{r})$ is now a periodic potential. Since the crystal has a quasi-infinite number of atoms, the number of electronic levels is also quasi-infinite. This generates a quasi-infinite secular equation if we solve for the electronic states by the molecular orbitals method. However, in the case of a crystal, we can use the fact that the crystal is a periodic structure based on a *unit cell* that repeats itself under the lattice vectors labeled by

$$\boldsymbol{R} = n\boldsymbol{a}_1 + m\boldsymbol{a}_2 + l\boldsymbol{a}_3\,, \tag{2.21}$$

where $\boldsymbol{a}_1, \boldsymbol{a}_2$, and $\boldsymbol{a}_3$ are the primitive vectors of the crystal lattice, and n, m, and l are integers. Since the potential $V(\boldsymbol{r})$ is periodic under an $\boldsymbol{R}$ translation ($V(\boldsymbol{r}) = V(\boldsymbol{r} + \boldsymbol{R})$), the solutions to Eq. (2.20) are wave functions that can be written as:

$$\Psi_k(\boldsymbol{r}) = e^{i\boldsymbol{k}\cdot\boldsymbol{r}} u_{\boldsymbol{k}}(\boldsymbol{r}) , \tag{2.22}$$

where

$$u_{\boldsymbol{k}}(\boldsymbol{r}) = u_{\boldsymbol{k}}(\boldsymbol{r} + \boldsymbol{R}) , \tag{2.23}$$

is periodic in accordance with Bloch's Theorem [31, 95]. Figure 2.5 illustrates the formation of Bloch states for a linear chain of atoms, and these Bloch states are defined by the unit cell wavefunction $u_{\boldsymbol{k}}$ (shown as s and p states in Figure 2.5) and e^{ikr} term, which modifies the sign and amplitude as a phase factor. Here $\boldsymbol{k}$ is the wavevector whose length is given by

$$k = 2\pi/\lambda , \tag{2.24}$$

where λ is the wavelength of the wavefunction.

Since $\boldsymbol{k}$ is a good quantum number (or a variable that is conserved under translation by $\boldsymbol{R}$), the electronic structures of crystals are displayed in a plot of the electron energy $E_{\boldsymbol{k}}$ vs. electron wavevector $\boldsymbol{k}$, called the *energy dispersion relations,* which consist of quasi-continuous states within a finite region of energy called *electronic energy bands.* From an atomic orbital in a unit cell, we can make a Bloch function and thus produce an energy band which can occupy two electrons per unit cell. Being quasi-continuous, these energy bands account for a quasi-infinity of electronic levels.

The *real coordinate space* (x, y, z) is the space where the atoms are displayed (Figure 1.2), and where the probability for finding an electron with a wave function Ψ is given by taking the square of Ψ, shown in Figure 2.5. The so-called

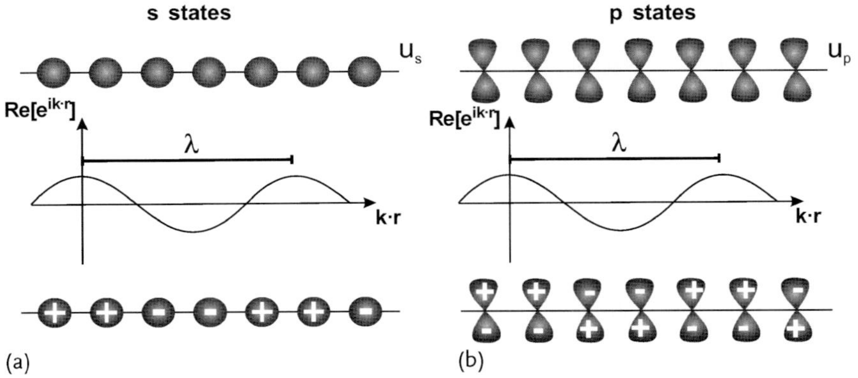

Figure 2.5 Schematic 1D Bloch orbitals formed by s (a) and p_z (b) atomic orbitals. The top shows the wavefunction of each atom, the middle shows the phase of the Bloch orbital e^{ikr}, and the bottom shows the amplitudes of the Bloch orbitals.

reciprocal space (k_x, k_y, k_z) is the space of the wavevectors k for the wavefunctions, and the electronic dispersion relations are representations of the electron energy in reciprocal space.

Consider a hypothetical one-dimensional (1D) crystal made of a quasi-infinite repetition of –N–O–N–O– atoms. Considering $a_{\text{-N-O}}$ as the primitive translation vector in real space, given by the distance between –N–O units, the allowed values for λ (or k, see Eq. (2.24)) are $a_{\text{-N-O}}, 2a_{\text{-N-O}}, 3a_{\text{-N-O}}, \ldots, \mathcal{N} a_{\text{-N-O}}$, where $\mathcal{N}$ is the number of –N–O units in this 1D crystal ($\mathcal{N} \sim 10^{23}$). Compared to the NO molecule in Section 2.1.3, instead of having the 11 electronic levels made of combinations of the N: $2s^2, 2p^3$ and O: $2s^2, 2p^4$ electrons, there will be $11 \times \mathcal{N}$ electronic levels. The energy dispersion relation ($E_{\boldsymbol{k}}$ vs. $\boldsymbol{k}$ plot) in reciprocal space will then be made of 11 energy bands, where the number of energy bands is given by the total number of atomic orbitals in the unit cell. Each energy band accepts $2\mathcal{N}$ electrons or 2 electrons per unit cell. While there will be 11 energy bands (similar to the 11 levels of the NO molecule), the lower 5 energy bands of the 11 energy bands are fully or partially occupied.

Like in the case of molecules, the electronic levels will be filled from the lowest to the highest energy. Because of the odd number of electrons in the unit cell, we expect metallic behavior in which the highest occupied energy band is half occupied. The electronic wavefuction can now change phase when moving along the one-dimensional crystal, with $\mathcal{N}$ wavevectors associated with each electronic band. The energy defining the boundary between the occupied and unoccupied levels is called the *Fermi energy*. If the Fermi energy falls within an electronic band, no energy will be required to take an electron from the occupied to the unoccupied state, and the material is therefore *metallic*.[6] If the Fermi energy falls within an energy gap between the valence (the highest occupied) and the conduction (the lowest unoccupied) energy bands, then the material will be *semiconducting* (with an energy gap on the order of 1 eV) or an *insulator* (gap on the order of 10 eV).[7] Graphene is an interesting system where the energy separation between the valence and conduction bands is zero, imposed by crystal symmetry. Therefore, graphene is a zero gap semiconductor (or metal) with a symmetry-imposed degeneracy between the valence and conduction bands at specific points in two-dimensional (k_x, k_y) reciprocal state [94].

Finally, we define the so-called *Brillouin zone*, which is a symmetry-based unit cell in reciprocal space providing a representation for a wavevector appropriate to a given crystal. The number of allowed $\boldsymbol{k}$ wavevectors inside a Brillouin zone is always limited by the number $\mathcal{N}$ of unit cells in the crystal, where $-\pi/a \le k \le \pi/a$ defines the so-called bounding region of the Brillouin zone. Figure 2.5 displays the case for $\lambda = 2a_{\text{-A-B}}$, where $k = \pi/a_{\text{-A-B}}$ at the boundary of the first Brillouin zone. Outside the first Brillouin zone ($k_{\text{out}} > \pi/a$, or $\lambda < 2a_{\text{-A-B}}$) the electronic structure repeats the electronic levels inside the first Brillouin zone with $k = k_{\text{out}} - K$,

6) This would likely happen to this hypothetical NO crystal, since there is one electron in the π^* state in the molecule (see Section 2.1.3).

7) This would likely happen to the hypothetical NO^+ crystal, since the LUMO is empty in the NO^+ molecule (see Section 2.1.3).

where $K = 2\pi/a_{-A-B}$ gives the reciprocal lattice vector in what is called the extended Brillouin zone [95]. This periodicity in reciprocal space is equally applicable to phonons, and we use this periodicity for phonons in Chapter 3. All the concepts briefly summarized in this section are broadly used in the study of sp^2 nanocarbons, as shown in the next sections and chapters. The reader who is not familiar with solid state physics may want to consult a more tutorial solid state physics presentation, such as the introductory text by Kittel [95].

2.2 Electrons in Graphene: the Mother of sp^2 Nanocarbons

The discussion of Section 2.1.5 is now applied to sp^2 nanocarbons. Graphene provides a simple illustration showing that the number of branches in the dispersion relations corresponds to the number of electrons in the unit cell. Graphene has two C atom sites per unit cell, which means 2 sets of $2s$ and $2p$ states (a total of 8 states per unit cell), so that there will be eight electronic energy bands, derived from the 3σ, $3\sigma^*$, 1π and $1\pi^*$ levels. The 8 electrons per unit cell will fill the 4 lower 3σ and 1π bonding energy bands with spin up and spin down electrons, and the 4 higher $3\sigma^*$ and $1\pi^*$ energy bands will be unoccupied.[8)]

2.2.1 Crystal Structure of Monolayer Graphene

The fundamental crystal structure that constitutes the basis for sp^2 carbon nanostructures is graphene, which is a two-dimensional (2D) planar structure based on a unit cell containing two carbon atoms A and B, as shown by the unit vectors $\boldsymbol{a}_1$ and $\boldsymbol{a}_2$ in Figure 2.6a. The carbon atoms in monolayer graphene are located at the vertices of the hexagons where $\boldsymbol{a}_1$ and $\boldsymbol{a}_2$ are unit vectors.

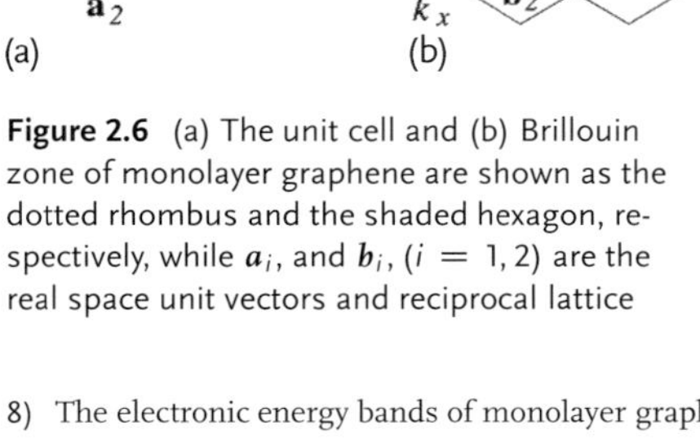

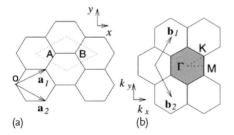

Figure 2.6 (a) The unit cell and (b) Brillouin zone of monolayer graphene are shown as the dotted rhombus and the shaded hexagon, respectively, while $\boldsymbol{a}_i$, and $\boldsymbol{b}_i$, ($i = 1, 2$) are the real space unit vectors and reciprocal lattice vectors, respectively. Energy dispersion relations are usually displayed along the perimeter of the dotted triangle connecting the high symmetry points, Γ, K and M (see inset to Figure 2.7).

8) The electronic energy bands of monolayer graphene are displayed in Figure 2.10.

As shown by Figure 2.6a, the real space unit vectors $\boldsymbol{a}_1$ and $\boldsymbol{a}_2$ of the hexagonal lattice are expressed in Cartesian coordinates as:

$$\boldsymbol{a}_1 = \left(\frac{\sqrt{3}}{2}a, \frac{a}{2}\right), \quad \boldsymbol{a}_2 = \left(\frac{\sqrt{3}}{2}a, -\frac{a}{2}\right), \tag{2.25}$$

where $a = |\boldsymbol{a}_1| = |\boldsymbol{a}_2| = 1.42 \times \sqrt{3} = 2.46$ Å is the lattice constant of monolayer graphene. Likewise, the unit cell in reciprocal space is shown by the shaded hexagon in Figure 2.6b and is described by the unit vectors $\boldsymbol{b}_1$ and $\boldsymbol{b}_2$ of the reciprocal lattice given by

$$\boldsymbol{b}_1 = \left(\frac{2\pi}{\sqrt{3}a}, \frac{2\pi}{a}\right), \quad \boldsymbol{b}_2 = \left(\frac{2\pi}{\sqrt{3}a}, -\frac{2\pi}{a}\right), \tag{2.26}$$

corresponding to a lattice constant of length $4\pi/\sqrt{3}a$ in reciprocal space. The unit vectors $\boldsymbol{b}_1$ and $\boldsymbol{b}_2$ of the reciprocal hexagonal lattice (see Figure 2.6b) are rotated by 30° from the unit vectors $\boldsymbol{a}_1$ and $\boldsymbol{a}_2$ in real space, respectively. The three high symmetry points of the Brillouin zone, Γ, K and M are the center, the corner, and the center of the edge of the hexagon, respectively. Other high symmetry points or lines are along ΓK (named T), KM (named T′) and ΓM (named Σ).

In monolayer graphene, three of the electrons form σ bonds which hybridize in a *sp*^{2} configuration, and the fourth electron of the carbon atom forms the $2p_z$ orbital, which is perpendicular to the graphene plane, and makes π covalent bonds. In Section 2.2.2 we use the tight-binding approximation to treat the covalent π energy bands for graphene which are the simplest for determining the solid state properties of graphene, reflecting the strong coupling of the in-plane carbon atoms. In Section 2.2.3 we review the σ-bands which, together with the π-bands, give the electronic structure of graphene.

2.2.2
The π-Bands of Graphene

In this section we review the derivation of the electronic π-bands of graphene based on the tight-binding model which is used here to provide an approximate description of the π-bands of monolayer graphene because of the very strong in-plane bonding between the carbon atoms in graphene. For a more detailed development of the tight-binding model applied to graphene and other *sp*^{2} carbon systems, see [31, 32].

Within the tight-binding method, the unperturbed eigenvectors are represented by atomic orbitals, and the crystalline potential is treated as a perturbation, thus forming the crystalline electronic states which are represented by Bloch states. Two Bloch functions (Φ_A and Φ_B), constructed from p_z atomic orbitals (φ, with $\Phi_{\mathrm{A,B}} \propto \sum_R e^{ik\cdot R}\varphi(r - R)$) for the two nonequivalent carbon atoms at A and B sites in Figure 2.6a, provide the basis functions for describing the electronic structure of monolayer graphene (1-LG). The secular equation is derived from a 2×2 Hamiltonian matrix, $H_{ij} = \langle \Phi_i | H | \Phi_j \rangle$, containing four matrix elements coupling Φ_A and Φ_B. When we consider only nearest neighbor interactions, then $\mathcal{H}_{AA} = \mathcal{H}_{BB} = \epsilon_{2p}$ for the diagonal matrix elements where ϵ_{2p} is the atomic $2p$ level energy of an isolated carbon atom. For the off-diagonal matrix element $\mathcal{H}_{AB}$, we must consider the three nearest neighbor B atoms relative to an A atom, which are denoted by the vectors $\boldsymbol{R}_1$, $\boldsymbol{R}_2$, and $\boldsymbol{R}_3$ connecting the A atom to its three nearest neighbor B atoms to obtain:

$$2\mathcal{H}_{AB} = t\left(e^{ik\cdot R_1} + e^{ik\cdot R_2} + e^{ik\cdot R_3}\right) = t\, f(k)\,, \tag{2.27}$$

where t is the nearest neighbor transfer integral ($\langle \varphi_A | \mathcal{H} | \varphi_B \rangle$) which is often called $-\gamma_0$ ($t = -\gamma_0$) in the literature, where γ_0 is given a positive value. The function $f(k)$ in Eq. (2.27) is a function of the sum of the phase factors of $e^{ik\cdot R_j}$ ($j = 1, \cdots, 3$). Using the x, y coordinates of Figure 2.6a, $f(k)$ is given by

$$f(k) = e^{ik_x a/\sqrt{3}} + 2e^{-ik_x a/2\sqrt{3}} \cos\left(\frac{k_y a}{2}\right). \tag{2.28}$$

Since $f(k)$ is a complex function, and the Hamiltonian forms a Hermitian matrix, we write $\mathcal{H}_{BA} = \mathcal{H}^*_{AB}$ in which * denotes the complex conjugate. Using Eq. (2.28), the overlap integral matrix, $S_{ij} = \langle \Phi_A | \Phi_B \rangle$ is given by $\mathcal{S}_{AA} = \mathcal{S}_{BB} = 1$, and $\mathcal{S}_{AB} = s\, f(k) = \mathcal{S}^*_{BA}$, with the nearest neighbor overlap integral for p_z wavefunctions, $s = \langle \varphi_A | \varphi_B \rangle$. The explicit forms for $\mathcal{H}$ and $\mathcal{S}$ can be written as:

$$\mathcal{H} = \begin{pmatrix} \epsilon_{2p} & t\, f(k) \\ t\, f(k)^* & \epsilon_{2p} \end{pmatrix}, \quad \mathcal{S} = \begin{pmatrix} 1 & s\, f(k) \\ s\, f(k)^* & 1 \end{pmatrix}. \tag{2.29}$$

Solving the secular equation $\det(\mathcal{H} - E\mathcal{S}) = 0$ (where "det" denotes the determinant) and using $\mathcal{H}$ and $\mathcal{S}$ as given in Eq. (2.29), the eigenvalues $E(\boldsymbol{k})$ for the graphene π-bands are obtained as a function $\boldsymbol{k} = (k_x, k_y)$:

$$E(\boldsymbol{k}) = \frac{\epsilon_{2p} \pm t w(\boldsymbol{k})}{1 \pm s w(\boldsymbol{k})}\,, \tag{2.30}$$

where the + signs in the numerator and denominator go together giving the bonding π energy band, and likewise for the − signs, which give the antibonding π^*-band as symmetric and antisymmetric combinations of Φ_A and Φ_B, respectively, (see Section 2.1.2), while the function $w(\boldsymbol{k})$ is given by

$$w(\boldsymbol{k}) = \sqrt{|f(\boldsymbol{k})|^2} = \sqrt{1 + 4\cos\frac{\sqrt{3}k_x a}{2}\cos\frac{k_y a}{2} + 4\cos^2\frac{k_y a}{2}}\,. \tag{2.31}$$

In Figure 2.7, the electronic energy dispersion relations for the π-bands of monolayer graphene are shown throughout the two-dimensional first Brillouin zone and the inset shows the energy dispersion relations along the high symmetry axes along

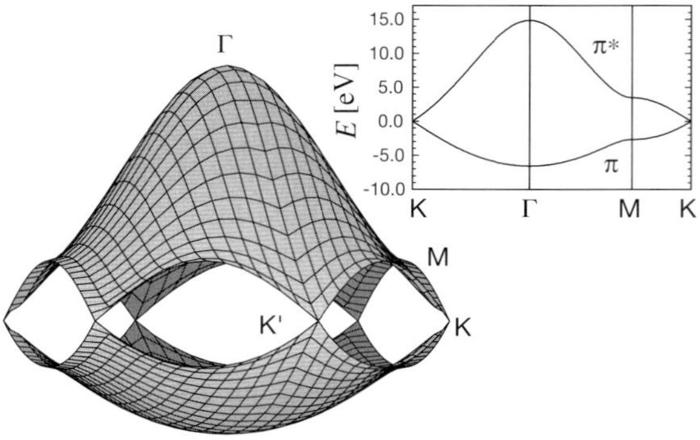

Figure 2.7 The energy dispersion relations for 2D graphite are shown throughout the whole region of the first Brillouin zone [31]. The valence and conduction bands of graphene touch at the points K and K', which are related by time reversal symmetry [94]. The inset shows the energy dispersion $E(k)$ along the high symmetry directions of the triangle $\Gamma M K$, shown in Figure 2.6b.

the perimeter of the triangle shown in Figure 2.6b. Here we use values of the parameters $\epsilon_{2p} = 0$, $t = -3.033$ eV, and $s = 0.129$ in order to reproduce the first principles calculation of the graphite energy bands [16, 96]. The upper half of the energy dispersion curves describes the π^*-energy "antibonding" band, and the lower half is the π-energy bonding band[9]. Since there are two π electrons per unit cell, these two π electrons fully occupy the lower π-band. Therefore, the π-band is filled by spin up and spin down electrons, while the π^*-band is empty. The upper π^*-band and the lower π-band are degenerate at the K (K') points through which the Fermi energy passes for an undoped monolayer graphene sample.

The existence of a zero gap at the K (K') points comes from the symmetry requirement that the two carbon sites A and B in the hexagonal lattice are distinct but equivalent to each other by symmetry. If the A and B sites had different atoms, such as B and N, then the site energy ϵ_{2p} would be different for B and N, and therefore the calculated energy dispersion would show an energy gap between the π and π^*-bands ($E_g = 3.5\,\text{eV} = \epsilon_{2p}^{\text{B}} - \epsilon_{2p}^{\text{N}}$ for BN).

When the graphene overlap integral s becomes zero, the π and π^*-bands become symmetrical around $E = \epsilon_{2p}$, which can be understood from Eq. (2.30). The energy dispersion relations in the case of $s = 0$ are commonly used as a simple approximation for the electronic structure of a graphene layer near $E = \epsilon_{2p}$:

$$E(k_x, k_y) = \pm t \left\{ 1 + 4\cos\left(\frac{\sqrt{3}k_x a}{2}\right)\cos\left(\frac{k_y a}{2}\right) + 4\cos^2\left(\frac{k_y a}{2}\right) \right\}^{1/2} . \tag{2.32}$$

In this case, the electronic energies have values of $\pm 3t$, $\pm t$ and 0, respectively, at

9) The bonding and antibonding assignment is not strictly correct for graphene because of the hexagonal symmetry.

the high symmetry points, Γ, M and K in the Brillouin zone, and the band width is $6t$, which is consistent with the three π bonds per atom.

It should be noted that the valence and conduction bands come into the K (K') point with a linear $E(k)$ relation. The $E(k)$ relation about the K and K' points was already shown to have a linear dependence of $E(k)$ in the early work of Wallace [53]. Most of the graphene literature makes use of the relation Eq. (2.32) for $s = 0$, and uses the lowest order term in the expansion of this equation around the K and K' points in the Brillouin zone, which are related by time inversion symmetry. This yields

$$E^{\pm}(k) = \pm\hbar v_{\mathrm{F}}|k| \,, \tag{2.33}$$

where v_{F} is the Fermi velocity of π electrons ($\sim 10^6$ km/s) given by

$$v_{\mathrm{F}} = \sqrt{3}(\gamma_0 a/2\hbar) \,, \tag{2.34}$$

and $a = \sqrt{3}a_{\text{C–C}}$ is the lattice constant of graphene and $a_{\text{C–C}} = 1.42$ Å is the nearest neighbor carbon–carbon distance [53].

It is interesting to point out that the linear dispersion given by Eq. (2.33) is the solution to the massless Dirac Hamiltonian at the $K(K')$ point [97]:

$$\mathcal{H} = \hbar v_F(\sigma \cdot \kappa) \,, \tag{2.35}$$

where $\kappa = -i\nabla$, and the σ are the Pauli matrices operating in the space of the electron wave function amplitude on the A,B sublattices of graphene (pseudo spin). Equation (2.35) gives a "chiral" nature to the quasi-particles defined by Eq. (2.33) [52]. The Dirac Hamiltonian of Eq. (2.35) (or the effective mass approximation model) gives good insights into the relativistic nature of electrons in monolayer graphene, and has been important for describing transport effects near the Fermi level. However, its accuracy is limited to low energies and care should be taken when using this expression to analyze optical phenomena. Nevertheless, in the visible range the linear k dispersion relation (see Figure 2.7) is usually accurate enough to explain most experimental results.

2.2.3
The σ-Bands of Graphene

Let us next consider the σ-bands of graphene. There are three atomic orbitals of sp^2 covalent bonding per carbon atom, $2s, 2p_x$ and $2p_y$. We thus have six Bloch orbitals in the 2 atom unit cells, yielding six σ-bands for the 6×6 Hamiltonian matrix. We calculate the electronic structure for these six σ-bands using this 6×6 Hamiltonian and the corresponding (6×6) overlap matrix, and we then solve the secular equation for each $\boldsymbol{k}$ point. Since the planar geometry of graphene satisfies the even symmetry of the Hamiltonian $\mathcal{H}$ and of the symmetry operators $2s, 2p_x$ and $2p_y$ upon mirror reflection about the xy plane, and the odd symmetry of the operator $2p_z$, the σ and π energy bands can be solved separately, because the matrix elements of different symmetry types do not couple in the Hamiltonian. For the eigenvalues thus obtained, three of the six σ-bands are bonding σ-bands which appear below the Fermi energy, and the other three σ-bands are antibonding σ^*-bands which appear above the Fermi energy.

The calculation of the Hamiltonian and overlap matrix is performed analytically, using a small number of parameters. Hereafter we arrange the matrix elements in accordance with their atomic identity for the free atom: $2s^{\mathrm{A}}, 2p_x^{\mathrm{A}}, 2p_y^{\mathrm{A}}, 2s^{\mathrm{B}}, 2p_x^{\mathrm{B}}, 2p_y^{\mathrm{B}}$. Then the matrix elements coupling the same atoms (for example A and A) can be expressed by a 3×3 small matrix, which is a sub-matrix of the 6×6 matrix. Within the nearest neighbor site approximation, the small Hamiltonian and overlap matrices are diagonal matrices as follows:

$$\mathcal{H}_{AA} = \begin{pmatrix} \epsilon_{2s} & 0 & 0 \\ 0 & \epsilon_{2p} & 0 \\ 0 & 0 & \epsilon_{2p} \end{pmatrix}, \quad \mathcal{S}_{AA} = \begin{pmatrix} 1 & 0 & 0 \\ 0 & 1 & 0 \\ 0 & 0 & 1 \end{pmatrix}, \tag{2.36}$$

where ϵ_{2s} and ϵ_{2p} denote the orbital energy of the $2s$ and $2p$ levels.

The matrix element for the Bloch orbitals between the A and B atoms can be obtained by taking the components of $2p_x$ and $2p_y$ in the directions parallel or perpendicular to the σ bond. In Figure 2.8, we show how to rotate the $2p_x$ atomic orbital and how to obtain the σ and π components for the rightmost bonds of this figure.[10] In Figure 2.8 the wavefunction of $|2p_x\rangle$ is decomposed into its σ and π components as follows:

$$|2p_x\rangle = \cos\frac{\pi}{3}|2p_\sigma\rangle + \sin\frac{\pi}{3}|2p_\pi\rangle\,. \tag{2.37}$$

This type of decomposition is called the Slater–Koster method [94].

By rotating the $2p_x$ and $2p_y$ orbitals in the directions parallel and perpendicular to the desired bonds, the matrix elements appear in only 8 patterns as shown in Figure 2.9, where shaded and nonshaded regions denote positive and negative

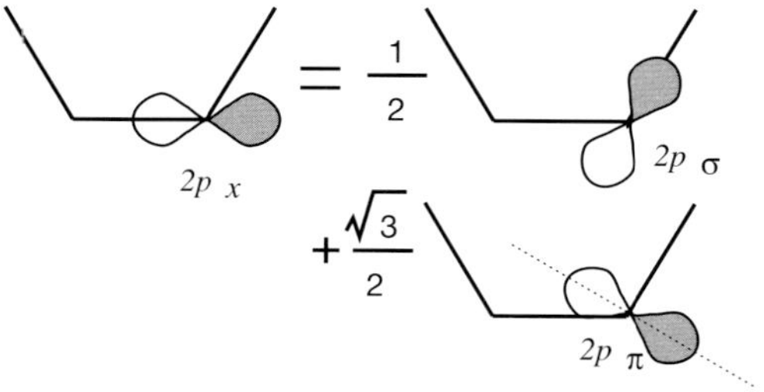

Figure 2.8 The rotation of $2p_x$. The figure shows how to project $2p_x$ into its σ and π components in the direction of the right C–C bond. This method is valid only for p orbitals [31].

10) Here the π component (in-plane) has nothing to do with the π orbital (out-of-plane) discussed in Section 2.2.2. The π component is named π because it is perpendicular to the considered σ orbital.

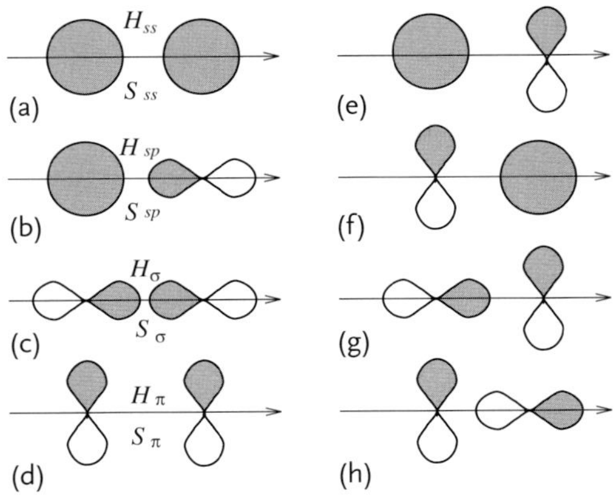

Figure 2.9 The band parameters for σ-bands. The four cases from (a) to (d) correspond to matrix elements having nonvanishing values and the remaining four cases from (e) to (h) correspond to matrix elements with zero values [31].

amplitudes of the wavefunctions, respectively. The four cases from Figure 2.9a to d correspond to nonvanishing matrix elements and the remaining four cases from Figure 2.9e to h correspond to matrix elements that vanish because of symmetry. The corresponding parameters for both the Hamiltonian and the overlap matrix elements are also shown in Figure 2.9.

When all the matrix elements of the 6×6 Hamiltonian and overlap matrices are calculated [31, 96], the energy dispersion of the σ-bands can be obtained from solution of the secular equation. Since the analytic solution of the 6×6 Hamiltonian is too complicated for practical use, we solve the Hamiltonian numerically by using, for example, the Lapack software package.[11] The results thus obtained for the calculated σ and π energy bands are shown in Figure 2.10, which result from a fit of the functional form of the energy bands imposed by symmetry to the energy values obtained from the first principles band calculations at the high symmetry points [31, 96].

2.2.4
N-Layer Graphene Systems

When joining graphene layers to form N-layer graphene (N-LG) with the Bernal AB stacking structure, the unit cell will be formed by $2N$ atoms. Consequently, the π and π^*-bands will split into symmetric and antisymmetric combinations of the graphene states. Figure 2.11a,b shows the unit cell for $N = 2$, that is, bilayer graphene (2-LG), and Figure 2.11c shows its π-band electronic structure. For 3-LG

11) Lapack is a linear algebra package written using Fortran or C languages. You can download the library as free software and the programs have been used and checked to be correct by many groups. We do not need to use a sub-program for matrix calculations but just call this library. For further details, search for "LAPACK" on the Internet. There are several versions of the Lapack library. The Intel compiler supports Lapack under the name of Math Kernel Library.

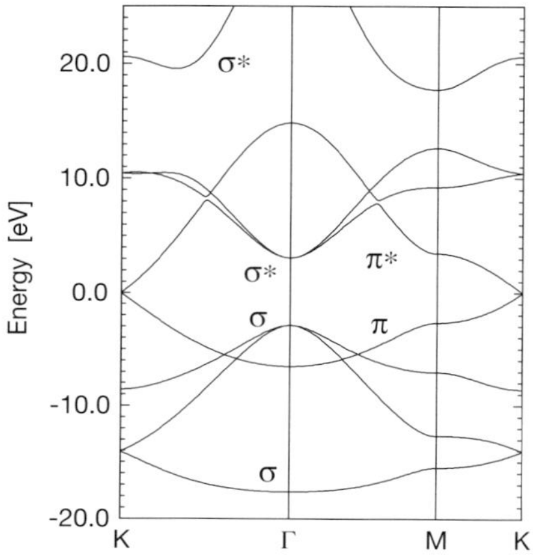

Figure 2.10 The energy dispersion relations along high symmetry directions for σ and π-bands of monolayer graphene [31]. The Fermi level (E_F) was chosen as the zero energy.

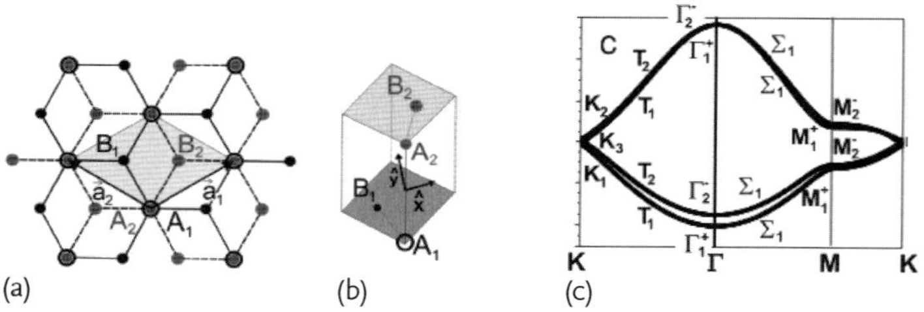

Figure 2.11 (a) The real space top-view of the setting for the unit cell for bilayer graphene (2-LG) with AB Bernal stacking, showing the non-equivalent A_1 and B_1 carbon atoms in the first layer and the A_2 and B_2 carbon atoms in the second layer. The unit cell vectors $\boldsymbol{a}_1$ and $\boldsymbol{a}_2$ are shown, considering the origin to be half way between atoms A_1 and A_2. The A atoms are above one another on adjacent layers, but the B atoms are staggered on adjacent layers, as shown in a 3D view in panel (b). (c) The electronic dispersion for the 2-LG π electrons calculated by DFT (density functional theory) along the KΓMK directions. The energy band labeling comes from group theory and will be discussed in Chapter 6 [98].

with AB stacking, atoms A_3 and B_3 would be placed on top of A_1 and B_1 in the top-view of Figure 2.11a and it would exhibit 3π and $3\pi^*$-bands. The stacking of 4-LG would look exactly like two 2-LG blocks on top of each other, and so on. The electronic structure of 2-LG can be described by the phenomenological Slonczewski–Weiss–McClure (SWM) model [16, 99, 100]. Since the unit cell of 2-LG with the Bernal AB stacking structure is the same as for graphite, which also has the same layer stacking structure, we can denote the electronic spectrum of bilayer graphene in terms of a model closely related to the SWM model for graphite. A larger set of parameters (γ_0, γ_1, γ_3, and γ_4),[12] that are associated with overlap and transfer integrals calculated for nearest neighbors atoms up to adjacent layers will be need-

12) γ_2 and γ_5 are transfer integrals for next-nearest layers.

ed to describe the electronic structure (more about this in Section 11.3). However, even in 3D graphite, the interaction between two adjacent layers is small compared with intralayer interactions, since the layer-layer separation of 3.35 Å is much larger than the nearest neighbor distance between two carbon atoms, $a_{\text{C–C}} = 1.42$ Å. Thus the electronic structure of graphene provides a building block for the electronic structure for N-LG and 3D graphite.

One important fact for N-LG is that the linear energy dispersion of 1-LG appears for odd-number LG near the Fermi energy, while parabolic energy dispersion appears for even-number LG. Koshino and Ando [101] explain this fact by showing that the Hamiltonian can be decoupled into 2×2 sub-matrices if we consider only γ_1 for the interlayer interaction [101]. Thus depending on whether we have an odd or even number of graphene layers, the effective mass of the carriers of N-LG becomes zero or finite, respectively, which is analogous to elementary particle physics in which two kinds of particles exist, such as massless (photon, neutrino) Bosons, and finite mass (electron, proton) Fermions depending on symmetry. When we consider the Fermi velocity $v_F = 1 \times 10^6$ m/s, which is $\sim c/300$ as the velocity of light, we can make an analogy between graphene and particle physics.

2.2.5 Nanoribbon Structure

When going from a bulk material to a low-dimensional structure, the electronic states are constrained by quantum effects in the nanoscale directions. If the low-dimensional system has the same crystal structure as the parent higher-dimensional material, the electronic states of the low-dimensional system can be considered as a subset of the electronic states of the bulk material. When we move from the two-dimensional graphene sheet to the one-dimensional carbon nanoribbon (or nanotube), the wave vector components in the nanoscale directions can only take on discrete values in order to maintain an integral number of wave function nodes, that is, these wave vector components then become quantized. The number of quantized states for a given orbital of each atom (such as $2s, 2p_x \ldots$) is equal to the number of unit cells of the parent higher-dimensional material in the nanoscale directions of the lower-dimensional structure.

The general procedure for confining the two-dimensional electronic structure of graphene into a one-dimensional structure will be discussed in detail for carbon nanotubes in Section 2.3. But before discussing nanotubes in detail, let us briefly mention graphene nanoribbons. Such nanoribbons consist of graphene with a finite width and infinite length, as shown in Figure 2.12. Thus the unit cell of a nanoribbon consists of 2N carbon atoms[13] in the direction of the width, while periodicity appears in the length direction. The wavevectors are quantized in the direction of the width, and 2N 1D energy sub-bands for the π ($2p_z$) band appears if we simply adopt the zone-folding method (see Figure 2.13).

13) Here N is the number of CC dimers along the ribbon width. Here we use *N* (italic font) for the number of graphene layers.

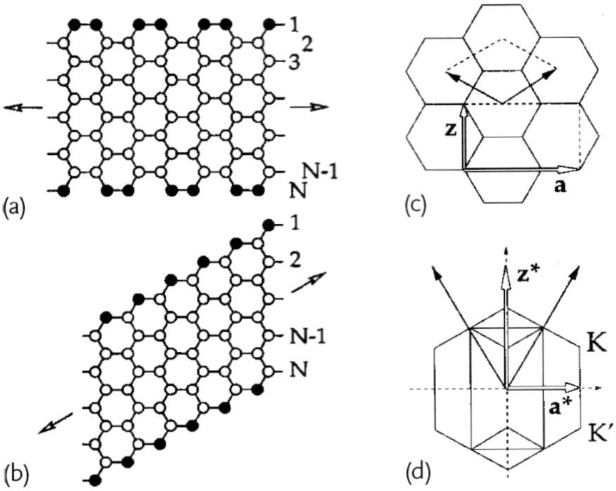

Figure 2.12 (a,b) The network skeleton of two nanoribbons. (a) N = 10 lines of C atoms from one edge to the other, and an armchair-like edge structure. (b) N = 5 and a zigzag-like edge structure. The arrows indicate the translational directions of the graphene ribbons. Unit cells in real space (c) and reciprocal space (d) of 2D graphite. The vectors $\boldsymbol{a}$ and $\boldsymbol{a}^*$ (z and z^*) relate to armchair (zigzag) ribbons [102].

The method of constructing 1D electronic energy sub-bands by *cutting* the 2D electronic dispersion relations with these lines is known as the "zone-folding scheme" [31]. The cutting lines represent the allowed k vectors for the 1D nanorib-

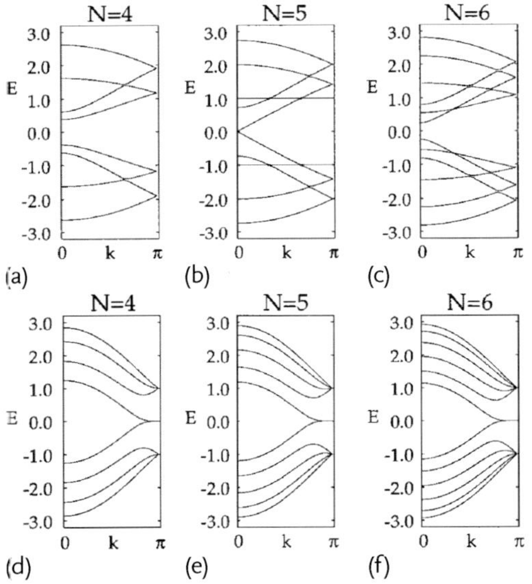

Figure 2.13 Band structure $E(k)$ of graphene nanoribbons of various widths obtained from the zone-folding procedure. Armchair nanoribbons with N = 4 (a), N = 5 (b) and N = 6 (c), and zigzag nanoribbons with N = 4 (d), N = 5 (e), and N = 6 (f) [102].

bon represented in the 2D Brillouin zone of graphene, which are continuous along the ribbon axis and discrete along the tube width. The length of each cutting line is $2\pi/T$, where T is the 1D unit vector in the translational direction of the nanoribbon or nanotube axis. The separation between two adjacent cutting lines is inversely proportional to the nanoribbon width (nanotube diameter). The orientation of the cutting lines in 2D reciprocal space is determined by the cutting direction, that is, the relative orientation of the nanoribbon axis with respect to the principal axes of graphene (the unrolled flat layer of the 2D parent graphene material [31]). The structure of the edges is very important. We can consider two possible edges, armchair and zigzag edges, which are more stable than the other shape of edges [103] and whose structures are shown in Figure 2.12. The nanoribbons with armchair and zigzag edges are called, respectively, armchair nanoribbons (A-NR) and zigzag nanoribbons (Z-NR). For both cases, the edge carbon atoms have two σ bonds and one π bond, while the remaining σ bonds exist as either being terminated by H atoms or by dangling bonds.

We see that for the π-band for nanoribbons, a flat energy band appears around the Fermi energy in the electronic energy dispersion from the K to M points for Z-NRs (see Figure 2.13d–f), while no edge states appear for A-NRs (see Figure 2.13a–c). Thus the density of states near the Fermi energy is singular for Z-NRs. In the case of A-NRs, the energy gap is oscillating as a function of N, and for $N = 3n - 1$ the A-NRs become metallic, while they are semiconducting in the other cases (see Figure 2.13a–c).

While the zone-folding procedure works as a first approximation to the electronic structure for nanoribbons, the presence of edge states can significantly alter their fundamental electronic properties. For example, different from what is shown in Figure 2.13, *ab initio* calculations [104] and experiments [105] show that because of the localized edge states, all nanoribbons are semiconducting materials with an energy gap magnitude depending on the ribbon width, which depends on N. The electron amplitude ratio between sites A and B is expressed by the pseudo-spin, and the edge states in Z-NRs can be understood as pseudo-spin polarized states. From this we can derive many interesting physical phenomena, such as half-metallicity (only one of two spin currents exist at the Fermi energy) and the occurrence of magnetism at the zigzag edges [104, 106]. Once you close the ribbon structure into a carbon nanotubes, this complex edge physics is gone, as discussed in the next section.

2.3 Electrons in Single-Wall Carbon Nanotubes

In this section we review the structure of carbon nanotubes (Section 2.3.1), their electronic dispersion relations (Section 2.3.2), and their density of electronic states (Section 2.3.3). In Section 2.3.4 we explain the importance of both the carbon nanotube electronic structure and the laser excitation energy on the details of the observed Raman spectra.

2.3.1 Nanotube Structure

A single-wall carbon nanotube (SWNT) is constructed starting from a graphene layer by rolling it up into a seamless cylinder [31]. The graphene layer is oriented with respect to the coordinate system in such a way that the armchair direction lies along the x-axis and the zigzag direction is along the y-axis, as shown in Figure 2.14. The nanotube structure is uniquely determined by the chiral vector $\boldsymbol{C}_h$ which spans the circumference of the cylinder when the graphene layer is rolled up into a tube. The chiral vector can be written in the form $\boldsymbol{C}_h = n\boldsymbol{a}_1 + m\boldsymbol{a}_2$, where n and m are integers and where the vectors $\boldsymbol{a}_1$ and $\boldsymbol{a}_2$ bounding the unit cell of the graphene layer with the two distinct carbon atom sites A and B are shown in Figure 2.14. In the shortened (n, m) form, the chiral vector is written as a pair of integers, and the same notation is widely used to characterize the geometry of each distinct (n, m) nanotube.

The nanotube can also be characterized by its diameter d_t and chiral angle θ from a zigzag direction, which determine the length $C_h = |\boldsymbol{C}_h| = \pi d_t$ of the chiral vector and its orientation on the graphene layer (see Figure 2.14). Both d_t and θ are expressed in terms of the indices n and m by the relations $d_t = a\sqrt{n^2 + nm + m^2}/\pi$ and $\tan\theta = \sqrt{3}m/(2n + m)$, as one can derive from Figure 2.14, where $a = |\boldsymbol{a}_1| = |\boldsymbol{a}_2| = \sqrt{3}a_{C\text{–}C} = 0.246$ nm is the lattice constant for the graphene layer and $a_{C\text{–}C} = 0.142$ nm is the nearest neighbor C–C distance [31]. As an example, the chiral vector $\boldsymbol{C}_h$ shown in Figure 2.14 is given by $\boldsymbol{C}_h = 4\boldsymbol{a}_1 + 2\boldsymbol{a}_2$, and thus the corresponding nanotube can be identified by the integer pair (4, 2). Due to the six-fold symmetry of one graphene layer, all non-equivalent nanotubes can be characterized by the (n, m) pairs of integers where

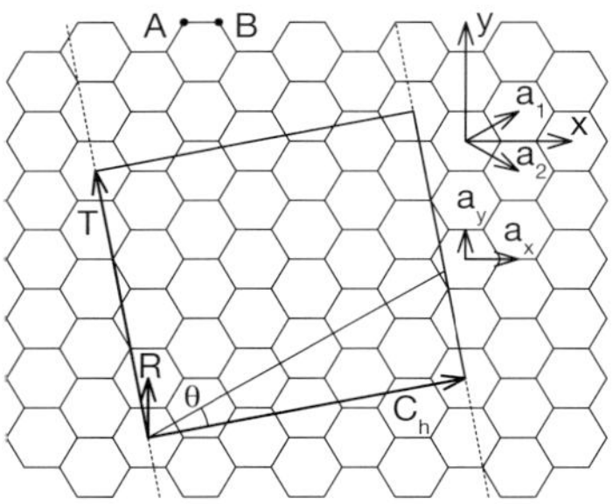

Figure 2.14 An unrolled nanotube projected on the graphene layer. When the nanotube is rolled up, the chiral vector $\boldsymbol{C}_h$ turns into the circumference of the cylinder, and the translation vector $\boldsymbol{T}$ is aligned along the cylinder axis. $\boldsymbol{R}$ is the symmetry vector and θ is the chiral angle (see text). The unit vectors $(\boldsymbol{a}_1, \boldsymbol{a}_2)$ of the graphene layer are indicated.[14)] The non-equivalent A and B sites within the unit cell of the graphene layer are shown at the top [31].

14) Notice the origin of the $(\boldsymbol{a}_1, \boldsymbol{a}_2)$ in this figure are chosen differently from that described in Figure 2.6a.

$0 \leq m \leq n$. The nanotubes are classified as chiral ($0 < m < n$) and achiral ($m = 0$ or $m = n$), which in turn are known as zigzag ($m = 0$) and armchair ($m = n$) nanotubes. A (4, 2) chiral nanotube is one of the smallest diameter nanotubes ever synthesized [107], requiring special calculational treatment because of its large curvature [108].

The unit cell of an unrolled nanotube on a graphene layer is a rectangle bounded by the vectors $\boldsymbol{C}_h$ and translational vector $\boldsymbol{T}$ (see the rectangle shown in Figure 2.14 for the (4, 2) nanotube). $\boldsymbol{T}$ is given by $t_1\boldsymbol{a}_1 + t_2\boldsymbol{a}_2$, where integers t_1 and t_2 are obtained by $\boldsymbol{C}_h \cdot \boldsymbol{T} = 0$ and $\text{gcm}(t_1, t_2) = 1$. Here gcm is an integer function of the greatest common multiplier of (n, m). The area of the nanotube unit cell can be easily calculated as a vector-product of these two vectors, $|\boldsymbol{C}_h \times \boldsymbol{T}| = \sqrt{3}a^2\left(n^2 + nm + m^2\right)/d_R$, where $d_R = \text{gcm}(2n + m, 2m + n)$. Using d_R, then t_1 and t_2 are given by $t_1 = (2m + n)/d_R$ and $t_2 = -(2n + m)/d_R$.

Dividing the cross product $|\boldsymbol{C}_h \times \boldsymbol{T}|$ by the area of the unit cell of a graphene layer $|\boldsymbol{a}_1 \times \boldsymbol{a}_2| = \sqrt{3}a^2/2$, one can get the number of hexagons in the unit cell of a nanotube, $N = 2\left(n^2 + nm + m^2\right)/d_R$. For the (4, 2) nanotube we have $N = 28$, so that the unit cell of the (4, 2) nanotube (see the rectangle shown in Figure 2.14) contains 28 hexagons, or $2 \times 28 = 56$ carbon atoms (see Table 2.1) [31].

The unit cell of a graphene layer is defined by the vectors $\boldsymbol{a}_1$ and $\boldsymbol{a}_2$. The graphene reciprocal lattice unit vectors $\boldsymbol{b}_1$ and $\boldsymbol{b}_2$ can be constructed from $\boldsymbol{a}_1$ and $\boldsymbol{a}_2$ using the standard definition $\boldsymbol{a}_i \cdot \boldsymbol{b}_j = 2\pi\delta_{ij}$, where δ_{ij} is the Kronecker delta symbol. The resulting reciprocal lattice unit vectors, $\boldsymbol{b}_1 = \boldsymbol{b}_x + \boldsymbol{b}_y$ and $\boldsymbol{b}_2 = \boldsymbol{b}_x - \boldsymbol{b}_y$, where $\boldsymbol{b}_x = 2\pi\hat{\boldsymbol{k}}_x/\left(\sqrt{3}a\right)$ and $\boldsymbol{b}_y = 2\pi\hat{\boldsymbol{k}}_y/a$, form the unit vectors for the hexagonal reciprocal lattice, as shown in Figure 2.15. Note the rotation of the hexagons in real space (Figure 2.14) and in reciprocal space (Figure 2.15).

In a similar fashion, the reciprocal space of a nanotube can be constructed [31]. The unrolled unit cell of the nanotube on a graphene layer is defined by the vectors $\boldsymbol{C}_h$ and $\boldsymbol{T}$, and therefore the reciprocal space vectors for the nanotube, $\boldsymbol{K}_1$ and $\boldsymbol{K}_2$, can be constructed using the standard definition, $\boldsymbol{C}_h \cdot \boldsymbol{K}_1 = \boldsymbol{T} \cdot \boldsymbol{K}_2 = 2\pi$ and $\boldsymbol{C}_h \cdot \boldsymbol{K}_2 = \boldsymbol{T} \cdot \boldsymbol{K}_1 = 0$. The vector $\boldsymbol{K}_1$ can be written in the form $\boldsymbol{K}_1 \propto t_2\boldsymbol{b}_1 - t_1\boldsymbol{b}_2$ to provide its orthogonality to the vector $\boldsymbol{T}$, taking into account that $\boldsymbol{a}_i \cdot \boldsymbol{b}_j = 2\pi\delta_{ij}$. Similarly, $\boldsymbol{K}_2 \propto m\boldsymbol{b}_1 - n\boldsymbol{b}_2$ is orthogonal to $\boldsymbol{C}_h$. The normalization conditions

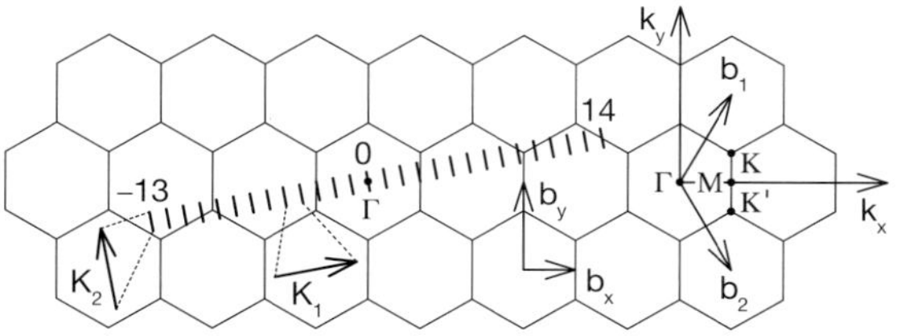

Figure 2.15 Reciprocal space of the graphene layer. Parallel equidistant lines represent the cutting lines for the (4, 2) nanotube, labeled by the cutting line index μ, which assumes values from $1 - N/2 = -13$ to $N/2 = 14$. The reciprocal lattice unit vectors ($\boldsymbol{b}_1$, $\boldsymbol{b}_2$) are indicated in this figure along with the (zoomed) reciprocal lattice unit vectors ($\boldsymbol{K}_1$, $\boldsymbol{K}_2$) of the nanotube [31].

$C_h \cdot K_1 = T \cdot K_2 = 2\pi$ are used to calculate the proportionality coefficients, yielding the magnitudes of the reciprocal space vectors, $|K_1| = 2/d_t$ and $|K_2| = 2\pi/|T|$. This results in the following expressions for the reciprocal space vectors, $K_1 = -(t_2 b_1 - t_1 b_2)/N$ and $K_2 = (m b_1 - n b_2)/N$ (see Table 2.1). Using the reciprocal space vectors K_1 and K_2, we can now construct the cutting lines for the nanotube as shown in Figure 2.15. The vectors K_1 and K_2 are orthogonal, and K_2 is directed along the nanotube axis, so that the cutting lines are also aligned along the tube axis.

The unrolled nanotube is extended in the direction of the translation vector T and has a nanoscale size in the direction of the chiral vector C_h (see Figure 2.14). Since the translation vector T is collinear with the wave vector K_2, and the chiral vector C_h corresponds to the wave vector K_1, the unrolled reciprocal space of the nanotube (see Figure 2.15) is quantized along the K_1 direction and is continuous along the K_2 direction.

Consequently, the N wave vectors μK_1, where μ is an integer number varying from $(1 - N/2)$ to $N/2$ (note that N is always even), form the N quantized states in the direction K_1 of the unrolled reciprocal space of the nanotube. Each of these N quantized states gives rise to a line segment of length $K_2 = |K_2|$ along the direction K_2 in the unrolled reciprocal space of the nanotube. These N line segments, defined by the wave vectors K_1 and K_2, represent the cutting lines in the unrolled reciprocal space of the nanotube. The length and orientation of each cutting line in reciprocal space is given by the wave vector K_2, while the separation between two adjacent cutting lines is given by the wave vector K_1. In the case of our model (4, 2) nanotube, the $N = 28$ cutting lines are shown in Figure 2.15 numbered by the index μ varying from $1 - N/2 = -13$ to $N/2 = 14$, where the middle cutting line $\mu = 0$ crosses the Γ point, the center of the first Brillouin zone of the graphene layer. In the case of an ideal infinitely long nanotube, the wave vectors along the nanotube axis (along the K_2 vector) would be continuous. If the nanotube length L is small enough, yet still much larger than the unit cell length $T = |T|$, the wave vector along the nanotube axis also becomes quantized, $\xi(T/L)K_2$, where ξ is an integer number ranging from $(2T - L)/(2T)$ to $L/(2T)$. Such quantization effects in short carbon nanotubes have been observed experimentally [109]. The SWNT parameters are summarized in Table 2.1.

2.3.2
Zone-Folding of Energy Dispersion Relations

The electronic structure of a single-wall nanotube can be obtained simply from that of two-dimensional graphite. By using periodic boundary conditions in the circumferential direction denoted by the chiral vector C_h, the wave vector associated with the C_h direction becomes quantized, while the wave vector associated with the direction of the translational vector T (or along the nanotube axis) remains

Table 2.1 Parameters for single-wall carbon nanotubes.[a]

Symbol	Name	Formula
a	Graphene lattice constant	$a=\sqrt{3}a_{\text{C–C}}=0.246\,\text{nm}$
$\boldsymbol{a}_1,\ \boldsymbol{a}_2$	Graphene unit vectors	$\left(\frac{\sqrt{3}}{2},\frac{1}{2}\right)a,\ \left(\frac{\sqrt{3}}{2},-\frac{1}{2}\right)a$
$\boldsymbol{b}_1,\ \boldsymbol{b}_2$	Graphene reciprocal lattice vectors	$\left(\frac{1}{\sqrt{3}},1\right)\frac{2\pi}{a},\ \left(\frac{1}{\sqrt{3}},-1\right)\frac{2\pi}{a}$
$\boldsymbol{C}_h$	Nanotube chiral vector	$\boldsymbol{C}_h=n\boldsymbol{a}_1+m\boldsymbol{a}_2\equiv(n,m)$
C_h	Length of $\boldsymbol{C}_h$	$C_h=\lvert\boldsymbol{C}_h\rvert=a\sqrt{n^2+m^2+nm}$
d_t	Nanotube diameter	$d_t=C_h/\pi$
θ	Nanotube chiral angle	$\tan\theta=\frac{\sqrt{3}m}{2n+m}$
d	$\gcd(n,m)$[b]	
d_R	$\gcd(2n+m,2m+n)$[b]	$d_R=\begin{cases} d & \text{if } (n-m) \text{ is not a multiple of } 3d \\ 3d & \text{if } (n-m) \text{ is a multiple of } 3d \end{cases}$
N	Number of hexagons in the nanotube unit cell	$N=\frac{2(n^2+m^2+nm)}{d_R}$
$\boldsymbol{T}$	Translational vector	$\boldsymbol{T}=t_1\boldsymbol{a}_1+t_2\boldsymbol{a}_2$
$t_1,\ t_2$	along nanotube axis	$t_1=\frac{2m+n}{d_R},\quad t_2=-\frac{2n+m}{d_R}$
T	Length of $\boldsymbol{T}$	$T=\lvert\boldsymbol{T}\rvert=\frac{\sqrt{3}C_h}{d_R}$
$\boldsymbol{R}$	Symmetry vector	$\boldsymbol{R}=p\boldsymbol{a}_1+q\boldsymbol{a}_2$
$p,\ q$	of the nanotube	$t_1q-t_2p=1,\quad 1\le mp-nq\le N$
τ	Pitch of $\boldsymbol{R}$	$\tau=\frac{(mp-nq)T}{N}=\frac{MT}{N}$
ψ	Rotation angle of $\boldsymbol{R}$	$\psi=\frac{2\pi}{N}$
M	Number of $\boldsymbol{T}$ in $N\boldsymbol{R}$	$N\boldsymbol{R}=\boldsymbol{C}_h+M\boldsymbol{T},\quad M=mp-nq$
$\boldsymbol{K}_1$	Nanotube reciprocal	$\boldsymbol{K}_1=-(t_2\boldsymbol{b}_1-t_1\boldsymbol{b}_2)/N$
$\boldsymbol{K}_2$	lattice vectors	$\boldsymbol{K}_2=(m\boldsymbol{b}_1-n\boldsymbol{b}_2)/N$
		$K_1=\lvert\boldsymbol{K}_1\rvert=2/d_t\,K_2=\lvert\boldsymbol{K}_2\rvert=2\pi/T$
		$\boldsymbol{K}_1\parallel\boldsymbol{C}_h$
		$\boldsymbol{K}_2\parallel\boldsymbol{T}$
	Translational vectors for the $\boldsymbol{K}_1$-extended representation of the cutting lines	$N\boldsymbol{K}_1=-t_2\boldsymbol{b}_1+t_1\boldsymbol{b}_2$ $\boldsymbol{K}_2-M\boldsymbol{K}_1=\frac{m+Mt_2}{N}\boldsymbol{b}_1-\frac{n+Mt_1}{N}\boldsymbol{b}_2$
	Translational vectors	$(N/Q)\boldsymbol{K}_2=\frac{m}{Q}\boldsymbol{b}_1-\frac{n}{Q}\boldsymbol{b}_2$
	for the $\boldsymbol{K}_2$-extended	$Q\boldsymbol{K}_1-W\boldsymbol{K}_2=r_1\boldsymbol{b}_1+r_2\boldsymbol{b}_2$
Q	representation of the	$Q=\gcd(M,N)$[b]
W	cutting lines	$W=r_2t_2-r_1t_1$
$r_1,\ r_2$		$nr_1-mr_2=Q,\quad 1\le t_2r_2-t_1r_1\le\frac{N}{Q}$

a In this table $n, m, t_1, t_2, r_1, r_2, p$, and q are integers and d, d_R, N, M, Q, and W are integer functions of these integers.

b $\gcd(n,m)$ denotes the greatest common divisor of the two integers n and m.

continuous for a nanotube of infinite length.[15] Figure 2.16 shows the reciprocal space of the (4,2) nanotube ($\boldsymbol{\kappa}_1 = 28\boldsymbol{K}_1$ and $\boldsymbol{\kappa}_2 = \boldsymbol{K}_2 - 6\boldsymbol{K}_1$) as compared to the graphene reciprocal space. The first Brillouin zone, which is shown in dark gray, can be translated to the adjacent Brillouin zones, shown in light gray, by applying reciprocal lattice vectors, as shown in Figure 2.16. Thus the energy bands consist of a set of one-dimensional energy dispersion relations which are cross-sections of those for two-dimensional graphite (see Figure 2.17a).

When the energy dispersion relations of two-dimensional graphite are folded, N pairs of 1D energy dispersion relations $E_\mu(k)$ are obtained (see Figure 2.17b).

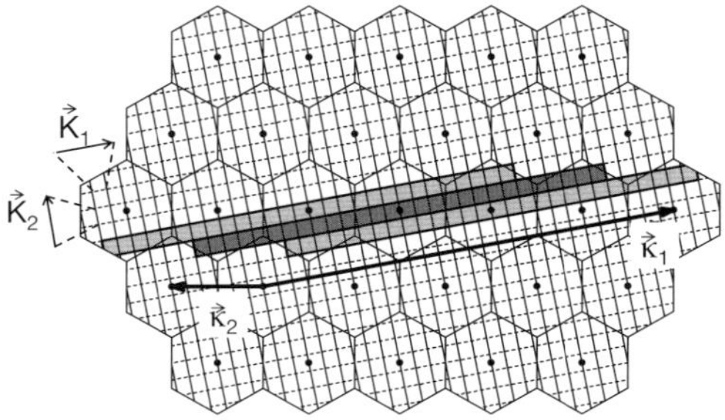

Figure 2.16 Reciprocal space of the graphene layer, showing the $\boldsymbol{K}_1$ and $\boldsymbol{K}_2$ reciprocal lattice vectors. Parallel equidistant lines represent the cutting lines for the (4, 2) nanotube. The first Brillouin zone is shown in dark gray. The light gray rectangles are the Brillouin zones obtained by the unit vector $\boldsymbol{\kappa}_2$ of the reciprocal space structures [110].

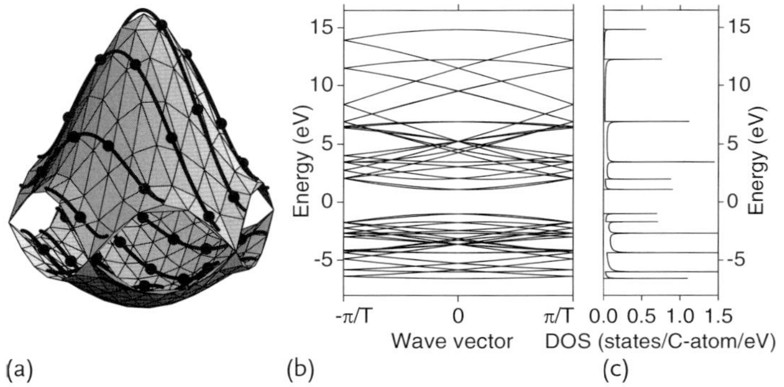

Figure 2.17 (a) The conduction and valence bands of the graphene layer in the first Brillouin zone calculated according to the π-band nearest neighbor tight-binding model [31]. Solid curves show the cutting lines for the (4, 2) nanotube in the fully reduced representation. Solid dots show the ends of the cutting lines in the fully $\boldsymbol{K}_1$-extended representation. (b) Electronic energy band diagram for the (4, 2) nanotube obtained by zone-folding from (a). (c) Density of electronic states for the energy band diagram shown in (b) [110].

15) For real carbon nanotubes, if the length of a nanotube (L_{CN}) is on the order of a micrometer or less, discrete k vectors ($\Delta k = 2\pi/L_{\mathrm{CN}}$) can be expected.

These 1D energy dispersion relations are given by

$$E_\mu(k) = E_{g2D}\left(k\frac{K_2}{|K_2|} + \mu K_1\right), \quad \left(\mu = 0, \cdots, N-1, \quad \text{and} \quad -\frac{\pi}{T} < k < \frac{\pi}{T}\right), \tag{2.38}$$

corresponding to the energy dispersion relations of a single-wall carbon nanotube, where E_{g2D} comes from Eq. (2.30). The N pairs of energy dispersion curves given by Eq. (2.38) correspond to the cross-sections of the two-dimensional energy dispersion surface shown in Figure 2.17a, where cuts are made on the lines of $k\boldsymbol{K}_2/|\boldsymbol{K}_2| + \mu \boldsymbol{K}_1$.

If for a particular (n, m) nanotube, a cutting line passes through a K point of the 2D Brillouin zone, where the π and π^* energy bands of two-dimensional graphite are degenerate by symmetry, then the one-dimensional energy bands have a zero energy gap, and are metallic. If, however, no cutting line passes through a K point, then the carbon nanotube is expected to show semiconducting behavior, with a finite energy gap between the valence and conduction bands.

The condition for obtaining a metallic energy band is that the ratio of the length of the vector $\boldsymbol{YK}$ to that of $\boldsymbol{K}_1$ in Figure 2.18 is an integer.[16] Since the vector $\boldsymbol{YK}$ is given by

$$\boldsymbol{YK} = \frac{2n+m}{3}\boldsymbol{K}_1\,, \tag{2.39}$$

the condition for metallic nanotubes is that $(2n + m)$ or equivalently $(n - m)$ is a multiple of 3.[17] In particular, the armchair nanotubes denoted by (n, n) are always

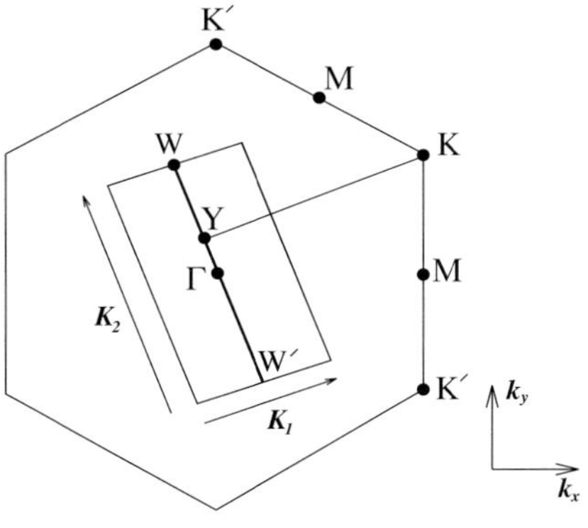

Figure 2.18 The condition for metallic energy bands: if the ratio of the length of the vector $\boldsymbol{YK}$ to that of $\boldsymbol{K}_1$ is an integer, metallic energy bands are obtained [31].

16) There are two non-equivalent K and K' points in the Brillouin zone of graphite as is shown in Figure 2.18 and thus the metallic condition can also be obtained in terms of K'. However, the results in that case are identical to the case specified by YK in Figure 2.18, since K and K' are a time-reversal pair in the k space.

17) Since $3n$ is a multiple of 3, the two remainders of $(2n + m)/3$ and $(n - m)/3$ are identical.

metallic, and the zigzag nanotubes $(n, 0)$ are only metallic when n is a multiple of 3.

The cutting lines in the vicinity of the K point are shown in Figure 2.19 for three different cases, $n - m = 3\ell$, $n - m = 3\ell + 1$, and $n - m = 3\ell + 2$, where ℓ is an integer. The first case $n - m = 3\ell$ corresponds to the cutting line crossing the K point, resulting in metallic behavior, as discussed above. The other two cases $n - m = 3\ell + 1$ and $n - m = 3\ell + 2$ (or, equivalently, $2n + m = 3\ell + 2$ and $2n + m = 3\ell + 1$, respectively, using a different notation) correspond to the K point being located at one third and at two thirds of the distance between two adjacent cutting lines, resulting in semiconducting behavior, since the cutting lines do not go through the K and K' points. We thus conclude that the number of semiconducting nanotubes is roughly twice that of metallic nanotubes.

The two cases of semiconducting nanotubes, $n - m = 3\ell + 1$ and $n - m = 3\ell + 2$, are also different from each other, depending on which side of the K point in the unfolded two-dimensional Brillouin zone of the nanotube the first van Hove singularity (vHS) in the electronic density of states (DOS) appears, as discussed further in Section 2.3.3. We classify these two types of semiconducting nanotubes as mod1 and mod2, in accordance with the number $(n - m - 3\ell)$ being equal to either 1 or 2, respectively. In the other notation, for $2n + m = 3\ell + 1$ and $2n + m = 3\ell + 2$, we call respectively, type I and type II (or S1 and S2). We must be careful for different notations used in the literature [110] where mod 1 (mod 2) corresponds to S2 (S1). In a similar fashion, we can classify metallic nanotubes by the ratio that the K point divides the cutting line in the $\boldsymbol{K}_1$-extended representation.

2.3.3 Density of States

As shown in the previous section, when 1D nanotubes are rolled up from 2D sheets of graphene, different sub-bands in the 1D reciprocal space of the nanotube can be

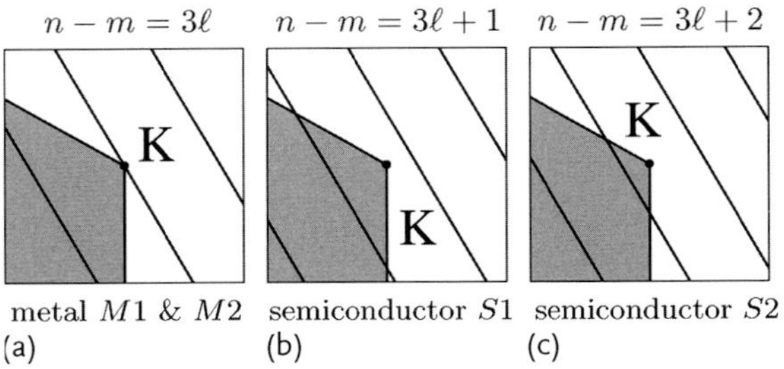

Figure 2.19 Three different configurations of the cutting lines in the vicinity of the K point. (a) Configuration $n - m = 3\ell$ corresponds to the case of metallic nanotubes of both $M1$ and $M2$ types. (b, c) Configurations $n - m = 3\ell + 1$ and $n - m = 3\ell + 2$ correspond to the case of semiconducting nanotubes of types mod 1 (or type 2 $S2$) and mod 2 (or type 1 $S1$), respectively [110].

extended into the 2D reciprocal space of a single sheet of the parent bulk layered material as a set of parallel equidistant lines or cutting lines. This procedure is shown in Figure 2.20a for states near the K point.

Figure 2.20b shows the electronic density of states (DOS) related to the nanotube electronic band structure plotted schematically in Figure 2.20a (see also Figure 2.17c). Each of the cutting lines in Figure 2.20a (except for the one that crosses the degenerate K point) gives rise to a local maximum in the DOS in Figure 2.20b, known as a (one-dimensional) van Hove singularity (vHS), given by

$$g(E) = \frac{2}{N} \sum_{\mu=1}^{N} \int \left[\frac{\partial E_\mu(k)}{\partial k} \right]^{-1} \delta[E_\mu(k) - E] dk \, . \tag{2.40}$$

The four vHS in Figure 2.20b are labeled by $E_i^{(v)}$ and $E_i^{(c)}$ for the electronic subbands in the valence and conduction bands, correspondingly. The presence of vHSs in the DOS of 1D structures makes these structures behave differently from their related 3D and 2D materials, as can be seen in Figure 2.21.

More generally, the DOS profiles for systems of different dimensionality (3D, 2D, 1D, and 0D) are very different from one another, as shown in Figure 2.21. The typical DOS dependence on energy near an energy band extremum, $g(E)$ is given by $g \propto (E - E_0)^{[(D/2)-1]}$, where D is an integer, denoting the spatial dimension and D

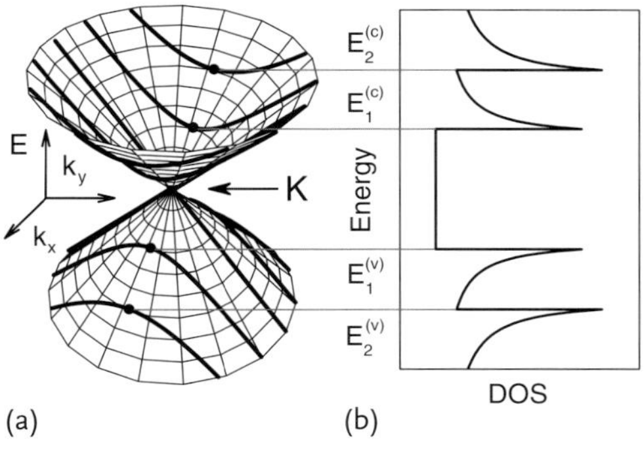

Figure 2.20 (a) The energy-momentum contours for the valence and conduction bands for a 2D system, each band obeying a linear dependence for $E(k)$ and forming a degenerate point K where the two bands touch to define a zero gap semiconductor. The cutting lines of these contours denote the dispersion relations for the 1D system derived from the 2D system. Each cutting line gives rise to a different energy sub-band. The energy extremum E_i for each cutting line at the wave vector k_i is known as the van Hove singularity. The energies $E_i^{(v)}$ and $E_i^{(c)}$ for the valence and conduction bands and the corresponding wave vectors $k_i^{(v)}$ and $k_i^{(c)}$ at the van Hove singularities are indicated on the figure by the solid dots. (b) The 1D density of states (DOS) for the conduction and valence bands in (b) corresponding to the $E(k)$ dispersion relations for the sub-bands shown in (a) as thick curves. The DOS shown in (b) is for a metallic 1D system, because one of the cutting lines in (a) crosses the degenerate Dirac point (the K point in the graphite Brillouin zone). For a semiconducting 1D system, no cutting line crosses the degenerate point, which results in a band gap opening up in the DOS between the van Hove singularities $E_1^{(v)}$ and $E_1^{(c)}$ [110].

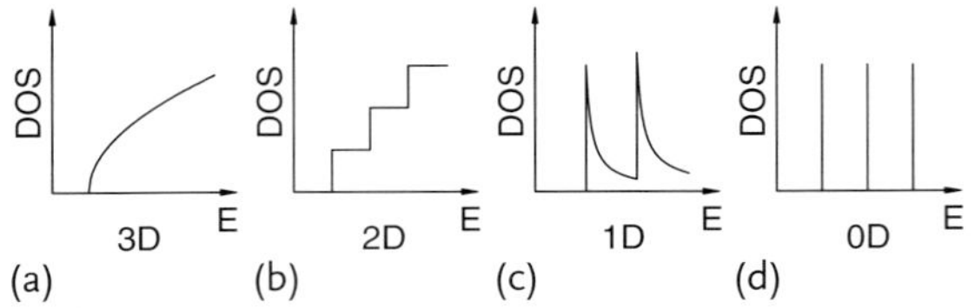

Figure 2.21 Typical electronic density of states for (a) 3D, (b) 2D, (c) 1D and (d) 0D systems [90].

assumes the values 1,2, and 3, respectively, for 1D, 2D, and 3D systems [90]. Here E_0 denotes the energy band minimum (or maximum) for the conduction (valence) energy bands. For a 1D system, E_0 would correspond to a vHS in the DOS occurring at each sub-band edge, where the magnitude of the DOS becomes very large. One can see from Figure 2.21 that 1D systems exhibit DOS profiles, which are quite similar to the case of 0D systems, with both 0D and 1D systems having very sharp maxima at certain energies, in contrast to the DOS profiles for 2D and 3D systems, which show a more monotonic increase with energy (see Figure 2.21). However, the 1D DOS is different from the 0D DOS (δ function at energy levels) in that the 1D DOS has a sharp threshold and a decaying tail, so that the 1D DOS does not go to zero between the sharp maxima, as the 0D DOS does (see Figure 2.21). The extremely high values of the DOS at the vHS allow us to observe physical phenomena for individual 1D objects in various experiments, as discussed in Section 2.3.4.

For metallic nanotubes, a cutting line crosses the Fermi level at the K point. It follows that the density of states per unit length along the nanotube axis is a constant given by

$$N(E_F) = \frac{8}{\sqrt{3}\pi a|t|}, \tag{2.41}$$

where a is the lattice constant of the graphene layer and $|t|$ is the nearest neighbor C–C tight-binding overlap energy usually denoted by γ_0 in the graphite literature [111].

For semiconducting nanotubes the DOS is zero up to the first van Hove singularities, and their energy gap depends roughly on $1/d_t$, the reciprocal nanotube diameter d_t [18)]

$$E_g = \frac{|t|a_{C-C}}{d_t}, \tag{2.42}$$

where $a_{C-C} = a/\sqrt{3}$ is the nearest neighbor C–C distance on a graphene sheet.

18) The energy gap also has a weak dependence on the chiral angle of the semiconducting nanotube.

2.3.4
Importance of the Electronic Structure and Excitation Laser Energy to the Raman Spectra of SWNTs

Figure 2.22a–c shows the density of electronic states (DOS) for three different SWNTs, and since SWNTs are one-dimensional (1D) systems, their DOS is characterized by their van Hove singularities (vHSs). The sharp vHSs define narrow energy ranges where the DOS intensity becomes very large. Therefore, in practice, a single carbon nanotube exhibits a "molecular-like" behavior, with well-defined electronic energy levels at each vHS. The three DOS curves in Figure 2.22a–c come from different SWNTs as labeled by their (n, m) indices [112] (see figure caption). An observable Raman signal from a carbon nanotube can be obtained when the laser excitation energy is equal to the energy separation between vHSs in the valence and conduction bands (e. g., see E_{11}^S, E_{22}^S and E_{11}^M in Figure 2.22), but restricted to the selection rules for optically allowed electronic transitions (see Chapter 6). A plot of the density of valence-conduction states fulfilling these selection rules that are available for optical transitions as a function of the excitation photon energy is called a joint density of states (JDOS) plot.

For the characterization of nanotubes by Raman spectroscopy, it is useful to consider plots of the energies E_{ii} vs. the nanotube diameter, d_t, as shown in Figure 2.22d [113]. Each point in this plot represents one optically allowed electronic transition energy (E_{ii}) from a given (n, m) SWNT. Crosses come from semiconducting SWNTs, and circles from metallic SWNTs. This plot should be considered as a guide for answering the question "if I use a given E_{laser} to excite my sample,

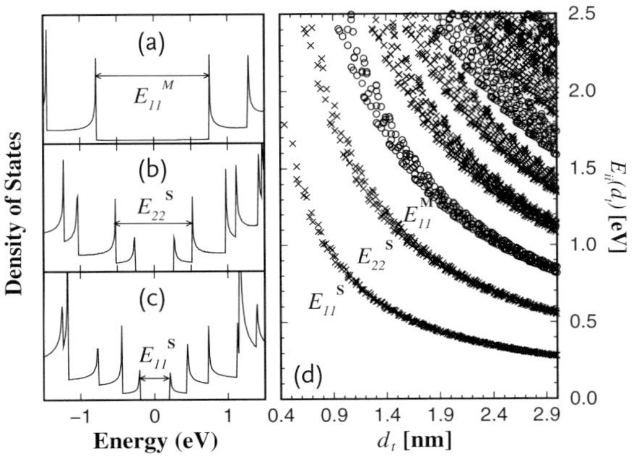

Figure 2.22 Density of electronic states for (a) an armchair (10, 10) SWNT, (b) a chiral (11, 9) SWNT, and (c) a zigzag (22, 0) SWNT obtained with the tight-binding model from [31]. (d) A plot (called a Kataura plot [113]) of the electronic transition energies E_{ii} for all the (n, m) SWNTs with diameters from 0.4 nm to 3.0 nm using a simple first-neighbor tight-binding model [31]. Deviations from this simple one-electron model are expected for the lower energy transitions E_{11}^S and for SWNTs with $d_t \lesssim 1$ nm. Other corrections due to many-body interactions are also important [112].

which (n, m) carbon nanotubes can be in resonance with my laser line?" In other words, since the observable Raman spectra come predominantly from tubes in resonance with E_{laser}, Figure 2.22 specifies the nanotubes that will be observable for a given laser line.

Because of this resonance process, Raman spectra at the single nanotube level allow us to study the electronic and phonon structure of SWNTs in great detail. The E_{ii} vs. d_t plot shown in Figure 2.22d is called the "Kataura plot" in the literature (from [113]), and the early plots were made using a first-neighbor tight-binding model and the zone-folding procedure described in this chapter [31]. As we will see later, resonance Raman spectroscopy shows that the general trends seen in Figure 2.22d are correct, but for an accurate determination of the E_{ii} for each (n, m) SWNT, other effects must also be considered, as discussed briefly in Section 2.4, and in more detail in Chapter 10.

2.4
Beyond the Simple Tight-Binding Approximation and Zone-Folding Procedure

The simple tight-binding method used to describe electrons in a sp^2 system is pedagogic and gives a very good first-order approximation for the electronic structure of graphene and the other sp^2 nanocarbons. However, for an accurate description of the physical properties, other effects also have to be considered:

- Many-body effects, like electron–electron (e-e) and electron–hole (e–h) interactions have to be considered to describe the energy of excited states of graphene.
- These many-body effects such as excitonic (electron–hole) interactions become extremely important when one-dimensional quantum confinement takes place, thus having a strong influence on the optical properties and Raman spectra of carbon nanotubes and nanoribbons.
- Curvature in nanotubes and edge effects in nanoribbons will substantially change the electronic properties of these materials when their dimensions (diameter for tubes, width for ribbons) reach the nanometer scale.

These concepts will be discussed in detail elsewhere in the book. Here we expand briefly on the failure of the description given in Section 2.3 for small diameter SWNTs such as the (4, 2) carbon nanotube. Beyond this simple description, deviations from the simple linear approximation for the electronic dispersion relations of the graphene layer arise from the $\sigma - \pi$ hybridization of the electronic orbitals caused by the curvature of the nanotube wall [114]. Curvature-related effects introduce a band gap in the electronic dispersion relations at the Fermi level near the K and K' points in the Brillouin zone and this band gap affects the electronic band structure and physical properties in many ways. The curvature effect scales inversely with the square of nanotube diameter, being especially important for small diameter nanotubes, such as for the (5, 0) nanotube which becomes metallic. The simple π-band nearest neighbor tight-binding approximation used in Figure 2.17

is not able to describe curvature effects, since the description of nanotube curvature requires at least four orbitals per carbon atom, $2s, 2p_x, 2p_y, 2p_z$, as discussed in Chapter 10. Furthermore, next-nearest neighbor interactions also affect the electronic band structure of a graphene layer, not only in the vicinity of the K and K' points, as shown by comparing the nearest neighbor and the third nearest neighbor approximations with the results of *ab initio* electronic structure calculations [115]. Thus, Figure 2.17 does not reflect the real electronic band structure of a (4, 2) nanotube which still remains semiconducting, being a lowest order approximation in the limit of small diameter SWNTs where more detailed calculations are necessary. *Ab initio* calculations for a (4, 2) nanotube [108] yield an electronic band structure that is substantially different from the results obtained by the zone-folding scheme, as discussed in Chapter 10.

Problems

[2-1] Using Eq. (2.5), estimate the $1s$ energy of B, C and N atoms. In order to observe the core state energy, we usually use XPS (X-ray photoelectron spectroscopy) measurements. Explain how to measure $1s$ states for these atomic species by XPS with some illustrative figures.

[2-2] An electronic p orbital has angular momentum $\ell = 1$. Obtain the spherical harmonic $Y_{\ell m}(\theta, \phi)$ for $\ell = 1$ and $m = -1, 0, 1$. By combining these three functions, construct the p_x, p_y and p_z functions. Plot the shape of the p_x, p_y and p_z functions in three dimensions.

[2-3] Plot the rough shape of $R_{n\ell}(r)$ in Eq. (2.2) for $1s$, $2s$ and $2p$ states as a function of r. Explain how these functions are orthogonal to each other.

[2-4] When we change Z to $Z - 2$ in Eq. (2.5) for expressing the screening between two $1s$ electrons, estimate the $2s$ energy of B, C and N atoms.

[2-5] In order to solve Eq. (2.1), express the wavefunction as $\Psi(\mathbf{r}) = R(r)\Theta(\theta)\Phi(\phi)$ and obtain the equations for the variables of r, θ, and ϕ. Explain qualitatively why $2p$ orbitals have a higher energy than $2s$ orbitals. It is only for the case of the hydrogen atom that the $2s$ and $2p$ levels have an identical energy.

[2-6] Using the unitary matrix U in Eq. (2.16), diagonalize the Hamiltonian in Eq. (2.14).

[2-7] Solve the Schrödinger equation for the H molecule for the case that $s = \langle \Psi_1 | \Psi_2 \rangle$ is not zero. Obtain both eigenvalues and wavefunctions.

[2-8] Use the tight-binding model to obtain the electronic band structure for π electrons in polyacetylene. In the case of polyacetylene, there are two kinds of structures, the cis- and trans-structures (see Figure 2.23). In the case of trans-polyacetylene, the transfer integral t for π electrons should have two

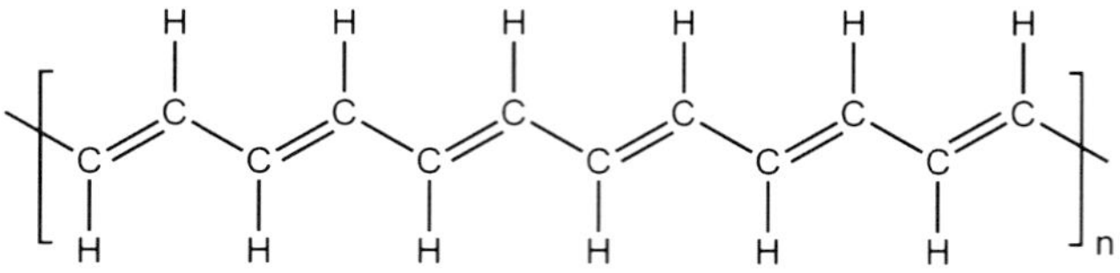

Figure 2.23 The structure of trans-polyacetylene, with alternative C–C bonds in the trans-configuration. In cis-polyactylene (not shown), an armchair edge shaped C–C chain appears.

values t_1 and t_2 depending on the double and single bond nature. Show that an energy gap which is proportional to $|t_1 - t_2|$ appears in the case of trans-polyacetylene.

[2-9] Obtain the coordinates in k space for the K, M, and Γ points and obtain the energy values for the simplest tight-binding energy for graphene at these high symmetry points in the Brillouin zone (Eq. (2.32)).

[2-10] Obtain Eqs. (2.33) and (2.34). Evaluate the value of the Fermi velocity.

[2-11] Show that the density of states is proportional to E when the energy dispersion is $E = a\sqrt{k_x^2 + k_y^2}$ in two-dimensional materials. How about for the linear dispersion in one-dimensional materials?

[2-12] Plot the density of states for the parabolic energy band, $E = a(k_x^2 + k_y^2)$.

[2-13] By expanding $f(k)$ of Eq. (2.28) near the K point, show that the Hamiltonian matrix is written by Eq. (2.35).

[2-14] Explain how to obtain the density of states for the cosine energy band, $E = a(\cos(k_x a) + \cos(k_y a))$ numerically.

[2-15] Calculate the diameter for an (n, m) SWNT. What are the values of the diameters for (5,5), (9,0), (10,5) SWNTs? What are possible (n, m) values for SWNTs having a 1.5 ± 0.02 nm diameter?

[2-16] For a given (n, m) SWNT, show the expression of $\boldsymbol{T} = (t_1, t_2)$ as a function of n, m. What is the length of $\boldsymbol{T}$?

[2-17] Show that $|\boldsymbol{a}_1 \times \boldsymbol{a}_2| = \sqrt{3}a^2/2$ and that the number of hexagons in the nanotube unit cell is $N = 2\left(n^2 + nm + m^2\right)/d_R$.

[2-18] Show the relation for an (n, m) SWNT $|\boldsymbol{C}_h \times \boldsymbol{T}| = \sqrt{3}a^2\left(n^2 + nm + m^2\right)/d_R$, where $d_R = \text{gcm}(2n + m, 2m + n)$ and gcm is an integer function of the greatest common multiplier.

[2-19] Show the reciprocal lattice unit vectors $(\boldsymbol{K}_1, \boldsymbol{K}_2)$ as a function of n, m in 2D reciprocal space.

[2-20] Show that $|\boldsymbol{K}_1| = 2/d_t$ and $|\boldsymbol{K}_2| = 2\pi/|\boldsymbol{T}|$.

[2-21] Give the 1D electronic structure for the (10, 10) and (18, 0) SWNTs. Plot the cutting lines (1D Brillouin zone of SWNTs) in the 2D reciprocal space of the 2D Brillouin zone of graphene for these two SWNTs.

[2-22] Show that in Figure 2.18, $\boldsymbol{YK} = [(2n + m)/3]\boldsymbol{K}_1$, which is Eq. (2.39).

[2-23] Plot the cutting lines for type-I, type-II semiconducting and metallic SWNTs and show which cutting lines correspond to E_{11}^S, E_{22}^S, E_{33}^S, and E_{11}^M transitions for each case.

3
Vibrations in sp^2 Nanocarbons

Although the inelastic process of Raman scattering of light can originate from the creation or annihilation of polaritons, plasmons, magnons or any elementary excitations in molecules and solids, it is the phonons, which are the quanta of atomic vibration that are the main source of Raman spectra in the literature, and are the focus of this chapter. In sp^2 nanocarbons, the phonons, like the electrons, depend on the atomic structure and the Raman phonon spectra can be used to study the similarities and differences between the various materials within the sp^2 nanocarbon family, thereby providing a sensitive tool to distinguish one member of the sp^2 nanocarbon family from another.

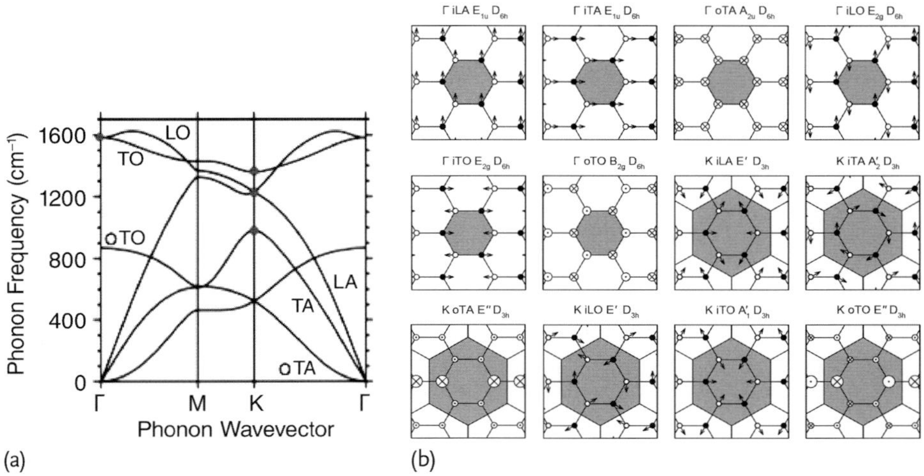

Figure 3.1 (a) Phonon dispersion relations for graphene and (b) the eigenvectors for the in-plane phonons relevant to the high symmetry Γ point and K (K') points of the Brillouin zone. Each of these twelve modes is labeled and their atom displacements are indicated [117]. i/o stands for in-plane/out-of-plane; T/L stands for transversal/longitudinal; A/O stands for acustic/optical. The other symbols are symmetry assignments according to group theory, as discussed in Chapter 6.

Raman Spectroscopy in Graphene Related Systems. Ado Jorio, Riichiro Saito, Gene Dresselhaus, and Mildred S. Dresselhaus

ISBN: 978-3-527-40811-5

Figure 3.1a shows the phonon dispersion relations $\omega(q)$ of monolayer graphene, the building block of many of the other carbon sp^2 nanostructures, calculated within the so-called *force constant model* [31, 116].

There are six branches in these phonon dispersion relations because the crystal has a unit cell with two distinct atom sites (A and B). The six eigenvectors *at* the Γ point ($q = 0$, that is the wavelength, $\lambda \to \infty$) consist of three translations of the crystal along x, y, z, which have no restoring force and, consequently, zero frequency, plus three vibrational modes, two of which are degenerate. The Γ point phonons shown in Figure 3.1 include the stretching of the C–C bond in the graphene unit cell, and the phonons are denoted by LO or TO according to whether the atomic vibrations are along or perpendicular to the direction of the wave vector $\boldsymbol{q}$ (not shown). Graphene is a nonionic crystal, since the unit cell is formed by two atoms of the same type, and because of the nonionic behavior of the graphene crystal, the LO and TO phonon modes are degenerate at the Γ point.[1)] For the K point phonons, $q \neq 0$ and there is a phase factor in the phonon eigenfunctions from one unit cell relative to the neighboring unit cell.

The brief description above introduces the picture of how to treat phonons in a crystal, and the goal of this chapter is to describe the vibrational structure of sp^2 carbons. Similar to what was covered in Chapter 2, we aim in this chapter to capture the fundamental concepts for the phonon dispersion from the energy levels for a very simple molecule and then to build the phonon dispersion structure of a crystal, as presented in Figure 3.1. Therefore, we begin in Section 3.1 by briefly reviewing some basic concepts: the harmonic oscillator, which describes the fundamental physics of molecular vibrations (Section 3.1.1); the concept of normal modes from molecules to crystals (Section 3.1.2); the force constant model (Section 3.1.3), in which interatomic forces are represented by spring constants to calculate the phonon dispersion relations in crystals. With these basic concepts in place, we develop the force constant model for graphene (Section 3.2). Like the tight-binding approximation for the electronic structure (Chapter 2), the force constant model is simple enough to be solved analytically, thus giving an understanding about how to build the phonon structure of graphene. Again similar to electrons, phonon confinement takes place in carbon nanotubes and can be described, as a first approximation, by the zone-folding procedure, which is developed in Section 3.4. Finally, both the zone-folding and the force constant model have their limitations which are discussed in Sections 3.4.2 and 3.5. The force constant model, with connections also made to 3D graphite, can be expressed quite accurately by increasing the number of neighbors and the number of force constants, but it fails when effects like the electron–phonon coupling are important. Such a failure is actually very important for interpreting the Raman spectroscopy of sp^2 carbon nanotubes, and is briefly discussed in this chapter. Further discussion is given in more detail throughout the book whenever new aspects of the electron–phonon interaction are needed to explain specific phenomena associated with the Raman spectra of sp^2

1) In the case of an ionic crystal, like NaCl, the LO mode has a higher frequency than the TO mode because of the Coulomb interaction between the ions. The LO-TO splitting is then related to the dielectric constant through the Lydane–Sacks–Teller (LST) theory [118].

carbons. Many basic textbooks on Raman spectroscopy are available to supplement the presentation provided in this book which is focused on sp^2 carbons [119–122].

3.1
Basic Concepts: from the Vibrational Levels in Molecules to Solids

To provide an overview of the basics, we first review the use of a harmonic oscillator (Section 3.1.1), which is important for defining phonons and describing phonon amplitudes and displacements that are needed for Raman intensity calculations. Then we discuss the normal vibrational modes in molecules and how the number of normal modes evolves from small molecules to a crystal with an "infinite" number of atoms, thereby building the phonon dispersion relations in crystals (Section 3.1.2). Finally we describe the force constant model for vibrations in general terms (Section 3.1.3).

3.1.1
The Harmonic Oscillator

The atomic motion in molecules and solids is described in terms of the normal modes of vibration, which are represented by an orthogonal set of harmonic oscillators. Classically, a harmonic oscillator can be described by

$$m\frac{d^2(x - x_{\text{eq}})}{dt^2} = -K(x - x_{\text{eq}}) \,, \tag{3.1}$$

where m, x_{eq} and K are, respectively, the mass, the equilibrium position (see Figure 3.2) and the force constant for the harmonic oscillator. A solution of Eq. (3.1) is given by

$$x(t) = x_{\text{eq}} + A\cos(\omega t + \phi), \quad \text{and} \quad \omega \equiv \sqrt{\frac{K}{m}} \,, \tag{3.2}$$

where A, ω and ϕ are, respectively, the amplitude, the frequency and the phase of the vibration. A and ϕ are determined by the initial conditions at time $t = 0$.

However, the description above does not take into account the quantum nature of the atoms, which can be described by solving the time-dependent Schrödinger

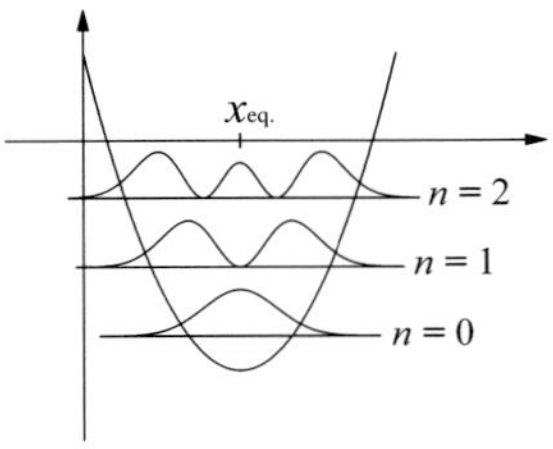

Figure 3.2 The potential energy for a harmonic oscillator showing the quantized energy levels for a molecular mode vibration. The waves associated with each energy level display the probability to find the interatomic distance at a given x value.

equation for the quantum harmonic oscillator (HO)

$$\left[-\frac{\hbar^2}{2m}\frac{\partial^2}{\partial x^2}+\frac{1}{2}Kx^2\right]\Psi_n = E_n\Psi_n, \quad (n=0,1,2,\ldots)\,, \tag{3.3}$$

in which Ψ_n is a wave function of the harmonic oscillator and n labels the quantum state of the harmonic oscillator. The vibrational amplitude for Ψ_n, which is proportional to $\sqrt{n}$, is quantized and the different vibrational levels are described by the number of vibrational quanta, called phonons, and quantified by the quantum number n (see Figure 3.2).

To account for the quantum nature of phonons and their energies, we introduce the annihilation operator a and the creation operator $a^\dagger$ which, respectively, annihilates one phonon or creates one phonon of energy $\hbar\omega$:

$$a=\frac{p-i\omega m x}{\sqrt{2\hbar\omega m}}, \quad \text{and} \quad a^\dagger=\frac{p+i\omega m x}{\sqrt{2\hbar\omega m}}\,. \tag{3.4}$$

The annihilation operator a lowers the quantum number of the state $|n\rangle \equiv \Psi_n$ to $|n-1\rangle$, while the creation operator $a^\dagger$ increases the quantum number of the state $|n\rangle$ to $|n+1\rangle$. Since the commutator $[p,x]=\hbar/i$, it follows that

$$[a,a^\dagger]=1\,, \tag{3.5}$$

so that the Hamiltonian for the harmonic oscillator in Eq. (3.3) can be written in terms of the operators a and $a^\dagger$ as:

$$\mathcal{H}=\frac{1}{2m}\left[(p+i\omega m x)(p-i\omega m x)+m\hbar\omega\right] \tag{3.6}$$

$$=\hbar\omega[a^\dagger a+1/2]\,. \tag{3.7}$$

Considering $N=a^\dagger a$ to be the number operator, the Hamiltonian in Eq. (3.7) can be written as

$$\mathcal{H}|n\rangle=\hbar\omega[N+1/2]|n\rangle=\hbar\omega(n+1/2)|n\rangle\,. \tag{3.8}$$

Thus the eigenvalues for the harmonic oscillator are written as:

$$E=\hbar\omega(n+1/2), \qquad (n=0,1,2,\ldots) \tag{3.9}$$

as shown in Figure 3.2. Here n corresponds to the number of phonons with frequency ω. The phonon amplitude will be related to the number of phonons n, which depends on the phonon energy and temperature, as given by the Bose–Einstein distribution function (see Section 4.3.2.1).

3.1.2
Normal Vibrational Modes from Molecules to a Periodic Lattice

In a molecule with N atoms, there exist $3N-6$ degrees of freedom for vibrations. Of these $3N-6$ modes, 6 correspond to the degrees of freedom for the translation and rotation of the center of mass, which either have no restoring force (zero frequency for translations) or have very small frequencies (rotations).[2] Any atomic motion of the molecule can be expressed by a linear combination of these $3N-6$ independent, orthogonal vibrations, which we call normal modes.

As a simple example, we discuss the molecular vibrations of the NO molecule, which is an exception to the $3N - 6$ rule, but has $3N - 5$ normal modes. Having two atoms, the NO molecule has 6 degrees of freedom for atomic motion in three dimensions, but has only one vibrational mode. Three degrees of freedom are associated with translations of the molecule along x, y and z, and only two degrees of freedom involve molecular rotations around the x and y axes. Rotation around the z (N–O bond) axis does not represent motion for a linear molecule, which therefore has $3N - 5 = 1$ vibrational modes. The vibrational mode is represented by the N–O bond stretching vibration, which is a breathing mode that does not alter the symmetry of the molecule but only changes the bond length (dipole moment).

Increasing the size of the molecule from two to three atoms, we now consider the CO_2 molecule, which is also a linear molecule with the carbon atom at the center and each of the oxygen atoms (along the z direction, for example) located at distances $\pm z_0$ from the carbon atom. In this case, there are 9 degrees of freedom, 4 of which are vibrational. This gives rise to a symmetric breathing mode with the C atom remaining static and the oxygen moving in $\pm z$ directions to preserve the center of mass. A second mode is the antisymmetric stretch mode with the carbon atom moving for example in the $+z$ direction when the two oxygen atoms move in the $-z$ direction to preserve the center of mass. In addition, there is a doubly degenerate bending mode where the two oxygen and the carbon atoms vibrate in the directions normal to the molecular axis (i. e., the $\pm x$ and $\pm y$) directions. In this case the bending and antisymmetric stretch modes that create a dipole moment are infrared-active and the symmetric stretching mode that transforms as a symmetric second rank tensor are Raman-active. These symmetry concepts behind Raman activity can be described by group theory and will be discussed in Chapter 6.

Finally, finding the normal modes of large and complex molecules, such as proteins, is not an easy task because of the large number of degrees of freedom. In cases like that, it is common to find the spectral features identified with the stretching and bending of local bonds (e. g., C=C, C–H, C=O, etc.) rather than the complete molecular normal modes. Crystals have a large number of atoms (ideally infinite), but periodic systems are, again, quite simple to describe in terms of molecules in real and reciprocal space, and require the use of *phonon dispersion relations*, $\omega(q)$ where q is the phonon wave vector.

The Bloch theorem developed in Section 2.1.5 for electrons can be used to describe the vibrational structure of crystalline solids. The same concepts, such as the unit cell in real and reciprocal space, the Brillouin zone, etc., are used to describe the vibrational structure, but the name *branches* is used in place of bands to designate the phonon dispersion relations. Consider N as the number of atoms in the unit cell, and $N_\Omega \sim 10^{23}$ per mole, as the number of unit cells in a mole of crystal.

2) Typical rotational energies are on the order of $\sim$ 1 meV and occur at far-infrared frequencies. The vibrational modes of molecules are observed in the mid-IR range, typically in the range 20–100 meV, and are the usual subject of study for molecular Raman spectroscopy.

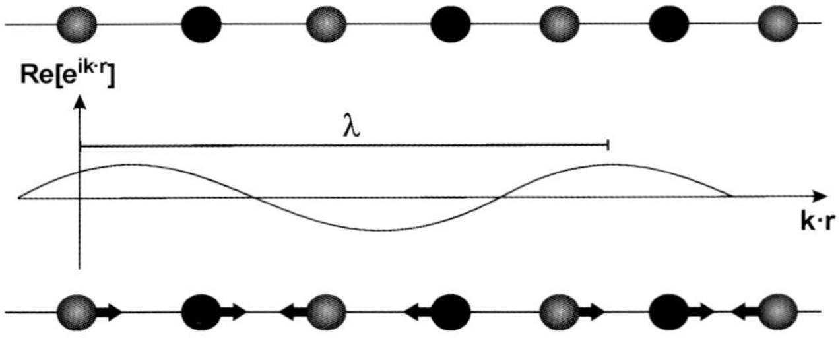

Figure 3.3 Schematic phonon eigenvector for a 1D crystal with two atoms in the unit cell. The phonon mode is the longitudinal acoustic (LA) phonon branch with $q = 8\pi/9a$ ($\lambda = 9a/4$).

There will then be $3 \times N_\Omega \times N$ vibrational modes, which is infinite for a crystal of infinite size, but the quasi-infinite number of vibrational modes are grouped into the various phonon branches. While the number of electronic bands is defined by the number of electrons in the unit cell, the number of phonon branches is defined by the number of atoms (N) in the unit cell. The difference between one phonon to another within the branch is not given by the atomic motion within the unit cell, but rather by the change in the phase of the vibration from one unit cell to the next. A schematic phonon eigenvector for a hypothetical 1D crystal with $N = 2$ is shown in Figure 3.3, which should be compared to Figure 2.5 for s and p electrons. This difference among phonons within the unit cell is described by phonon wave vectors, which are usually labeled $\boldsymbol{q}$ (where $\boldsymbol{k}$ is used for the electron wave vector), and their energies are given by $E_q = \hbar\omega_q$. A plot of ω_q vs. $\boldsymbol{q}$, such as in Figure 3.1a, gives the phonon dispersion relations for graphene [31].

There are $3 \times N_\Omega$ modes related to the translation of N_Ω unit cells along the 3 directions of real space. However, only the translations along x, y and z with infinite wavelength λ (i. e., for the null wave vector $q = 0$ at the Γ point) represent a crystal translation (no restoring force, zero frequency). All the other $3N_\Omega - 3$ modes actually have a restoring force from the neighboring unit cells, and they are all vibrational modes grouped in three branches called *acoustic branches,* since they are related to the transport of acoustic waves at long wavelengths.[3)] In general, the sound velocity for *longitudinal waves* is faster than that for *transverse waves.* In high symmetry crystals such as 2D graphene and 3D graphite we define the wave propagation direction by the wave vector $\boldsymbol{q}$, and for a given $\boldsymbol{q}$ we can define one longitudinal acoustic (LA) branch, where the vibrational amplitude is parallel to the wave propagation direction $\boldsymbol{q}$, and two transverse acoustic (TA) branches, whose amplitudes are perpendicular to $\boldsymbol{q}$ (Figure 3.4).

The rotation of a unit cell in a crystal is not allowed. All the other $3N_\Omega N - 3N_\Omega$ modes are also vibrational modes, and they group into $3N - 3$ branches named *optical branches,* making reference to the fact that they are usually studied using optics. Like for the acoustic branches, the optical branches can be classified into longitudinal and transversal modes, using the general nomenclatures LO and TO, respectively (with "O" for optical replacing "A" for acoustic).

3) A long wavelength phonon would have $\lambda > 50$ cm if the velocity of sound is 10 km/s and if the highest frequency of sound is 20 kHz.

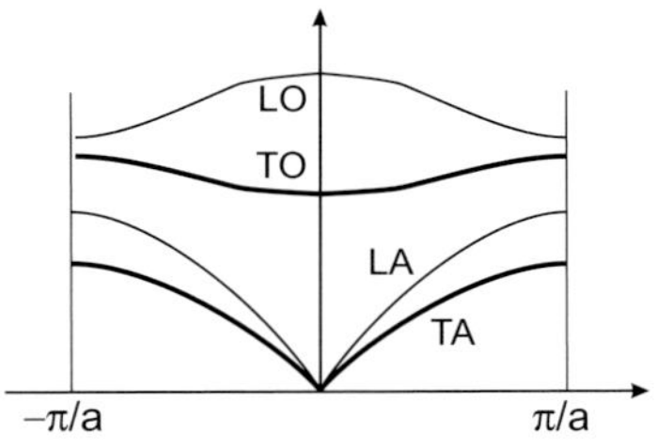

Figure 3.4 Schematic phonon dispersion relations for a hypothetical 1D crystal made by the repetition of two different atoms. The thicker lines indicate doubly degenerate branches for both the acoustic and optical branches.

As an illustration, Figure 3.4 shows the phonon dispersion relations schematically for a hypothetical 1D crystal made of two different alternating atoms (Figure 3.3). The two lower energy branches are for the acoustic phonons and two higher energy branches are for the optical phonons. Up to six branches (3 acoustic and 3 optical) are expected in 3D graphite, since we have two atoms unit cell. However, in the case of the hypothetical 1D crystal, two acoustic and two optical modes are degenerate, since atomic vibrations along x or y will have the same energies.

Another important concept is the *zone boundary of the phonon dispersion*, already discussed in Section 2.1.5, which is given in Figure 3.4 by $q = \pm\pi/a$. This boundary which we call the first Brillouin zone boundary is defined by the largest possible value for $q = 2\pi/\lambda = 2\pi/2a$. Any value larger than that can be folded back inside the $\pm\pi/a$ boundaries of the first Brillouin zone. For example, if we draw in Figure 3.3 the motion for $q = 0$ ($\lambda \to \infty$) and for $q = 2\pi/a$ ($\lambda = a$), we will see that the motion of equivalent atoms is identical.

Understanding the meaning of a phonon dispersion relation like Figures 3.1 and 3.4 is very important for Raman spectroscopy and for materials science in general, and this is the goal of Section 3.1. In the next section we introduce a model used to calculate such phonon dispersion relations for actual materials.

3.1.3 The Force Constant Model

In general, the equations of motion for the displacement of the ith atom measured from the equivalent position, $\boldsymbol{u}_i = (x_i, y_i, z_i)$ for N atoms in the unit cell is given by

$$M_i \ddot{\boldsymbol{u}}_i = \sum_j K^{(ij)}(\boldsymbol{u}_j - \boldsymbol{u}_i), \quad (i = 1, \ldots, N)\,, \tag{3.10}$$

where M_i is the mass of the ith atom and $K^{(ij)}$ represents the 3×3 force constant tensor[4] between the ith and the jth atoms. The sum over j in Eq. (3.10) is normally

4) A second rank tensor is defined by a 3×3 matrix whose elements ($K_{xx}, K_{xy}, \ldots, K_{zz}$) can be transformed as $U^{-1}KU$, where U is a unitary matrix which transforms the x, y, z coordinates into another orthogonal x', y', z' coordinate system without changing the length scale.

taken over only a few neighbor distances relative to the ith site, which for a 2D graphene sheet has been carried out up to fourth nearest neighbor interactions in [123]. In order to reproduce the experimental results, up to twentieth nearest neighbor interactions have been considered [116, 124]. In a periodic system we can perform a Fourier transform of the displacement of the ith atom with the wave number $\boldsymbol{k}'$ to obtain the normal mode displacements $\boldsymbol{u}_{\boldsymbol{k}^{(i)}}$

$$\boldsymbol{u}_i = \frac{1}{\sqrt{N_\Omega}} \sum_{q'} e^{-i(\boldsymbol{q}'\cdot \boldsymbol{R}_i - \omega t)} \boldsymbol{u}_{q'}^{(i)} , \quad \text{or} \quad \boldsymbol{u}_{\boldsymbol{q}^{(i)}} = \frac{1}{\sqrt{N_\Omega}} \sum_{\boldsymbol{R}_i} e^{i(\boldsymbol{q}\cdot \boldsymbol{R}_i - \omega t)\boldsymbol{u}_i} , \tag{3.11}$$

in which the sum is taken over all (N_Ω) wave vectors $\boldsymbol{q}'$ in the first Brillouin zone[5] and $\boldsymbol{R}_i$ denotes the atomic position of the ith atom in the crystal. When we assume the same eigenfrequencies ω for all $\boldsymbol{u}_i$, that is $\ddot{\boldsymbol{u}}_i = -\omega^2 \boldsymbol{u}_i$, then Eq. (3.10) can be formally written by defining a $3N \times 3N$ dynamical matrix $\mathcal{D}(\boldsymbol{q})$

$$\mathcal{D}(\boldsymbol{q}) \boldsymbol{u}_q = 0 . \tag{3.12}$$

To obtain the eigenvalues $\omega^2(\boldsymbol{q})$ for $\mathcal{D}(\boldsymbol{q})$ and nontrivial eigenvectors $\boldsymbol{u}_q \neq \boldsymbol{0}$, we solve the secular equation $\det \mathcal{D}(\boldsymbol{q}) = 0$ for a given $\boldsymbol{q}$ vector. It is convenient to divide the dynamical matrix $\mathcal{D}(\boldsymbol{q})$ into small 3×3 matrices $\mathcal{D}^{(ij)}(\boldsymbol{q})$, $(i, j = 1, \cdots, N)$, where we denote $\mathcal{D}(\boldsymbol{q})$ by $\{\mathcal{D}^{(ij)}(\boldsymbol{q})\}$, and from Eq. (3.12) it follows that $\mathcal{D}^{(ij)}(\boldsymbol{q})$ is expressed as:

$$\mathcal{D}^{(ij)}(\boldsymbol{q}) = \left(\sum_{j''} K^{(ij'')} - M_i \omega^2(\boldsymbol{q}) I \right) \delta_{ij} - \sum_{j'} K^{(ij')} e^{i\boldsymbol{q}\cdot\Delta \boldsymbol{R}_{ij'}} , \tag{3.13}$$

in which I is a 3×3 unit matrix and $\Delta \boldsymbol{R}_{ij} = \boldsymbol{R}_i - \boldsymbol{R}_j$ is the relative coordinate of the ith atom with respect to the jth atom. The vibration of the ith atom is coupled to that of the jth atom through the $K^{(ij)}$ force constant tensor. The sum over j'' is taken for all neighbor sites from the ith atom with $K^{(ij'')} \neq 0$, and the sum over j' is taken over the equivalent sites to the jth atom. The first two terms[6] of Eq. (3.13) have nonvanishing values only when $i = j$, and the last term appears only when the j'th atom is coupled to the ith atom through $K^{(ij')} \neq 0$.

In a periodic system, the dynamical matrix elements are given by the product of the force constant tensor $K^{(ij)}$ and the phase difference factor $e^{i\boldsymbol{q}\cdot\Delta \boldsymbol{R}_{ij}}$. This situation is similar to the case of the tight-binding calculation for the electronic structure where the matrix element is given by the product of the atomic matrix element and the phase difference factor (see Section 2.2.2).

5) N_Ω is the number of unit cells in the solid and thus $N_\Omega \sim 10^{23}$/mole.

6) These terms correspond to the diagonal block of the dynamical matrix. The last term in Eq. (3.13) is in the off-diagonal (ij) block of the dynamical matrix. When the ith atom has equivalent neighbor atoms in the adjacent unit cells, the last term can appear in the diagonal block of the dynamical matrix.

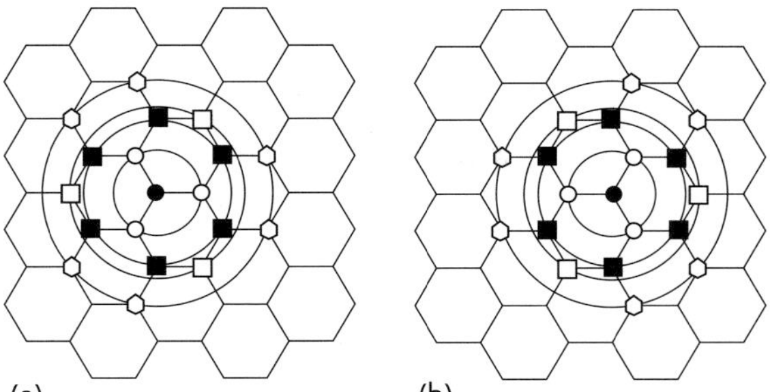

Figure 3.5 Neighbor atoms of graphene up to fourth nearest neighbors for (a) an A atom and (b) a B atom at the center denoted by solid circles. From the first to the fourth neighbor atoms, we plot 3 open circles (first neighbor), 6 solid squares (second), 3 open squares (third), and 6 open hexagons (fourth), respectively. Circles connecting the same neighbor atoms are for guides to the eye [31].

3.2 Phonons in Graphene

Now we describe the force constant model applied to graphene (two-dimensional graphite). In graphene, since there are two distinct carbon atoms, A and B, in the unit cell, we must consider six coordinates $\boldsymbol{u}_k$ (or 6 degrees of freedom) in Eq. (3.12). The secular equation to be solved is thus a 6×6 dynamical matrix $\mathcal{D}$. The dynamical matrix $\mathcal{D}$ for graphene is written in terms of the 3×3 matrices: (1) D^{AA}, (2) D^{AB}, (3) D^{BA}, and (4) D^{BB} for the coupling between (1) A and A, (2) A and B, (3) B and A and several (4) B and B atoms in the various unit cells

$$\mathcal{D} = \begin{pmatrix} D^{AA} & D^{AB} \\ D^{BA} & D^{BB} \end{pmatrix} . \tag{3.14}$$

When we consider an A atom, the three nearest neighbor atoms (see Figures 3.5 and 3.6) are B_1, B_2, and B_3 whose contributions to $\mathcal{D}$ are contained in D^{AB}, while the six next-nearest neighbor atoms denoted by solid squares in Figure 3.5a are all A atoms, with contributions to $\mathcal{D}$ that are contained in D^{AA} and so on. In Figure 3.5a, b, we show neighbor atoms up to fourth nearest neighbors for the A and B atoms, respectively. It is important to note that the A and B sites do not always appear alternately for the nth neighbors. In fact the third and the fourth neighbor atoms in Figure 3.5 belong to equivalent atoms.

The remaining problem is how to construct the force constant tensor $K^{(ij)}$. Here we show a simple way to obtain $K^{(ij)}$.[7] First we consider the force constant between

7) Since the determinant of the dynamical matrix is a scalar variable, the determinant should be invariant under any operation of the point group for the unit cell. Thus the proper combination of terms in the product of the force constant tensor $K^{(ij)}$ and the phase difference factor $e^{i\boldsymbol{q}\cdot\Delta\boldsymbol{R}_{ij}}$ is determined by group theory, which gives block-diagonalization in accordance with the irreducible representations of the symmetry groups of periodic structures (see Chapter 6). Further details are given in the literature for Si and Ge in [125], and for graphite in [126].

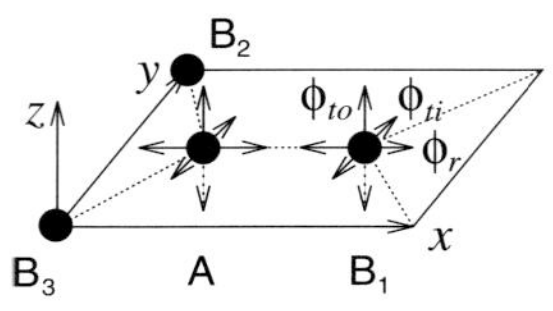

Figure 3.6 Force constants between the A and B_1 atoms on a graphene sheet. Here ϕ_r, ϕ_{ti}, and ϕ_{to} represent forces for the nearest neighbor atoms in the radial (bond-stretching), in-plane and out-of-plane tangential (bond-bending) directions, respectively. B_2 and B_3 are the nearest neighbors equivalent to B_1, whose force constant tensors are obtained by appropriately rotating the $K^{(ij)}$ tensor for A and B_1 [31].

an A atom and a nearest neighbor B_1 atom on the x axis as shown in Figure 3.6 (see also Figure 2.6a). The force constant tensor is given by

$$K^{(A,B_1)} = \begin{pmatrix} \phi_r^{(1)} & 0 & 0 \\ 0 & \phi_{ti}^{(1)} & 0 \\ 0 & 0 & \phi_{to}^{(1)} \end{pmatrix} , \tag{3.15}$$

where $\phi_r^{(n)}$, $\phi_{ti}^{(n)}$, and $\phi_{to}^{(n)}$ represent the force constant parameters in the radial (bond-stretching), in-plane and out-of-plane tangential (bond-bending) directions of the nth nearest neighbors, respectively. Here the graphene plane is the xy plane, the radial direction (x in the case of Figure 3.6) corresponds to the direction of the σ bonds (dotted lines), and the two tangential directions (y and z) are taken to be perpendicular to the radial direction. Since graphite is an anisotropic material, we introduce two parameters to describe the in-plane (y) and out-of-plane (z) tangential phonon mode, and the corresponding phase factor, $e^{i\mathbf{q}\cdot\Delta\mathbf{R}_{ij}}$, becomes $\exp(-iq_x a/\sqrt{3})$ for the B_1 atom at $(a/\sqrt{3}, 0, 0)$.

The force constant matrices for the two other nearest neighbor atoms, B_2 and B_3 are obtained by rotating the matrix in Eq. (3.15) according to the rules for a second-rank tensor

$$K^{(A,Bm)} = U_m^{-1} K^{(A,B_1)} U_m , \quad (m = 2, 3) , \tag{3.16}$$

where the unitary matrix U_m is here defined by a rotation matrix around the z axis in Figure 3.6, taking the B_1 atom into the Bm atom,[8)]

$$U_m = \begin{pmatrix} \cos\theta_m & \sin\theta_m & 0 \\ -\sin\theta_m & \cos\theta_m & 0 \\ 0 & 0 & 1 \end{pmatrix} . \tag{3.17}$$

8) The formulation should be in terms of the rotation of the axes connecting an atom A to its various equivalent neighbors. However, for easy understanding, we present in Eq. (3.16) the rotation of atoms. The matrix for the rotation of the *axes* is the transpose matrix of the matrix for the rotation of *atoms*.

To make the method explicit, we show next the force constant matrix for the B_2 atom at $[-a/(2\sqrt{3}), a/2, 0]$, and U_2 is evaluated assuming $\theta_2 = 2\pi/3$,

$$K^{(A,B_2)} = \frac{1}{4}\begin{pmatrix} \phi_r^{(1)} + 3\phi_{ti}^{(1)} & \sqrt{3}(\phi_{ti}^{(1)} - \phi_r^{(1)}) & 0 \\ \sqrt{3}(\phi_{ti}^{(1)} - \phi_r^{(1)}) & 3\phi_r^{(1)} + \phi_{ti}^{(1)} & 0 \\ 0 & 0 & \phi_{to}^{(1)} \end{pmatrix}, \quad (3.18)$$

and the corresponding phase factor is given by $\exp[-iq_x a/(2\sqrt{3}) + iq_y a/2]$.

In the case of the phonon dispersion relations calculation for monolayer graphene, the interactions between two nearest-neighbor atoms are not sufficient to reproduce the experimental results, and we generally need to consider contributions from long-distance forces, such as from the nth neighbor atoms, $(n = 1, 2, 3, 4\ldots)$.[9] To describe the twisted motion of four atoms, in which the outer two atoms vibrate around the bond of the two inner atoms as shown in Figure 3.7, contributions up to at least the fourth nearest neighbor interactions are necessary [127]. Values for the force constants [123] (see Table 3.1) are obtained by fitting the 2D phonon dispersion relations over the Brillouin zone as determined experimentally, as for example from electron energy loss spectroscopy [128], inelastic neutron scattering [123] or inelastic X-ray scattering [129, 130].

In Figure 3.8a the phonon dispersion curves for a monolayer graphene sheet, denoted by solid lines, are shown using the set of force constants in Table 3.1. In Figure 3.8b the corresponding density of phonon states is plotted per C atom per cm^{-1}, where the energy is in units of cm^{-1}. The calculated phonon disper-

Table 3.1 Force constant parameters for 2D graphene out to fourth neighbors in units of 10^4 dyn/cm [123]. Here the subscripts r, ti, and to refer to radial, transverse in-plane and transverse out-of-plane, respectively. See Figures 3.5 and 3.6.

Radial		Tangential			
$\phi_r^{(1)}$ =	36.50	$\phi_{ti}^{(1)}$ =	24.50	$\phi_{to}^{(1)}$ =	9.82
$\phi_r^{(2)}$ =	8.80	$\phi_{ti}^{(2)}$ =	−3.23	$\phi_{to}^{(2)}$ =	−0.40
$\phi_r^{(3)}$ =	3.00	$\phi_{ti}^{(3)}$ =	−5.25	$\phi_{to}^{(3)}$ =	0.15
$\phi_r^{(4)}$ =	−1.92	$\phi_{ti}^{(4)}$ =	2.29	$\phi_{to}^{(4)}$ =	−0.58

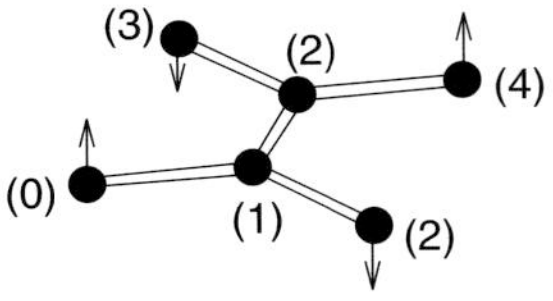

Figure 3.7 In order to describe the twisted motion of four atoms, it is necessary to consider up to at last fourth-nearest neighbor interactions. The numbers shown in the figure denote the nth nearest neighbor atoms from the leftmost zeroth atom.

9) When we consider the force constant matrix of the nth neighbor atoms, these atoms are not always located on the x (or y) axis. In that case it does not seem that we can build an initial force constant matrix as given by Eq. (3.15). This happens at the fourth neighbor atoms in graphene. However, if we consider a virtual atom on the x axis, and if we then rotate the matrix, we can get the force constant matrix without any difficulty.

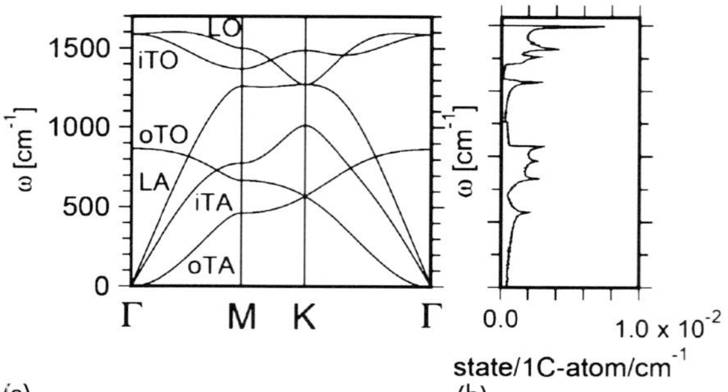

(a) (b)

Figure 3.8 (a) The phonon dispersion curves, plotted along high symmetry directions, for a 2D monolayer graphene sheet, using the set of force constants in Table 3.1 [123]. (b) The corresponding density of states vs. phonon energy for phonon modes in units of states/1C-atom/cm^{-1} × 10^{-2} [31].

sion curves of Figure 3.8a reproduce the experimental points obtained by electron energy loss spectroscopy in general [127, 128], but are not very accurate for the optical phonons near the K point (see the difference between Figures 3.8 and 3.1 and [129]). Thus, to a first approximation the inclusion of fourth-neighbor interactions is sufficient for reproducing the phonon dispersion relations of 2D graphite, but for a very accurate description of the phonon structure near the K point, other effects have to be considered, as discussed briefly in Section 3.5.

The three phonon dispersion branches, which originate from the Γ point of the Brillouin zone (see Figure 3.8a), correspond to acoustic modes: an out-of-plane mode (oTA), an in-plane tangential (bond-bending) mode (iTA) and an in-plane longitudinal (or radial, bond-stretching) mode (iLA),[10] listed in order of increasing energy, respectively. The remaining three branches correspond to optical modes: one out-of-plane mode (oTO) and two in-plane modes (iTO and iLO).

It is noted that the oTA branch shows a q^2 energy dispersion relation around the Γ point, while the other two in-plane acoustic branches show a linear q dependence, as is normally seen for acoustic modes. One reason why we get a q^2 dependence for the out-of-plane mode is simply because this branch corresponds to a two-dimensional phonon mode and because graphite has three-fold rotational symmetry. It is clear in Eq. (3.16) that all rotations U are within the x, y plane in the case of monolayer graphene. Thus the force constant matrix can be decomposed into a 2×2 matrix of x, y components and a 1×1 matrix of z components. The 1×1 force constant tensor $K_{zz}^{(ij)}$ for the nth neighbor atoms does not depend on the coordinates, and $\omega(q)$ thus becomes an even function of k which is obtained from

10) Since the longitudinal modes are always in-plane phonon modes, we can omit "i" from iLA or iLO.

the sum of the differential phase factors $e^{i\boldsymbol{q}\cdot\Delta\boldsymbol{R}_{ij}}$.[11] If we consider only the three nearest neighbor atoms, the sum of the differential phase factors is nothing but $f(k)$ obtained in Eq. (2.28) when discussing the electronic structure. The energy dispersion relation thus obtained (see Eq. (2.31)) is an even function of q around the Γ point. The optical out-of-plane transverse branch ($\sim 865\,\mathrm{cm}^{-1}$ at the Γ point in Figure 3.8a) shows a q^2 dependence for the same reason. Thus, there is neither a phase velocity nor a group velocity for the z component of the vibrations at the Γ point, and the phonon density of states shows a step function which is known as a two-dimensional van Hove singularity (see Figure 3.8b). Finally, remember that while Figure 3.8 introduces all the basic concepts of the phonon dispersion in graphene, it is not accurate enough for describing the experimental measurements, mainly for the in-plane optical modes near the Γ and K points. This will be discussed further in Section 3.4.2 and discussed in depth in later chapters of this book.

3.3 Phonons in Nanoribbons

The phonon dispersion for nanoribbons consists of many one-dimensional phonon dispersion relations which can be obtained, as a first approximation, by the zone-folding technique that we discussed in Section 2.2.5. Like for electrons in Chapter 2, we will discuss such a zone-folding procedure when applying it to carbon nanotubes in Section 3.4, which represents a structure where such a procedure is fully applicable because of the cyclic boundary condition. Nanoribbons are terminated at their boundaries, and the zone-folding procedure has to be applied with care.

Some special modes are of interest to Raman spectroscopy. The width breathing phonon mode in which the ribbon width vibrates is a Raman-active phonon mode with A symmetry. This mode originates from the LA phonon mode in graphene.

11) In general, the phase factor $e^{i\boldsymbol{q}\cdot\Delta\boldsymbol{R}_{ij}}$ goes into its complex conjugate if we change $\boldsymbol{q}$ to $-\boldsymbol{q}$. Thus when we change $\boldsymbol{q}$ to $-\boldsymbol{q}$, the dynamical matrix for the z components in a two-dimensional system becomes its complex conjugate. It is clear that $|D^*| = |D|$ for the Hermitian matrix $\mathcal{D}$, and thus the eigenvalues are even functions of $\boldsymbol{q}$ around $\boldsymbol{q} = 0$ (the Γ point). Even though $\omega(\boldsymbol{q})$ is an even function of $\boldsymbol{q}$, a term proportional to $|\boldsymbol{q}|$ might appear in $\omega(\boldsymbol{q})$. For example, for a one-dimensional spring constant model with the force constant, K, we get $\omega(q) = 2\sqrt{K/M}|\sin qa| \propto |q|$, for ($q \sim 0$). The absence of a linear q term in the phonon dispersion relations along the z axis of graphite comes from the three-fold rotational axis, C_3 along the z direction. Because of this symmetry, $\omega(q_x, q_y)$ should have three-fold rotational symmetry around the C_3 axis. However, no linear combination of q_x and q_y, such as $aq_x + bq_y$ (with constant values for a, b), can be invariant under a $2\pi/3$ rotation around the q_z axis. The simplest invariant form is a constant, and the quadratic form of $q_x^2 + q_y^2$ is also invariant. This is why we get a q^2 dependence for $\omega(q)$ for the out-of-plane branch. When the force constant matrix depends on the atom locations, such as for the in-plane modes, this invariant condition applies to the product of the force constant matrix and the phase difference factor, which generally has a linear q term in $\omega(q)$.

Furthermore, according to the edge structure of graphene nanoribbons, we expect edge-localized phonon modes to appear [131]. Some calculations show that we can see such modes at 1450 cm^{-1} and 2060 cm^{-1} for zigzag and armchair edges, respectively. The reason why we get a relatively lower frequency 1450 cm^{-1} compared to the G-band frequency of 1585 cm^{-1} for the zigzag edge structure is that the edge atoms have only two chemical bonds. In the case of the armchair edge, the dangling bonds of A and B edge atoms form another π bond which makes the C–C bond at the armchair edge a triple bond whose optical phonon modes are around 2000 cm^{-1}. Similar Raman spectra are observed in the polyene C_nH_2, ($n = 8, 10, 12$) encapsulated in a SWNT in the frequency region around 2000 cm^{-1}[4]. When the dangling bonds are terminated by hydrogen atoms, a triple bond becomes a double bond, whose frequency may appear at around 1530 cm^{-1}[132]. The downshift in frequency from 1585 cm^{-1} to 1530 cm^{-1} can be understood by considering the weight of the hydrogen atom. Since the H mass is much lighter than the C mass and since the C–H bond is much stiffer compared with the C–C bond, we may consider that the mass of the edge carbon atoms changes from 12 to 13. In fact, when we multiply $\sqrt{12/13}$ by 1585 cm^{-1}, we get 1530 cm^{-1}. Thus by measuring the micro-Raman modes associated with the edge, we can get information about the edge structure of graphene and related functionalized graphene materials.

3.4 Phonons in Single-Wall Carbon Nanotubes

The vibrational structure of carbon nanotubes is obtained by rolling up the graphene nanoribbon into a cylinder. In this section, we review the zone-folding picture for obtaining the first-approximation to the phonon dispersion relations for nanotubes (Section 3.4.1), while the effect of nanotube curvature is discussed in Section 3.4.2.

3.4.1 The Zone-Folding Picture

As a first approximation, the phonon structure of carbon nanotubes can be obtained using a similar procedure to that used for electrons (see Section 2.3), by superimposing the N cutting lines in the $\boldsymbol{K}_1$-extended representation on the six phonon frequency surfaces in the reciprocal space of the graphene layer [31, 110]. The corresponding one-dimensional phonon energy dispersion relation $\omega_{1D}^{m\mu}(q)$ for the nanotubes is given by:

$$\omega_{1D}^{m\mu}(q) = \omega_{2D}^{m}\left(q\frac{\boldsymbol{K}_2}{|\boldsymbol{K}_2|} + \mu\boldsymbol{K}_1\right), \quad \left(\begin{array}{l} m = 1,\dots,6, \\ \mu = 0,\dots,N-1, \end{array} \text{ and } -\frac{\pi}{T} < q \le \frac{\pi}{T}\right), \tag{3.19}$$

where $\omega_{2D}^{m}(\boldsymbol{q})$ denotes the two-dimensional phonon dispersion relations for a monolayer graphene sheet, q is a one-dimensional wave vector, T is the magnitude of the one-dimensional translation vector $\boldsymbol{T}$, and μ is a cutting line index.

According to the zone-folding scheme, this procedure yields $6N$ phonon modes for each carbon nanotube. The $6(N/2-1)$ pairs of the phonon modes arising from the cutting lines of the indices μ and $-\mu$, where $\mu = 1, \dots, (N/2-1)$, are expected to be doubly degenerate, similar to the case of the electronic sub-bands, while the phonon modes arising from the cutting lines for the indices $\mu = 0$ and $\mu = N/2$ are nondegenerate. The total number of distinct phonon branches is $3(N+2)$. For our prototype (4,2) nanotube, $N = 28$, so that there are 90 distinct phonon branches.

Spikes appear in the phonon density of states (DOS) of the carbon nanotube, similar to the spikes (VHSs) appearing in the electronic DOS (see Figure 3.9c), except for the presence of a much larger number of spikes in the phonon DOS than in the electronic DOS, due to the larger number of phonon modes relative to the number of electronic bands, and the more complex structure of the dispersion relations for phonons than for electrons in the graphene layer. However, the spikes in the phonon DOS do not play such an important role in the experimental outcomes as the spikes in the electronic DOS because of symmetry selection rules. Among the large number of the phonon modes in carbon nanotubes, only a few are Raman-active or infrared-active [134, 135]. Further details about the selection rules for the phonon modes are discussed in Chapter 6, where the relevant group theory discussion is presented. The phonon dispersion relations of the graphene layer calculated by the force constant model are shown in Figure 3.9b along with the cutting lines for our (4,2) sample nanotube in Figure 3.9a. The corresponding phonon density of states for the (4,2) nanotube are shown in Figure 3.9c.

3.4.2 Beyond the Zone-Folding Picture

The zone-folding scheme neglects the curvature of the nanotube wall. Meanwhile, the nanotube curvature couples the in-plane and out-of-plane phonon modes of the graphene layer to each other, especially affecting the low frequency acoustic phonon modes. Among the three acoustic phonon modes of the graphene layer, only one of the two in-plane modes results in the acoustic phonon mode of the nanotube corresponding to the vibrational motion along the nanotube axis. The two other in-plane and out-of-plane acoustic phonon modes give rise to the twisting mode (TW, the vibrational motion in the circumferential direction of the nanotube) and to the radial breathing mode (RBM, the vibrational motion in the radial direction of the nanotube), correspondingly, while the two related acoustic phonon modes of the nanotube (the vibrational motion in two orthogonal directions perpendicular to the nanotube axis) can be constructed as linear combinations of the acoustic modes with wave vectors $q = 2/d_t$ [31].

The zone-folding scheme predicts zero frequencies for both the TW and RBM phonon modes of the nanotube at the center of the Brillouin zone ($q = 0, \omega = 0$),

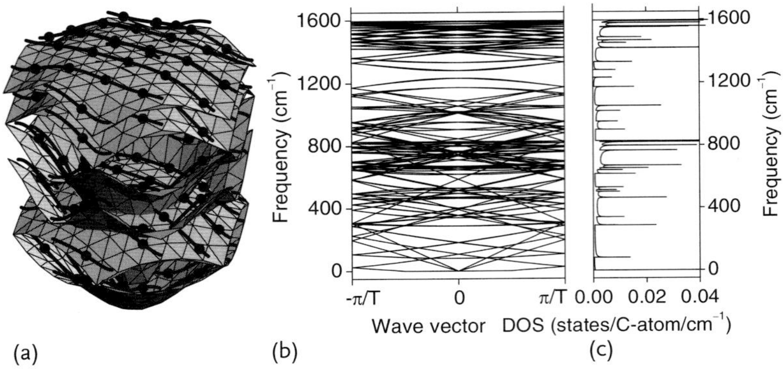

Figure 3.9 (a) The phonon dispersion relations of the graphene layer in the first Brillouin zone calculated with the force constants fitted to the Raman scattering data for various graphitic materials [133]. Solid curves show the cutting lines for the (4,2) nanotube in the fully reduced representation. Solid dots show the ends of the cutting lines in the K_1-extended representation. (b) Phonon dispersion relations for the (4,2) nanotube obtained by zone-folding from (a). (c) The density of phonon states for the phonon modes shown in (b) for monolayer graphene [110].

since they arise from the acoustic phonon modes of the graphene layer. However, the RBM phonon frequency cannot have a zero frequency since it is an in-plane, bond stretching phonon mode, though the corresponding TW phonon mode is a bond bending, in-plane phonon mode. In this sense, the RBM mode is not an acoustic phonon mode but an optical phonon mode. The frequency of the RBM is inversely proportional to the nanotube diameter, varying from around 60 to 450 cm^{-1} for typical diameters of 2.0 to 0.5 nm. This was first predicted within the force constant model [123], and then confirmed by resonance Raman scattering measurements [136], and by *ab initio* calculations [137], which also revealed a weak dependence of the RBM frequency on the nanotube chirality. The zone-folding scheme does not explain the characteristics of the RBM [123], although zone-folding does provide better results for the high frequency phonon modes (optical modes), as confirmed by *ab initio* calculations [138]. In order to avoid the limitations of the zone-folding scheme for the low frequency phonon modes, the force constant model can be used directly for the nanotube by constructing and solving the $6N \times 6N$ dynamical matrix for the unit cell of the nanotube, instead of using the 6×6 dynamical matrix for the unit cell of the graphene layer with subsequent zone-folding [31]. Alternatively, the first-principles methods can be used instead of the force constant models to calculate the phonon modes [116, 124], yet the size of the unit cell cannot be too large for *ab initio* methods. Therefore, *ab initio* calculations are presently limited to achiral nanotubes and to only a few chiral nanotubes with relatively small unit cells. Also, the accuracy of the experiments significantly exceeds what *ab initio* calculational methods can presently achieve.

3.5
Beyond the Force Constant Model and Zone-Folding Procedure

The phonon dispersion relations of the graphene layer can be calculated within a force constant model [31], or by tight-binding [139], or *ab initio* [116] methods. In the force constant model, interactions including as many nearest neighbors in the graphene layer can be considered in order to improve the agreement with experiment. The phonon dispersion relations for monolayer graphene can be measured along high symmetry directions in the Brillouin zone by electron energy loss spectroscopy [128], inelastic neutron scattering [123], and inelastic X-ray scattering [129, 130]. The force constants up to the fourth nearest-neighbor were fitted to the phonon frequencies measured by inelastic neutron scattering in graphite [123]. Furthermore, the force constant model using up to 20 nearest neighbor terms were fitted to inelastic X-ray scattering data [140].

Surprisingly, resonance Raman scattering in sp^2 carbons is not restricted to the Γ point ($q = 0$ selection rule), but Raman spectra are also sensitive to the regions around the high symmetry points K (K') in the first Brillouin zone, and this will be discussed in more detail in Chapters 12 and 13. Actually, Raman experiments on sp^2 carbons probing zone center modes provided evidence for a failure of all the force constant, tight-binding and *ab initio* calculational methods used to describe the phonon dispersion relations around these high symmetry points, until the phonon energy renormalization due to electron–phonon coupling was included in the calculations [141].

In the presence of electron–phonon coupling, the phonon lifetime is no longer infinite. When we consider the electron–phonon interaction as a perturbation, the phonon energy can be modified by a virtual excitation of an electron and this effect is significant both for certain Γ and K point phonons. Because of this virtual excitation of an electron, the phonon lifetime becomes finite, and the phonon frequency shows a broadening due to the Heisenberg uncertainty relation. This phenomena generally becomes very strong for the phonon wave vector $q = 2k_\mathrm{F}$ where k_F is the Fermi wave vector, and we call this effect the Kohn anomaly effect [141]. In the case of graphene and carbon nanotubes, the Kohn anomaly effect occurs for Γ and K point phonons and we will discuss the effect of the Kohn anomaly on the Raman spectra further in Chapter 8.

Problems

[3-1] Obtain the frequencies of two atoms (of mass M), which are connected to each other by a spring with spring constant K and also connected to the two walls (that is, wall-atom-atom-wall geometry). Here we consider only the longitudinal modes for which the vibration is along the bond axis. We consider that the walls have an infinite mass. Show the two normal modes in a figure, using arrows to show the atomic mode displacements.

[3-2] Obtain the vibrational frequencies of three atoms in similar geometries to the previous problem, which involves the wall-atom-atom-atom-wall geometry. Show the three normal modes using arrows in the figure to denote atom displacements. In order to solve this problem, you can use the fact that any normal mode should be either symmetric or antisymmetric under the inversion operation $x \to -x$.

[3-3] Next we consider $N-1$ atoms attached between the two walls. When we consider x_0 and x_n as the coordinates of the two walls, show that all the equations of motion are expressed by the same formula. Then considering that the Bloch theorem applies, substitute $x_\ell = A\exp(iq\ell a - i\omega t)$ into the equations of motion and obtain the dispersion of the phonon frequency $\omega(q)$. Plot the dispersion of the phonon within the first Brillouin zone.

[3-4] In the previous Problem 3-3, we can use the fixed boundary condition of $x_0 = x_n = 0$. Obtain the $N-1$ q-independent phonon frequency values from the boundary condition.

[3-5] What happens in the previous Problem 3-3, if we adopt the periodic boundary condition $x_0 = x_n$. Here we consider the center of mass motion to be zero.

[3-6] Show that the previous results satisfy the case of $N-1=2$ and $N-1=3$ by comparing your results with answers to earlier questions in Chapter 3.

[3-7] The problems above describe the phonon dispersion for a one-atom per unit cell 1D crystal. If you replace each even atom by a different atom, the system will become a two-atoms per unit cell 1D crystal, with a doubled-size unit cell. Show that the first Brillouin zone is reduced to half the size of the original one and show that the phonon dispersion will be represented by a zone-folding of the one-atom phonon dispersion.

[3-8] In Figure 3.3, choose any unit cell vibrational mode and show that the motion of the atoms for $q = 0$ ($\lambda \to \infty$) and for $q = 2\pi/a$ ($\lambda = a$) is the same.

[3-9] Consider the phonon dispersion of a two-dimensional square lattice with atoms of mass M and spring constant K. Plot the resulting phonon dispersion curve as a function of q_x and q_y.

[3-10] Consider the phonon dispersion of a two-dimensional honeycomb lattice with atoms of mass M and spring constant K. In this case, we have two atoms per unit cell. Show the first Brillouin zone and plot the phonon dispersion for the high symmetry directions within the first Brillouin zone.

[3-11] Consider a $2(N-1)$ linear chain of atoms in which two different atoms A and B with masses M_A and M_B are connected along a chain (wall-A-B-A-B-…-B-wall). Obtain and plot the phonon dispersion for this configuration.

[3-12] In the previous problem, show the normal modes for the Γ point and for the zone boundary for each phonon mode.

[3-13] Consider the $2(N-1)$ linear chain with one type of atom (mass M) with two different spring constants alternating in the sequence wall-(K1)-atom-(K2)-...-(K1)-wall. In this case, we have two phonon branches. Plot the phonon dispersion.

[3-14] When we consider a transverse phonon mode, how should we consider the force constant for a linear chain of similar atoms of mass M. Show that stretching a spring in such a linear chain does not give a deformation which is proportional to y or z when we consider the direction along the chain as x.

[3-15] How many normal modes exist for the CH_4, C_2H_2 and C_{60} molecules which, respectively, have the shapes of a regular tetrahedron, a linear chain, and a truncated icosahedron?

[3-16] Let us consider the H_2O molecule. Solve for the phonon normal modes by considering the spring constant K_1 for the H–O bond stretching force constant and K_2 for the H–O–H bond angle force constant. Obtain these force constants by using the experimental values for the mode frequencies.

[3-17] When we consider the two-dimensional square lattice for atoms of mass M and force constant K between nearest neighbor atoms, write an equation of motion for the transverse phonon mode and give a solution for its mode frequency.

[3-18] Let us consider the phonon modes of the C_{60} molecule. When we consider two spring constants for pentagonal and hexagonal C–C bonds, show how to construct the dynamical matrix and how to calculate the phonon modes.

[3-19] Consider zigzag and armchair graphene nanoribbons which have edges with zigzag and armchair shapes. Obtain discrete q vectors in the nanoribbon width direction by defining the width of the nanoribbon.

[3-20] Consider a ring which consists of N carbon atoms, each connected to its neighboring atoms by spring constant K and consider only the radial breathing phonon mode. Then show that the corresponding phonon frequency is inversely proportional to N.

[3-21] When we consider the hexagonal corners of the two-dimensional Brillouin zone of graphene, that is the K and K' points, show that the phonon eigenvectors have the periodicity of a $\sqrt{3} \times \sqrt{3}$ super-cell. How many atoms exist in this super-cell?

[3-22] Consider the previous Problem 3-21 for the case of the M point, which is the center of the hexagonal edge of the two-dimensional Brillouin zone.

[3-23] Study the general theory of the LO and TO phonon modes whose mode frequency ratio depends on the dielectric constant of the materials. In par-

ticular, show that the LO and in-plane TO phonons are degenerate at the Γ point ($q = 0$) in 3D graphite.

[3-24] The sound velocity of the LA phonon mode of graphite is about 21 km/s. Estimate the spring constant of the C–C chemical bond of graphite.

[3-25] Review the principles of inelastic neutron scattering and inelastic X-ray scattering. What is the merit of each of these experimental techniques for studying carbon systems such as graphene and carbon nanotubes?

[3-26] In order to get momentum and energy information from inelastic neutron scattering measurements, we need a monochromator to disperse the neutron beams. How do we get a neutron beam with a fixed kinetic energy?

[3-27] When we use 10 KeV X-rays for observing 0.1 eV phonons, we need high accuracy for observing the scattering angles. Estimate the accuracy of the angles needed for the X-ray detector in such an experiment.

4
Raman Spectroscopy: from Graphite to sp^2 Nanocarbons

This chapter gives a broad perspective on how Raman spectroscopy provides an especially sensitive characterization tool for carbon-based materials and even more so for sp^2 nanocarbon materials. Since many other optical effects can also be utilized in probing sp^2 carbons, to gain a clear understanding of the Raman scattering process, we start by contextualizing the optical processes under the broad heading of light–matter interaction phenomena. We then briefly review a number of photophysical phenomena in order to put the Raman effect into proper perspective (Sections 4.1 and 4.2). Following this discussion, we present the big picture of Raman spectra from sp^2 carbon nanomaterials (Section 4.3). The atomic vibrational nature of the modes related to each Raman peak is then introduced and the differences in the observed Raman spectra for the different sp^2 structures are addressed (Section 4.4). The presentation here is basic and follows a historic perspective, aiming to give the reader a broad picture of the Raman spectra from bulk graphite to sp^2 nanocarbons. In Chapters 5 and 6 we discuss the quantum description and selection rules for the Raman scattering process, respectively. In Part Two of the book we elaborate on the science of each Raman mode that is introduced in the present chapter, showing the great deal of information one can obtain from each of these Raman features for each specific sp^2 carbon system.

4.1
Light Absorption

When shining light into a material (molecule or solid), part of the energy simply passes through the sample (by transmission), while the remaining photons interact with the system through light absorption, reflection, photoluminescence or light scattering. The amount of light that will be transmitted, as well as the details for all the light–matter interactions will be determined by the electronic and vibrational properties of the material. Furthermore, different phenomena occur when shining light into a given material with different energy photons [1, 142], because different energies will be related to the different optical transitions occurring in the medium. As an example of the richness of light–matter interactions, a schematic optical absorption curve for a semiconducting material is shown in Figure 4.1. Using this

Raman Spectroscopy in Graphene Related Systems. Ado Jorio, Riichiro Saito, Gene Dresselhaus, and Mildred S. Dresselhaus

ISBN: 978-3-527-40811-5

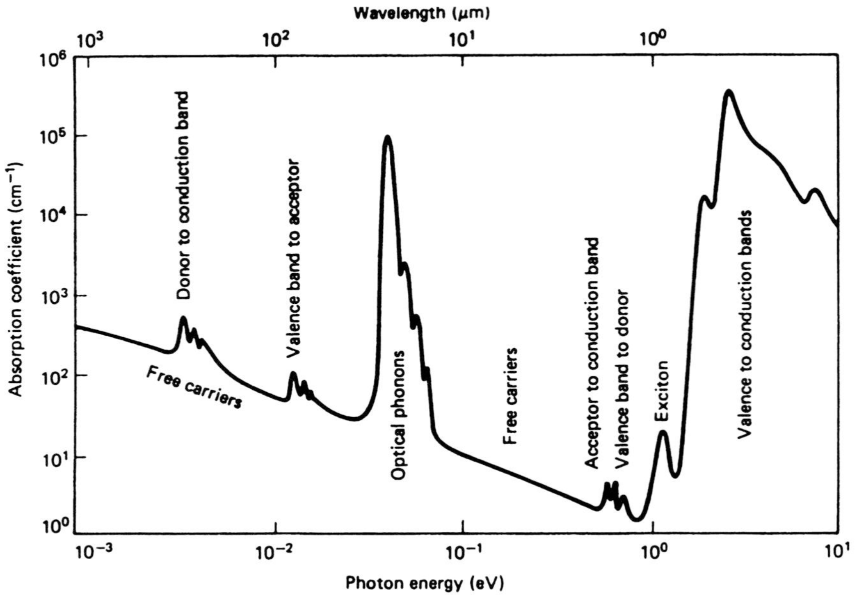

Figure 4.1 Photophysical mechanisms operative for various regions of the electromagnetic spectrum as photons in various energy ranges interact with materials.

figure as a guide, examples are given for the many different effects that might occur when light interacts with a material.

Starting from the high energy side of Figure 4.1:

- The photon (1–5 eV) can be absorbed by an electron making a transition from the valence band to the conduction band. Such a transition generates a free electron in the conduction band leaving behind a "hole" in the valence band, using the nomenclature of semiconductor physics.
- The photon with an energy smaller than the energy gap can generate an exciton level, which corresponds to an electron bound to a hole through the Coulomb interaction. Whereas excitonic levels in model semiconductor systems have excitonic levels of a few millielectron volts below the band gap, the energies in carbon nanotubes are much deeper (on the order of a few hundred millielectron volt range).
- If the semiconductor crystal contains impurities (foreign atoms), new energy levels appear in the energy band gap for an electron bound to such an impurity atom. If the impurity atom has more valence electrons than the atom it replaces, the impurity will act as an electron donor. If it has fewer electrons it will be an electron acceptor. Light can be absorbed, generating electronic transitions from the valence band to the donor impurity levels, or taking electrons from the acceptor level to the conduction band. The corresponding photon energy is 10 to 100 meV smaller than the energy gap.

- When the photon energy coincides with the energy of optical phonons (10 meV to 0.2 eV for first-order processes), light will be absorbed, thereby creating phonons. Harmonics and combination modes are observed also, extending the photon energy range to $\sim$ 350 meV. These processes occur in the infrared energy range (infrared absorption), and play a major role in the field of infrared spectroscopy.
- Shallow donor level transitions to the conduction band and valence band transitions to shallow acceptor levels can also be responsible for light absorption. The corresponding photon energy would be significantly lower than the stronger excitations from the dominant bright state. In some cases the lowest state is a dark state, which means it cannot be created by light absorption.
- The free carriers, which are electrons in metallic systems, and electrons and/or holes in doped semiconducting systems, can also absorb light, usually occurring over a broad energy range from 1 to 10 meV. Other free carrier processes not shown in the diagram also occur. In a much higher energy region (1 to 20 eV), the collective excitation of electrons also occurs. This is known as plasmon absorption.
  For the much higher energy region of ultraviolet and X-ray electron excitation (not shown in Figure 4.1), transitions from core levels also occur. In this case we observe the photoexcited electrons using experimental techniques known as ultra-violet photoelectron spectroscopy (UPS) and X-ray photoelectron spectroscopy (XPS).

One important aspect of optical absorption by crystals is related to wave vector or crystal momentum conservation. In the visible range, the wavelength of light is on the order of $\lambda_{\text{light}} \sim 500$ nm. The dimensions of the Brillouin zones are defined by the maximum value of the wave vector k, which is usually given by $k_{\text{BZ}} = \pi/a$ (see Section 2.1.5), where the primitive translation vector a in the unit cell is about 0.1 to 0.2 nm. Therefore, the photon wave vector is related to the maximum dimension of the Brillouin zone (k_{BZ}) by

$$k_{\text{light}} = \frac{2\pi}{\lambda_{\text{light}}} \approx \frac{k_{\text{BZ}}}{3000} \,. \tag{4.1}$$

Since $k_{\text{light}} \ll k_{\text{BZ}}$, we say that a photon excites one electron from the valence band to the conduction band with the same k. The transition is vertical in the electron energy dispersion, that is, there is no change of wave vector in the electronic energy dispersion.

4.2 Other Photophysical Phenomena

Section 4.1 describes schematically various mechanisms that are responsible for the optical absorption of light in a semiconducting material, as displayed in Fig-

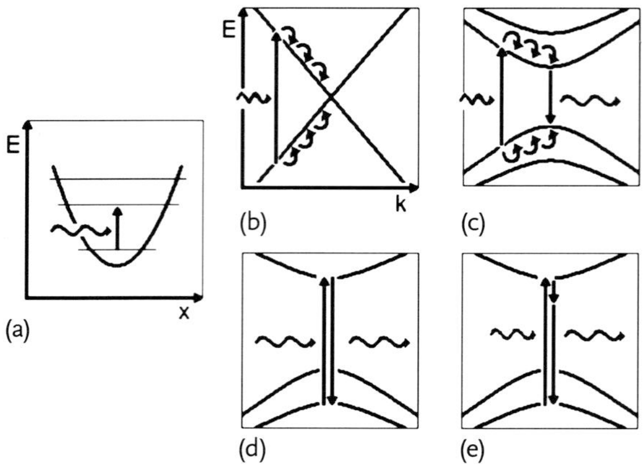

Figure 4.2 The light–matter interaction, showing the most commonly occurring processes. The waved arrows indicate incident and scattered photons. The vertical arrows denote the light-induced transitions between (a) vibrational levels (see Figure 3.2) and (b–e) electronic states. Curved arrow segments indicate electron–phonon (hole–phonon) scattering events. In (e) the shortest vertical arrow also indicates an electron–phonon transition in Raman scattering. In (d) and (e) the processes will be resonant if the incident (or scattered) light energy exactly matches the energy difference between initial and excited electronic states. When far from the resonance window where resonance occurs, the transition is called virtual. The intensity for resonance Raman scattering can be much larger for the vertical processes.

ure 4.1. Now we broadly discuss different phenomena occurring via light–matter interactions (see Figure 4.2):

- The photon energy that has been absorbed can in one case be transformed into atomic vibrations, that is, heat. When the light energy matches the energy for allowed phonon transitions, the photon can transfer energy directly to create an acoustic or optical phonon (see Figure 4.2a). This resonance process is called infrared (IR) absorption, since phonon energies occur at infrared frequencies.
- Even if the photon energy does not match the energy of the optical phonons, this photon energy can be transferred to the electrons. The photoexcited electrons then lose energy by creating multiple phonons of different frequencies by electron–phonon coupling. In a metal, such photoexcited electrons will decay down to their ground states through electron–phonon coupling in a process as shown in Figure 4.2b. If the material has an energy gap between the occupied (valence) and unoccupied (conduction) bands, the photoexcited electron can decay first to the bottom of the conduction band by an electron–phonon process and then to its ground state by emitting a photon with the band gap energy, (see Figure 4.2c). This process is called photoluminescence.
- A photon may be virtually absorbed by a material (not real absorption), which means the photon (oscillating electric field) just shakes the electrons, which will then scatter that energy back to another photon with the same energy as the incident one. In the case where the incident and scattered photons have the

same energy, the scattering process is said to be "elastic" and is named Rayleigh scattering (see Figure 4.2d).

- The photon may shake the electrons, again with no real absorption, thereby causing vibrations of the atoms at their natural vibrational frequencies (by generating phonons). In this case, when the electrons scatter the energy back into another photon, this photon will have lost (gained) energy to (from) the vibration of the atoms. This is an inelastic scattering process that creates or absorbs a phonon and it is named Raman scattering (see Figure 4.2e). When the photon loses energy in creating a phonon we call this a Stokes process. When the photon gains energy by absorbing a phonon we call the process an anti-Stokes process. Since an additional phonon energy E_{ph} is needed for the anti-Stokes process, this process is temperature-dependent and is proportional to $\exp(-\hbar\omega_{ph}/kT)$.
- In a solid, a further distinction is made between inelastic scattering by acoustic phonons (called Brillouin scattering) and by optical phonons (called Raman scattering). This concept does not apply to molecular systems where the acoustic phonon would represent a translation of the molecule (see Section 3.1.2). It is important to remember that Raman and Brillouin scattering also denote light scattering processes due to other elementary excitations in solids and molecules, but in this book we restrict the discussion of Raman scattering to the most general inelastic scattering processes that occur by optical phonons.

Many other processes not listed here, usually related to nonlinear optics, may also occur. They are usually less important for the energetic balance of the light–matter interaction and are not treated in this book. Besides, it is important to draw a distinction between Raman scattering (Figure 4.2e) and photoluminescence (Figure 4.2c), as illustrated in Figure 4.3 [143]. Several light scattering peaks are highlighted by circles in Figure 4.3b. The vertical gray band in Figure 4.3a denotes photoluminescence emission at the band gap $E_{PL} = E_{11} = 1.26\,\text{eV}$. The horizontal gray bands denote nearly continuous luminescence emission associated with thermally excited processes involving different phonon branches. The cutoff energy at 1.06 eV is marked in Figure 4.3b by a vertical dotted line. Slanted dotted lines denote emission from resonant Raman scattering processes for three different phonons named G-band, M-band, and G′-band in Figure 4.3b. Notice in Figure 4.3a the strong emission spots at E_{11} where these Raman lines cross E_{11}. These intersection points are denoted by circles in Figure 4.3b. They are associated with a mixture of photoluminescence and resonance Raman scattering processes, which differ in linewidth (Raman peaks are much sharper) and by the fact that, when changing the excitation laser line, the photoluminescence (PL) emission is fixed at E_{11}, while the Raman peaks change in absolute frequency, keeping fixed the energy shift from the excitation laser line. Occurring at the same energy, these two processes are sometimes confused in the literature, and the major reason is that Raman scattering in solids often has a much greater (say, 10^3 times larger) intensity when the photon energy is equal to an energy band gap, and this effect is called *resonant Raman scattering* (RRS) [144]. To differentiate between these processes one can just look at what happens when changing the excitation laser energy. Alterna-

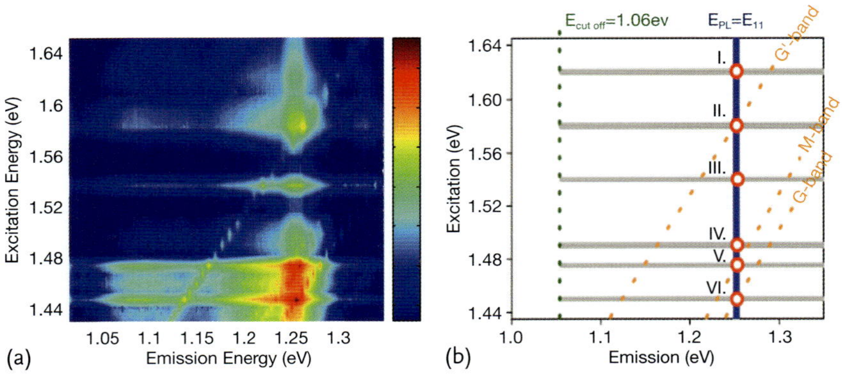

Figure 4.3 (a) A 2D excitation vs. emission contour map for a (6,5)-enriched DNA-SWNT sample. The spectral intensity is plotted using the log scale shown on the right. (b) A schematic view of the observed light emissions plotted as the excitation energy vs. photon emission energy. See the description of the different processes in the text [143].

tively, PL bands are usually broad (hundreds of cm^{-1} or larger) while Raman peaks are sharp (tens of cm^{-1} or sharper).

The difference in linewidth between Raman and PL arises because, in Raman scattering, the intermediate states that are excited between the initial state (incident photon plus the energy of the system before light absorption) and the final state (emitted photon plus the energy of the system after light emission) are "virtual" states. These virtual states do not have to correspond to real states (that is eigenstates) of the physical "system" – any optical excitation frequency will, in principle, suffice. In photoluminescence, on the other hand, the optically excited state must be a real state of the system and, in this case, a real absorption of light occurs, followed by a real emission at a different frequency. Here "real absorption" means that the photoexcited electron can be in the excited states for a sufficient time for measurement, for example, 1 ns.

4.3 Raman Scattering Effect

Within the optical processes, light scattering techniques provide an exceedingly useful tool to study fundamental excitations in solids and molecules, because light can be scattered inelastically so that the incident and scattered photons have different frequencies and this frequency difference is related to the properties of each material. The inelastic scattering of light is called the Raman effect, named in honor of the discoverer of the Raman effect in 1927, is commonly attributed to Sir Chandrasekhara Venkata Raman (1988–1970), an Indian scientist for whom the effect is named.[1]

1) Sir C. V. Raman was awarded the Nobel Prize of Physics in 1930 for his work on "the scattering of light" and for the discovery of the effect named after him.

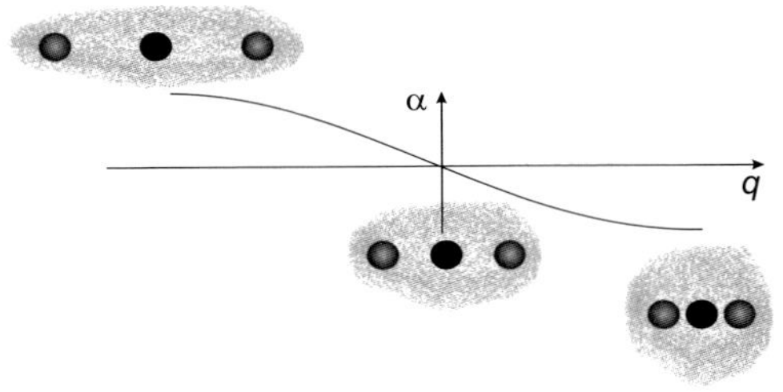

Figure 4.4 Schematics showing how the bond stretching in the CO_2 molecule changes the polarizability α of the molecule. When the oxygen atoms move farther from and closer to the carbon atom, it will be respectively easier and harder for the electric field to mode the electron clouds and thus polarize the molecule. Therefore the polarizability undergoes an oscillatory behavior with the molecule vibration.

In the Raman process, an incident photon with energy $E_i = E_{\text{laser}}$ and momentum $k_i = k_{\text{laser}}$ reaches the sample and is scattered, resulting in a photon with a different energy E_s and momentum k_s. For energy and momentum conservation:

$$E_s = E_i \pm E_q \quad \text{and} \quad k_s = k_i \pm q \,, \tag{4.2}$$

where E_q and q are the energy and momentum change during the scattering event induced by this excitation of the medium. Although different excitations can result from Raman scattering, the most usual scattering phenomenon involves phonons (see Chapter 3), so that E_q and q can be considered to be the energy and the momentum of the phonon that is created or annihilated in the inelastic Raman scattering event.

In Raman scattering, the photon shakes the electrons. The inelastic scattering by phonons occurs because at different atomic positions within the vibrational mode displacements of the atom, the ability of the photon to shake the electrons will be different. This "ability" to shake the electrons is measured by the polarizability (see Figure 4.4 and Section 4.3.1 for a classical description of the Raman effect). These characteristic vibrational modes are called normal modes and are related to the chemical and structural properties of materials. Since every material has a unique set of such normal modes, Raman spectroscopy can be used to probe materials properties in detail and to provide an accurate characterization of certain Raman-active phonon modes in specific materials.

4.3.1 Light–Matter Interaction and Polarizability: Classical Description of the Raman Effect

As an optical phenomena, the basic concepts of Raman spectroscopy can be introduced within the framework of classical electromagnetism. Concepts like dielectric constant and susceptibility are more familiar when describing optical phenomena, but the introduction of the concept of the *polarizability* of a material is needed to discuss inelastic scattering. In this section we briefly discuss the polarizability of materials and give a classical description of the Raman effect.

In describing the polarizability α of an atom in a material, we start by defining α in terms of the polarization vector of the atom p and the local electric field E_{local} at the position of the atom:

$$p = \alpha E_{\text{local}} \,. \tag{4.3}$$

The polarizability α is an atomic property, and the dielectric constant ϵ of a material will depend on the manner in which the atoms are assembled to form a crystal. For a nonspherical atom, α is described by a tensor.[2)] The polarization P of a crystal or of a molecule may be approximated by summing the product of the polarizability of the individual atoms in the crystal (or of the molecule) times the local electric fields

$$P = \sum_j N_j p_j = \sum_j N_j \alpha_j E_{\text{local}}(j) \,, \tag{4.4}$$

where N_j is the atomic concentration of each species and we sum over all the atoms in the crystal using the atomic polarization p_j given by Eq. (4.3). If the local field $E_{\text{local}}(j)$ is given by the Lorentz relation $E_{\text{local}}(j) = E + (4\pi/3)P$, then we obtain

$$P = \sum_j N_j \alpha_j \left(E + \frac{4\pi}{3} P \right) . \tag{4.5}$$

Solving for the susceptibility χ we then obtain

$$\chi \equiv \frac{P}{E} = \frac{\sum_j N_j \alpha_j}{1 - \frac{4\pi}{3} \sum_j N_j \alpha_j} \,. \tag{4.6}$$

Using the definition for the dielectric constant ϵ which relates the displacement vector D to the electric field E through the relation $\epsilon = 1 + 4\pi\chi$, one obtains the Clausius–Mossotti relation

$$\frac{\epsilon - 1}{\epsilon + 2} = \frac{4\pi}{3} \sum_j N_j \alpha_j \,, \tag{4.7}$$

which relates the dielectric constant ϵ to the electronic polarizability α, but only for crystal structures for which the Lorentz local field relation applies.

Light scattering can then be understood simply on the basis of classical electromagnetic theory. When an electric field $\boldsymbol{E}$ is applied to a solid, a polarization $\boldsymbol{P}$ results

$$\boldsymbol{P} = \overset{\leftrightarrow}{\alpha} \cdot \boldsymbol{E} \,, \tag{4.8}$$

2) In this case, the second-rank tensor α is a 3×3 matrix. When we consider p and E_{local} using another coordinate system, each component of the p and E_{local} vectors is transformed by a unitary matrix U, representing rotation, inversion, etc., such as Up and UE_{local}. Then α is transformed by the unitary transformation $U\alpha U^{-1}$, which preserves length scales. A matrix which transforms as $U\alpha U^{-1}$ for a transformation of coordinates is defined as a second-rank tensor. p and E_{local} are both vectors and are called first-rank tensors since U appears once under a coordinate or more general unitary transformation.

where $\overleftrightarrow{\alpha}$ is the polarizability tensor of the atom in the solid, indicating that positive charge moves in one direction and negative charge moves in the opposite direction under the influence of the applied field. We now use these results to obtain a classical description for the Raman effect.

In light scattering experiments, the electric field of the light is oscillating at an optical frequency ω_i

$$E = E_0 \sin \omega_i t \ . \tag{4.9}$$

The lattice vibrations in the solid with a frequency ω_q modulate the polarizability of the atoms α, where

$$\alpha = \alpha_0 + \alpha_1 \sin \omega_q t \ , \tag{4.10}$$

and ω_q is a normal mode frequency of the solid that couples to the optical field so that the polarization which is induced by the applied electric field becomes:

$$\begin{aligned} P &= E_0(\alpha_0 + \alpha_1 \sin \omega_q t) \sin \omega_0 t \\ &= E_0 \left[\alpha_0 \sin(\omega_i t) + \frac{1}{2}\alpha_1 \cos(\omega_i - \omega_q)t - \frac{1}{2}\alpha_1 \cos(\omega_i + \omega_q)t \right] . \end{aligned} \tag{4.11}$$

Thus, we see from Eq. (4.11) that light will be scattered both elastically at a frequency ω_i (Rayleigh scattering) and also inelastically, being downshifted by the natural vibration frequency ω_q of the atom (i. e., by the Stokes process for the emission of a phonon) or upshifted by the same frequency ω_q (by the anti-Stokes process for the absorption of a phonon). For a good appreciation of α_1, it is necessary to introduce quantum theory, and this is the subject of Chapter 5. In the present chapter we simply introduce the basic concept of Raman spectroscopy. In Section 4.3.2, we describe its general characteristics.

4.3.2 Characteristics of the Raman Effect

In this section we very briefly summarize a number of the important characteristics of the Raman Effect that will be broadly used throughout this book.

4.3.2.1 Stokes and Anti-Stokes Raman Processes

In the inelastic scattering process, the incident photon can decrease or increase its energy by creating (Stokes process) or destroying (anti-Stokes process) a phonon excitation in the medium. The plus (minus) signs in Eqs. (4.2) and inside parenthesis in Eq. (4.11) apply when energy has been received from (transferred to) the medium excited by the Raman signal. The probability for the two types of events depends on the excitation photon energy E_i in the scattered photon energy E_s and on the temperature.

The probability to annihilate or create a phonon depends on the phonon statistics, which is given by the Bose–Einstein distribution function. At a given temperature, the average number of phonons n with energy E_q is given by:

$$n = \frac{1}{e^{E_q/k_B T} - 1} \ , \tag{4.12}$$

where k_B is the Boltzmann constant and T is the temperature. Since the vibrational energy E_q of the harmonic oscillator with n phonons is given by $E_q(n + 1/2)$, consequently the scattering event for having n phonons depends on the temperature. The probability for the Stokes (S) and anti-Stokes (aS) processes differs because in the Stokes process the system goes from n phonons to $n + 1$, while in the anti-Stokes process the opposite occurs. Using time reversal symmetry, the matrix elements for the transition $n \rightarrow n + 1$ (Stokes) and $n + 1 \rightarrow n$ (anti-Stokes) are the same, and the intensity ratio between the Stokes and anti-Stokes signals from one given phonon can be obtained by

$$\frac{I_S}{I_{aS}} \propto \frac{n+1}{n} = e^{E_q/k_B T} , \tag{4.13}$$

where I_S and I_{aS} denote the measured intensity for the Stokes and anti-Stokes peaks, respectively.

Since Eq. (4.13) is obtained through time reversal symmetry, this relation should be applicable only when the Stokes and anti-Stokes processes are measured using different incident photon energies (different E_{laser}), that is, when the incident energy for the Stokes process matches the scattered energy for the anti-Stokes process (time reversal). However, usually this is not important, since the phonon energy is much smaller than the laser energy, so that incident and scattered energies are very close to each other. The previous assumption does not hold when resonance Raman scattering with sharp energy levels takes place. In this case, strong deviations from Eq. (4.13) can be obtained if I_S and I_{aS} are taken with the same E_{laser}. The resonance condition might not occur at the same E_{laser} for S and aS Raman scattering.

Finally, because the anti-Stokes signal is usually weaker than the Stokes process, it is usual that people only care about the Stokes spectra. Therefore, in this book, when not referring explicitly to the type of scattering process, it is the Stokes process that is being addressed.

4.3.2.2 The Raman Spectrum

A Raman spectrum is a plot of the scattered intensity I_s as a function of $E_{laser} - E_s$ (Raman shift, see Figure 4.5), and the energy conservation relation given by Eq. (4.2) is a very important aspect of Raman spectroscopy. The Raman spectra will show peaks at a phonon energy $\pm E_q$, where E_q is the energy of the excitation associated with the Raman effect. By convention, the energy of the Stokes process occurs at positive energy while the anti-Stokes process occurs at a negative energy. Thus, in the spectrometer (grating) which divides the scattered light into different directions, the anti-Stokes signal appears in the opposite position relative to the Stokes signal with respect to the central Rayleigh signal.

4.3.2.3 Raman Lineshape and Raman Spectral Linewidth Γ_q

The Raman spectrum exhibits a peak at $E_{laser} - E_s = E_q$ where the phonon excitation can be represented by a harmonic oscillator damped by the interaction with other excitations in the medium (similar to a mass-spring system inside a liquid).

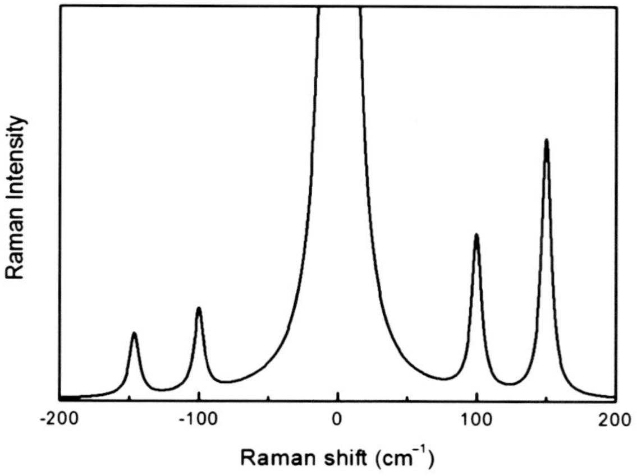

Figure 4.5 Schematics showing the Rayleigh (at 0 cm^{-1}) and the Raman spectrum. The Rayleigh intensity is always much stronger and it has to be filtered out for any meaningful Raman experiment. The Stokes processes (positive frequency peaks) are usually stronger than the anti-Stokes processes (negative frequency peaks) due to phonon creation/annihilation statistics.

Therefore, the shape of the Raman peak will be the response of a damped harmonic oscillator with eigenfrequency ω_q that is forced by an external field with a frequency ω. Considering the damping energy given by Γ_q, the power dissipated by a forced damped harmonic oscillator is a Lorentzian curve

$$I(\omega) = \frac{I_0}{\pi \Gamma_q} \frac{1}{(\omega - \omega_q)^2 + \Gamma_q^2} , \tag{4.14}$$

in the limit where the frequency $\omega_q \gg \Gamma_q$.[3] The full width at half maximum intensity is given by FWHM = $2\Gamma_q$. The center of the Lorentzian gives the natural vibration frequency ω_q, and Γ_q is related to the damping or the energy uncertainty or the lifetime of the phonon [145]. Therefore, when the damping of the amplitude occurs as the scattered light energy is varied, as characterized by Γ_q, the corresponding phonon has a finite life time, Δt. The uncertainty principle $\Delta E \Delta t \sim \hbar$ gives an uncertainty in the value of the phonon energy, as measured in the Raman spectrum, which corresponds to the spectral FWHM of $2\Gamma_q$. Therefore, Γ_q is the inverse of the lifetime for a phonon, and Raman spectra in this way provide information on phonon lifetimes.

There are two main reasons for the finite phonon lifetime:

- Anharmonicity of the potential for the phonon so that, for large q far from the potential minimum, q_{eq} is no longer a good quantum number and phonon scattering occurs by emitting a phonon (third-order process) or by phonon–phonon

3) The solution of a forced damped harmonic oscillator is not a Lorentzian. It approaches the Lorentzian function for $\omega_q \gg \Gamma_q$. If ω_q approaches Γ_q, the lineshape departs from a Lorentzian shape.

scattering (fourth-order anharmonicity). Anharmonicity is a main contribution to the thermal expansion (third-order process) and to the thermal conductivity (fourth-order process).

- Another possible interaction is the electron–phonon interaction in which a phonon excites an electron in the valence band to the conduction band or scatters a photoexcited electron to other unoccupied states. The former electron–phonon process works for electrons in the valence band while the latter electron–phonon process associated with anharmonicity works for electrons in excited states. Thus the origin of these two electron–phonon processes are different from each other.

In specific cases, the Raman feature can deviate from the simple Lorentzian shape. One obvious case is when the feature is actually composed of more than one phonon contribution. Then the Raman peak will be a convolution of several Lorentzian peaks, depending on the weight of each phonon contribution. Another case is when the lattice vibration couples to electrons, that is, when the electron–phonon interaction takes place. In this case, additional line broadening and even distorted (asymmetric) lineshapes can result and this effect is known as the Kohn anomaly. In cases where phonons are coupled to electrons, the Raman peak may exhibit a so-called Breit–Wigner–Fano (BWF) lineshape, given by [147]:

$$I(\omega) = I_0 \frac{[1 + (\omega - \omega_{\mathrm{BWF}}/q_{\mathrm{BWF}}\Gamma_{\mathrm{BWF}})]^2}{1 + [(\omega - \omega_{\mathrm{BWF}}/\Gamma_{\mathrm{BWF}})]^2} , \tag{4.15}$$

where $1/q_{\mathrm{BWF}}$ is a measure of the interaction of a discrete level (the phonon) with a continuum of states (the electrons), ω_{BWF} is the BWF peak frequency at maximum intensity I_0, and Γ_{BWF} is the half width of the BWF peak. Such effects are observed in certain metallic sp^2 carbon materials and are discussed in Chapter 8, in connection with metallic carbon nanotubes.

4.3.2.4 Energy Units: cm^{-1}

The energy axis in the Raman spectra is usually displayed in units of cm^{-1}. 1 cm^{-1} is the energy of a photom whose wavelength is 2π cm. Lasers are usually described by the wavelength of the light, that is, in nanometers, but the phonon energies are usually too small a number when displayed in nanometers, and the Raman shifts are thus given in units of cm^{-1} (1 cm^{-1} is equivalent to 10^{-7} nm^{-1}). Furthermore, the accuracy of a common Raman spectrometer is on the order of 1 cm^{-1}. It is important to note that the wave number is expressed in units of cm^{-1} but the definition of the wavenumber in this case is $k = 1/\lambda$ (where λ is the wavelength of light) which is different from the definition of the wavenumber in solid state physics, $k = 2\pi/\lambda$. The energy conversion factors are: $1\,\mathrm{eV} = 8065.5\,\mathrm{cm}^{-1} = 2.418 \times 10^{14}\,\mathrm{Hz} = 11\,600\,\mathrm{K}$. Also 1 eV corresponds to a wavelength of 1.2398 µm.

4.3.2.5 Resonance Raman Scattering and Resonance Window Linewidth γ_r

The laser excitation energies are usually much higher than the phonon energies. Therefore, although the exchange in energy between light and the medium is transferred to the atomic vibrations, the light–matter interaction is mediated by electrons. Usually the photon energy is not large enough to achieve a real electronic transition, and the electron that absorbs the light is said to be excited to a "virtual state", from where it couples to the lattice, generating the Raman scattering process. However, when the excitation laser energy E_{laser} matches the actual energy gap between the valence and conduction bands E_g in a semiconducting medium (or between an occupied initial state and an unoccupied final state more generally), the probability for the scattering event to occur increases by many orders of magnitude ($\sim 10^3$), and the process is then called a resonance Raman process (non-resonant otherwise). The same happens if the scattered light ($E_{\text{laser}} \pm E_q$, where "+" denotes the anti-Stokes and "−" denotes the Stokes process) is equal to the electronic transition E_g. Therefore, by varying E_{laser} through a discrete level E_g, the Raman intensity should increase when $E_{\text{laser}} \to E_g$ (resonance with incident light) and when $E_{\text{laser}} \to E_g \pm E_q$ (resonance with scattered light). A plot of the Raman intensity vs. E_{laser} gives the Raman excitation profile, according to

$$I(E_{\text{laser}}) = \left| \frac{A}{(E_{\text{laser}} - E_g - i\gamma_r)(E_{\text{laser}} - (E_g \pm E_q) - i\gamma_r)} \right|^2 . \qquad (4.16)$$

The FWHM of each peak in such a plot of the Raman excitation profile is the resonance window width γ_r, and is related to the lifetime of the excited states, that is, the lifetime for the Raman scattering process, which is the time delay between absorption of the incoming photon and emission of the outgoing photon. In other words, γ_r is the inverse of the lifetime for the photoexcited carrier. The photoexcited carrier can be relaxed from the excited states by

- The electron–phonon interaction for all possible phonons (with lifetimes < 1 ps);
- The electron–photon interaction (with lifetimes < 1 ns);
- Other excitations such as the Auger process (Coulomb interaction) (lifetime range not known yet for sp^2 carbon).

Thus the electron–phonon interaction from k (a photoexcited state) to the energy-momentum conserved $k + q$ (phonon emitting electron state) is dominant. Notice γ_r (resonance window width) is different from the Γ_q (width on the Raman intensity vs. scattered light energy Raman spectrum, where Γ_q is related to the phonon lifetime, see Section 4.3.2.3). These two experimental widths may be related or not, depending on the electron–phonon coupling. The physics behind the connection between these widths is further discussed in Part Two of this book.

The resonance effect is extremely important in nano-scale systems, since the Raman signal from nanomaterials is generally very weak because of the very small sample size. Thus the large resonance enhancement by the resonance Raman effect allows the observation of measurable Raman signals from nanostructures.

For example, the large enhancement associated with the resonance Raman scattering (RRS) process provides a means to study the Raman spectrum from a single graphene sheet, a single graphene ribbon or an individual carbon nanotube, as discussed further in Section 4.4.

4.3.2.6 Momentum Conservation and Backscattering Configuration of Light

As discussed in this section, the $\boldsymbol{q}$ vector of the phonon carries information about the wavelength of the vibration ($q = 2\pi/\lambda$) and the direction along which the oscillation occurs. In an inelastic scattering process, momentum conservation is required as given by Eq. (4.2). Different scattering geometries of light are possible by the appropriate placement of the detector of the scattered light relative to the direction of the incident light. If we select a specific choice of this geometry, we can select different phonons due to the anisotropy of the scattering event by the selection rules for Raman scattering.

In a general scattering geometry in which the scattered light wave vector $\boldsymbol{k}_s$ makes an angle ϕ with the incident $\boldsymbol{k}_i$, the modulus of the phonon wave vector q will be given by the law of cosines:

$$q^2 = k_i^2 + k_s^2 \pm 2k_i k_s \cos\phi \,. \tag{4.17}$$

The backscattering configuration of the light, for which k_i and k_s for the incident and scattered light, respectively, have the same direction and opposite signs, gives the largest possible q vector and is the most common scattering geometry when working with nanomaterials, because a microscope is usually needed to focus the incident light onto small samples and the scattered light is also collected by the same microscope.

For the Raman-allowed one-phonon scattering process, the momentum transfer is usually neglected, that is, $k_s - k_i = q \sim 0$. The momenta associated with the first-order light scattering process are on the order of $k_i = 2\pi/\lambda_{\text{light}}$, where λ_{light} is in the visible range (800–400 nm). Therefore, k_i is very small when compared to the dimensions of the first Brillouin zone, which is limited to vectors no longer than $q = 2\pi/a$, and where the unit cell vector a in real space is on the order of a tenth of a nanometer and $a = 0.246$ nm for graphene and carbon nanotubes. This discussion explains why the first-order Raman process can only access phonons at $q \to 0$, that is, very near to the Γ point. Thus, the phonon momentum $q \neq 0$ becomes important only in defect-induced or higher-order Raman scattering processes.

4.3.2.7 First and Higher-Order Raman Processes

The order of the Raman process is given by the number of scattering events that are involved in the Raman process. The most usual case is the first-order Stokes Raman scattering process, where the photon energy exchange creates one phonon in the crystal with a very small momentum ($q \approx 0$). If two, three or more scattering events occur in the Raman process, the process is called second, third, or higher-order, respectively. The first-order Raman process gives the basic quantum of vibration, while higher-order processes give very interesting information about overtones and combination modes. In the case of overtones, the Raman signal appears at $n E_q$ ($n = 2, 3, \ldots$) and the Raman signal from combination modes appears at the sum of different phonon energies ($E_{q1} + E_{q2}$, etc.). What is an interesting point in the higher-order Raman signal in a solid material is that the restriction for $q \approx 0$ in first-order Raman scattering is relaxed. The photoexcited electron at k can be scattered to $k + q$ and can go back to its original position at k after the second scattering event by a phonon with wave vector $-q$, which allows the recombination of photoexcited electrons with their corresponding holes. The probability for selecting a pair of q and $-q$ phonons is usually small and not very important for solids. However, we will see in Chapters 12 and 13 that, under special resonance conditions (multiple resonance condition) common in sp^2 nanocarbons, we can expect a distinct Raman signal from $q \neq 0$ scattering events.

For example, in Figure 1.5, the one-phonon Raman bands in sp^2 carbons go up to frequencies of 1620 cm^{-1}, and the spectra above 1620 cm^{-1} are composed of overtone ($G' = 2D, 2G$) and combination ($D + D'$) modes. The disorder-induced one-phonon D and D′ features are due to second-order Raman scattering process, since both features involve a scattering event by a $q \neq 0$ phonon and another scattering event induced by a symmetry breaking elastic scattering process such as a defect which contributes with a wave vector $q_{\text{defect}} = -q$, to achieve momentum conservation. The effect of defects is discussed in Chapter 13.

4.3.2.8 **Coherence**

It is not trivial to define whether a real system is large enough to be considered as effectively infinite and therefore to exhibit a continuous phonon (or electron) energy dispersion relation. Whether or not a dispersion relation can be defined indeed depends on the process that is under evaluation and the characteristics of this process. In the Raman process, we ask how long does it take for an electron excited by the incident photon to decay? Considering this scattering time, what is the distance probed by an electron wave function? These issues are discussed in condensed matter physics textbooks under the concept of *coherence*. The coherence time is the time the electron takes to experience an event such as a scattering process that changes its state. Thus, the coherence length is the size over which the electron maintains its quantum state identity and its coherence. The coherence length is defined by the electron speed and the coherence time, both of which can be measured experimentally. The Raman process is an extremely fast process, in the range of femtoseconds (10^{-15} s). Considering the speed of electrons in graphite and graphene (10^6 m/s), this electron speed gives a coherence length on the order of nanometers. Interestingly this number is much smaller than the wavelength of visible light. On the other hand, this is a particle picture for the scattering process and consideration of both the particle and wave aspects of electrons and phonons are important for carbon nanostructures. Playing with these concepts is actually quite interesting and important when dealing with local processes induced by defects, as is discussed in Chapter 13.

4.4 General Overview of the sp^2 Carbon Raman Spectra

Next, we provide an overview of the Raman spectra of sp^2 carbon-based materials following the basic concepts of Raman spectroscopy described above. Figure 1.5 shows the Raman spectra from different crystalline and disordered sp^2 carbon nanostructures in comparison to graphite and amorphous carbon. In the present section we will introduce the spectral features observed in the Raman spectra of many sp^2 carbons, following a historic perspective, that is, starting from the precursor material graphite, going through nanotubes and ending with the most fundamental material, graphene.

4.4.1 Graphite

The Raman spectrum of crystalline graphite is marked by the presence of two strong peaks centered at $1580\,\text{cm}^{-1}$ and $2700\,\text{cm}^{-1}$, being named the G and G′ bands, respectively, where the G label comes from graphite (see Figure 1.5). In 1970, Tuinstra and Koenig proposed that the lowest frequency peak (G-band, see Figure 4.6) is a first-order Raman-allowed feature originating from the in-plane stretching of the C–C bond [148, 149]. The highest frequency peak (G′ band) was reported by Nemanich and Solin [150, 151] and was then assigned as a second-order (two-phonon) feature with $q \neq 0$.

In the Raman spectra obtained from samples with small crystallite size L_a ($< 0.5\,\mu\text{m}$, that is, smaller than the wavelength of light), the presence of an additional peak centered at $\sim 1350\,\text{cm}^{-1}$ was observed (see Figure 4.7a). Tuinstra and Koenig assigned this feature to the breathing of the carbon hexagons that "achieves Raman activity at the borders of the crystallite areas due to loss of translational symmetry" [148, 149]. Since the frequency of this feature is about half of

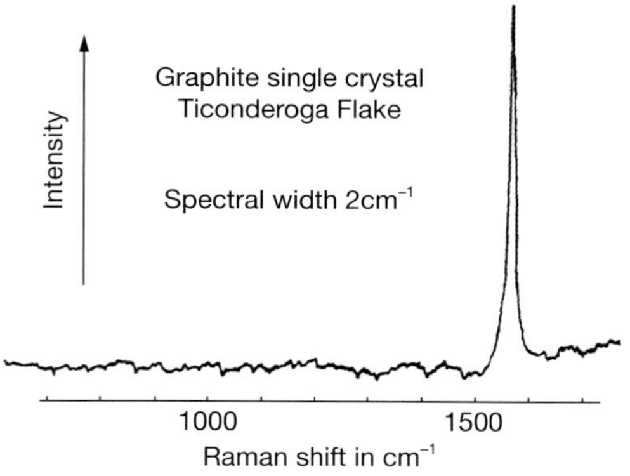

Figure 4.6 Raman spectrum from a single crystal of graphite, in which the presence of the one-phonon allowed G-band is observed [148].

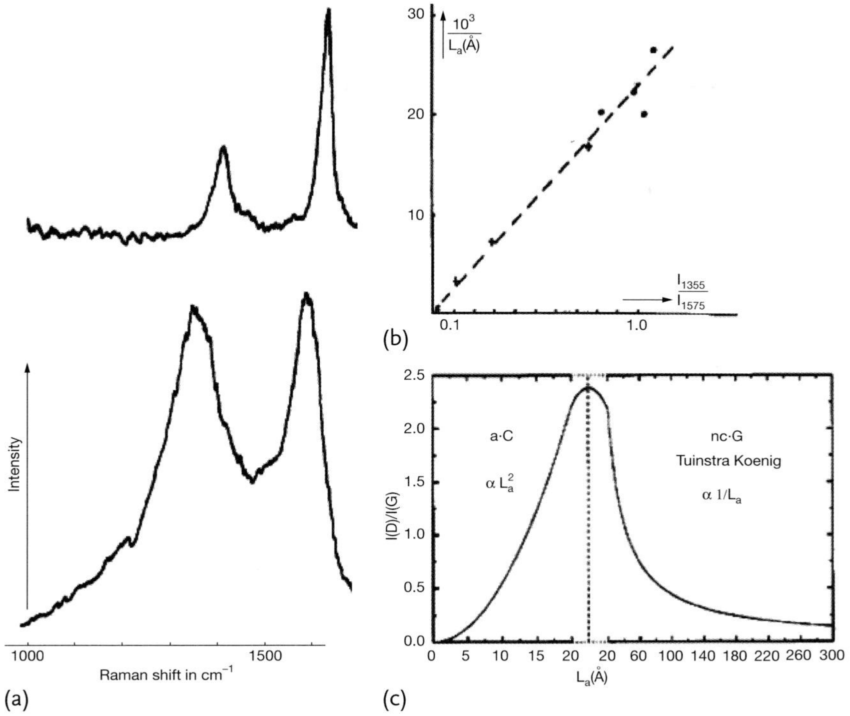

Figure 4.7 (a) Raman spectrum obtained from nanographite. The top spectrum comes from commercial graphite, and the bottom spectrum from activated charcoal. The *x*-axis gives the Raman shift in cm^{-1}. (b) X-ray data for L_a plotted as a function of the measured intensity ratio I_D/I_G between the disorder-induced (D) and the Raman-allowed (G) peaks [148]. (c) The proposed amorphitization trajectory for I_D/I_G over a wide range of L_a (Å) values [88].

the second-order G′ frequency, it was identified as the first-order of the G′ peak. Since the $1350\,\mathrm{cm}^{-1}$ peak is observed in the presence of defects in an otherwise perfect infinite graphite structure, it has been named the D-band (D for defect or disorder). Based on the assumption that this D-band is also associated with boundaries and interfaces, they proposed that its intensity should be proportional to the amount of crystallite boundary in the sample, and showed that the ratio between the intensities of the disorder-induced D-band and the first-order graphite G-band (I_D/I_G) is linearly proportional to the inverse of the crystallite size L_a (see Figure 4.7b) [148, 149]

$$I_\mathrm{D}/I_\mathrm{G} = \frac{\mathrm{A}}{L_a}\,, \tag{4.18}$$

where A is a constant for a fixed Raman excitation frequency. This relation can be applied to large enough carbon sp^2 crystallites, while the complete amorphitization trajectory for I_D/I_G going down to small L_a values was proposed in 2000 by Ferrari and Robertson [88] (see Figure 4.7c). As proposed by Ferrari and Robertson, I_D/I_G starts to decrease for sp^2 carbon hexagonal structure starts to disappear. Furthermore, this I_D/I_G intensity ratio was further shown to be excitation laser energy-dependent [152]. Another disorder-induced band centered at 1620 cm^{-1} is usually observed in the Raman spectra of disordered graphitic materials, although with smaller intensity as compared to the D-band. This feature, reported in 1978 by Tsu *et al.* [153], has been named the D′-band and also depends on L_a and E_{laser} [153].

In 1981, Vidano *et al.* [154] showed that the D and G′-band are dispersive, that is, their frequencies change with the incident laser energy E_{laser} [154] with $\Delta\omega_D/\Delta E_{laser} \sim 50\,cm^{-1}/eV$ and $\Delta\omega_{G'}/\Delta E_{laser} \sim 100\,cm^{-1}/eV$. The out-of-plane stacking order has also been shown to affect the G′ Raman spectra [155–157]. Baranov *et al.* [158] proposed, in 1987, that the dispersive behavior of the D-band comes from the coupled resonance between the excited electron and the scattered phonons, as previously discussed in semiconductor physics [91, 142]. The full appreciation of the double resonance model came in 2000, as discussed by Thomsen and Reich [159], and extended to explain the mechanism behind many other dispersive Raman peaks usually observed in the literature [160], yielding an explanation for all the features observed in the Raman spectra of ordered and disordered graphite [88]. Many of the weak Raman features are dispersive and can be used to measure the graphite phonon dispersion [160], that is, the atomic vibrations with different wave vectors, usually obtained only with inelastic neutron scattering due to momentum conservation requirements. Another interesting result obtained by the double resonance model (see Chapter 13) was the definition of the atomic structure at graphite edges [161].

In parallel to the solid state physics approach for exhibiting the Raman spectra of sp^2 carbons, a molecular approach to the Raman spectroscopy of graphite has developed based on polycyclic aromatic hydrocarbons (PAH) [162–164]. Here PAH denotes a class of planar two-dimensional π-conjugated structures consisting of condensed aromatic rings, with a structure similar to graphene (see Figure 4.8). The PAHs can be synthesized with well-defined size and shape [165, 166], allowing study of the effects of the confinement and de-localization of π-electrons using quantum chemistry calculations. The quantum chemical studies show two collective vibrational displacements characteristic of PAHs giving rise to strong Raman signals, the first appearing in the frequency range 1200–1400 cm^{-1}, with a large projection of a totally symmetric atomic vibration related to the D-band, and the second appearing in the frequency range 1600–1700 cm^{-1}, with a large projection on the G-band displacement (see Figure 4.8). The D-like band is active in PAHs due to the relaxation of the structure correlated with the confinement of π electrons. The confinement and de-localization effects can be induced in the presence of finite-size graphite domains or edges. These effects change progressively with the molecular size in connection with disordered and nanostructured sp^2 carbon materials, and exhibit a clear size-dependent resonance phenomenon. These concepts have been behind the development of long-range electron–phonon coupling models for the periodic graphene structure, such as for the Kohn anomaly, discussed in Chapter 8 [141].

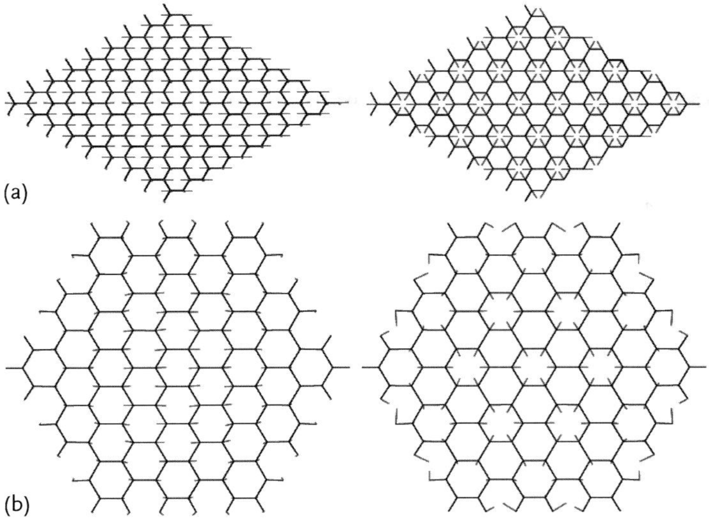

Figure 4.8 (a) nuclear displacements associated with the G-band (at Γ point) and D-band phonons (at the K point) of a perfect 2D graphene monolayer. (b) comparison with selected normal modes of $C_{114}H_{30}$, as obtained from semi-empirical quantum consistent-force-field π electron method (QC FF/PI) calculations [167].

At this point, the main ingredients responsible for the large impact of Raman spectroscopy on sp^2 nanocarbons are already in place. The G-band gives the first-order signature. As we will see, the G-band exhibits a rich behavior for nanostructured systems due to quantum confinement effects, curvature effects and electron–phonon coupling. The G′-band is sensitive to small changes in both the electronic and vibrational structures, and acts as a probe for electrons and phonons and their uniqueness in each distinct sp^2 nanocarbon. The disorder-induced D-band (and also the smaller intensity D′-band) appears due to symmetry-breaking effects, and, together with theory, can provide valuable information about local disorder. The D-band is a second-order process that includes an inelastic scattering event and thus D-band is not the first-order process of the G′-band which only involves two phonon scattering processes of the same phonon as for the D-band. Finally, many other small features can be observed and related to specific physical properties.

To conclude this brief review of the Raman spectra of graphite, we note that when strong amorphitization takes place, substantially increasing the sp^3 carbon sample content, significant changes in the Raman lineshapes are observed [88]. Amorphous carbon and diamond-like materials have been intensively studied and these materials have been broadly utilized in surface coating applications. This book will discuss in depth the disorder-induced features related to symmetry breaking of the sp^2 configuration, but discussion of the Raman spectra associated with amorphitization is beyond the scope of this volume. For more discussion on sp^3 carbon systems, see for example [88].

4.4.2
Carbon Nanotubes – Historical Background

The use of Raman spectroscopy to characterize carbon materials generally [90, 168] motivated researchers to apply this technique also to SWNTs shortly after the early synthesis of SWNTs in 1993 [25, 26]. Even though only $\sim$ 1% of the carbonaceous material in the sample used in the 1994 experiments was estimated to be due to SWNTs [169], a unique Raman spectrum was observed, different from any other previously observed spectrum, namely a double-peak G-band structure. These findings motivated further development of this noninvasive characterization technique for SWNTs, and the laser ablation technique, developed in 1996 to synthesize large enough quantities of high purity SWNT material [170], opened up this field of investigation [136].

The first phase explorations of Raman spectroscopy on SWNTs spanned the four year period 1997–2001. Figure 4.9 gives a general view of the Raman spectra from a typical SWNT bundle sample. There are two dominant Raman signatures in these Raman spectra that distinguish a SWNT from other forms of carbon. The first relates to the low frequency feature, usually in the range 100–300 cm^{-1}, arising from phonon scattering by the radial breathing mode (RBM), which corresponds to symmetric in-phase displacements of all the carbon atoms in the radial direction

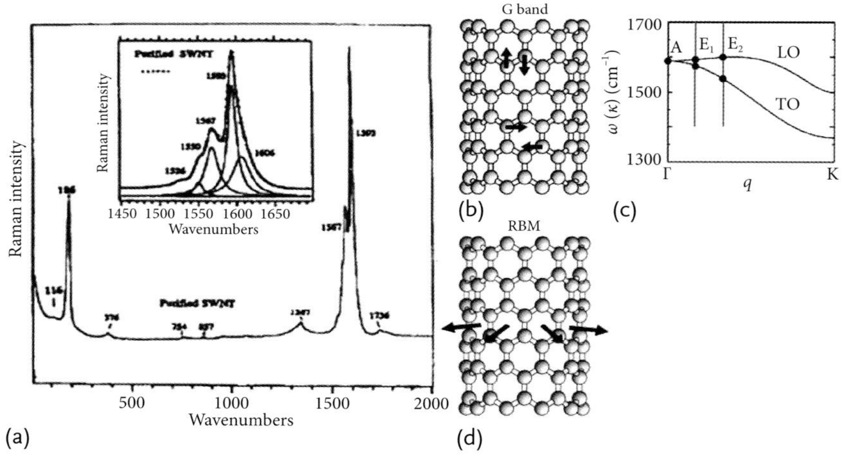

Figure 4.9 (a) Room temperature Raman spectrum from a SWNT bundle grown by the laser vaporization method. The inset shows a Lorentzian fit to the G-band multiple feature [112, 136]. (b) The G-band eigenvectors for the C–C bond stretching mode. The G-band is composed of up to six peaks that are allowed in the first-order Raman spectra. The vibrations are tangential to the tube surface, three along the tube axis (LO) and three along the circumference (TO), having two modes with A, E_1 and E_2 symmetries. Within the set of three peaks, what changes is the relative phase of the vibration, containing 0, 2 or 4 nodes along the tube circumference, which is related to increasing q in the unfolded graphene phonon dispersion $\omega(q)$ along ΓK in the Brillouin zone (c). (d) Eigenvector for the radial breathing mode (RBM) appearing around 186 cm^{-1} in (a).

(see Figure 4.9d). The second signature relates to the multi-component higher frequency features (around 1500–1600 cm^{-1}) associated with the tangential (G-band) vibrational modes of SWNTs and is related to both the early observations in [169] and to the Raman-allowed feature appearing in Raman spectra for graphite (see Figure 4.9b,c). Neither the RBM feature nor the multi-component G-band features in the Raman spectra for SWNTs were previously observed in any other sp^2 bonded carbon material.

Like in graphite, several studies were carried out on the D, G, G′ bands of carbon nanotubes, as well as on other combination modes and overtones [112]. The power of Raman spectroscopy for studying carbon nanotubes was in particular revealed through exploitation of the resonance Raman effect, which is greatly enhanced by the singular density of electronic states, which comes from one-dimensional confinement of the electronic states. Soon after the discovery of the resonance Raman effect in SWNTs [136], it was found that the resonance lineshape could be used to distinguish metallic from semiconducting SWNTs. Specifically, the lineshape for the lower frequency G-band feature (G^-) is broader for a metallic tube, and this feature follows a Breit–Wigner–Fano lineshape, and downshifted in frequency from that for a semiconducting tube of similar diameter (see Figure 4.10a) [147, 171]. The RBMs from SWNT bundles were found to exhibit an oscillatory behavior depending on the excitation laser energy, due to the confinement of the electronic structure [172], and a similar effect was observed for the G′ band [173]. The interpretation of all these resonance Raman effects was systematized by the develop-

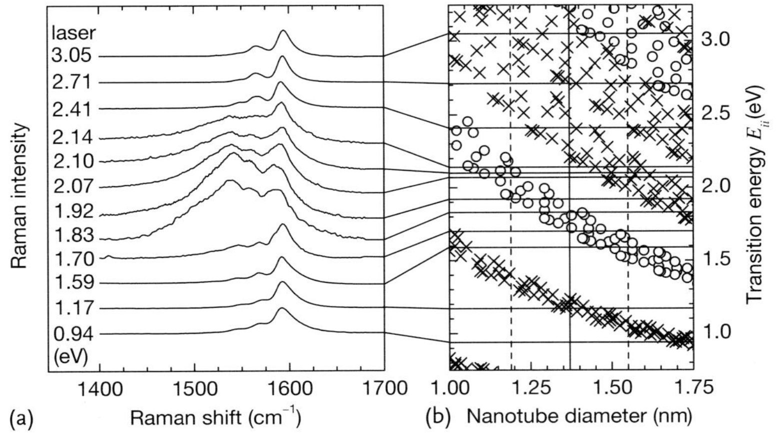

Figure 4.10 (a) Raman spectra of the tangential G-band modes of carbon nanotube bundles measured with several different laser lines showing lineshapes typical of semiconducting and metallic carbon nanotubes, taken from [171]. (b) Optical transition energies for the nanotube diameter distribution in the bundles (the vertical solid line is the average nanotube diameter and the vertical dashed lines define the full width at half maximum diameter distribution of the nanotube sample). Crosses denote optical transition energies for semiconducting nanotubes and open circles are for metallic SWNTs. The G-band gets broad and downshifted when the laser is in resonance with optical transitions from metallic SWNTs, as shown in (a) [110].

ment, in 1999, of the so-called Kataura plot, where Kataura and coworkers displayed the optical transition energies for each (n, m) SWNT as a function of tube diameter [174], as shown in Figure 4.10b. Advances in the synthesis of carbon nanotubes also took place in the 1997–2001 time frame, allowing the use of catalyzed chemical vapor deposition (CVD) techniques in the growth of samples of isolated individual SWNTs on an insulating substrate, such as oxidized silicon (Si/SiO_2) [175].

The second phase explorations of the Raman spectroscopy of SWNTs spanned the four year period 2001–2004 and this phase was opened by the observation of Raman spectra from a single isolated SWNT [176], as shown in Figure 4.11. Measurements of the frequency of the radial breathing mode and the resonant energy E_{ii} at the single nanotube level allow a determination to be made of the (n, m) in-

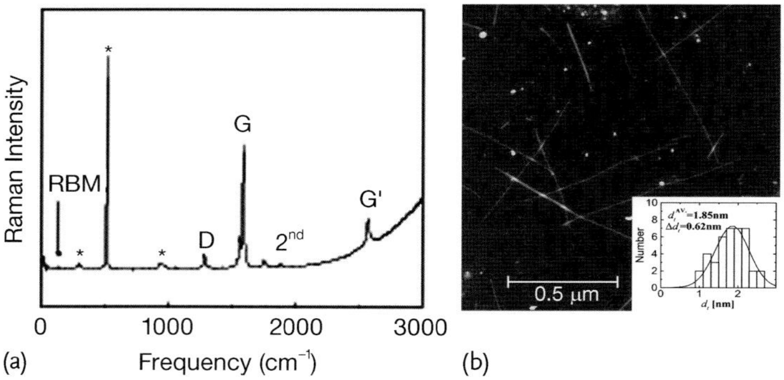

Figure 4.11 (a) Raman spectra from SWNTs at the single nanotube level. The Raman features denoted by "*" come from the Si/SiO_2 substrate on which the nanotube is mounted. (b) Atomic force microscopy image of the SWNT sample used in the experiment. The inset (lower right) shows the diameter distribution of the sample in (b) [176].

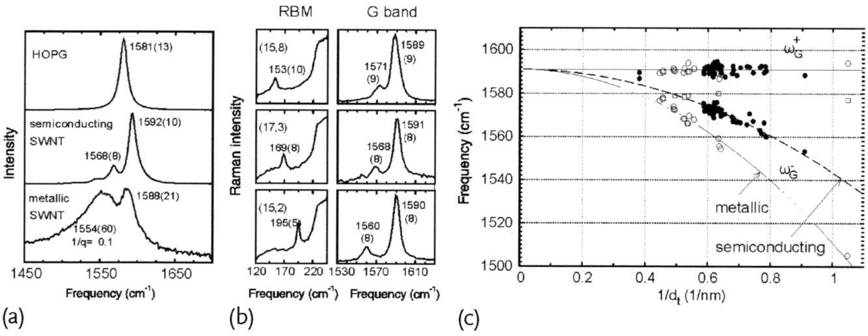

Figure 4.12 (a) The G-band for highly oriented pyrolytic graphite (HOPG), one semiconducting SWNT and one metallic SWNT. For C_{60} fullerenes a peak in the Raman spectra is observed at 1469 cm^{-1}, but it is not considered a G-band. (b) The radial breathing mode (RBM) and the G-band Raman spectra for three semiconducting isolated SWNTs with the indicated (n, m) values. (c) Frequency vs. $1/d_t$ for the two most intense G-band features (ω_{G^-} and ω_{G^+}) from isolated semiconducting and metallic SWNTs [179].

tegers of a SWNT as a function of nanotube diameter d_t and chiral angle θ [31]. Once the (n, m) of a given tube is identified, then the dependence of the frequency, width and intensity of each of the features in the Raman spectra on diameter and chiral angle can be found, including the RBM, G-band, D-band, G′-band, and various overtone and combination modes within the first-order and higher-order frequency regimes (see Figure 4.12 for the ω_G dependence on $1/d_t$) [80, 179]. Such studies have played a major role in advancing both the fundamental understanding of Raman spectroscopy for 1D systems and in characterizing actual SWNT samples [80, 177, 178].

Third phase explorations of Raman spectroscopy on SWNTs started in 2004. It was initiated by both the development of the high pressure CO (HiPCO) and compact synthesis processes for nanotubes, that generated a large amount of SWNTs in the low diameter limit of 0.7–1.3 nm, and the development of a method for dispersing the nanotube bundles to produce isolated tubes in solution. The presence of isolated tubes with relatively small diameters allowed detailed studies to be made of the first electronic transition E_{11}^S for semiconducting nanotubes using mostly photoluminescence experiments [180] ($E_{ii} \propto 1/d_t$, E_{11}^S being in the far-infrared spectral range for usual tube diameters $d_t > 1.3$ nm). Studies on the small diameter SWNTs led to an observation of the departure from the tight-binding model which predicted the "ratio rule", that is, $E_{22}^S/E_{11}^S = 2$ for armchair tubes [181]. The observation of $2n + m$ family effects [180, 182] led to much more sophisticated first principles and tight-binding models for treating many-body effects in 1D systems.

In order to accurately study the optical transition energies, resonance Raman experiments with many laser lines and tunable laser systems (based on dye lasers and Ti:Sapphire laser systems) started to be made [183, 184]. Using several closely spaced excitation laser energies, 2D plots were made from the RBM spectra obtained from Stokes resonance Raman measurements as a function of E_{laser}. Figure 4.13a presents a map of Stokes resonance Raman measurements of carbon nanotubes grown by the HiPCO process, dispersed in aqueous solution and wrapped with sodium dodecyl sulfate (SDS) [185], in the frequency region of the RBM features, built from 76 values of E_{laser} for $1.52 \leq E_{laser} \leq 2.71$ eV. Several RBM peaks appear in Figure 4.13a, each peak corresponding to an (n, m) SWNT in resonance with E_{laser}, thereby delineating for each nanotube both the resonance spectra as a function of E_{laser} and the resonance window (Raman intensity as a function of the energy in the range where the RBM feature can be observed).

Figure 4.13b shows a comparison between the experimental and theoretical Kataura plots. The filled circles are experimental E_{ii} vs. ω_{RBM} obtained by Telg *et al.* [184] from analysis of an experiment very similar to the one performed by Fantini *et al.* [183] and shown in Figure 4.13a. The open circles come from a third-neighbor tight-binding calculation [184]. Black and gray open circles represent, respectively, M-SWNTs and S-SWNTs. The geometrical patterns for carbon nanotube families with $(2n + m)$ = constant (dashed gray lines) for E_{22}^S, E_{11}^M and E_{33}^S are also shown along with the (n, m) values assigned for some SWNTs. The energies do not match very well due to the simplicity of the TB method, even going up to third-neighbor interactions. However, the observed geometrical patterns can be

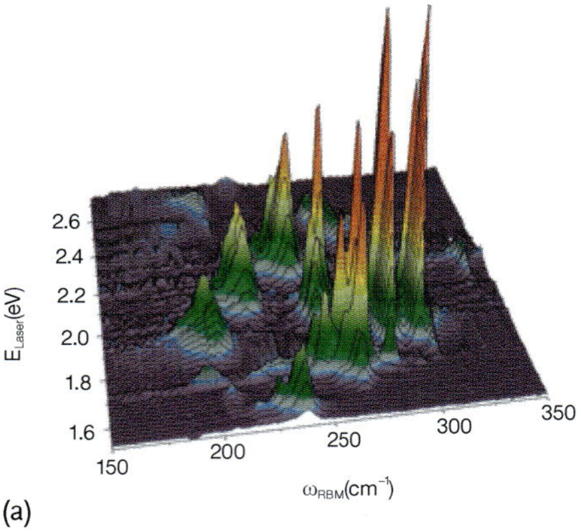

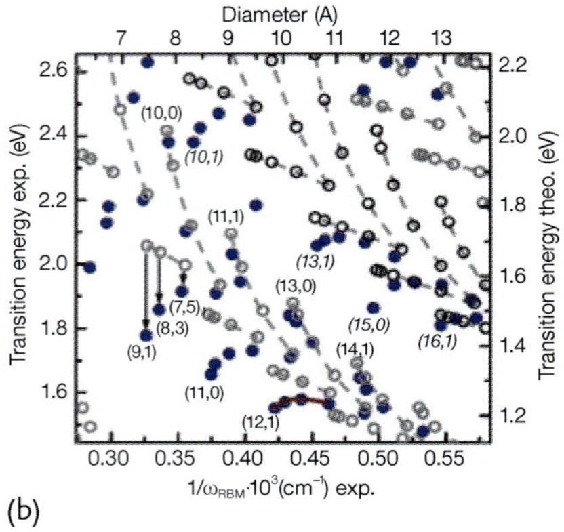

Figure 4.13 (a) RBM Raman measurements of HiPCO SWNTs dispersed in an SDS aqueous solution [185], measured with 76 different laser lines E_{laser} [183]. The nonresonance Raman spectrum from a separated CCl_4 solution is acquired after each RBM measurement, and used to calibrate the spectral intensities and to check the frequency calibration. (b) Filled circles are experimental E_{ii} vs. ω_{RBM} data obtained by Telg *et al.* [184] from analysis of an experiment very similar to the one shown in (a). The label "Transition energy exp." actually indicates the excitation laser energy (E_{laser}). Open circles come from third-neighbor tight-binding calculations, showing that even the addition of interactions with more neighbors in the π-band based tight-binding model is not enough to accurately describe the experimental results. Gray and black circles indicate calculated optical transitions from semiconducting (E_{22}^S and E_{33}^S) and metallic (E_{11}^M) tubes, respectively.

compared with the predicted patterns, and the comparison leads to the (n, m) assignment of individual SWNTs. Based on these results, an extended tight-binding picture, which includes σ–π hybridization due to tube curvature, strong excitonic effects of the excited electron–hole pair and the Coulomb repulsion of electrons was developed [186]. These theoretical advances allowed very accurate calculations of (n, m)-dependent resonance Raman cross-sections [187, 188]. Present work in this field has been devoted to studying effects of both the nanotube environment [189, 190] and nanotube defects [191].

4.4.3
Graphene

Among the sp^2 carbon systems, monolayer graphene is the simplest and has, consequently, the simplest Raman spectra (see Figure 4.14). Intensive graphene science study started quite recently (in 2004 [51]), but the carbon-Raman community was ready for almost instantaneous appreciation of the important achievements made in characterizing sp^2 carbons with the Raman technique [5, 86]. One interesting aspect of the graphene studies was the sudden realization of a perfect prototype system for studying sp^2 carbons. The dispersion of the G′-band by changing the excitation laser energy in bilayer graphene was used to study the effects of interlayer coupling on the *electronic* structure [192]. The effect of different substrates

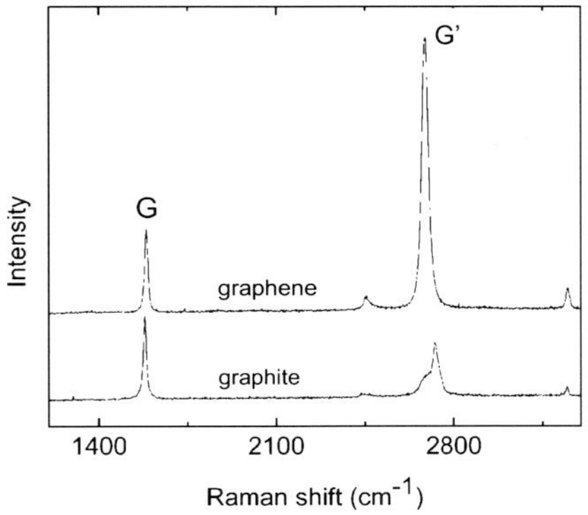

Figure 4.14 Raman spectrum of single-layer graphene in comparison to graphite measured with a 514 nm ($E_{laser} = 2.41$ eV) laser. The two most intense features are the first-order Raman-allowed G-band and the second-order G′-band. The spectrum of pristine single-layer graphene is unique in sp^2 carbons for exhibiting a very intense G′-band as compared to its G-band feature.

and deposited top gates on the G and G′-band of graphene due to strain, charge transfer effects and disorder have been studied with unprecedented detail [193–195]. As a result of all these characterization studies, Raman spectroscopy is now being used commonly as a tool for measuring doping in graphene electronic devices [196] as well as in carbon nanotubes [191]. This is one of the many examples of how findings in graphene helped with understanding experimental results on nanotubes, and vice-versa. Because of the peculiar and unique electronic dispersion of graphene, being a zero gap semiconductor with a linear $E(k)$ dispersion relation near the Fermi level, the G-band phonons (energy of 0.2 eV) can promote electrons from the valence to the conduction band. The electron–phonon coupling in this system is quite strong, and gives rise to a renormalization of the electronic and phonon energies, including a sensitive dependence of the electronic structure on electron or hole doping [196].

Furthermore, interesting confinement and polarization effects can be observed in the G-band of a graphene nanoribbon, as shown in Figure 4.15 [83]. The lower frequency G_1-band coming from the nanoribbon on top of an HOPG substrate can be separated from the substrate higher frequency G_2-band by use of laser heating of the nanoribbon (see Figure 4.15). The temperature rise of the ribbon due to laser heating is greater than that of the substrate and therefore ω_{G_1} for the ribbon decreases more than for the substrate because of the higher thermal conductivity of the substrate relative to the graphene ribbon. The ribbon G_1-band shows a clear antenna effect, where the Raman signal disappears when crossing the light polarization direction with respect to the ribbon axis, in accordance with theoretical predictions [80].

The high sensitivity of the G′-band feature to the electronic structure was used to differentiate between 1-LG, 2-LG and many-layer graphene (see Figure 4.14) [86, 197], as well as folded (but not stacked in accordance with AB Bernal stacked

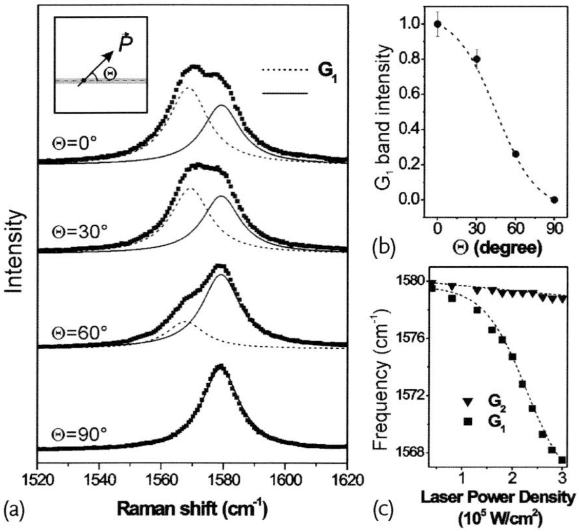

Figure 4.15 (a) The G-band Raman spectra from a graphene nanoribbon (G_1) and from the graphite substrate on which the nanoribbon sits (G_2). (b) The G_1 peak intensity dependence on the light polarization direction with respect to the ribbon axis, including experimental points on the dark curve and theoretical predictions by the dashed curve. (c) Frequency of the G-band peaks as a function of incident laser power for the graphene ribbon (G_2) in contrast to the results for the HOPG substrate (G_1) [83].

graphite) layers [198]. The Raman technique soon became the fingerprint for quick characterization of few-layer graphene samples. Raman imaging also differentiates between the number of layers in different locations of a large graphene flake, due to the dependence of the Raman intensity on the number of scattering layers [197], although this information has not yet provided an accurate characterization tool for determining the number of layers in few-layer graphene. The epitaxial growth of graphene on a SiC substrate, among others, and the effect of other chemically induced environmental interactions has also been studied using Raman spectroscopy [199].

Problems

[4-1] This problem tests your understanding of the difference between resonance Raman scattering (RRS) and hot luminescence (PL) processes. Consider a material with a discrete optical level with energy E_g and a phonon with energy E_q. Build a plot of E_i vs. E_s and show how the PL and the RRS should appear in such a plot. Make schematic pictures to show you can differentiate the two processes when they overlap in energy.

[4-2] Obtain Eq. (4.7) from Eq. (4.6).

[4-3] Consider the equation for a damped harmonic oscillator driven by a force F_0

$$m\frac{\partial^2 x}{\partial t^2} + \gamma\frac{\partial x}{\partial t} + Kx = F_0 \exp i(\omega t + \phi) . \qquad (4.19)$$

a. Obtain and plot the oscillator response x as a function of the driving frequency ω.
b. Show that the power response ($|x|^2$) exhibits a Lorentzian lineshape when the natural oscillator frequency is much larger than the peak full width at half maximum intensity. Obtain an expression for the time dependence of the full width at half maximum intensity.

[4-4] Show that 1 eV corresponds to 8065 cm^{-1}. What is the energy in eV for the G-band Raman spectrum of graphite at 1580 cm^{-1}?

[4-5] Obtain the energy in eV of a laser with a wavelength of 633 nm. What is the conversion formula from nm to eV, and from eV to nm?

[4-6] What is the wavelength of the Stokes and anti-Stokes scattered light for the graphene (1590 cm^{-1}) G-band for 514.5 nm laser light?

[4-7] When we consider the room temperature Raman spectrum for a carbon nanotube at 300 K, what is the intensity ratio of the anti-Stokes and Stokes intensities $I_{\text{aS}}/I_{\text{S}}$ for the G-band (1590 cm^{-1}) and for the radial breathing mode for a 1nm diameter SWNT, with $\omega_{\text{RBM}} = 227\,\text{cm}^{-1}$? When the temperature is changed to 4 K and to 2000 K, what are the new values of $I_{\text{aS}}/I_{\text{S}}$ for these cases?

[4-8] Derive the relation between the Raman peak FWHM and phonon lifetime.

[4-9] Calculate the wave vectors for 488 nm, 514 nm, 633 nm lasers and compare them with the wave vector of the K and M points of graphene.

[4-10] Consider the linear dispersion of energy for graphene,

$$E(k) = \pm\frac{\sqrt{3}\gamma_0}{2}\left(\sqrt{k_x^2 + k_y^2}\right) a ,$$

where $\pm$ denote the valence and conduction bands, respectively, and $\gamma_0 = 2.9$ eV (nearest neighbor transfer energy for an optical transition) and $a = 0.246$ nm. Show that this energy dispersion has the shape of two cone structures with the apexes of their Dirac cones that touch each other at the K (Dirac) point. Calculate the $|\boldsymbol{k}|$ value (distance from the Γ point in k space) for an optical transition observed with a 514 nm laser.

[4-11] In Problem 4-10, when an optical phonon (LO branch) with energy 1580 cm^{-1} is emitted, show the energy-momentum conserved final states

in the energy dispersion and show the possible $\boldsymbol{q}$ vectors in a figure. (Hint: the final states should be on an equi-energy contour of $E_{\text{laser}} - E_q$).

[4-12] In the case of graphene, the Dirac cone shape appears not only at the K point but also at the K' point. Explain in what ways K and K' are not equivalent to each other. If we consider the inelastic scattering from $\boldsymbol{k}$ in the K Dirac cone to $\boldsymbol{k} - \boldsymbol{q}$ in the K' Dirac cone, plot the possible $\boldsymbol{q}$ vectors measured from the Γ point in the two-dimensional Brillouin zone.

[4-13] In the previous problem, for a given laser excitation, the $\boldsymbol{k}$ wave vector for photoexcited electrons exists on an equi-energy contour. Plot the possible $\boldsymbol{q}$ vectors in the two-dimensional Brillouin zone.

[4-14] The condition for the resonance Raman effect is given by $E_{\text{laser}} = E_{\text{gap}}$, where E_{gap} denotes the energy separation between the conduction and valence energy bands. If the resonance condition is applied for the scattered resonance for phonon energy E_q, show the resonance condition for the scattered light, which we call the scattered light resonance condition. Explain why the scattered light resonance condition depends on the phonon energy while the incident light resonance condition does not.

[4-15] When we plot the Raman intensity as a function of E_{laser} (Raman excitation profile), we expect two peaks: one for the incident resonance and one for the scattered resonance conditions. When we have two different phonons with different phonon energies, illustrate the expected Raman excitation profiles.

[4-16] Explain the resonance conditions for anti-Stokes shifts. Illustrate two Raman excitation profiles for the Stokes and anti-Stokes Raman signals. Explain why the two Raman excitation profiles do not appear at the same excitation energy values.

[4-17] If we consider second-order Raman scattering for a $q \neq 0$ phonon for graphene, we have two distinct $\boldsymbol{q}$ vectors for a $q \neq 0$ scattering event. Explain this situation by using the two Dirac cone model. We denote these two scattering events by intravalley scattering and intervalley scattering.

[4-18] Explain the results in Problem 4-17 with an illustration of intervalley scattering for the forward (q) and backward ($-q$) scattering geometries.

[4-19] Explain in the phonon dispersion of graphene (Figure 3.1), which phonons are selected for intravalley and intervalley $q \neq 0$ scattering. Explain what happens for the case of forward and backward scattering.

[4-20] For an intravalley process, when we consider two laser energies $E_1 < E_2$, show that the $\boldsymbol{q}$ vectors are larger for $E_{\text{laser}} = E_2$. When the laser energy increases, how is the scattered $\boldsymbol{q}$ vector selected in the phonon energy dispersion?

[4-21] If E_{laser} is selected in the Raman measurement of single-wall carbon nanotubes (SWNTs), we can get a Raman signal only from SWNTs, which satis-

fy the resonance condition $E_{\text{laser}} = E_{ii}$. When we observe the radial breathing mode (RBM) whose frequency is inversely proportional to the diameter, how can we determine (n, m) values from the known calculated values of E_{ii} and ω_{RBM}? Explain a procedure that could be used to get (n, m) values.

[4-22] In order to obtain all the (n, m) values for SWNTs in a given sample, we need an almost continuously tunable energy output from E_{laser}. Suppose that you have a laser system with continuously tunable energy, how should you arrange the experimental schedule to obtain all these (n, m) values? Explain your purpose and the method you will use.

[4-23] Suppose that we have only one SWNT on a Si substrate. The position of the SWNT cannot be seen by optical microscopy. Explain why obtaining (n, m) for the specified SWNT by Raman spectroscopy is difficult. In order to overcome this difficulty, how should we prepare our samples or arrange our experimental set up?

[4-24] Try to correlate the modifications in the G′ feature, as shown in Figure 4.14b with the changes in the electronic structure, as discussed in Section 2.2.4.

[4-25] How should the I_D/I_G intensity ratio depend on the defect density in a graphene layer? Discuss a physical picture for some defect density condition with use of some typical lengths of the system.

5
Quantum Description of Raman Scattering

The classical description of the Raman effect developed in Chapter 4 provides an account of the frequencies observed in the Raman spectra for carbon nanotubes, including quantitative descriptions of some of the main spectral features. The observed Raman lines are shifted in energy from the laser line in accordance with their phonon energies, with energy conservation occurring for each scattering process. For a quantitative description of the Raman intensities, a quantum description of the scattering processes is needed. In the case of sp^2 nanocarbons, a quantum treatment is essential because of the resonance Raman scattering process. Even for the main spectral features, their observed Raman frequencies depend strongly on the quantum description of the internal scattering events. The goal of this chapter is to introduce a quantum description of the Raman scattering process, starting with the Fermi Golden Rule, which provides a theoretical basis for the scattering process, ending with the form of the electron–photon and electron–phonon interaction Hamiltonians.

5.1
The Fermi Golden Rule

In this section we review a few of the results of time-dependent perturbation theory and the use of the Fermi Golden Rule to provide the background for a quantum mechanical description of the Raman effect in Section 5.2. A detailed discussion of these topics can be found in standard quantum mechanics text books [92, 200], since the development of the formalism of Raman spectroscopy depends strongly on the use of time-dependent electromagnetic fields. The most important case of interest is the one where the external field is a sinusoidal function of time. For most practical applications, the external fields are sufficiently weak, so that their effect can be handled within the framework of perturbation theory, where the unperturbed wavefunctions serve as a basis for describing the perturbed system. If the perturbation has an explicit time dependence, it must be handled by *time-dependent* perturbation theory.

Raman Spectroscopy in Graphene Related Systems. Ado Jorio, Riichiro Saito, Gene Dresselhaus, and Mildred S. Dresselhaus

ISBN: 978-3-527-40811-5

In doing time-dependent perturbation theory, we solve the time-dependent form of the Schrödinger equation, which is:

$$i\hbar\frac{\partial\psi}{\partial t} = \mathcal{H}\psi = (\mathcal{H}_0 + \mathcal{H}'(t))\psi \, , \tag{5.1}$$

where $\mathcal{H}'(t)$ is a time-dependent perturbation. We then expand the time-dependent wavefunctions $\psi(\mathbf{r}, t)$ in terms of the complete set of eigenfunctions of $\mathcal{H}_0$ $u_n(\mathbf{r})e^{-iE_nt/\hbar}$,

$$\psi(\mathbf{r}, t) = \sum_n a_n(t)u_n(\mathbf{r})\, e^{-iE_nt/\hbar} \, , \tag{5.2}$$

where the $a_n(t)$ are the time-dependent expansion coefficients. Combining Eq. (5.1) and Eq. (5.2) we obtain a relation

$$\dot{a}_m(t) = \frac{1}{i\hbar}\sum_n a_n(t)e^{i\omega_{mn}t}\langle m|\mathcal{H}'(t)|n\rangle \, , \tag{5.3}$$

where ω_{mn} is the Bohr frequency proportional to the energy difference between states m and n

$$\omega_{mn} = (E_m - E_n)/\hbar \, , \tag{5.4}$$

and $\langle m|\mathcal{H}'(t)|n\rangle$ is the time-dependent matrix element given by

$$\langle m|\mathcal{H}'(t)|n\rangle = \int u_m^*(\mathbf{r})\mathcal{H}'(t)u_n(\mathbf{r})d^3r \, . \tag{5.5}$$

Since $\mathcal{H}'(t)$ is time-dependent, so too are its matrix elements time-dependent.

In applying perturbation theory, we consider the matrix element $\langle m|\mathcal{H}'(t)|n\rangle$ to be small, and we write each time-dependent amplitude as an expansion in perturbation theory

$$a_m = a_m^{(0)} + a_m^{(1)} + a_m^{(2)} + \cdots = \sum_{i=0}^{\infty} a_m^{(i)} \, , \tag{5.6}$$

where the superscript (i) gives the order of each term in perturbation theory. Thus $a_n^{(0)}$ is the zeroth-order term and $a_n^{(i)}$ is the ith order correction to a_n. From Eq. (5.3), we see that $a_m(t)$ changes its value with time only because of the time-dependent perturbation. Thus, the unperturbed situation (zeroth-order perturbation theory) must give no time dependence in zeroth-order and has a value only for the initial state labeled ℓ

$$\dot{a}_m^{(0)} = 0, \quad \text{and} \quad a_m^{(0)} = \delta_{m\ell} \, , \tag{5.7}$$

where $\delta_{m\ell} = 1$ for $m = \ell$ and $\delta_{m\ell} = 0$ for $m \neq \ell$ (Kronecker's delta function). Then the first-order correction becomes:

$$\dot{a}_m^{(1)} = \frac{1}{i\hbar}\sum_n a_n^{(0)}\langle m|\mathcal{H}'(t)|n\rangle e^{i\omega_{mn}t} = \frac{1}{i\hbar}a_\ell^{(0)}\langle m|\mathcal{H}'|\ell\rangle e^{i\omega_{m\ell}t} \, . \tag{5.8}$$

For our interest here, if the perturbation $\mathcal{H}'(t)$ has a sinusoidal time dependence with frequency ω, which is the situation for all resonant phenomena, we can write

$$\mathcal{H}'(t) = \mathcal{H}'(0)e^{\pm i\omega t} . \tag{5.9}$$

This shows the explicit time dependence, so that upon integration of Eq. (5.8), and after some manipulation of terms, we obtain the probability for finding an electron in the state m, that is

$$|a_m^{(1)}(t)|^2 = \frac{|\langle m|\mathcal{H}'|\ell\rangle|^2}{\hbar^2} \frac{4\sin^2((\omega_{m\ell} \pm \omega)t/2)}{(\omega_{m\ell} \pm \omega)^2} , \quad (m \neq \ell) , \tag{5.10}$$

where ω is the applied frequency and $\omega_{m\ell}$ is the resonant frequency for the transition. Here the explicit time dependence is contained in an oscillatory term of the form $[\sin^2(\omega' t/2)/\omega'^2]$ where $\omega' = \omega_{m\ell} \pm \omega$. This function is also encountered in diffraction theory and looks like that shown in Figure 5.1.

Of special interest here is the fact that the main contribution to this function comes from $\omega' \cong 0$, with the height of the main peak proportional to $t^2/4$ and the width proportional to $1/t$. This means that the area under the central peak is proportional to $t/4$. If ω' becomes zero, the system makes a selective transition from a state ℓ to the corresponding state m with a transition probability proportional to t. If we then wait long enough, a system in an energy state ℓ will eventually make a transition to a state m, if photons of the resonant frequency $\omega_{\ell m}$ are present.[1)]

Since the transition probability is proportional to t, it is therefore useful to introduce the quantity called the *transition probability per unit time* and the relation giving this quantity is called Fermi's Golden Rule (named for Enrico Fermi who first introduced this rule to calculate such transition probabilities).

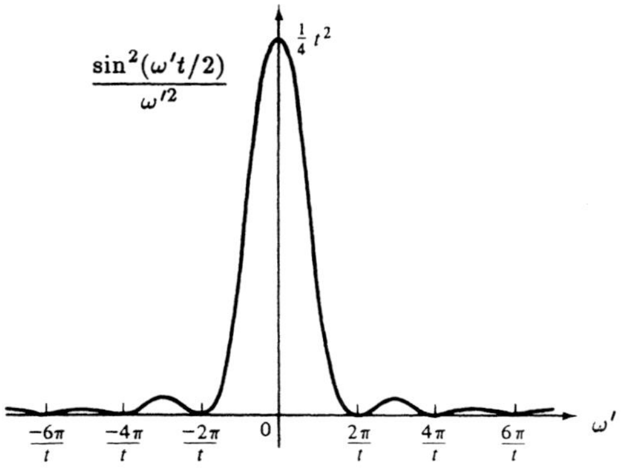

Figure 5.1 Plot of $\sin^2(\omega' t/2)/\omega'^2$ vs. ω', a function which enters the calculation of typical time-dependent perturbation problems [200].

1) When $|a_m^{(1)}|^2 \sim 1$, perturbation treatment can no longer be used. The increase of $|a_m^{(1)}|^2$ discussed here is valid when $|a_m^{(1)}|^2 \ll 1$.

In deriving Fermi's Golden Rule, we must consider the system to be exposed to the perturbation for a time sufficiently long, so that we can make a meaningful measurement within the framework of the Heisenberg uncertainty principle:

$$\Delta E \Delta t \sim \hbar \,, \tag{5.11}$$

so that the uncertainty in energy (or frequency) during the time that the perturbation acts is

$$\Delta E \sim h/t \tag{5.12}$$

or

$$\Delta \omega_{\ell m} \sim 2\pi/t \,. \tag{5.13}$$

But this is precisely the period of the oscillatory function shown in Figure 5.1. In this context, we must think of the concept of the transition probability/unit time as encompassing a range of energies and times consistent with the uncertainty principle. In the case of solids, it is quite natural to do this in any case, because the wave vector $\boldsymbol{k}$ is a quasi-continuous variable. That is, there are a large number of k states which have energies close to a given energy since the quantum states labeled by wave vector $\boldsymbol{k}$ are close together in a solid having about 10^{22} atoms/cm^3. Since the photon source itself has a bandwidth, we would automatically want to consider a range of energy differences $\hbar\delta\omega'$. From this point of view, we introduce the transition probability/unit time W_m for making a transition to a state m

$$W_m = \frac{1}{t} \sum_{m' \approx m} | \, a_{m'}^{(1)}(t) \, |^2 \,, \tag{5.14}$$

where the summation is carried out over a range of energy states consistent with the uncertainty principle; $\Delta\omega_{mm'} \sim 2\pi/t$.

Substituting for $|a_{m'}^{(1)}(t)|^2$ from Eq. (5.10), we obtain

$$|a_m^{(1)}(t)|^2 = \frac{4|\langle m|\mathcal{H}'|\ell\rangle|^2}{\hbar^2} \frac{\sin^2(\omega' t/2)}{\omega'^2} \tag{5.15}$$

and the summation in Eq. (5.16) is replaced by an integration over a narrow energy range weighted by the density of states $\rho(E_m)$ which denotes the number of states per unit energy range. We thus obtain

$$W_m = \frac{4}{\hbar^2 t} \int |\mathcal{H}'_{m'\ell}|^2 \frac{\sin^2(\omega_{m'\ell} t/2)}{\omega^2_{m'\ell}} \rho(E_{m'}) \, dE_{m'} \tag{5.16}$$

where we have written $\mathcal{H}'_{m'\ell}$ for the matrix element $\langle m'|\mathcal{H}'|\ell\rangle$. But, by hypothesis, we are only considering energies within a small energy range $E_{m'}$ around E_m and over this energy range the matrix elements and density of final states will not be varying much. However, the function $[\sin^2(\omega' t/2)/\omega'^2]$ will be varying rapidly, as can be seen from Figure 5.1. Therefore, it is adequate to integrate Eq. (5.16) only over the rapidly varying function $[\sin^2(\omega t/2)]/\omega^2$. Writing $dE = \hbar d\omega'$, we then obtain

$$W_m \simeq \frac{4|\mathcal{H}'_{m\ell}|^2 \rho(E_m)}{t\hbar^2} \int \frac{\sin^2 \frac{\omega' t}{2}}{\omega'^2} d\omega' . \tag{5.17}$$

The most important contribution to the integral in Eq. (5.17) comes from values of ω close to ω'. On the other hand, we know how to do this integral between $-\infty$ and $+\infty$, since

$$\int_{-\infty}^{\infty} \frac{\sin^2 x}{x^2} dx = \pi . \tag{5.18}$$

Therefore we can write an approximate relation based on Eq. (5.17) by setting $x = \omega' t/2$

$$W_m \cong \frac{2\pi}{\hbar} |\mathcal{H}'_{m\ell}|^2 \rho(E_m) . \tag{5.19}$$

The simple formula in Eq. (5.19) is called Fermi's Golden Rule, and is used to calculate transition probabilities per unit time when considering the optical properties of solids, including Raman scattering intensities.

If the initial state is a discrete level (such as a donor impurity level) and the final state is a continuum (such as the conduction band), then the Fermi's Golden Rule (Eq. (5.19)), as written, yields the transition probability per unit time and $\rho(E_m)$ is interpreted as the density of *final* states. Likewise if the final state is discrete and the initial state is a continuum, then W_m also gives the transition probability per unit time, only in this case $\rho(E_m)$ is now interpreted as the density of initial states. If the transitions of interest are between a continuum of initial states and a continuum of final states, then the Fermi Golden Rule must be interpreted in terms of a joint density of states, whereby the initial and final states are separated by the photon energy $\hbar\omega$ inducing the transition.

Our discussion up to this point introduces the basic concepts behind Fermi's Golden Rule, that is it provides an understanding of each term in W_m and its relation to the uncertainty principle. For further interpreting Raman spectra, we need to consider first-order one-phonon scattering processes as well as second-order two-phonon scattering processes, as shown in Figures 5.2 and 5.3. Therefore, to describe the Raman processes we typically consider second-order and higher-order perturbation theory, where we start in an initial state, scatter into one or more intermediate states before scattering back to a final state. The expression for the Raman intensity obtained by these higher-order perturbation processes is given in Section 5.2. We do not derive the second-order and higher-order perturbation theory in this book since it is laborious, it does not add new physical insights and such derivations can be found in standard quantum mechanics texts [92, 200, 201]. In short, by increasing the order in perturbation theory, one matrix element and one more term in the denominator will be added, with a summation over all possible intermediate states, as described in the next section.

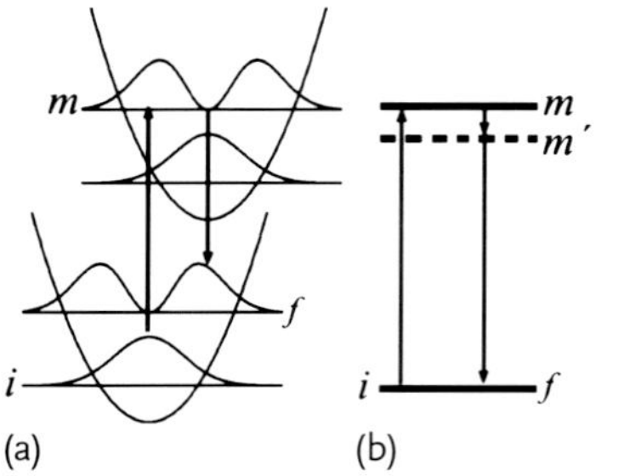

Figure 5.2 (a) Schematics showing the second-order Raman process for small molecules, where both the electronic (parabola) and vibrational levels are displayed, comparable to Figure 3.2. Notice the horizontal displacement of the two parabolas, indicating that the two different electronic levels have different atomic positions. The upwards arrow indicates vibronic state transition $i \rightarrow m$ mediated by the photon absorption, and the downwards arrow indicates vibronic state transition $m \rightarrow f$ mediated by the photon emission. The energy difference between the incident and scattered photons corresponds to a quantum of atomic vibration. (b) Schematics showing the third-order process often used to describe the Raman process in crystals. While the large upwards $i \rightarrow m$ and downwards $m' \rightarrow i$ arrows represent photon absorption and emission by the electron, the small downwards arrow $m \rightarrow m'$ represents the electron losing energy to the lattice through an electron–phonon scattering event. The vibrational levels are not displayed.

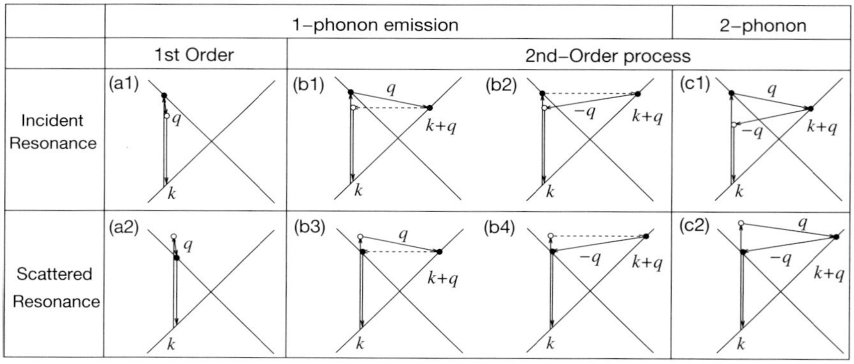

Figure 5.3 (a) One-phonon first-order, (b) one-phonon second-order, and (c) two-phonon second-order, resonance Raman spectral processes. Parts (a1), (b1), (b2), and (c1) show incident photon resonance conditions, and parts (a2), (b3), (b4), and (c2) show scattered photon resonance conditions. For one-phonon second-order transitions, one of the two scattering events is an elastic scattering event (dashed lines). Resonance points are shown as solid circles [80].

5.2
The Quantum Description of Raman Spectroscopy

In computing the Raman spectra related to totally symmetric phonons in small molecules, we make use of the Frank–Condon effect, whereby the excitation of one electron by a photon changes the atomic arrangement in a molecule, so that an overlap between the vibrational states n_q and $n_q + 1$ is possible (see Figure 5.2a). That is, since the excited states have different wavefunctions which give a different stable position for the atom in the molecule, the atom in the excited state moves from its original position in the ground state, thereby inducing an atomic vibration. Each of the four levels displayed in Figure 5.2a represents a different *vibronic* level, which designates a given electronic and vibrational level.

In larger molecules or crystals, however, exciting one electronic state does not change the atomic configuration, and second-order perturbation theory can only give rise to elastic (Rayleigh) scattering of the light. In these cases, it is necessary to go to third-order perturbation theory, whereby the excited electron perturbs the atom in creating a phonon through an electron–phonon interaction. Such a process is sketched in Figure 5.2b and is different from Figure 5.2a, where both the electronic and vibrational levels are displayed, showing a transition between vibronic states. In Figure 5.2b only the electronic energies are displayed, and the system is described as having the same initial and final electronic level, although one phonon has been left in the system, mediated by the electron–phonon scattering event represented by the small downwards arrow. Figure 5.2b is pictorial, and the excited energy levels m and m' do not necessarily represent real electronic states, but just the energy gained by the electron from its original energy E_i, as discussed below.

If the electron is initially in a state i with energy E_i, then light scattering can excite the electron to a higher energy state m with energy E_m (see Figure 5.2b) by the absorption of an excitation energy $(E_m - E_i)$ from a laser (E_{laser}). If m is a real electronic state (we usually draw a solid line), the light absorption is a resonant process. This electron will be further scattered by a $q \approx 0$ phonon to a "virtual" state m', and decay back to state i by emitting the scattered light. Virtual states are usually displayed by dashed lines and, within perturbation theory, they are described by a linear combination of the electron eigenstates of the system with a large energy uncertainty and a small lifetime to compensate for the uncertainty principle. The initial "system", therefore, has an electron in the state i and a photon with energy $(E_m - E_i)$, while the final "system" has an electron in the state i, a phonon of energy E_q and a photon with energy $(E_m - E_i - E_q)$. Alternatively, the incident photon can excite the electron to a virtual state m with higher energy E_m and light scattering can serve to bring the system to a final state of lower energy E_i by the emission of energy $(E_m - E_i)$. In this case, it is the photon emission that is resonant, rather than the photon absorption. This alternative process is shown in Figure 5.3a2. While in Figure 5.2b we show discrete electronic levels, in Figure 5.3 we show Raman scattering processes within the continuum electronic dispersion near the Fermi level at the K point of graphene (see Section 2.2.2). Figure 5.3a1 is analogous to Figure 5.2b, and Figure 5.3a2 shows first-order Raman scattering with a resonant process involving the scattered light, while the other cases in Figure 5.3 show other possible processes discussed later.

Therefore, the first-order Raman intensity as a function of phonon energy $E_q = \hbar\omega_q$ and of the incident laser energy E_{laser} is given by third-order perturbation theory by [91, 142]

$$I(\omega_q, E_{\text{laser}}) = \sum_f \left| \sum_{m,m'} \frac{M^{\text{op}}(\boldsymbol{k}-\boldsymbol{q}, im')M^{\text{ep}}(\boldsymbol{q}, m'm)M^{\text{op}}(\boldsymbol{k}, mi)}{(E_{\text{laser}} - \Delta E_{mi})(E_{\text{laser}} - \hbar\omega_q - \Delta E_{m'i})} \right|^2 , \tag{5.20}$$

in which

$$\Delta E_{m^{(\prime)}i} \equiv (E_{m^{(\prime)}} - E_i) - i\gamma_r , \tag{5.21}$$

and i, m, m' and f denote, respectively, the initial state, the two excited intermediate states, and the final state of an electron, while γ_r denotes the broadening factor of the resonance event (see Section 4.3.2.5). The physical process is described by an electron at wave vector $\boldsymbol{k}$ that is (1) excited by an electric dipole interaction $M^{\text{op}}(\boldsymbol{k}, mi)$ with the incident photon to make a transition from state i to m, and the electron is then (2) scattered by emitting a phonon with energy $\hbar\omega_q$ and wave vector $\boldsymbol{q}$ through an electron–phonon interaction, $M^{\text{ep}}(\boldsymbol{q}, m'm)$, and finally (3) the electron in state m' emits a photon by an electric dipole transition, through the interaction $M^{\text{op}}(\boldsymbol{k}-\boldsymbol{q}, im')$ to reach the final electronic state $f = i$. For momentum conservation, $q \approx 0$. For an energy separation E_{im} between the i and m states, the resonance conditions are either with the incident photon, $E_{\text{laser}} = E_{mi}$, or with the scattered photon, $E_{\text{laser}} = E_{mi} + \hbar\omega_q$. To reach a given final state, the sum in Eq. (5.20) is taken over all possible intermediate states m and m'. The intermediate states m are determined by specifying the initial state i with use of energy-momentum conservation. In order to take the sum over the intermediate states, we need to know the electric dipole matrix elements of the electron–photon interaction, M^{op}, and of the electron–phonon interaction, M^{ep}, which will be discussed in Section 5.4. In the scattering process, energy-momentum conservation for an electron and phonon holds, but this is not explicitly written in Eq. (5.20).

Moving to a higher-order Raman process, the various diagrams in Figure 5.3b1–b4 and 5.3c1,c2 show inelastic scattering processes which have to be described by fourth-order perturbation theory. In Figure 5.3b1–b4, there is an internal electron scattering process by a phonon, and another by a lattice defect or impurity, which can cause an elastic scattering event. Both phonon emission and absorption are possible and the order of the elastic and inelastic scattering processes can be interchanged. These processes will be discussed in Chapter 13. Figure 5.3c1,c2 shows two second-order two-phonon Raman scattering processes. In this case, the intensity as a function of E_{laser} and the sum of the two phonon energies $\omega = \omega_1 + \omega_2$ is given by a similar formula,

$$I(\omega, E_{\text{laser}}) \propto \sum_i \left| \sum_{m',m'',\omega_1,\omega_2} J_{m',m''}(\omega_1, \omega_2) \right|^2 , \tag{5.22}$$

where

$$J_{m',m''}(\omega_1, \omega_2) = \frac{M^{\text{op}}(\boldsymbol{k}, im'')M^{\text{ep}}(-\boldsymbol{q}, m''m')M^{\text{ep}}(\boldsymbol{q}, m'm)M^{\text{op}}(\boldsymbol{k}, mi)}{(E_{\text{laser}} - \Delta E_{mi})(E_{\text{laser}} - \hbar\omega_1 - \Delta E_{m'i})(E_{\text{laser}} - \hbar\omega_1 - \hbar\omega_2 - \Delta E_{m''i})} . \tag{5.23}$$

Now we have two-phonon scattering processes with phonon wave vectors $\boldsymbol{q}$ and $-\boldsymbol{q}$, so that momentum conservation is possible with $q \neq 0$. Due to momentum conservation for the whole process, most often the m'' and m states will be the same, since all the others will generally be much farther in energy. In order to get two resonance conditions at the same time, an intermediate electronic state $E_{m'i}$ is always in resonance ($E_{\text{laser}} = \Delta E_{m'i} + \hbar\omega_1$), and either the incident resonance condition ($E_{\text{laser}} = \Delta E_{mi}$) or the scattered resonance condition ($E_{\text{laser}} = \Delta E_{m''i} + \hbar\omega_1 + \hbar\omega_2$) is satisfied. On the other hand, for a second-order one-phonon process, the Raman intensity is calculated by replacing one of the two phonon scattering processes by an elastic impurity scattering process. Another point to mention concerns the energy uncertainty γ_r, which enters Eq. (5.21) and is operative in Eqs. (5.20) and (5.23). Actually, the different denominators may have different γ_r values since they are related to different scattering events. However it is usual to consider the same γ_r for simplicity when there is no experimental information distinguishing them. The physical origin of γ_r has been discussed in Section 4.3.2.5.

5.3
Feynman Diagrams for Light Scattering

Feynman diagrams are useful for keeping track of various processes that may occur in an inelastic scattering process that absorbs or creates an excitation, such as the six scattering processes shown in Figure 5.4a–f for creating an excitation (e. g., the Raman Stokes process). The basic notation used in drawing Feynman diagrams consists of propagators, such as electrons, phonons or photons and vertices where interactions occur, as shown in Figure 5.4g. Time goes from left to right in the diagrams in Figure 5.4. The basic diagram for the Raman process is given in Figure 5.4a and is taken from the Yu and Cardona book "Fundamentals of Semiconductors" [202]. The other permutations of Figure 5.4a obtained by different orders of the vertices are given in Figure 5.4b–f and are also found in [202]. We then use the Fermi Golden Rule (Eq. (5.19)) for each diagram, by multiplying the contributions from each vertex. For example, the first vertex in Figure 5.4a contributes a term to the scattering probability per unit time of the form

$$\frac{\langle n|\mathcal{H}_{eR}(\omega_{\text{laser}})|i\rangle}{[E_{\text{laser}} - (E_n - E_i)]} ,$$

where the Hamiltonian $\mathcal{H}_{eR}(\omega_{\text{laser}})$ denotes the interaction between the electron and the incident electromagnetic radiation field ($\omega_i = \omega_{\text{laser}}$) taking the system from an initial state i to an intermediate state n. The interaction energy for the second vertex $\mathcal{H}_{\text{e-ion}}(\omega_{\text{laser}})$ is between the electron and the lattice vibrations of the ion (or the electron–phonon interaction) and the corresponding energy denominator is

$$E_{\text{laser}} - (E_n - E_i) - \hbar\omega_q - (E_{n'} - E_n) = [E_{\text{laser}} - \hbar\omega_q - (E_{n'} - E_i)] .$$

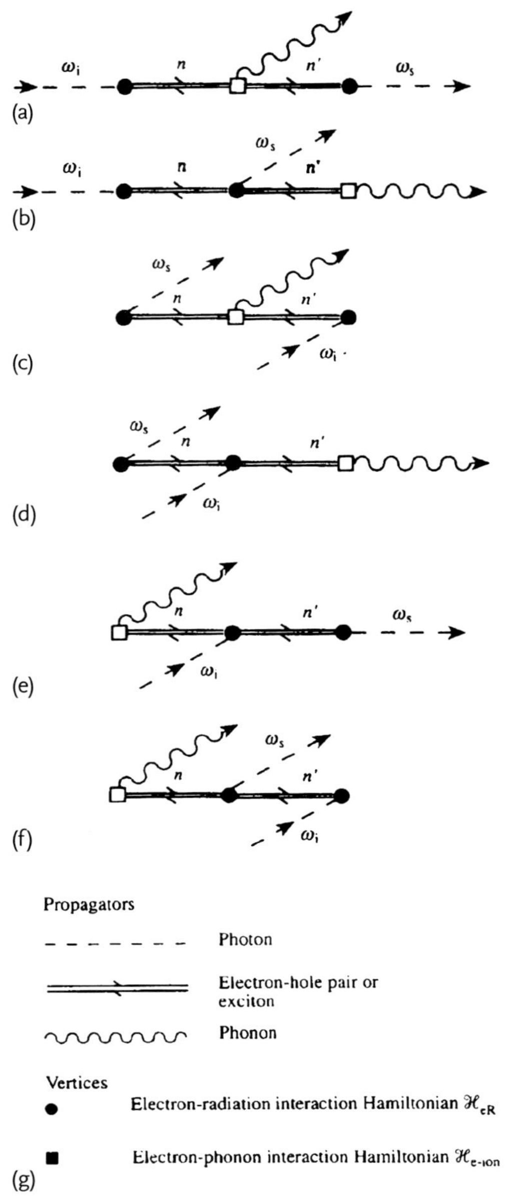

Figure 5.4 (a–f) Feynman diagrams for the six scattering processes that contribute to one-phonon (Stokes) Raman scattering. Here $\omega_i = \omega_{\text{laser}}$ and ω_s is the scattered light frequency. (g) Symbols used in drawing Feynman diagrams to represent Raman scattering [202].

Here $E_{\text{laser}} = \hbar\omega_{\text{laser}}$, $\hbar\omega_q$ is the phonon energy, E_i is the electron energy in the initial state and $E_{n(')}$ is the electron energy in the intermediate state. For the third vertex the denominator becomes $[E_{\text{laser}} - \hbar\omega_q - \hbar\omega_s - (E_f - E_i)]$, but since the initial and final electron energies are the same, energy conservation requires the δ function $\delta(\hbar\omega_{\text{laser}} - \hbar\omega_q - \hbar\omega_s)$ to yield the probability per unit time for Raman scattering for the diagram in Figure 5.4a:

$$P_{\text{ph}}(\omega_s) = \frac{2\pi}{\hbar} \left| \sum_{n,n'} \frac{\langle i|\mathcal{H}_{eR}(\omega_s)|n'\rangle\langle n'|\mathcal{H}_{\text{e-ion}}|n\rangle\langle n|\mathcal{H}_{eR}(\omega_{\text{laser}})|i\rangle}{[E_{\text{laser}} - (E_n - E_i)][E_{\text{laser}} - \hbar\omega_q - (E_{n'} - E_i)]} \right|^2 \times \delta(E_{\text{laser}} - \hbar\omega_q - \hbar\omega_s) , \quad (5.24)$$

in which $\hbar\omega_s$ denotes the photon energy for the scattered light.

The rules in drawing Feynman diagrams are:

- Excitations such as photon, phonons and electron–hole pairs that occur in Raman scattering are represented by lines (or propagators). These propagators can be labeled by the properties of the excitations, such as their wave vectors, frequencies and polarizations.
- The interaction between two excitations is represented by an intersection of their propagators. This intersection is known as a *vertex* and is sometimes highlighted by a symbol such as a filled circle (electron–photon interaction) or an empty square (electron–phonon interaction).
- Propagators are drawn with an arrow to indicate whether excitations are created or annihilated in an interaction. Arrows pointing towards a vertex represent excitations which are annihilated. Those pointing away from the vertex are for excitations that are created.
- When several interactions are involved, they are always assumed to proceed sequentially from the left to the right as a function of time.
- Once a diagram has been drawn for a certain process, other possible processes are derived by permuting the time order in which the vertices occur in the Feynman diagram.

Then summing over the six diagrams in Figure 5.4 yields the result

$$P_{\text{ph}}(\omega_s) = \frac{2\pi}{\hbar} \left| \sum_{n,n'} \frac{\langle i|\mathcal{H}_{eR}(\omega_s)|n'\rangle\langle n'|\mathcal{H}_{\text{e-ion}}|n\rangle\langle n|\mathcal{H}_{eR}(\omega_{\text{laser}})|i\rangle}{[\hbar\omega_i - (E_n - E_i)][\hbar\omega_i - \hbar\omega_q - (E_{n'} - E_i)]} \right.$$
$$+ \frac{\langle i|\mathcal{H}_{\text{e-ion}}|n'\rangle\langle n'|\mathcal{H}_{eR}(\omega_s)|n\rangle\langle n|\mathcal{H}_{eR}(\omega_{\text{laser}})|i\rangle}{[\hbar\omega_i - (E_n - E_i)][\hbar\omega_i - \hbar\omega_s - (E_{n'} - E_i)]}$$
$$+ \frac{\langle i|\mathcal{H}_{eR}(\omega_{\text{laser}})|n'\rangle\langle n'|\mathcal{H}_{\text{e-ion}}|n\rangle\langle n|\mathcal{H}_{eR}(\omega_s)|i\rangle}{[-\hbar\omega_s - (E_n - E_i)][-\hbar\omega_s - \hbar\omega_q - (E_{n'} - E_i)]}$$
$$+ \frac{\langle i|\mathcal{H}_{\text{e-ion}}|n'\rangle\langle n'|\mathcal{H}_{eR}(\omega_{\text{laser}})|n\rangle\langle n|\mathcal{H}_{eR}(\omega_s)|i\rangle}{[-\hbar\omega_s - (E_n - E_i)][-\hbar\omega_s + \hbar\omega_i - (E_{n'} - E_i)]}$$

$$+\frac{\langle i|\mathcal{H}_{eR}(\omega_s)|n'\rangle\langle n'|\mathcal{H}_{eR}(\omega_{\text{laser}})|n\rangle\langle n|\mathcal{H}_{\text{e-ion}}|i\rangle}{[-\hbar\omega_q-(E_n-E_i)][-\hbar\omega_q+\hbar\omega_i-(E_{n'}-E_i)]}$$

$$+\left.\frac{\langle i|\mathcal{H}_{eR}(\omega_{\text{laser}})|n'\rangle\langle n'|\mathcal{H}_{eR}(\omega_s)|n\rangle\langle n|\mathcal{H}_{\text{e-ion}}|i\rangle}{[-\hbar\omega_q-(E_n-E_i)][-\hbar\omega_q-\hbar\omega_s-(E_{n'}-E_i)]}\right|^2$$

$$\times\,\delta(\hbar\omega_i-\hbar\omega_s-\hbar\omega_q)\,. \tag{5.25}$$

Here we used $\hbar\omega_i$ instead of E_{laser}. Notice that having $\hbar\omega_i$ in the denominators or not depends on the position (time ordering) at which the optical absorption takes place in the Feynman diagram, and not all terms can be resonant, that is, when one of the terms in the denominator vanishes, the transition probability diverges and a resonance process takes place. Therefore, while all the terms in Eq. (5.25) play a role in nonresonant Raman scattering, when considering resonance Raman scattering with a fixed energy between the valence and conduction bands, only one of the terms in Eq. (5.25) will dominate. Interestingly, this is exactly the term where scattering occurs in the most intuitive order of events, that is, light absorption, electron–phonon scattering, light emission (first term in Eq. (5.25)). For this reason, the simpler Eq. (5.20) is enough for the calculation of Raman intensities under resonance conditions. This concept is developed further in the problem set for this chapter.

5.4 Interaction Hamiltonians

In this section we discuss the form of the interaction Hamiltonian $\mathcal{H}_{eR}$, which denotes the interaction between the electron and the electromagnetic radiation field (the electron–photon interaction), and the interaction Hamiltonian $\mathcal{H}_{\text{e-ion}}$ is between the electron and the lattice vibrations of the ion (or the electron–phonon interaction). Together with information about the electronic and vibrational states, plus the exciting field, these Hamiltonians can be used to calculate the matrix elements M^{op} and M^{ep}.

5.4.1 Electron–Radiation Interaction

The Hamiltonian $\mathcal{H}_{eR}$ that can be used to obtain the Lorentz force for an electron in an electromagnetic field is given by:

$$\mathcal{H}_{eR}=\frac{1}{2m}(\boldsymbol{p}-e\boldsymbol{A})^2+V(\boldsymbol{r})\,, \tag{5.26}$$

where m and e are the electron mass and charge, $\boldsymbol{p}$, $\boldsymbol{A}$ and $V(\boldsymbol{r})$ are, respectively, the momentum, vector potential and crystal potential. This is known from classical electromagnetism theory, that in the presence of an electromagnetic field, the canonical momentum has to be substituted by $\boldsymbol{p} \to \boldsymbol{p} - e\boldsymbol{A}$ written here in S.I. units. In the Coulomb gauge ($\nabla \cdot \boldsymbol{A} = 0$), then $\boldsymbol{p}$ and $\boldsymbol{A}$ commute and Eq. (5.26) becomes

$$\mathcal{H}_{eR} = \left[\frac{\boldsymbol{p}^2}{2m} + V(\boldsymbol{r})\right] - \frac{e}{m}\boldsymbol{p} \cdot \boldsymbol{A} + \frac{e^2 \boldsymbol{A}^2}{2m} . \tag{5.27}$$

The term between brackets gives the Hamiltonian $\mathcal{H}_0$ for an electron in the potential $V(\boldsymbol{r})$. Considering electromagnetic fields that are not so intense, the $\boldsymbol{A}^2$ term can be neglected (weak field approximation), and the electron-electromagnetic field interaction is given by:

$$\mathcal{H}_{eR} = -\frac{e}{m}\boldsymbol{p} \cdot \boldsymbol{A} . \tag{5.28}$$

For light scattering by a crystal, the wavelength of light is much larger than the unit cell dimensions, and an electron basis wavefunction $\varphi(\boldsymbol{r})$ (for example, a tight-binding wavefunction) is localized around the position of an atom $\boldsymbol{r}_0$ (a situation that is also valid for light scattering by a molecule). Considering monochromatic plane waves ($\boldsymbol{A}(\boldsymbol{r}, t) \propto e^{i\boldsymbol{k}\cdot\boldsymbol{r}}$), the interaction Hamiltonian $\mathcal{H}_{eR}$ can be considered within the dipole approximation, which is represented by $\boldsymbol{A}(\boldsymbol{r}, t)|\Psi\rangle \approx \boldsymbol{A}(\boldsymbol{r}_0, t)|\Psi\rangle$.

Finally, by considering $\boldsymbol{p} \equiv m(d\boldsymbol{r}/dt)$, we can write $\mathcal{H}_{eR}$ in terms of the electric field and the position vector by

$$\mathcal{H}_{eR} = -e\boldsymbol{r} \cdot \boldsymbol{E}(\boldsymbol{r}_0, t) . \tag{5.29}$$

A third approximation has been considered, that is, the contribution from the derivative $\partial[\boldsymbol{r} \cdot \boldsymbol{A}(\boldsymbol{r}_0, t)]/\partial t$ vanishes, considering the time average over a complete field oscillation period.

5.4.2
Electron–Phonon Interaction

The Hamiltonian $\mathcal{H}_{\text{e-ion}}$ for the electron–phonon interaction describes how the energy of the atoms change when they move through the so-called deformation potential [203]

$$\mathcal{H}^{\sigma}_{\text{e-ion}}(\boldsymbol{R}_{S'}, \boldsymbol{R}_S) = \int \phi(\boldsymbol{r} - \boldsymbol{R}_{S'})\nabla v(\boldsymbol{r} - \boldsymbol{R}_\sigma)\phi(\boldsymbol{r} - \boldsymbol{R}_S)d^3 r , \tag{5.30}$$

where $\phi(\boldsymbol{r} - \boldsymbol{R}_S)$ is the electron wave function at site $\boldsymbol{R}_S$ and $v(\boldsymbol{r} - \boldsymbol{R}_\sigma)$ is the atomic potential.[2] Calculation of the electron–phonon interaction in nanocarbons will be treated in Section 11.7, but considering here a simplified picture, the electron–phonon matrix element for optical phonons can be obtained from the shift of the

2) For completeness, when calculating these matrix elements for determining the electron–phonon coupling, both electrons and holes have to be taken into account.

electronic bands under the deformation of the atomic structure corresponding to the phonon-pattern by [204, 205]

$$M^{\mathrm{ep}}(\boldsymbol{k}-\boldsymbol{q}, mm') = \sqrt{\frac{\hbar}{2N_{\Omega} M \omega_q}} \sum_a \varepsilon_a^q \frac{\partial E_m}{\partial u_a} , \tag{5.31}$$

where the sum over a runs over all atoms in the unit cell. The wave vector and band index of the electronic state are here denoted by k and m, respectively. The q index denotes the phonon with polarization vector ε_a^q while E_m is the electronic energy and u_a is the atomic displacement. N_{Ω}, M and ω_q are the number of unit cells in the system, the atomic mass and the phonon eigenvalue, respectively.

5.5
Absolute Raman Intensity and the E_{laser} Dependence

In Eqs. (5.20) and (5.22) the Raman intensity is given as proportional to the transition probability. Different authors present different proportionality constants, and measuring the absolute Raman intensity is not an easy task, since it depends on several experimental details. The simplest procedure is calibrating the Raman intensities experimentally by measuring the Raman spectra of a well-established Raman scatterer, such as a standard reference material. For example, the dependence of the absolute Raman cross-sections for the cyclohexane liquid (C_6H_{12}) are known from the literature [206].

Of special interest in this proportionality constant is a ω_s^4-dependence predicted by Raman scattering theory [207–215]. This ω_s^4 dependence is not a special result of Raman spectroscopy, but it comes from the general theory for dipole radiation [216]. In short, let's define the dipole moment by $\boldsymbol{d} = e\boldsymbol{r}$, where e is the electric charge and $\boldsymbol{r}$ is the vector connecting the negative and positive charges in the dipole. It is known that radiation occurs only when the electric charge exhibits acceleration. For this reason, the fields (H and E) are proportional to the second time-derivative of the dipole momentum, that is H and $E \propto \ddot{\boldsymbol{d}} = e\ddot{\boldsymbol{r}}$ describing $\boldsymbol{r} = \boldsymbol{r}_0 e^{i\omega_s t}$, $\ddot{\boldsymbol{d}} = -\omega_s^2 \boldsymbol{d}$. In addition, the scattering intensity I can be related to the energy flux given by the Poynting vector S which, for plane waves, is related to the squared fields ($I \propto S \propto H^2$ or $E^2 \propto (\ddot{\boldsymbol{d}})^2$). This gives rise to a ω_s^4 dependence for the light emission.[3)] It is true that in Raman spectroscopy the incident and scattered light have different energies from one another but, since the phonon energies involving $\hbar\omega_q$ are usually much smaller than the excitation laser energy, $\hbar\omega_i \sim \hbar\omega_s$ is a good approximation and we can thus say that the absolute Raman intensity should increase with E_{laser}^4.

3) This is also the reason why the sky is blue. For more details, see [216].

Problems

[5-1] Using Maxwell's equations, explain why we need a gauge field for the vector potential **A** and a static potential ϕ. Consider some gauges explicitly and explain under what kind of situation such gauges are useful by giving some explanation of the physical phenomena.

[5-2] In a vacuum, show that the electric field is expressed by the vector potential **A**.

[5-3] When we consider the Hamiltonian in the presence of a vector potential, expand the Hamiltonian and retain the linear term in **A**. This corresponds to a perturbation Hamiltonian for the electron–photon coupling constant. Use the Coulomb gauge $\text{div}\,\mathbf{A} = 0$ when you obtain this result.

[5-4] In the previous problem, we also have a term which is proportional to $\mathbf{A}^2$. In order to neglect this term, this term should be at least 1/10 smaller than the linear **A** term. What is the corresponding value of the electric field? If **A** gives an electric field above this value, we should then consider the nonlinear $\mathbf{A}^2$ effect of light.[4)]

[5-5] The Poynting vector, $\mathbf{S} = \mathbf{E} \times \mathbf{H}$ is the power density per unit area of the electromagnetic field. In a typical micro-Raman measurement system, the diameter of the light beam is about 1 μm and the laser power is 1 mW. Estimate the power density of this micro-Raman setup and calculate **E**. Show that the electric field thus obtained is not strong enough to be in the nonlinear regime.

[5-6] Show that there is C_2 rotational symmetry in graphene. C_2 rotational symmetry means that the lattice structure does not change under a 180° rotation about a point. Specify the symmetry axis of this C_2 rotation.

[5-7] By the C_2 rotation, the A and B carbon atoms in the unit cell are exchanged with each other. Show that all A and B carbon atoms in the lattice are exchanged for any C_2 rotation.

[5-8] When we put the origin of the coordinate system at the axis point of the C_2 rotation of graphene, show that the perturbation Hamiltonian for the interaction of graphene with an incident light beam is an odd function of the coordinates.

[5-9] When we solve a simple 2×2 tight-binding Hamiltonian for the π-band of graphene (Eq. (2.29)), we get the wavefunction $\Psi(\mathbf{k})$ as a linear combination of the Bloch functions $\Phi_A(\mathbf{k})$ and $\Phi_B(\mathbf{k})$ consisting of A and B terms:

$$\Psi(\mathbf{k}) = C_A(\mathbf{k})\Phi_A(\mathbf{k}) + C_B(\mathbf{k})\Phi_B(\mathbf{k}) \,.$$

4) The general nonlinear optical effect in materials occurs at a much lower power level for light since the physical properties are saturated as a function of E. For example, the polarization vector **P** can be expanded as $\alpha E + \alpha^{(2)} E^2 + \ldots$, etc.

Obtain the explicit form of $C_A(\boldsymbol{k})$ and $C_B(\boldsymbol{k})$ for $s = 0$ in Eq. (2.29) with the use of $f(k)$ and $w(k)$ as defined in Eqs. (2.28) and (2.31), respectively.

[5-10] In the calculation for Problem 5-9, show $C_A = C_B$ and $C_A = -C_B$ for the valence and conduction band, respectively, for any $\boldsymbol{k}$ point. Combining this result with the fact that the perturbation Hamiltonian is an odd function of the coordinates for exchanging A and B, obtain the dipole transition matrix element which is proportional to $\langle \pi^* | \nabla | \pi \rangle$ as a function of $\boldsymbol{k}$.

[5-11] Using time-dependent perturbation theory, obtain the result for $|a_m^{(1)}(t)|^2$ in Eq. (5.10).

[5-12] Plot $|a_m^{(1)}(t)|^2$ in Eq. (5.10) for the case of $\omega_{m\ell} = \omega/2$, $2\omega/3$, $3\omega/4$, and ω. Show that the peak height of the oscillation is increasing as $\omega_{m\ell}$ increases. This means that we can select only the closest m states for the calculation of $|a^{(1)}(t)|^2$ as a first approximation.

[5-13] A delta function $\delta(x)$ has two significant properties: (1) The values of $\delta(x) = \infty$ for $x = 0$, but $= 0$ for $x \neq 0$. (2) When we integrate $\delta(x)$ over any region which includes $x = 0$, the integrated value is 1. Using these definitions, obtain the following formula:

$$\lim_{t \to \infty} \frac{\sin^2(\alpha t)}{\pi \alpha^2 t} = \delta(\alpha) ,$$

and then, using this formula, obtain Fermi's Golden Rule directly.

[5-14] The uncertainty relation of $\Delta E \Delta t \sim \hbar$ (Eq. (5.11)) is important in spectroscopy since most optical processes have a finite lifetime. If an excited state has a lifetime Δt, the energy of the excited states has an energy uncertainty value ΔE. In this case, time-dependent theory in which we consider a definite energy, should be modified. Explain how the resonance condition can be relaxed in the case of such a lifetime.

[5-15] For a photoexcited electron in a carbon nanotube, the electron can emit a phonon within 1 ps. Estimate the uncertainty energy value for this electron. For the photoluminescence process for which a photon is emitted, the excited electron has a relatively slower lifetime, on the order of 1 ns. Is the accuracy of the spectrometer sufficient to observe the energy of a photoluminescent phonon?

[5-16] Analyze the first-order processes depicted in the Feynman diagrams of Figure 5.4 within the framework of a resonant process, which is obtained by having null terms in Eq. (5.25). By starting with a given $\hbar\omega_i$ and the material in the ground state, which processes can be and which cannot be resonant? Consider $n = n'$ for simplicity, which happens when the phonon does not break the symmetry of the system.

[5-17] In the previous problem, derive a quantitative analysis to understand which of the six processes in Eq. (5.25) dominate the total intensity. Choose values

for $E_m - E_i$ and for $\hbar\omega_q$ and introduce the damping term $i\gamma_r$ with a specific value for γ_r. You will realize that processes which seem "reasonable" within our own "common sense" are closely related to whether or not the resonance condition is achieved.

[5-18] When we consider one optical-absorption, one optical-emission, and one phonon-emission vertex, we can obtain six possible Feynman diagrams. How many Feynman diagrams are expected if we change one phonon-emission vertex into a two phonon-emission vertex? Illustrate using some diagrams where the process starts from the optical absorption.

[5-19] Demonstrate Eqs. (5.28) and (5.29) making clear where each approximation is introduced.

[5-20] Show that for a periodic motion of charged particles forming a dipole d, and c is the velocity of light, the intensity of radiation with frequency ω is given by:

$$I = \frac{4\omega^2}{3c^3}|d|^2 . \tag{5.32}$$

6
Symmetry Aspects and Selection Rules: Group Theory

Symmetry is an important concept in physics. Momentum (angular momentum) conservation, for example, goes hand in hand with the translational (rotational) symmetry of space, and in a periodic lattice, the same can be said for crystal momentum for an electron or phonon state. Many optical processes are governed by conserved physical variables or selection rules that are the consequence of symmetry requirements. The selection rules governing the Raman scattering are derived from group theory which is the central focus of this chapter.

Group theory is a branch of mathematics whose beauty and strength, when applied to physics, resides in the transformation of many complex symmetry operations into a very simple linear algebra. A deep understanding of group theory requires a devoted study [94] and cannot be gained on the basis of this chapter. However, we can in this chapter provide a taste of the basic concepts, the power and usage of group theory in the field of Raman spectroscopy, giving examples of the useful information which comes from its application to the Raman spectroscopy of sp^2 nanocarbons. Therefore, be aware that this chapter may be difficult for a reader without a background in group theory, but with the help of a knowledgeable mentor, the beauty of symmetry can be introduced.

First, in Section 6.1, we briefly present the basic concepts of group theory [94] as they are applied to Raman spectroscopy. In Section 6.2 we give a group theoretical treatment of the Raman scattering selection rules. Section 6.3 summarizes a group theory analysis of optical processes for electrons and phonons in monolayer, bilayer and trilayer graphene, and extending all the way to N layer graphene, distinguishing the cases of even N, odd N and very large N, the latter corresponding to graphite [98, 217]. The symmetry of graphene is chosen for the introduction to this topic because the graphene family provides the fundamental building blocks for all sp^2 carbons, and monolayer graphene (1-LG) is the building block for all graphenes. In the second half of the chapter, we summarize in Section 6.4 the symmetry properties of single-wall carbon nanotubes (SWNTs) [135, 218], including an interesting connection given in Section 6.4.6 between the nature of the selection rules for the scattering processes in one dimension and those pertinent to the unfolded two-dimensional wave vector space [110].

Raman Spectroscopy in Graphene Related Systems. Ado Jorio, Riichiro Saito, Gene Dresselhaus, and Mildred S. Dresselhaus

ISBN: 978-3-527-40811-5

6.1 The Basic Concepts of Group Theory

6.1.1 Definition of a Group

The structure of sp^2 nanocarbons can be constructed by considering one single C atom and applying successively all the symmetry operations that take one atom into another, which include rotations, reflections, inversion, translations and combined operations. The set of symmetry operations that a molecule or crystal exhibits form a group in the group theory sense, and a group is defined as the following. A collection of elements $A, B, C, \ldots$ form a group when the following four conditions are satisfied:

1. The product of any two elements of the group is itself an element of the group. For example, relations of the type $AB = C$ are valid for all members of the group.
2. The associative law is valid, that is, $(AB)C = A(BC)$.
3. There exists a unit element E (also called the identity element) such that the product of E with any group element leaves that element unchanged $AE = EA = A$.
4. For every element A there exists an inverse element A^{-1} such that $A^{-1}A = AA^{-1} = E$.

As a simple example of a group, consider the permutation group for three numbers, $P(3)$. Below are listed the $3! = 6$ possible permutations that can be carried out; the top row denotes the initial arrangement of the three numbers and the bottom row denotes the final arrangement. Each permutation is an element of $P(3)$.

$$E = \begin{pmatrix} 1 & 2 & 3 \\ 1 & 2 & 3 \end{pmatrix} \quad A = \begin{pmatrix} 1 & 2 & 3 \\ 1 & 3 & 2 \end{pmatrix} \quad B = \begin{pmatrix} 1 & 2 & 3 \\ 3 & 2 & 1 \end{pmatrix}$$
$$C = \begin{pmatrix} 1 & 2 & 3 \\ 2 & 1 & 3 \end{pmatrix} \quad D = \begin{pmatrix} 1 & 2 & 3 \\ 3 & 1 & 2 \end{pmatrix} \quad F = \begin{pmatrix} 1 & 2 & 3 \\ 2 & 3 & 1 \end{pmatrix} . \tag{6.1}$$

We can also think of the elements in Eq. (6.1) in terms of the three points of an equilateral triangle (see Figure 6.1). Again, the top row denotes the initial state and the bottom row denotes the final position of each number as the effect of the six distinct symmetry operations that can be performed on these three points (see caption to Figure 6.1). The element D is a clockwise rotation of $2\pi/3$ and F is a counter-clockwise rotation of $2\pi/3$. We can call each symmetry operation an *element* of the group. This group is, therefore, identical with the group $P(3)$.

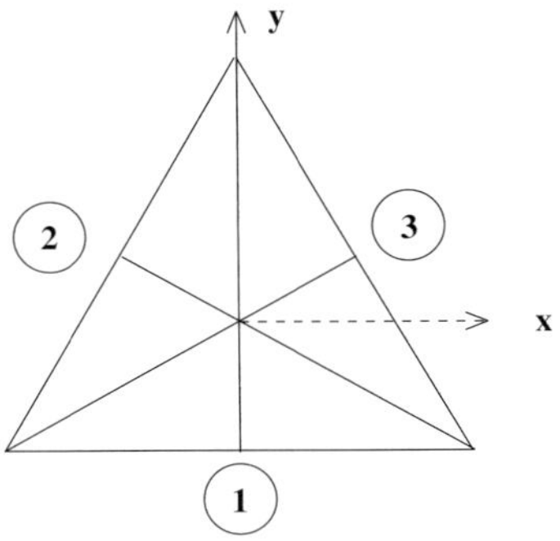

Figure 6.1 The symmetry operations on an equilateral triangle, are the rotations by $\pm 2\pi/3$ about the origin 0 and the rotations by 180° (π) about the three two-fold axes. Here the three two-fold axes are denoted by numbers in circles.

We illustrate the use of the notation by verifying the *associative law* $(AB)C = A(BC)$ for a few elements:

$$\begin{aligned}(AB)C &= DC = B \\ A(BC) &= AD = B\ .\end{aligned} \tag{6.2}$$

Point groups are groups without translations. There is at least one point which does not move under all the operations of a point group. The groups where translations are included as elements are named space groups.

6.1.2
Representations

Two groups are *isomorphic* or *homomorphic* if there exists a correspondence between their elements, such that $A \to \mathcal{A}$, $B \to \mathcal{B}$ and $AB \to \mathcal{AB}$, where the plain letters denote elements in one group and the other letters denote elements in the second group. If the two groups have the same order (same number of elements, such as $P(3)$ and the symmetries of an equilateral triangle), then they are *isomorphic* (one-to-one correspondence). Otherwise they are *homomorphic* (many-to-one correspondence).

The *representation*[1)] for an element R (e.g., $R = A, B, C, \ldots$) denoted by $D(R)$ is given by a square matrix in which a set of basis functions $\boldsymbol{u} \equiv (u_1, u_2, \cdots, u_m)$ are transformed by R as

$$R\boldsymbol{u} = D(R)\boldsymbol{u}\ . \tag{6.3}$$

In this way we assign a matrix $D(A)$ to each element A of the group such that $D(RR') = D(R)D(R')$.

1) The representation of an abstract group is a substitution group (matrix group with square matrices) such that the substitution group is homomorphic (or isomorphic) to the abstract group.

Consider the following group of matrices:

$$E = \begin{pmatrix} 1 & 0 \\ 0 & 1 \end{pmatrix} \quad A = \begin{pmatrix} -1 & 0 \\ 0 & 1 \end{pmatrix} \quad B = \begin{pmatrix} \frac{1}{2} & -\frac{\sqrt{3}}{2} \\ -\frac{\sqrt{3}}{2} & -\frac{1}{2} \end{pmatrix}$$
$$C = \begin{pmatrix} \frac{1}{2} & \frac{\sqrt{3}}{2} \\ \frac{\sqrt{3}}{2} & -\frac{1}{2} \end{pmatrix} \quad D = \begin{pmatrix} -\frac{1}{2} & \frac{\sqrt{3}}{2} \\ -\frac{\sqrt{3}}{2} & -\frac{1}{2} \end{pmatrix} \quad F = \begin{pmatrix} -\frac{1}{2} & -\frac{\sqrt{3}}{2} \\ \frac{\sqrt{3}}{2} & -\frac{1}{2} \end{pmatrix} . \tag{6.4}$$

The matrix corresponding to the identity operation is always a unit matrix. The matrices in Eq. (6.4) constitute a matrix representation of the group that is isomorphic to $P(3)$ and to the symmetry operations on an equilateral triangle, since they obey the same multiplication rules. The A matrix represents a rotation by $\pm\pi$ about the y axis while the B and C matrices, respectively, represent rotations by $\pm\pi$ about axes 2 and 3 in Figure 6.1. D and F, respectively, represent rotation of $-2\pi/3$ and $+2\pi/3$ around the center of the triangle (see Section 6.1.1).

6.1.3
Irreducible and Reducible Representations

When $D(R)$ for all the elements in a group can be blocked into similar sub-matrices by one unitarity transformation,[2] the $D(A)$ is called a reducible representation. In this case, the space of $\boldsymbol{u}$ can be decomposed into smaller spaces of basis functions with smaller dimensions. If $D(A)$ cannot be reduced any further into smaller blocks, then $D(A)$ is called an *irreducible representation* (IR).

Three irreducible representations for the permutation group $P(3)$ are:

$$\begin{array}{llll} & E & A & B \\ \Gamma_1: & (1) & (1) & (1) \\ \Gamma_{1'}: & (1) & (-1) & (-1) \\ \Gamma_2: & \begin{pmatrix} 1 & 0 \\ 0 & 1 \end{pmatrix} & \begin{pmatrix} -1 & 0 \\ 0 & 1 \end{pmatrix} & \begin{pmatrix} \frac{1}{2} & -\frac{\sqrt{3}}{2} \\ -\frac{\sqrt{3}}{2} & -\frac{1}{2} \end{pmatrix} \\ & C & D & F \\ \Gamma_1: & (1) & (1) & (1) \\ \Gamma_{1'}: & (-1) & (1) & (1) \\ \Gamma_2: & \begin{pmatrix} \frac{1}{2} & \frac{\sqrt{3}}{2} \\ \frac{\sqrt{3}}{2} & -\frac{1}{2} \end{pmatrix} & \begin{pmatrix} -\frac{1}{2} & \frac{\sqrt{3}}{2} \\ -\frac{\sqrt{3}}{2} & -\frac{1}{2} \end{pmatrix} & \begin{pmatrix} -\frac{1}{2} & -\frac{\sqrt{3}}{2} \\ \frac{\sqrt{3}}{2} & -\frac{1}{2} \end{pmatrix} \end{array} . \tag{6.5}$$

$\Gamma_1, \Gamma_{1'}$ and Γ_2 obey the same multiplication rules as $P(3)$ and as the equilateral triangle. However, Γ_1 and $\Gamma_{1'}$ are one-dimensional representations having only one and two elements, respectively, being homomorphic to $P(3)$ and to the equilateral triangle. The one-dimensional Γ_1 is named the totally symmetric representation and it exists for all groups. The basis function of the irreducible representation

2) A unitary transformation is an operation such as $U D(R) U^{-1}$, where U is an unitary matrix. For space symmetry operations, such transformations are equivalent to a rotation (without deformation) of the coordinate axes.

Γ_1 does not change the form for any of the operations of the group, such as 1 or $r = x^2 + y^2$ for a bidimensional point group using the spacial coordinates. Thus the dimension of Γ_1 is one and the matrix representation is a 1×1 matrix whose matrix element is 1. For $\Gamma_{1'}$, z is a suitable basis function. By considering the equilateral triangle in the xy plane, the z coordinate goes into $-z$ when applying the symmetry operations A, B and C. Thus the dimension of $\Gamma_{1'}$ is one and the matrix representation is a 1×1 matrix whose matrix element now is either 1 or -1. The bidimensional Γ_2 has six elements and is isomorphic to $P(3)$, as stated previously. A suitable set of basis functions (two orthogonal functions are needed) using the spacial coordinates is (x, y). A reducible representation containing these three irreducible representations is:

$$\Gamma_R : \quad \overset{E}{\begin{pmatrix} 1 & 0 & 0 & 0 \\ 0 & 1 & 0 & 0 \\ 0 & 0 & 1 & 0 \\ 0 & 0 & 0 & 1 \end{pmatrix}} \quad \overset{A}{\begin{pmatrix} 1 & 0 & 0 & 0 \\ 0 & -1 & 0 & 0 \\ 0 & 0 & -1 & 0 \\ 0 & 0 & 0 & 1 \end{pmatrix}} \quad \overset{B}{\begin{pmatrix} 1 & 0 & 0 & 0 \\ 0 & -1 & 0 & 0 \\ 0 & 0 & \frac{1}{2} & -\frac{\sqrt{3}}{2} \\ 0 & 0 & -\frac{\sqrt{3}}{2} & -\frac{1}{2} \end{pmatrix}} \text{, etc.,} \tag{6.6}$$

where Γ_R is of the block form[3)]

$$\left(\begin{array}{c|c|c} \Gamma_1 & 0 & \mathcal{O} \\ \hline 0 & \Gamma_{1'} & \mathcal{O} \\ \hline \mathcal{O} & \mathcal{O} & \Gamma_2 \end{array}\right) . \tag{6.7}$$

It is customary to list the irreducible representations (IR) contained in a reducible representation Γ_R as:

$$\Gamma_R = \Gamma_1 + \Gamma_{1'} + \Gamma_2 . \tag{6.8}$$

In quantum mechanics, the matrix representation of a group, $D(R)$, is important for several reasons. First of all, an eigenfunction for a quantum mechanical operator is a basis function of an IR of the group of the system.[4)] Thus, without actually calculating the eigenvalue problem, we can know from group theory the form of functions which belong to each IR.[5)] Secondly, quantum mechanical operators are usually written in terms of a matrix representation which is generally a reducible representation of the group, and thus by selecting the basis function belonging to the IR, the matrix can be put into block form. Thus matrix algebra becomes much easier to manipulate than the original symmetry operations.

3) $\mathcal{O}$ indicates a null matrix.

4) Any operator of the group of the system, A commutes with the quantum operator $\mathcal{O}$ as $[A, \mathcal{O}] = 0$. This is a requirement of the group. Then the wavefunction of $A\Psi$ has the same eigenvalue O of $\mathcal{O}$ for Ψ ($\mathcal{O}A\Psi = A\mathcal{O}\Psi = OA\Psi$). Thus Ψ is also an eigenfunction of A (if Ψ is not degenerate) in the group and Ψ is the basis function of an IR. This proof is also valid when E is degenerate.

5) This does not mean that group theory solves the eigenvalue problem. Group theory, however, helps solving the eigenvalue problem by reducing the dimensions of the space to be considered.

6.1.4 The Character Table

The *character* of the representation matrix $D(R)$, is the trace $\chi(R)$ of $D(R)$.[6] The table of $\chi(R)$ for all R and for all IRs constitutes a *character table*, schematically represented in Table 6.1 in which $\chi(R)$ for an IR is listed in a row. The first row of the character table is a totally symmetric IR, denoted by Γ_1 (or $A, A_1, A_g \ldots$ depending on the group and notation) in which all $\chi(R) = 1$. This row is present in the character table for all point groups.

A column of Table 6.1 pertains to a *class* which is defined by a set of operations $(R_1, R_2, \cdots, R_c)$ that transform into one another by any operations O_j in the group such as $R_i = O_j^{-1} R O_j$. For example, if three C_2 rotational operators are equivalent to one another, the three C_2 belong to a class of $3C_2$, where 3 denotes the dimension of the class. The characters for operators within a class are the same. Class 1 in Table 6.1 consists of the identity element E and this element is also present in all groups. The characters for Class 1 for an IR is the dimension of the IR since no basis function changes under the operation E and thus the diagonal matrix elements of $D(E)$ are all 1. Notice that in a group the number of IRs is equal to the number of *classes* in the group.

The character table for the permutation group $P(3)$ is shown in Table 6.2.

Table 6.1 Schematics for a group theory character table. The IR denoted by 1 represents the totally symmetric IR, and is always present in all character tables. $\chi_{\mathrm{IR}j}^{\mathcal{C}_k}$ represents the characters for the symmetry operations in class $\mathcal{C}_k$, belonging to the IR j, where $j = 1, 2, 3$. Also N_k is the number of elements in $\mathcal{C}_k$.

Basis functions		$N_1\mathcal{C}_1$	$N_2\mathcal{C}_2$	$N_3\mathcal{C}_3$
Function 1	IR 1	1	1	1
Function 2	IR 2	$\chi_{IR2}^{C_1}$	$\chi_{IR2}^{C_2}$	$\chi_{IR2}^{C_3}$
Function 3	IR 3	$\chi_{IR3}^{C_1}$	$\chi_{IR3}^{C_2}$	$\chi_{IR3}^{C_3}$

Table 6.2 Character table for the permutation group $P(3)$, for the symmetry operations of the equilateral triangle or, more generally, for the so-called group D_3, using the Schoenflies notation [94].

Class →	$\mathcal{C}_1$	$3\mathcal{C}_2$	$2\mathcal{C}_3$
IR ↓	$\chi(E)$	$\chi(A, B, C)$	$\chi(D, F)$
Γ_1	1	1	1
$\Gamma_{1'}$	1	-1	1
Γ_2	2	0	-1

6) The trace of a matrix is the sum of diagonal matrix elements $\chi(R) = \mathrm{Tr}(R) = \sum_i R_{ii}$.

Table 6.3 Classes for group D_3 or equivalently for the permutation group $P(3)$ and for the symmetry operations of the equilateral triangle.

Notation for each class	D_3	**Equilateral triangle**	$P(3)$
Class 1 E ($N_k = 1$)	$1C_1$	(Identity class)	(1)(2)(3)
Class 2 A, B, C ($N_k = 3$)	$3C_2$	(Rotation of π about two-fold axis)	(1)(23)
Class 3 D, F ($N_k = 2$)	$2C_3$	(Rotation of 120° about three-fold axis)	(123)

This point group is named D_3 (Schoenflies notation) [94]. In Table 6.2 the notation $N_k C_k$ is used in the character table to label each class C_k, and N_k is the number of elements in C_k. The classes for group D_3 and $P(3)$ are listed in Table 6.3, showing different ways that the classes of a given group are presented.

6.1.5 Products and Orthogonality

When we define the *inner product* of characters for two IRs weighted by the dimension of the class divided by the normalization factor of the dimension of the group, we get the rules for the ortho-normal conditions for IRs, which is generally referred to as the *Wonderful Orthogonality Theorem for Characters.* To illustrate the meaning of the Wonderful Orthogonality Theorem for Characters, consider the character table for the group $D(3)$ shown in Table 6.2 or given in Table 6.4 in its most commonly used form. Let $\Gamma_j = \Gamma_1$ and $\Gamma_{j'} = \Gamma_{1'}$. Then calculate:

$$\sum_k N_k \chi^{(\Gamma_j)}(C_k)\left[\chi^{(\Gamma_{j'})}(C_k)\right]^* = \underbrace{(1)(1)(1)}_{\text{class of } E} + \underbrace{(3)(1)(-1)}_{\text{class of } A,B,C} + \underbrace{(2)(1)(1)}_{\text{class of } D,F}$$

$$= 1 - 3 + 2 = 0. \tag{6.9}$$

It can likewise be verified that the Wonderful Orthogonality Theorem works for all possible combinations of Γ_j and $\Gamma_{j'}$ in Table 6.4.

Since an eigenfunction Ψ for an electron (or phonon) state pertains to a given IR, we identify in the character table the set of symmetry operations that Ψ exhibits, thus describing the effect of the symmetry operations on Ψ based on the simple linear algebra of matrices. The dimension of each IR, d, gives the degeneracy of an

Table 6.4 Character table for group D_3 (rhombohedral).

	D_3 (32)		E	$2C_3$	$3C_2'$
x^2+y^2, z^2		A_1	1	1	1
	R_z, z	A_2	1	1	−1
(xz, yz), (x^2-y^2, xy)	(x, y), (R_x, R_y)	E	2	−1	0

energy level which belongs to the IR. If the system has m IR eigenstates given by group theoretical considerations as below, we expect to have m, d-fold eigenvalues (this means $m \times d$ eigenfunctions). As shown later, for a chiral nanotube, we have five doubly degenerate Raman-active phonon modes for a given IR (named E_1).

6.1.6
Other Basis Functions

For each symmetry group, one can build a character table whose structure is shown in Table 6.1. We remind readers that the basis function can be obtained by considering the representation matrices and vice-versa. In the leftmost column, the corresponding simple basis functions belong to the IR, such as x, y, z for translation, R_x, R_y, R_z for rotation along the x, y, z axes, or quadratic functions xx, yy, zz, xy, yz, zx and so on. An example is given in Table 6.4 for the D_3 group. This information on the basis functions is useful for knowing which IR belongs to the vibration along the x direction, the rotation along the x axis, and the Raman-active modes (see Section 6.2). The known groups such as point groups, space groups, symmetry groups, are all listed in crystallography tables and their character tables can be found in group theory books.

6.1.7
Finding the IRs for Normal Modes Vibrations

To find the normal modes for the vibration problem, we carry out the following steps:

1. Identify the symmetry operations that define the point group G of the crystal unit cell in its equilibrium configuration.
2. Find the characters for the equivalence representation, $\Gamma_{\text{equivalence}} = \Gamma^{\text{a.s.}}$ (a.s. stands for atom sites). These characters represent the number of atoms that are invariant under the symmetry operations of the group. Since $\Gamma^{\text{a.s.}}$ is, in general, a reducible representation of the group G, we must decompose $\Gamma^{\text{a.s.}}$ into its irreducible representations (e. g., see Eq. (6.8)).
3. We next use the concept that a molecular vibration involves the transformation properties of a vector. In group theoretical terms, this means that the molecular vibrations are found by taking the direct product of $\Gamma^{\text{a.s.}}$ with the irreducible representations for a radial vector (such as (x, y, z)). The representation for the molecular vibrations $\Gamma_{\text{lat. mode}}$ is thus found according to the relation[7]

$$\Gamma_{\text{lat. mode}} = (\Gamma^{\text{a.s.}} \otimes \Gamma_{\text{vec}}) \, . \qquad (6.10)$$

7) If working with molecules, Γ_{trans} and Γ_{rot} have to be subtracted from Eq. (6.10), where Γ_{trans} and Γ_{rot} denote the representations for the simple translations and rotations of the molecule about its center of mass.

The symbol $\otimes$ denotes the direct product of two IRs.[8] The characters found from Eq. (6.10), in general, correspond to a reducible representation of Group G. We therefore express $\Gamma_{\text{lat. mode}}$ in terms of the *irreducible* representations of group G to obtain the normal modes. Each eigenmode is labeled by one of these irreducible representations, and the degeneracy of each eigenfrequency is the dimensionality of the corresponding irreducible representation. The characters for Γ_{trans} are found by identifying the irreducible representations of the group G corresponding to the basis functions (x, y, z) for the radial vector $\boldsymbol{r}$. The characters for Γ_{rot} are found by identifying the irreducible representations corresponding to the basis functions (R_x, R_y, R_z) for the axial vector (e. g., angular momentum which, for example, corresponds to $\boldsymbol{r} \times \boldsymbol{p}$). Since the radial vector $\boldsymbol{r}(x, y, z)$ and the axial vector $\boldsymbol{r} \times \boldsymbol{p}$ (R_x, R_y, R_z) transform differently under the symmetry operations of group G, every standard character table normally lists the irreducible representations for the six basis functions for (x, y, z) and (R_x, R_y, R_z) (see Table 6.4).

4. From the characters for the irreducible representations for the molecular vibrations, we find the normal modes. The normal modes for a molecule as defined by Eq. (6.10) are constrained to contain only internal degrees of freedom, and no translations or rotations of the full molecule. Furthermore, the normal modes must be orthogonal to each other.
5. We use the techniques for selection rules (see Section 6.1.8) to find out whether or not each of the normal modes is Raman-active.

It is important to recall that $\Gamma_{\text{vec}}(R)$ is obtained by summing the irreducible representations to which the x, y, and z basis functions belong. If (x, y, z) are the partners of a three-dimensional irreducible representation for T translations, then $\Gamma_{\text{vec}}(R) = \Gamma^T(R)$. If, instead, x, y, and z belong to the same one-dimensional irreducible representation A, then $\Gamma_{\text{vec}}(R) = 3\Gamma^A(R)$. If the x, y, and z basis functions are not given in the character table, $\Gamma_{\text{vec}}(R)$ can be found directly from the trace of the matrix representation for R. All the point group operations are rotations or combination of rotations with inversion. For proper rotations by an angle θ, $\chi_{\text{vec}}(R) = 1 + 2\cos\theta$, so that the trace for the rotation matrix can be always be found directly from the rotation matrix

$$\begin{pmatrix} \cos(\theta) & \sin(\theta) & 0 \\ -\sin(\theta) & \cos(\theta) & 0 \\ 0 & 0 & 1 \end{pmatrix} . \tag{6.11}$$

Improper rotations consist of a rotation followed by a reflection about a horizontal plane resulting in the character $-1 + 2\cos\theta$ where the +1 for a proper rotation goes into -1 for an improper rotation, since z goes into $-z$ upon reflection.

8) The direct product of two IRs means that the characters of the two IRs for each class are multiplied with one another. The representation thus obtained is generally a reducible representation which is then decomposed into IRs.

Considering group D_3, imagine a hypothetical molecule made of three identical atoms at the vertices of the equilateral triangle. We have

	E	$2C_3$	$3\sigma_v$	
$\Gamma^{\text{a.s.}}$	3	0	1	$\Rightarrow A_1 + E$

So that

$$\Gamma_{\text{lat. mode}} = \Gamma^{\text{a.s.}} \otimes \Gamma_{\text{vec}}$$
$$\Gamma_{\text{lat. mode}} = (A_1 + E) \otimes (A_2 + E) = A_1 + 2A_2 + 3E \,. \tag{6.12}$$

From these modes, three modes $(A_2 + E)$ are molecular rotations, three modes $(A_2 + E)$ are molecular translations (see respective basis functions in Table 6.4) and the other three $(A_1 + E)$ are vibrational modes.

6.1.8
Selection Rules

In considering selection rules we always involve some interaction matrix $\mathcal{H}'$ that couples two states ψ_α and ψ_β. If $\mathcal{H}'\psi_\beta$ is orthogonal to ψ_α, then the matrix element $\langle\psi_\alpha|\mathcal{H}'|\psi_\beta\rangle$ vanishes by symmetry;[9] otherwise, the matrix element need not vanish, and the transition from state α to β may occur via $\mathcal{H}'$. Group theory is often invoked to decide whether or not $\langle\psi_\alpha|\mathcal{H}'|\psi_\beta\rangle$ vanishes by symmetry, and this information can be extracted from the character tables. First, we identify the IRs for $\psi_\alpha, \mathcal{H}'$ and ψ_β. Then we multiply their respective characters $\chi_{\mathcal{H}'}(R) \otimes \chi_{\psi_\beta}(R)$. Such a multiplication process can be described by a linear combination of characters coming from different IRs of the group. If this linear combination contains the IR for the ψ_α state,[10] the matrix is nonvanishing by symmetry. Otherwise, $\mathcal{H}'\psi_\beta$ is orthogonal to ψ_α. This rule is applied to obtain the selection rules for the matrix elements occurring in Raman scattering, as discussed in this chapter.

6.2
First-Order Raman Scattering Selection Rules

In Section 4.3.2.7 we described the momentum conservation requirement ($q \sim 0$) for first-order Raman scattering, which goes hand in hand with the translational symmetry in the periodic lattice. Here we derive other symmetry requirements for first-order symmetry-allowed Raman scattering processes for phonons related to the other symmetry elements (rotations, reflections, etc.) of the crystal lattice. Suppose that we have a group G with symmetry elements R and symmetry opera-

9) This means that the integrated function is an odd function of the variables so the implied integration gives a zero value.
10) Or alternatively the representation whose character is the direct product $\psi_\alpha(R) \otimes \chi_{\mathcal{H}'}(R) \otimes \chi_{\psi_\beta}(R)$ contains the totally symmetric IR Γ_1.

tors $\hat{P}_R$. We denote the IRs by Γ_n, where n labels the IR. We can then define a set of basis functions denoted by $|\Gamma_n j\rangle$, with $j = 1, 2, \dots \ell_j$, where ℓ_j is the dimension of the IR.

As shown in Section 5.4.1, the electromagnetic interaction for electric dipole transitions is given by:

$$\mathcal{H}_{eR} = -\frac{e}{m}\boldsymbol{p} \cdot \boldsymbol{A} , \tag{6.13}$$

in which $\boldsymbol{p}$ is the momentum operator of the electron and $\boldsymbol{A}$ is the vector potential of an external electromagnetic field.[11] In the dipole approximation, $\boldsymbol{A}$ (or $\boldsymbol{E}$) which have a much longer wavelength than the unit cell size is considered to be a constant vector within the unit cell, which means that $\boldsymbol{A}$ can be taken out of the matrix element $\langle a|\boldsymbol{p} \cdot \boldsymbol{A}|i\rangle = \langle a|\boldsymbol{p}|i\rangle \cdot \boldsymbol{A}$.

As discussed in the previous section (Section 6.1.8), to have a nonvanishing matrix element $\langle a|\boldsymbol{p}|i\rangle$, we need

$$\Gamma_a \subset \Gamma_p \otimes \Gamma_i , \tag{6.14}$$

where Γ_i, Γ_a and Γ_p are the IRs for the initial and intermediate electronic states and for the electron radiation Hamiltonian interaction, respectively. The symbol $A \subset B$ is defined so that A is a subset of B. That is, in the reducible representation of B, we can find the IR for A.

Similarly, the electron–phonon matrix element $\langle b|H_{e\text{-ion}}|a\rangle$ is nonvanishing if

$$\Gamma_b \subset \Gamma_{H_{e\text{-ion}}} \otimes \Gamma_a \subset \Gamma_{H_{e\text{-ion}}} \otimes \Gamma_p \otimes \Gamma_i , \tag{6.15}$$

if we consider that state $|a\rangle$ here was generated from $|i\rangle$ by $\mathcal{H}_{eR}$. In sequence, the symmetry for the final state in $\langle f|\boldsymbol{p}|b\rangle\langle b|H_{e\text{-ion}}|a\rangle\langle a|\boldsymbol{p}|i\rangle$ has to obey

$$\Gamma_f \subset \Gamma_p \otimes \Gamma_b \subset \Gamma_p \otimes \Gamma_{H_{e\text{-ion}}} \otimes \Gamma_p \otimes \Gamma_i , \tag{6.16}$$

which gives the selection rules for state $|f\rangle$being generated from state $|i\rangle$ by the third-order Raman scattering process. The Raman process ends with the electron decaying back to its original state, which means the initial and final electronic states are the same ($|f\rangle \equiv |i\rangle$). Therefore,

$$\Gamma_{\psi_i} \subset \Gamma_p \otimes \Gamma_{H_{e\text{-ion}}} \otimes \Gamma_p \otimes \Gamma_{\psi_i} , \tag{6.17}$$

with Γ_{ψ_i} being the IR for the initial electronic state. This condition is the same as saying that a Raman-active mode has to obey

$$\Gamma_p \otimes \Gamma_{H_{e\text{-ion}}} \otimes \Gamma_p \supset \Gamma_1 , \tag{6.18}$$

11) Alternatively, $\mathcal{H}_{eR}$ can also be given by $-e\boldsymbol{r} \cdot \boldsymbol{E}$ (see Section 5.4.1), and then $\boldsymbol{r}$ transforms as a vector.

where Γ_1 is the totally symmetric IR. Since Γ_p pertains to the same IR as the basis functions x, y and z, the condition above is valid if $\Gamma_{H_{e\text{-ion}}}$ belongs to the same IR as the symmetric combinations of the biquadratic basis functions xx, yy, zz, xy, xz and yz.[12] Therefore, one can identify the Raman-active modes as those pertaining to the same IRs as the biquadratic functions, which are often listed in character tables. Furthermore, the first and second letters in the biquadratic functions denote the polarization directions that the incident and scattered light must have, respectively, in order to observe the phonon pertaining to that IR. For the hypothetical triangular molecule discussed in Section 6.1.7, both the A_1 and E symmetry vibrational modes are Raman-active (see Table 6.4). For this molecule, only the A_1 mode can be seen for incident and scattered light polarized parallel to each other (basis functions $x^2 + y^2$ or z^2). Only the E modes can be seen when the incident and scattered light are cross polarized (basis functions $(x^2 - y^2, xy)$ or (xz, yz)).

6.3 Symmetry Aspects of Graphene Systems

In this section the group theory of graphene systems is summarized to provide a fundamental basis for using group theory to describe the electronic and vibrational properties of sp^2 carbons, previously discussed in Chapters 2 and 3, respectively.

6.3.1 Group of the Wave Vector

Figure 6.2a shows the hexagonal real space structure for monolayer graphene (1-LG) with two nonequivalent atoms per unit cell. The origin in real space is set at the highest symmetry point, that is, at the center of a hexagon, and Figure 6.2a shows the unit vectors defining the rhombic in-plane unit cell, containing the two inequivalent carbon atom sites A and B. Monolayer graphene is an isotropic planar medium described by the 2D space group P6/mm[13] in the Hermann–Mauguin notation.[14] Electrons and phonons at the Γ point both exhibit the symmetries of the point group D_{6h}. The character table for group D_6 is given in Table 6.5 and $D_{6h} = D_6 \otimes C_i$. Here the symbol $\otimes$ represents the so-called direct product,[15]

12) The Raman tensor needs to be a symmetric second rank tensor, and the symmetrized forms of xy, xz and yz are here needed, such as $(xy + yx)/2$, etc.

13) P6/mm denotes a primitive (or simple) lattice, with a six-fold rotational axis, and two mirror planes.

14) Both Hermann–Mauguin and Schoenflies notations are used to describe point group and space group symmetry operations [94].

15) Let $G_A = E, A_2, \ldots, A_{h_a}$ and $G_B = E, B_2, \ldots, B_{h_b}$ be two groups such that all operators A_R commute with all operators B_S. Then the direct product group is written as $G_A \otimes G_B = E, A_2, \ldots, A_{h_a}, B_2, A_2 B_2, \ldots, A_{h_a} B_2, \ldots, A_{h_a} B_{h_b}$, where the elements of the direct product group are also indicated.

Table 6.5 Character table for group D_6 (hexagonal)[a] [94].

	D_6 (622)		E	C_2	$2C_3$	$2C_6$	$3C_2'$	$3C_2''$
x^2+y^2, z^2		A_1	1	1	1	1	1	1
	R_z, z	A_2	1	1	1	1	−1	−1
		B_1	1	−1	1	−1	1	−1
		B_2	1	−1	1	−1	−1	1
(xz, yz)	(x, y) (R_x, R_y)	E_1	2	−2	−1	1	0	0
(x^2-y^2, xy)		E_2	2	2	−1	−1	0	0

a $D_{6h} = D_6 \otimes i$; $(6/mmm)$ (hexagonal).

which here indicates the direct product of the group D_6 and the group C_i.[16] This is equivalent to adding a horizontal plane of symmetry.[17]

Wave functions away from the Γ point exhibit symmetries lower than D_{6h}, and these lower symmetries are listed in Table 6.6. Character tables for the point groups for all the high symmetry points for monolayer graphene can be found in [94]. This reference also contains extensive information relevant to all types of graphene and sp^2 carbons. More explicitly, the point groups for electron and phonon wavefunctions at other high symmetry points for monolayer graphene are: $3m$[18] for K (K') points, mm for M points, m for T (T') and Σ points, where T (T') and Σ points lie on the line of ΓK (KM) and ΓM, respectively (see Figure 6.2c), and C_1 simply denotes the point group for the general points u in the Brillouin zone that have no special symmetry operations. The groups for wavefunctions at the k point in the Brillouin zone are usually named the *group of the wave vector* (GWV).

When graphene layers are stacked in the AB Bernal structure in real space, carbon atoms A_1 and A_2 on A sites are found one above the other on adjacent layers,

Table 6.6 The space groups and the group of the wave vector point groups for monolayer, *N*-layer graphene and graphite at all high symmetry points in the Brillouin zone [98].

	Space group	Γ	K (K')	M	T (T')	Σ	u
Monolayer	$P6/mm$	D_{6h}	D_{3h}	D_{2h}	C_{2v}	C_{2v}	C_{1h}
N even	$P\bar{3}m1$	D_{3d}	D_3	C_{2h}	C_2	C_{1v}	C_1
N odd	$P\bar{6}m2$	D_{3h}	C_{3h}	C_{2v}	C_{1h}	C_{2v}	C_{1h}
N infinite	$P6_3/mmc$	D_{6h}	D_{3h}	D_{2h}	C_{2v}	C_{2v}	C_{1h}

16) C_i has two symmetry elements, the identity E and the inversion i.

17) $C_2 i = \sigma_h$, and $C_2' i = \sigma_v$, where σ_h and σ_v are, respectively, horizontal and vertical mirror plane operations when we put the C_6 axis in the vertical direction.

18) $3m$ denotes three-fold rotational and mirror plane symmetry.

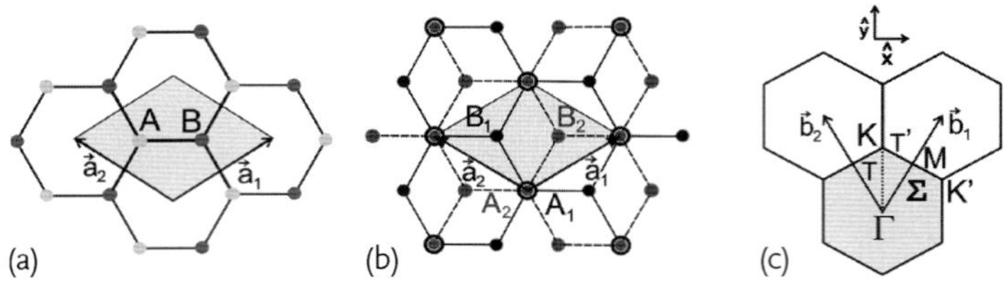

Figure 6.2 (a) The real space top-view of the setting for the unit cell for monolayer graphene, showing the nonequivalent A and B atoms and unit cell vectors $\boldsymbol{a}_1$ and $\boldsymbol{a}_2$. (b) The real space top-view of the setting for the unit cell for bilayer graphene. Light and dark gray dots in (a) denote A and B atoms in 1-LG. Large gray circles represent A atoms which are above one another in bilayer graphene. Small black and gray dots represent B atoms on the lower and upper layers, respectively. Thus the A atoms are above one another on adjacent layers, but the B atoms are instead staggered on adjacent layers for Bernal stacking. (c) The hexagonal reciprocal space showing high symmetry points and lines [98].

while the B atoms alternate between the B_1 and B_2 sites on adjacent layers, as seen in Figure 6.2b.[19]

The real space unit cell for bilayer graphene with AB Bernal stacking is shown in Figure 6.2b, and this is the fundamental unit cell for all N even-layer graphenes, including graphite in the large N limit. All N odd-layer graphenes can be considered as having a single graphene layer, flanked by bilayer unit cells on either side. The main symmetry operation distinguishing the point groups between even and odd layers is the horizontal mirror plane symmetry, which is absent for N even, and the inversion operation, which is absent for N odd. The space groups, and the group of the wave vector for all the high symmetry points in monolayer graphene, even and odd layer graphenes ($N > 1$), and for graphite in the large N limit are listed in Table 6.6. The GWV for N-layer graphenes are subgroups of the GWV for single-layer graphene. The direct product between the space group for N even with no translations and the space group for N odd with no translations gives the space group for the monolayer graphene GWV with no translations, that is,

$$\{G_{\text{even}}|0\} \otimes \{G_{\text{odd}}|0\} = \{G_{1-LG}|0\}, \tag{6.19}$$

which can be seen when carrying out the direct product operations between these groups. Graphite belongs to the $P6_3/mmc$ ($D_{6h}{}^4$) nonsymmorphic space group[20] while the space group for monolayer graphene is $P6/mm$ and is symmorphic. For graphite, the GWV is $P6_3/mmc$ at the Γ point. However, the group of the wave vector for graphite at high symmetry points in the BZ is isomorphic to the GWV of monolayer graphene, but they differ fundamentally for some classes where a

19) For three layers, the B_1 and B_3 appear at the same in-plane positions, while the B_2 atom is in a staggered location as shown in Figure 6.2b. Such stacking of adjacent planes is called ABAB stacking.

20) Depending on the existence or not of a spiral (or chiral) operation in the crystal symmetry, the space group is divided into nonsymmorphic and symmorphic space groups. Most of the space groups that we discuss in solid state textbooks are symmorphic space groups, for which the translation operations and the point group operations commute. For nonsymmorphic groups, translations and point group operations do not commute.

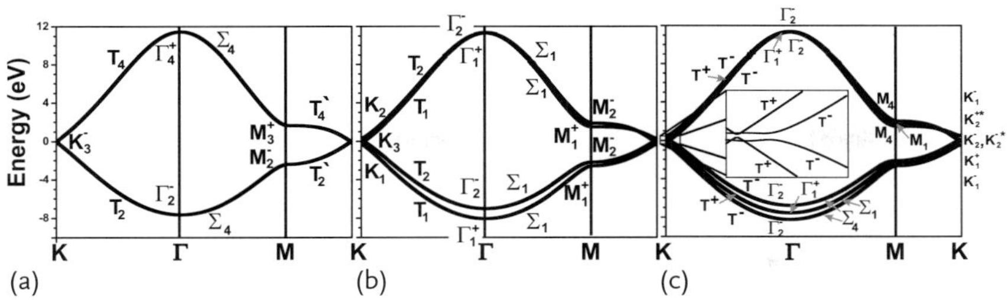

Figure 6.3 The electronic dispersion for the π electrons, calculated by DFT and using the irreducible representations (Γ_π), are shown for (a) monolayer, (b) bilayer and (c) trilayer graphene along the KΓMK directions [98].

translation of $c/2$ is present in the symmetry operations in graphite, and in this way graphite is homomorphic to monolayer graphene.

6.3.2 Lattice Vibrations and π Electrons

The group theoretical representations for lattice vibrations ($\Gamma_{\text{lat. mode}}$) and for π electron states (Γ_π) are given by the direct products as $\Gamma_{\text{lat. mode}} = \Gamma^{\text{a.s.}} \otimes \Gamma^{\text{vec}}$ and $\Gamma_\pi = \Gamma^{\text{a.s.}} \otimes \Gamma^z$, respectively, where $\Gamma^{\text{a.s.}}$ is the atom sites equivalence representation,[21] and Γ^{vec} is the representation for the vectors x, y and z [94].[22] For Γ_π we use only Γ^z, which is the irreducible representation for the vector z, since π electrons in graphene are formed by p_z electronic orbitals normal to the layer planes. The irreducible representations for all high symmetry points and lines in the first BZ for $\Gamma_{\text{lat. mode}}$ are found in Table 6.7. The corresponding results for Γ_π are found in Table 6.8 (see Section 6.3.6 for the notation conversion from space group to point group). Furthermore, Table 6.8 shows that the π electrons in monolayer graphene are degenerate at the K (Dirac) point, as obtained by theory [31]. Figure 6.3a, b and c show the electronic structure of monolayer, bilayer and trilayer graphene, respectively, calculated via density functional theory (DFT) [98, 217]. The symmetry assignments of the different electronic branches shown in Figure 6.3 were made according to the symmetries of the DFT projected electronic density of states [98].

21) The atom sites equivalence representation is a reducible representation of the point group of the unit cell, in which each value for $\Gamma^{\text{a.s.}}$ corresponds to the number of atoms that do not change their position by each symmetry operation. For the mirror symmetry operation, the number of atoms on the mirror corresponds to that value. For rotational operations, the number of atoms on the rotational axis corresponds to that value.

22) If you look at the character table, you sometimes see simple functions such as x, y, z or xy, etc., which can correspond to an irreducible representation. If this is not the case, you need to find by yourself which irreducible representation corresponds to x, y and z.

Table 6.7 The $\Gamma_{\text{lat. mode}}$ wave vector point-group irreducible representations for mono- and N-layer graphene at all distinct symmetry points in the BZ [98].

	Monolayer	N even	N odd
Γ	$\Gamma_2^- + \Gamma_5^- + \Gamma_4^+ + \Gamma_6^+$	$N(\Gamma_1^+ + \Gamma_3^+ + \Gamma_2^- + \Gamma_3^-)$	$(N-1)\Gamma_1^+ + (N+1)\Gamma_2^- + (N+1)\Gamma_3^+ + (N-1)\Gamma_3^-$
K	$K_1^+ + K_2^+ + K_3^+ + K_3^-$	$N(K_1 + K_2 + 2K_3)$	$NK_1^+ + NK_1^- + [f(N)+2]K_2^+ + [f(N-2)]K_2^{+*}$ $+NK_2^- + (N-1)K_2^{-*\text{a}}$
M	$M_1^+ + M_2^+ + M_3^+ + M_2^- + M_3^- + M_4^-$	$N(2M_1^+ + M_2^+ + M_1^- + 2M_2^-)$	$2NM_1 + (N-1)M_2 + (N+1)M_3 + 2NM_4$
$T\,(T')$	$2T_1 + T_2 + 2T_3 + T_4$	$3N(T_1 + T_2)$	$(3N+1)10T^+ + (3N-1)T^-$
Σ	$2\Sigma_1 + 2\Sigma_3 + 2\Sigma_4$	$N(4\Sigma_1 + 2\Sigma_2)$	$2N\Sigma_1 + (N-1)\Sigma_2 + (N+1)\Sigma_3 + 2N\Sigma_4$
u	$4u^+ + 2u^-$	$6Nu$	$(3N+1)u^+ + (3N-1)u^-$

a Where $f(N) = \sum_{m=0}^{\infty}[\Theta(N-4m-2) + 3\Theta(N-4m-4)]$, in which $\Theta(x)$ is equal to 0 if $x < 0$, and equal to 1 otherwise.

Table 6.8 The Γ_π wave vector point-group irreducible representations for mono- and N-layer graphene at all high symmetry points in the BZ [98].

	Monolayer	N even	N odd
Γ	$\Gamma_2^- + \Gamma_4^+$	$N(\Gamma_1^+ + \Gamma_2^-)$	$(N-1)\Gamma_1^+ + (N+1)\Gamma_2^-$
$K\,(K')$	K_3^-	$\frac{N}{2}(K_1 + K_2 + K_3)$	$(\frac{N-1}{2})K_1^+ + (\frac{N+1}{2})K_1^- + g(N)K_2^{+*}(K_2^+) + g(N-2)(K_2^+)(K_2^{+*}) + g(N)K_2^- + g(N+2)K_2^{-*}$ [a]
M	$M_3^+ + M_2^-$	$N(M_1^+ + M_2^-)$	$(N-1)M_1 + (N+1)M_4$
$T\,(T')$	$T_2 + T_4$	$N(T_1 + T_2)$	$(N-1)T^+ + (N+1)T^-$
Σ	$2\Sigma_4$	$2N\Sigma_1$	$(N-1)\Sigma_1 + (N+1)\Sigma_4$
u	$2u^-$	$2Nu$	$(N-1)u^+ + (N+1)u^-$

a Where $g(N) = \sum_{m=0}^{\infty} \Theta(N-4m-2)$, in which $\Theta(x)$ is equal to 0 if $x < 0$ and equal 1 otherwise.

Trilayer graphene with ABA Bernal stacking belongs to the D_{3h} point group, and Figure 6.3c shows its electronic dispersion. The group of the wave vector for the K point of trilayer graphene is isomorphic to the point group C_{3h}. In Tables 6.7 and 6.8, K_2^+ and K_2^{+*} are the two one-dimensional representations comprising the K_2^+ representation, and $*$ denotes the complex conjugate. The same applies to the K_2^- representation. The irreducible representations for the electronic bands are given by $\Gamma_\pi^K = K_1^+ + 2K_1^- + K_2^{+*} + K_2^- + K_2^{-*}$ for the K point and $\Gamma_\pi^{K'} = K_1^+ + 2K_1^- + K_2^+ + K_2^{-*} + K_2^-$ for the K$'$ point. Although time reversal symmetry can imply degeneracy between the complex conjugate representations for cyclic groups, in graphene the complex conjugation operation also takes K into the K' point and vice versa. Consequently, there are no degenerate bands at the K (K') point, in agreement with *tight-binding* calculations. These calculations include the γ_2 and γ_5 next-nearest layer coupling parameters in describing $E(k)$ for graphite [97, 221], which are necessary to describe the Fermi surface for graphite. An energy gap at the K point is obtained for both DFT calculations and *ab initio* calculations (see the inset of Figure 6.3c and more details in [98]).

6.3.3
Selection Rules for the Electron–Photon Interaction

In this section we discuss the selection rules for the electron–photon interaction in the dipole approximation, with emphasis given to the high symmetry lines T and T' in the electronic dispersion (see Figure 6.2c). These high symmetry T and T' lines along the ΓK and KM directions in reciprocal space, respectively, are important for Raman spectroscopy.[23] In fact, the light absorption up to 3 eV occurs mostly along the T, T' lines but there is also some absorption at general u points near the K (K') point in the case of graphene.

In Table 6.9, we show the direct products for three irreducible representations for the final state, the electron–photon perturbation, and the initial state $\Gamma_f \otimes \Gamma_p \otimes \Gamma_i$ in the column of the absorption matrix element $W(k)$. Knowing the symmetry of the initial and final states, and the irreducible representation that generates the basis function of the light polarization vector (x, y or z), group theory can be used to calculate whether $W(k)$ is zero or not by only using the character table. The results are summarized in Table 6.9 considering the graphene layers to be in the (x, y) plane and the light propagating along the z direction.

It is important to highlight some results given in Table 6.9 where the symbol $x \in T_3$ tells us that x belongs to the IR for T_3. The column for $W(k)$ gives the selection rules for the electron–photon interaction in terms of the direct products of the relevant irreducible representations. Along the T line, absorption by visible light has to couple T_2 and T_4 π electron symmetries which have the same basis functions of x and y as for monolayer graphene (see Figure 6.3a). For the T line

23) In the Raman spectroscopy of carbon nanotubes, the Raman signal or intensity depends on the corresponding one-dimensional Brillouin zone lying along the ΓK or KM directions. This gives a so-called type I and II dependence for semiconducting SWNTs, respectively, for Raman spectroscopy [222].

Table 6.9 Selection rules for the electron–photon interaction $W(k)$ with $\hat{x}$ and $\hat{y}$ light polarization in monolayer, bilayer and trilayer graphene (see Figure 6.2c for definitions of $\hat{x}$ and $\hat{y}$). For N even and N odd, the selection rules are the same as for bilayer and trilayer graphene, respectively. T_1 to T_4 are IRs for the GWV for the T point [98]. Here $x \in T_3$ means x transforms according to the IR T_3. For notation see Section 6.3.6.

	BZ point	Polarization	W(k)
Monolayer	T	$x \in T_3$	$T_2 \otimes T_3 \otimes T_4$ nonzero
		$y \in T_1$	$T_2 \otimes T_1 \otimes T_4$ zero
	u	$x, y \in u^+$	$u^- \otimes u^+ \otimes u^-$ nonzero
Gated	T	$x \in T_2$	$T_1 \otimes T_2 \otimes T_2$ nonzero
monolayer		$y \in T_1$	$T_1 \otimes T_1 \otimes T_2$ zero
	u	$x, y \in u$	$u \otimes u \otimes u$ nonzero
Bilayer	T	$x \in T_2$	$T_1 \otimes T_2 \otimes T_1$ zero
(N-even)			$T_1 \otimes T_2 \otimes T_2$ nonzero
			$T_2 \otimes T_2 \otimes T_2$ zero
		$y \in T_1$	$T_1 \otimes T_1 \otimes T_1$ nonzero
			$T_1 \otimes T_1 \otimes T_2$ zero
			$T_2 \otimes T_1 \otimes T_2$ nonzero
	u	$x, y \in u$	$u \otimes u \otimes u$ nonzero
Biased	T	$x, y \in T$	$T \otimes T \otimes T$ nonzero
bilayer			
	u	$x, y \in u$	$u \otimes u \otimes u$ nonzero
Trilayer	T	$x, y \in T^+$	$T^+ \otimes T^+ \otimes T^+$ nonzero
(N-odd)			$T^+ \otimes T^+ \otimes T^-$ zero
			$T^- \otimes T^+ \otimes T^-$ nonzero
	u	$x, y \in u^+$	$u^+ \otimes u^+ \otimes u^+$ nonzero
			$u^+ \otimes u^+ \otimes u^-$ zero
			$u^- \otimes u^+ \otimes u^-$ nonzero

direction along $\hat{y}$, the only allowed absorption is for light polarized along the $\hat{x}$ direction (T_3). For incident light polarization along the $\hat{y}$ direction (T_1), no absorption will occur along the KΓ direction along the k_y axis, giving rise to an optical absorption anisotropy for monolayer graphene [83, 220].

Bilayer graphene contains four electronic bands along the T line, belonging to two T_1 and two T_2 irreducible representations. The four possible transitions are illustrated in Figure 6.4 (a,b). In this case, both x and y polarized light can be absorbed. Trilayer graphene will have more possibilities for light-induced transitions, since there are more possibilities between the three π and three π^*-bands. Along the T (T') direction, there are two T^+ and four T^--bands for trilayer graphene, giving rise to five possible transitions (see Table 6.9), as shown in Figure 6.4c.

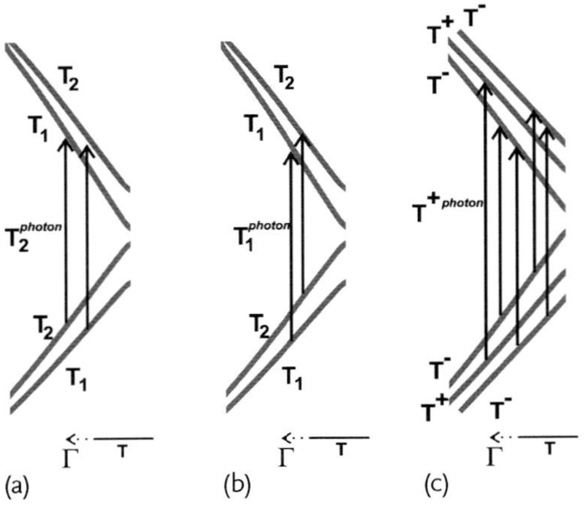

Figure 6.4 (a,b) Schematic electron dispersion of bilayer graphene along the KΓ direction showing the possible transition induced by (a) a photon with T_2 symmetry (x polarization) and (b) a T_1 photon (y polarization). (c) The electronic dispersion of trilayer graphene showing the five possible transitions by light absorption [98].

6.3.4
Selection Rules for First-Order Raman Scattering

The first-order Raman scattering process is limited to phonons at the center of the BZ (Γ point) due to the momentum conservation requirement (phonon wave vector $q = 0$). In monolayer graphene, although there are three optical modes at the Γ point, two (the LO and iTO) are degenerate and one (the oTO) is not Raman-active. The first-order Raman spectra is therefore composed of the G-band vibrational mode, which is doubly degenerate at the Γ point with Γ_6^+ (or E_{2g}) symmetry.[24] The Raman-active modes of N graphene layers which depend on N ($N > 1$) (without the acoustic modes) are:

$$\Gamma^{\text{Raman}} = N(\Gamma_3^+ + \Gamma_1^+), \text{ for } N \text{ even}$$
$$\Gamma^{\text{Raman}} = N\Gamma_3^+ + (N-1)(\Gamma_3^- + \Gamma_1^+), \text{ for } N \text{ odd.} \quad (6.20)$$

For an even number of graphene layers, the G-band belongs to the Γ_3^+irreducible representation. There is a low frequency Γ_3^+ mode with a frequency depending on the number of layers (35–53 cm^{-1}) [223]. Two new Raman-active modes near $\sim$ 80 cm^{-1} and $\sim$ 900 cm^{-1} appear belonging to the Γ_1^+ irreducible representations [223, 224]. For an odd number of graphene layers, the G-band is assigned to a combination of Γ_3^+ and Γ_3^- representations, and also the lower wavenumber component is Raman-active by a Γ_1^+ representation.

24) Although the space group notation, where the IRs are labeled by the BZ point label, is more complete, it is common in the Raman spectroscopy literature to use the notation from the isomorphic point groups, since only the Γ point ($q \sim 0$) is usually relevant. For the D_{6h} point group notation, Γ_6^+ corresponds to the E_{2g} IR. Information about the space group to point group notation conversion is found in Section 6.3.6.

6.3.5
Electron Scattering by $q \neq 0$ Phonons

The electron–phonon (el–ph) interaction is calculated from the coupling of the initial and final electron wave functions to the phonon eigenvector [8, 203] using a phonon-induced deformation potential. Therefore, the selection rules for the el–ph processes are obtained by taking the direct product of the symmetries of the initial and final electronic states and the symmetry of the phonon involved in the el–ph process. The allowed el–ph scattering processes for monolayer, gated monolayer, bilayer, biased bilayer and trilayer graphene along the $K\Gamma$ and KM directions (T and T' lines, respectively) and at a generic u point are summarized in Table 6.10.

6.3.6
Notation Conversion from Space Group to Point Group Irreducible Representations

Here we derive the Γ_π and $\Gamma_{\text{lat. mode}}$ representations for the electrons and phonons for all points in the first BZ of multi-layer graphene maintaining the notation of the space group (SG) for the irreducible representations. The conversion to point group (PG) representations is obtained by considering that (a) the superscript sign

Table 6.10 Allowed processes for electron–phonon scattering for monolayer, bilayer and trilayer graphene along the T and T' lines and at a generic u point for each phonon symmetry. For N even and N odd graphenes, the selection rules are the same as for bilayer and trilayer graphene, respectively. The table also includes entries for a gated monolayer and a biased bilayer [98].

	BZ point	Phonon	Allowed scattering
Monolayer	$T\ (T')$	T_1	$T_2 \to T_2$, $T_4 \to T_4$
		T_3	$T_2 \to T_4$
	u	u^+	$u^- \to u^-$
Gated monolayer	$T\ (T')$	T_1	$T_1 \to T_1$, $T_2 \to T_2$
		T_2	$T_1 \to T_2$
	u	u	$u \to u$
Bilayer (N-even)	$T\ (T')$	T_1	$T_1 \to T_1$, $T_2 \to T_2$
		T_2	$T_1 \to T_2$
	u	u	$u \to u$
Biased bilayer	$T\ (T')$	T	$T \to T$
	u	u	$u \to u$
Trilayer (N-odd)	$T\ (T')$	T^+	$T^+ \to T^+$, $T^- \to T^-$
		T^-	$T^+ \to T^-$
	u	u^+	$u^+ \to u^+$, $u^- \to u^-$
		u^-	$u^+ \to u^-$

Table 6.11 Example of the irreducible representation notation conversion from the Γ point space group (SG) to the D_{3h} and D_{3d} point groups (PG), and from the *K* point space group (SG) to the C_{3h} and D_3 point groups (PG) [98].

Γ point				*K* point			
D_{3h}		D_{3d}		C_{3h}		D_3	
SG	PG	SG	PG	SG	PG	SG	PG
Γ_1^+	A'_1	Γ_1^+	A_{1g}	$K_1{}^+$	A'	K_1	A_1
Γ_1^-	A''_1	Γ_1^-	A_{1u}	$K_1{}^-$	A''	K_2	A_2
Γ_2^+	A'_2	Γ_2^+	A_{2g}	$K_2{}^+$	E'	K_3	E
Γ_2^-	A''_2	Γ_2^-	A_{2u}	$K_2{}^{+*}$	E'^*		
Γ_3^+	E'	Γ_3^+	E_g	$K_2{}^-$	E''		
Γ_3^-	E''	Γ_3^-	E_u	$K_2{}^{-*}$	E''^*		

"+" or "−" applies if the character of the horizontal mirror plane (σ_h) or the inversion operation (i) is positive or negative, respectively; (b) the subscript number is given following the order of the point group irreducible representations; (c) two representations can only have the same subscript number if they both have superscripts with positive or negative signs. As an example, we give in Table 6.11 the Γ point space group notation conversion to the D_{3h} (N-odd) and D_{3d} (N-even) point groups and for the K point space group to the C_{3h} (N-odd) and D_3 (N-even) point groups.

6.4
Symmetry Aspects of Carbon Nanotubes

The nanotube physical properties depend on how the graphene sheet is rolled up, and from a symmetry point of view, two types of nanotubes can be formed, namely the symmorphic achiral armchair or zigzag tubes, as shown in Figure 6.5a,b, respectively, and the nonsymmorphic chiral tubes, shown in Figure 6.5c.[25] Simply put, for the symmorphic groups, the translations and rotations can be decoupled, while for nonsymmorphic the rotations also contain translations, such as screw rotations along the tube axis [94]. We note in Figure 6.5 that each nanotube has a cap at either end of the nanotube. Because of the small diameter of a carbon nanotube (~1 nm) and the large length-to-diameter ratio ($> 10^4$), it is assumed that the nanotube length is much larger than its diameter, so that the nanotube ends (see Figure 6.5) can be neglected when discussing the electronic and lattice properties of the nanotubes. Thus from a symmetry standpoint, a carbon nanotube is a one-dimensional crystal with a translation vector $\boldsymbol{T}$ along the cylinder axis and

25) If we take the smallest unit cell consisting of two carbon atoms for achiral nanotubes, we should consider the screw rotation. In general, the symmetry selection rule depends on the shape of the unit cell. The smallest unit cell is not always the best for understanding the relevant physics.

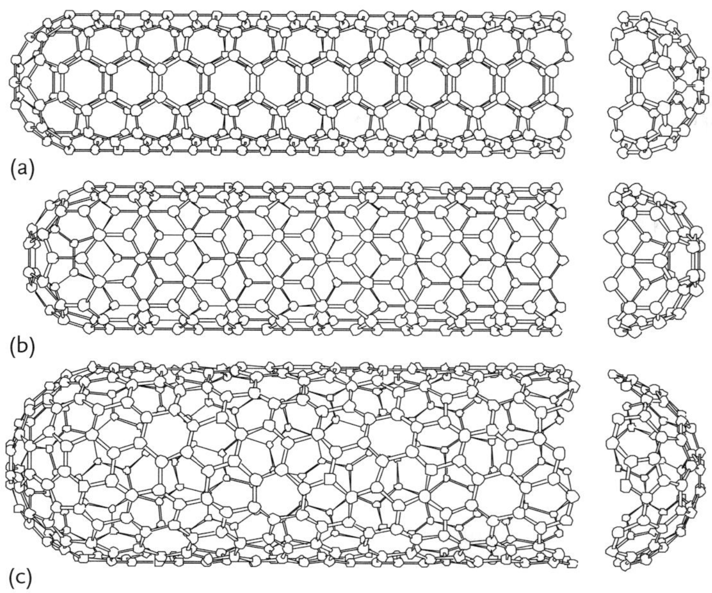

Figure 6.5 Schematic theoretical model for the three different types of single-wall carbon nanotubes: (a) the "armchair" nanotube, (b) the "zigzag" nanotube, and (c) the "chiral" nanotube. The actual nanotubes shown in the figure correspond to (n, m) values of (a) (5, 5), (b) (9, 0), and (c) (10, 5). See text for more information [31].

a small number of carbon hexagons associated with the circumferential direction. The nanotube structure can be considered as a folded graphene defined by the (n, m) indices, as discussed in Section 2.3.1. In this section we give more details that are important for the description of the eigenvectors and selection rules for nanotubes.

6.4.1
Compound Operations and Tube Chirality

All multiples of the translation vector $\boldsymbol{T}$ will be a translational symmetry operation of the nanotube [225]. However, to be more general, it is necessary to consider that any lattice vector

$$\boldsymbol{t}_{p,q} = p\boldsymbol{a}_1 + q\boldsymbol{a}_2, \tag{6.21}$$

with p and q integers, of the unfolded graphene layer will also be a symmetry operation of the nanotube. In fact, the symmetry operation that arises from $\boldsymbol{t}_{p,q}$ will appear as a screw translation of the nanotube. Screw translations are combinations of a rotation by an angle ϕ (R_ϕ) and a translation $\boldsymbol{\tau}$ in the axial direction of the nanotube. The screw translation can be written as $\{R_\phi|\tau\}$, using a notation common for space group operations [31, 94].

The translation vector from $\boldsymbol{t}_{p,q}$ can also be written in terms of components of the nanotube lattice vectors, $\boldsymbol{T}$ and $\boldsymbol{C}_h$, as

$$\boldsymbol{t}_{p,q} = \boldsymbol{t}_{u,v} = \frac{1}{N}\left(u\boldsymbol{C}_h + v\boldsymbol{T}\right), \tag{6.22}$$

where u and v are given by:

$$u = \frac{(2n+m)p + (2m+n)q}{d_R} \tag{6.23}$$

and

$$v = mp - nq\,. \tag{6.24}$$

N and d_R have been defined in Section 2.3.1 (see Table 2.1). Both u and v in Eqs. (6.23) and (6.24), respectively, are integer numbers which can assume either negative or positive values.

The screw translation of the nanotube $\boldsymbol{t}_{u,v}$ which is associated with the graphene lattice vectors can then be written using the space group notation as:

$$\boldsymbol{t}_{u,v} = \{C_N^u | vT/N\}\,, \tag{6.25}$$

where C_N^u is a rotation of u by $(2\pi/N)$ around the nanotube axis, and $\{E|vT/N\}$ is a translation of vT/N along the nanotube axis, with T being the magnitude of the primitive translation vector $\boldsymbol{T}$ along the tube axis. It is clear that if a screw vector $\{C_N^u|vT/N\}$ is a symmetry operation of the nanotube, then the vectors $\{C_N^u|vT/N\}^s$, for any integer value of s, are also symmetry operations of the nanotube. The number of hexagons in the unit cell N assumes the role of the "order" of the screw axis, since the symmetry operation $\{C_N^u|vT/N\}^N = \{E|vT\}$, where E is the identity operator, and $v\boldsymbol{T}$ is a pure translation of the nanotube.

The nanotube structure can be obtained from a small number of atoms (between 2 and $2N$) by using any choice of two independent noncolinear screw vectors, such as $\{C_N^{u_1}|v_1T/N\}$ and $\{C_N^{u_2}|v_2T/N\}$. Here noncolinear vectors are defined such that there does not exists a pair of integers s and l except for 1, which satisfy $lu_1 = su_2 + \lambda N$, and $lv_1 = sv_2 + \gamma T$, where λ and γ are two arbitrary integers [135].

When $\boldsymbol{T}$ is specified, a screw vector $C_N^u|0$ for p and q which satisfies $v = mp - nq = 1$ for an (n, m) nanotube, generates N carbon atoms in the 1D unit cell. This screw vector is called a *symmetry vector* R [31]. To better illustrate the action of the symmetry vector $\boldsymbol{R}$, we show in Figure 6.6 a diagram of the screw vector applied to the $(4, 2)$ nanotube. The dark atoms in the bottom represent a two-atom motif. We also show in Figure 6.6 another set of dark atoms which is equivalent to this motif due to a rotation of $2\pi/d$, with $d = 2$, around the nanotube axis. The dark gray helix of atoms is composed of the atoms in the nanotube unit cell which can be obtained by consecutive applications of the screw vector $\boldsymbol{R}$ to the atoms in the motif, while the other atoms are obtained by successive operations of the screw vector $\boldsymbol{R}$ followed by a pure translation which brings the motif back to the original unit cell [135].

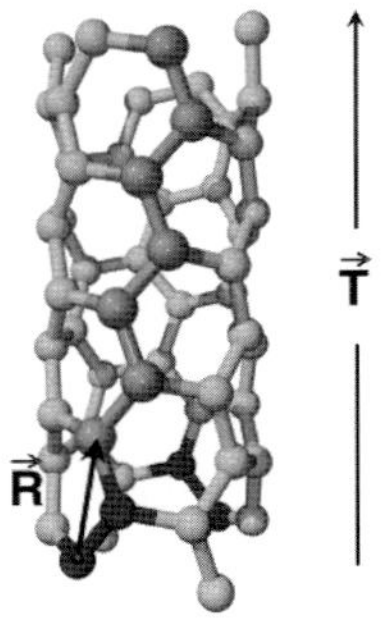

Figure 6.6 Unit cell of the (4,2) nanotube, with its 28 atoms. The dark gray atoms can be directly obtained by the application of the R vector, while the other atoms can be obtained from the latter by applying other symmetry operations, such as the translation vector T [135].

6.4.2
Symmetries for Carbon Nanotubes

The chiral nanotube symmetry operations can be separated into two sets [135]. The first set, which we shall call the symmorphic set, is formed by the translation operations of the nanotube and the point group operations. This symmorphic set forms a sub-group of the total space-group of the nanotube, and thus it can be used to obtain some of the symmetry-related properties. To obtain the point group of the nanotube, the nanotube can be rotated by an angle $2\pi/d$ (called the C_d operation) without changing its structure where $d = \text{gcm}(n, m)$ and gcm denotes the greatest common multiplier. Therefore, the C_d operation is a point group operation of the nanotube. Also, by choosing an axis perpendicular to the nanotube axis, the rotation by π around this axis (C_2' or C_2'') will also be a symmetry operation of the nanotube, as shown in Figure 6.7a. There are two different classes of rotations perpendicular to the nanotube axis (C_2' and C_2''). For one of the classes (C_2'), the axis goes through the center of bonds between two equivalent atoms (shown in Figure 6.7a). For the other class (C_2''), the axis goes through the centers of the hexagons.[26] The point group of the nanotube is thus obtained as the axial point group D_d.[27] The second set of symmetry operations, which we shall refer to as the nonsymmorphic set, is formed by the compound operations of the space group of the nanotube, which cannot be decomposed into pure translations of multiples of T and point group operations. All the screw vectors $t_{u,v}$, with the exception of multiples of T and C_h/d, are part of this set of operations [135].

Both armchair (n, n) and zigzag $(n, 0)$ carbon nanotubes exhibit all the symmetry operations that were observed for chiral nanotubes, namely the screw axes $\{C_N^u | vT/N\}^s$, where $N = 2n$, the rotation around the nanotube axis C_d, where $d = n$, and the rotations perpendicular to the nanotube axis C_2' and C_2''. However, achiral nanotubes also exhibit other symmetry operations, such as inversion centers as well as mirror planes and glide planes. The horizontal mirror plane σ_h and one of the vertical mirror planes σ_v are shown in Figure 6.7b,c, respectively. There

26) The C_2'' axis of the nanotube corresponds to the C_6 axis in the case of flat graphene.

27) The label D_d means that the group has one C_d rotational operation about the z axis and $2d$ C_2 rotational operations in the xy plane.

is also an inversion center at the intersection of the σ_h plane and the nanotube axis. The glide planes are represented by $\{\sigma_v|T/2\}$.

For chiral tubes, at the Γ point the GWV exhibits the symmetries of the D_N point group [135], for which the character table is shown in Table 6.12. The symmetry properties of general nanotube wave vectors $0 < k < \pi/T$ can be fully described by using the C_N group. In Table 6.13, we show the character table for the irreducible representations of the C_N point group. There are $[(N/2) - 1]$ representations which are doubly degenerate due to time reversal symmetry. For $k = \pi/T$ and $k = -\pi/T$, which can be translated into each other by a reciprocal lattice vector $K_2 = 2\pi/T$, the group of the wave vector is also isomorphic to D_N.

For achiral tubes, at the Γ point the GWV is isomorphic to D_{2nh}. The character table for group D_{2nh} is shown in Table 6.14, where the C_{2n} classes correspond to the screw vectors of the nanotube, while the σ'_v and σ''_v classes correspond, respectively, to mirror and glide planes containing the nanotube axis [135]. For $0 < k < \pi/T$ the only symmetry operations which maintain k invariant are the screw vectors and the mirror and glide planes which contain the nanotube axis (σ'_v and σ''_v). The GWV will then be isomorphic to the C_{2nv} point group, for which the character table is shown in Table 6.15. In the case of the (3,3) nanotube (see Figure 6.7b,c), the group of the wave vector at a general point $0 < k < \pi/T$ is isomorphic to the C_{6v} point group, while at $k = 0$ and $k = \pi/T$ the group of the wave vector is isomorphic to the D_{6h} point group [94, 135].

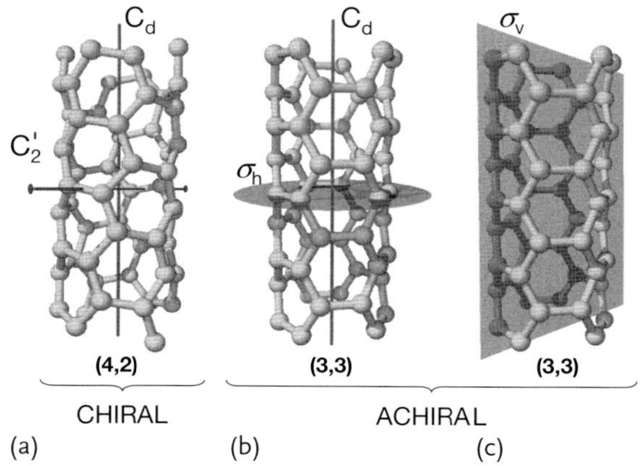

Figure 6.7 (a) Unit cell of the chiral (4,2) nanotube, showing the C_d rotation around the nanotube axis with $d = 2$, and one of the C'_2 rotations perpendicular to the tube axis. A different class of rotations (C''_2), which is also present in chiral and achiral nanotubes, is not shown here. (b) A section of an achiral armchair (3,3) nanotube is shown along with the horizontal mirror plane σ_h and the symmetry operation C_d with $d = 3$. (c) The same (3,3) armchair nanotube is shown along with one of the vertical mirror planes σ_v [135].

Table 6.12 Character table for the group of the wave vectors $k = 0$ and $k = \pi/T$ for chiral tubes. This group is isomorphic to the point group D_N.

D_N	$\{E\|0\}$	$2\{C_N^u\|\nu T/N\}$	$2\{C_N^u\|\nu T/N\}^2$	$\ldots$	$2\{C_N^u\|\nu T/N\}^{(N/2)-1}$	$\{C_N^u\|\nu T/N\}^{N/2}$	$(N/2)\{C_2'\|0\}$	$(N/2)\{C_2''\|0\}$
A_1	1	1	1	$\ldots$	1	1	1	1
A_2	1	1	1	$\ldots$	1	1	−1	−1
B_1	1	−1	1	$\ldots$	$(-1)^{(N/2-1)}$	$(-1)^{N/2}$	1	−1
B_2	1	−1	1	$\ldots$	$(-1)^{(N/2-1)}$	$(-1)^{N/2}$	−1	1
E_1	2	$2\cos 2\pi/N$	$2\cos 4\pi/N$	$\ldots$	$2\cos 2(N/2-1)\pi/N$	−2	0	0
E_2	2	$2\cos 4\pi/N$	$2\cos 8\pi/N$	$\ldots$	$2\cos 4(N/2-1)\pi/N$	2	0	0
$\vdots$	$\vdots$	$\vdots$	$\vdots$	$\vdots$	$\vdots$	$\vdots$	$\vdots$	$\vdots$
$E_{(N/2-1)}$	2	$2\cos 2(N/2-1)\pi/N$	$2\cos 4(N/2-1)\pi/N$	$\ldots$	$2\cos 2(N/2-1)^2\pi/N$	$2\cos(N/2-1)\pi$	0	0

Table 6.13 Character table for the group of the general wave vector $0 < k < \pi/T$ for chiral nanotubes. This group is isomorphic to the point group C_N. The $\pm$ signs label the different 1D irreducible representations ($\mathbb{E}$) with characters which are complex conjugates of each other[a]. These representations are degenerate due to time reversal symmetry [135].

C_N	$\{E\|0\}$	$\{C_N^u\|\nu T/N\}^1$	$\{C_N^u\|\nu T/N\}^2$	$\cdots$	$\{C_N^u\|\nu T/N\}^\ell$	$\cdots$	$\{C_N^u\|\nu T/N\}^{N-1}$
A	1	1	1	$\cdots$	1	$\cdots$	1
B	1	-1	1	$\cdots$	$(-1)^\ell$	$\cdots$	-1
$\mathbb{E}_{\pm 1}$	1	ϵ	ϵ^2	$\cdots$	ϵ^ℓ	$\cdots$	ϵ^{N-1}
	1	ϵ^*	ϵ^{*2}		$\epsilon^{*\ell}$		$\epsilon^{*(N-1)}$
$\mathbb{E}_{\pm 2}$	1	ϵ^2	ϵ^4	$\cdots$	$\epsilon^{2\ell}$	$\cdots$	$\epsilon^{2(N-1)}$
	1	ϵ^{*2}	ϵ^{*4}		$\epsilon^{*2\ell}$		$\epsilon^{*2(N-1)}$
$\vdots$	$\vdots$	$\vdots$	$\vdots$	$\vdots$	$\vdots$	$\vdots$	$\vdots$
$\mathbb{E}_{\pm(\frac{N}{2}-1)}$	1	$\epsilon^{\frac{N}{2}-1}$	$\epsilon^{2(\frac{N}{2}-1)}$	$\cdots$	$\epsilon^{\ell(\frac{N}{2}-1)}$	$\cdots$	$\epsilon^{(N-1)(\frac{N}{2}-1)}$
	1	$\epsilon^{*\frac{N}{2}-1}$	$\epsilon^{*2(\frac{N}{2}-1)}$		$\epsilon^{*\ell(\frac{N}{2}-1)}$		$\epsilon^{*(N-1)(\frac{N}{2}-1)}$

a The complex number ϵ in this table denotes $e^{i2\pi/N}$.

Table 6.14 Character table for the group of the wave vectors $k = 0$ and $k = \pi/T$ for achiral nanotubes [135]. This group is isomorphic to the point group D_{2nh}.[a]

D_{2nh}	$\{E\|0\}$	$\cdots$	$2\{C_{2n}^{u}\|\nu T/2n\}^s$	$\cdots$	$\{C_{2n}^{u}\|\nu T/2n\}^n$	$n\{C_2'\|0\}$	$n\{C_2''\|0\}$	$\{I\|0\}$	$\cdots$	$2\{IC_{2n}^{u}\|\nu T/2n\}^s$	$\cdots$	$\{\sigma_h\|0\}$	$n\{\sigma_v'\|0\}$	$n\{\sigma_v''\|T/2\}$
A_{1g}	1	$\cdots$	1	$\cdots$	1	1	1	1	$\cdots$	1	$\cdots$	1	1	1
A_{2g}	1	$\cdots$	1	$\cdots$	1	−1	−1	1	$\cdots$	1	$\cdots$	1	−1	−1
B_{1g}	1	$\cdots$	$(-1)^s$	$\cdots$	−1	1	−1	1	$\cdots$	$(-1)^s$	$\cdots$	−1	1	−1
B_{2g}	1	$\cdots$	$(-1)^s$	$\cdots$	−1	−1	1	1	$\cdots$	$(-1)^s$	$\cdots$	−1	−1	1
⋮	⋮	⋮	⋮	⋮	⋮	⋮	⋮	⋮	⋮	⋮	⋮	⋮	⋮	
$E_{\mu g}$	2	$\cdots$	$2\cos(\mu s\pi/n)$	$\cdots$	$2(-1)^\mu$	0	0	2	$\cdots$	$2\cos(\mu s\pi/n)$	$\cdots$	$2(-1)^\mu$	0	0
⋮	⋮	⋮	⋮	⋮	⋮	⋮	⋮	⋮	⋮	⋮	⋮	⋮	⋮	⋮
A_{1u}	1	$\cdots$	1	$\cdots$	1	1	1	−1	$\cdots$	−1	$\cdots$	−1	−1	−1
A_{2u}	1	$\cdots$	1	$\cdots$	1	−1	−1	−1	$\cdots$	−1	$\cdots$	−1	1	1
B_{1u}	1	$\cdots$	$(-1)^s$	$\cdots$	−1	1	−1	−1	$\cdots$	$-(-1)^s$	$\cdots$	1	−1	1
B_{2u}	1	$\cdots$	$(-1)^s$	$\cdots$	−1	−1	1	−1	$\cdots$	$-(-1)^s$	$\cdots$	1	1	−1
⋮	⋮	⋮	⋮	⋮	⋮	⋮	⋮	⋮	⋮	⋮	⋮			
$E_{\mu u}$	2	$\cdots$	$2\cos(\mu s\pi/n)$	$\cdots$	$2(-1)^\mu$	0	0	−2	$\cdots$	$-2\cos(\mu s\pi/n)$	$\cdots$	$-2(-1)^\mu$	0	0
⋮	⋮	⋮	⋮	⋮	⋮	⋮	⋮	⋮	⋮	⋮	⋮	⋮	⋮	⋮

a The values of s and μ span the integer values between 1 and $n-1$.

Table 6.15 Character table for the group of the wave vectors $0 < k < \pi/T$ for achiral nanotubes [135]. This group is isomorphic to the point group $C_{2n\nu}$.

$C_{2n\nu}$	$\{E\|0\}$	$2\{C^u_{2n}\|\nu T/2n\}^1$	$\{C^u_{2n}\|\nu T/2n\}^2$	...	$2\{C^u_{2n}\|\nu T/2n\}^{n-1}$	$\{C^u_{2n}\|\nu T/2n\}^n$	$n\{\sigma'_v\|0\}$	$n\{\sigma''_v\|T/2\}$
A'	1	1	1	...	1	1	1	1
A''	1	1	1	...	1	1	−1	−1
B'	1	−1	1	...	$(-1)^{(n-1)}$	$(-1)^n$	1	−1
B''	1	−1	1	...	$(-1)^{(n-1)}$	$(-1)^n$	−1	1
E_1	2	$2\cos\pi/n$	$2\cos 2\pi/n$	...	$2\cos 2(n-1)\pi/n$	−2	0	0
E_2	2	$2\cos 2\pi/n$	$2\cos 4\pi/n$	...	$2\cos 4(n-1)\pi/n$	2	0	0
⋮	⋮	⋮	⋮	⋮	⋮	⋮	⋮	⋮
$E_{(n-1)}$	2	$2\cos(n-1)\pi/n$	$2\cos 2(n-1)\pi/n$	...	$2\cos(n-1)^2\pi/n$	$2\cos(n-1)\pi$	0	0

Table 6.16 Irreducible representations for the electronic conduction and valence bands of chiral as well as armchair and zigzag achiral nanotubes [135].

		VALENCE		CONDUCTION	
		$k = 0, \pi/T$	$0 < k < \pi/T$	$k = 0, \pi/T$	$0 < k < \pi/T$
CHIRAL	$\mu = 0$	A_1	A	A_2	A
	$0 < \mu < N/2$	E_μ	$\mathbb{E}_{\pm\mu}$	E_μ	$\mathbb{E}_{\pm\mu}$
	$\mu = N/2$	B_1	B	2	B
ARMCHAIR	$\mu = 0$	A_{1g}	A'	A_{2g}	A''
	$0 < \mu < n$	$E_{\mu g}$	E_μ	$E_{\mu u}$	E_μ
	$\mu = n$	B_{1g}	B'	B_{2g}	B''
ZIGZAG	$\mu = 0$	A_{1g}	A'	A_{2u}	A'
	$0 < \mu < n$	$E_{\mu u,\mu g}$ [a]	E_μ	$E_{\mu g,\mu u}$ [a]	E_μ
	$\mu = n$	B_{1g}	B'	B_{2u}	B'

a For zigzag nanotubes, if $\mu < 2n/3$, the valence (conduction) band at $k = 0$ belongs to the $E_{\mu g}$ ($E_{\mu u}$) representation for μ even and to the $E_{\mu u}$ ($E_{\mu g}$) representation for μ odd, while if $\mu > 2n/3$ it is the opposite irreducible representations that apply.

6.4.3
Electrons in Carbon Nanotubes

Having shown the irreducible representations of the wave vector k, it is now possible to obtain the symmetries of the eigenvectors used to describe the electronic and vibrational properties for all the points of the first Brillouin zone. The irreducible representations of the electronic states of chiral nanotubes and achiral nanotubes are summarized in Table 6.16. In general, what defines the symmetry of wavefunctions in the quasi-one-dimensional carbon nanotubes are the number of nodes for the wavefunction phase along the tube circumference. The A modes are totally symmetric, while the E_μ modes exhibit 2μ nodes along the tube circumference, as depicted in Figure 6.8.

6.4.4
Phonons in Carbon Nanotubes

The phonon symmetries also obey the general picture displayed in Figure 6.8. For $k = 0$ phonons in achiral nanotubes (D_{2nh} group), $\Gamma_{\text{vec}} = A_{2u} + E_{1u}$ (z and x, y). The $\Gamma^{\text{a.s.}}$ for zigzag SWNTs is [134, 135]:

$$\Gamma^{\text{a.s.}}_{\text{zigzag}} = A_{1g} + B_{2g} + A_{2u} + B_{1u} + \sum_{j=1}^{n-1} (E_{jg} + E_{ju}) \,, \tag{6.26}$$

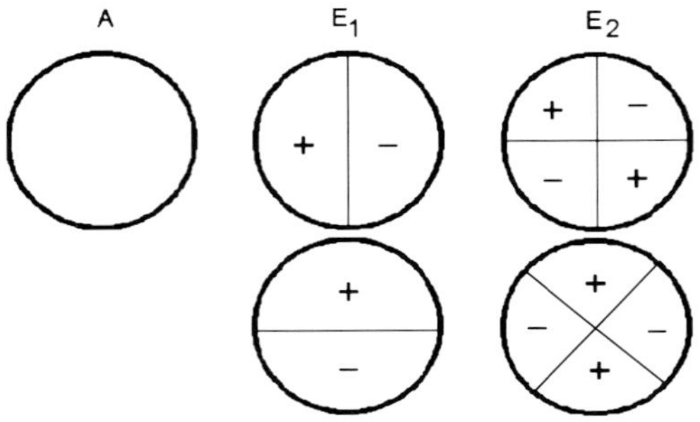

Figure 6.8 Schematics showing the phase change for SWNT wavefunctions with the one-dimensional totally symmetric A IR, and the two doubly degenerate E_1 and E_2 IRs.

giving rise to the following irreducible representations for the lattice modes [134, 135]:

$$\Gamma_{\text{zigzag}}^{\text{lat. mode}} = 2A_{1g} + A_{2g} + B_{1g} + 2B_{2g} + A_{1u} + 2A_{2u} + 2B_{1u} + B_{2u} + \sum_{j=1}^{n-1}(3E_{jg} + 3E_{ju}). \tag{6.27}$$

Finding Γ_{vec}, $\Gamma^{\text{a.s.}}$ and $\Gamma^{\text{lat. mode}}$ for armchair and chiral tubes is left as a problem for the reader.

6.4.5
Selection Rules for First-Order Raman Scattering

The optical activity of phonons in a first-order Raman scattering process is easily obtained from the basis functions in the character tables related to the irreducible representations that describe each of the lattice modes. The Raman-active modes are those transforming like symmetric combinations of quadratic functions (xx, yy, zz, xy, yz, zx).

The list of Raman-active modes are given below [134, 135]:

$$\Gamma_{\text{zigzag}}^{\text{Raman}} = 2A_{1g} + 3E_{1g} + 3E_{2g} \rightarrow 8 \text{ modes}, \tag{6.28}$$

$$\Gamma_{\text{armchair}}^{\text{Raman}} = 2A_{1g} + 2E_{1g} + 4E_{2g} \rightarrow 8 \text{ modes}, \tag{6.29}$$

$$\Gamma_{\text{chiral}}^{\text{Raman}} = 3A_1 + 5E_1 + 6E_2 \rightarrow 14 \text{ modes}. \tag{6.30}$$

A more detailed analysis of the Raman-active modes for chiral and achiral nanotubes is provided in [31, 135].

6.4.6
Insights into Selection Rules from Matrix Elements and Zone Folding

To illustrate the usage of the selection rules introduced by the electron–photon and electron–phonon interaction processes, we consider the first-order resonance Raman scattering process in carbon nanotubes. The first-order Raman scattering process involves the following steps: creation of an electron–hole pair, scattering by a phonon, and light emission by an electron–hole recombination process. The Raman signal is greatly enhanced when the electron scatters between VHSs in the valence and conduction band DOS, so that we can consider only the transitions between the two VHSs in the DOS as a first approximation. By utilizing the selection rules introduced above, we come up with the following five cases for allowed first-order resonance Raman scattering processes between the electronic energy VHSs in the valence and conduction bands denoted by $\mathbb{E}_\mu^{(v)}$ and $\mathbb{E}_{\mu'}^{(c)}$ [135, 226]:

$$\begin{array}{lllllllll}
\text{(I)} & \mathbb{E}_\mu^{(v)} & \xrightarrow{Z} & \mathbb{E}_\mu^{(c)} & \xrightarrow{A} & \mathbb{E}_\mu^{(c)} & \xrightarrow{Z} & \mathbb{E}_\mu^{(v)}, \\
\text{(II)} & \mathbb{E}_\mu^{(v)} & \xrightarrow{X} & \mathbb{E}_{\mu\pm1}^{(c)} & \xrightarrow{A} & \mathbb{E}_{\mu\pm1}^{(c)} & \xrightarrow{X} & \mathbb{E}_\mu^{(v)}, \\
\text{(III)} & \mathbb{E}_\mu^{(v)} & \xrightarrow{Z} & \mathbb{E}_\mu^{(c)} & \xrightarrow{\mathbb{E}_1} & \mathbb{E}_{\mu\pm1}^{(c)} & \xrightarrow{X} & \mathbb{E}_\mu^{(v)}, \\
\text{(IV)} & \mathbb{E}_\mu^{(v)} & \xrightarrow{X} & \mathbb{E}_{\mu\pm1}^{(c)} & \xrightarrow{\mathbb{E}_1} & \mathbb{E}_\mu^{(c)} & \xrightarrow{Z} & \mathbb{E}_\mu^{(v)}, \\
\text{(V)} & \mathbb{E}_\mu^{(v)} & \xrightarrow{X} & \mathbb{E}_{\mu\pm1}^{(c)} & \xrightarrow{\mathbb{E}_2} & \mathbb{E}_{\mu\mp1}^{(c)} & \xrightarrow{X} & \mathbb{E}_\mu^{(v)},
\end{array} \tag{6.31}$$

where A, E_1, and E_2 denote the symmetries of the phonon modes at $k = 0$, which are associated with the $\mu = 0, \mu = \pm1$, and $\mu = \pm2$ cutting lines (1D Brillouin zones in 2D k space), respectively. Thus, for a transition to occur between an electron in a state E_{μ_1} and a state E_{μ_2} it is necessary for the phonon which couples the two states to have $E_{\mu_2-\mu_1}$ symmetry. The XZ plane is parallel to the substrate on which the nanotubes lie, the Z axis is directed along the nanotube axis, and the Y axis is directed along the light propagation direction, so that Z and X in Eq. (6.31) stand for the light polarized parallel and perpendicular to the nanotube axis, respectively.

The five processes of Eq. (6.31) result in different polarization configurations for different phonon modes, ZZ and XX for A; ZX and XZ for E_1; and XX for E_2,[28] in perfect agreement with the basis functions predicted by group theory. Also, Eq. (6.31) predicts *different* resonance conditions for *different* phonon modes. While the A and E_1 modes can be observed in resonance for the $\mathbb{E}_\mu^{(v)} \to \mathbb{E}_\mu^{(c)}$ and the $\mathbb{E}_\mu^{(v)} \to \mathbb{E}_{\mu\pm1}^{(c)}$ processes, corresponding to E_{ii} and E_{ij} ($j = i \pm 1$) transitions, respectively, the E_2 modes can only be observed in resonance for the $\mathbb{E}_\mu^{(v)} \to \mathbb{E}_{\mu\pm1}^{(c)}$ process. Experimentally observed Raman scattering spectra do follow these predicted polarization configurations and resonance conditions [226–228].

It is also interesting to discuss how equivalent selection rules can be derived considering momentum conservation in the unfolded two-dimensional graphene-

28) ZX corresponds to the linear polarization directions of the incident (Z) and scattered (X) light.

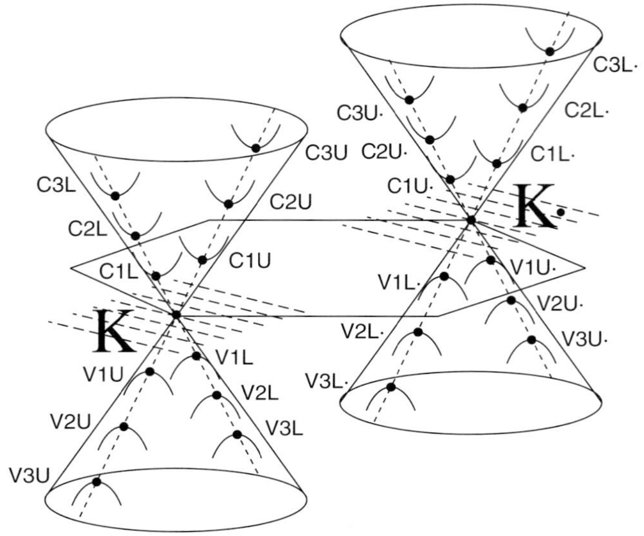

Figure 6.9 The electronic sub-bands for zigzag M-SWNTs in the vicinity of the K and K' points (near the Fermi energy) in the first Brillouin zone. The VHSs are labeled by three symbols: the first denotes the valence or conduction band (C/V), the second denotes the VHS index counted away from the Fermi energy or the cutting line index counted away from the K and K' points, and the third index denotes the lower and upper energy components (L/U) due to the trigonal warping near the $K(K')$ point that distort the cone and splits the energy of the VHSs for metallic SWNTs [229].

sheet, and considering the concepts of cutting lines. This consideration will give insights related to the importance of dimensionality in materials science.

The optical transition in the nanotube is vertical (momentum conserving) within the 1D Brillouin zone, that is, the electronic wave vector along the nanotube axis (along the $\boldsymbol{K}_2$ vector in the unfolded 2D Brillouin zone) does not change. In contrast to the case of the graphene layer, the polarization vector can be either parallel or perpendicular to the nanotube axis for light propagating perpendicular to the substrate on which the nanotubes lie. The dipole selection rules tell us that the optical transition in the nanotube conserves the electronic sub-band index (the cutting line index μ) for light polarized parallel to the nanotube axis. Conservation of both the 1D wave vector and the sub-band index implies conservation of the 2D wave vector in the Brillouin zone of the graphene layer (unfolded Brillouin zone of the nanotube).

As an example, we plot in Figure 6.9 the schematic band diagram of the nanotube in the unfolded 2D Brillouin zone. If the electron starts from the VHS in the valence sub-band V2U (see the solid dot on the sub-band V2U in Figure 6.9), this electron goes to the VHS in the conduction sub-band C2L, and the optical absorption is enhanced substantially because of the extremely high DOS at the VHSs in the valence and conduction sub-bands, V2U and C2L. If an electron in the valence sub-band V2U in Figure 6.9 absorbs a photon polarized perpendicular to the nanotube axis (i. e., polarized along the $\boldsymbol{K}_1$ vector), it can scatter to one of the two conduction sub-bands, either C1L or C3L, depending on the photon frequency and on the interband transition energies $E_{2,1}$ and $E_{2,3}$. This implies a different set of VHSs in the JDOS for perpendicular polarization, $E_{\mu,\mu\pm1}$, and these energies are located between the VHSs in the JDOS for parallel polarization, $E_{\mu\mu}$.

While the optical transition is vertical for the light polarized parallel to the nanotube axis, it involves a wave vector change of $\pm\boldsymbol{K}_1$ (the distance between two adjacent cutting lines) for the perpendicular polarization. This wave vector change can be understood by considering an unrolled nanotube, as shown in Figure 6.10. When the nanotube is unrolled into the graphene layer, the light polarized parallel to the nanotube axis transforms into light polarized parallel to the graphene layer, as shown in Figure 6.10a. This results in a vertical interband optical transition in the unfolded 2D Brillouin zone, which is equivalent to the optical transition within the same sub-band μ in the folded 1D Brillouin zone of the nanotube, as is predicted by the dipole selection rules.

However, perpendicular polarization in nanotubes becomes transformed into the in-plane and out-of-plane polarizations in the unfolded graphene layer, periodically modulated along the direction of the $\boldsymbol{C}_h$ (or $\boldsymbol{K}_1$) vector with the period $|\boldsymbol{C}_h| = \pi d_t$ (nanotube circumference), as shown in Figure 6.10b [110]. The optical transitions induced by the out-of-plane polarization are expected to be much weaker compared to those induced by the in-plane polarization and are usually ignored, because of the much stronger in-plane interaction in the graphene layer [220]. This implies that the light polarization in the unrolled nanotube shown in Figure 6.10b can be considered, as a first approximation, to be parallel to the graphene layer, with an additional phase factor describing oscillations of the in-plane polarization component, arising from the rotation of the polarization vector. The phase factor is given by $\cos(\boldsymbol{k}\cdot\boldsymbol{r})$ where the wave vector $\boldsymbol{k}$ has the direction of $\boldsymbol{K}_1$ and a magnitude of $2\pi/(\pi d_t) = 2/d_t$, that is, $\boldsymbol{k} = \boldsymbol{K}_1$. By assuming wave vector conservation in the unfolded 2D Brillouin zone for the optical transition process, we come up with the selection rules $\boldsymbol{k}_c = \boldsymbol{k}_v \pm \boldsymbol{K}_1$ for light absorption and $\boldsymbol{k}_v = \boldsymbol{k}_c \pm \boldsymbol{K}_1$ for light emission, which correspond to an electronic transition to the adjacent cutting line in the unfolded 2D Brillouin zone, or the electronic transition to the adjacent sub-band in the 1D Brillouin zone of the nanotube. It is interesting to note that the photon wave vector $\pm\boldsymbol{K}_1$ in the unrolled graphene layer is much larger in magnitude than the photon wave vector κ in free space, $K_1 = 2/d_t \gg \kappa = 2\pi/\lambda$, because the nanotube diameter d_t is much smaller than the optical wave length λ. Therefore, an

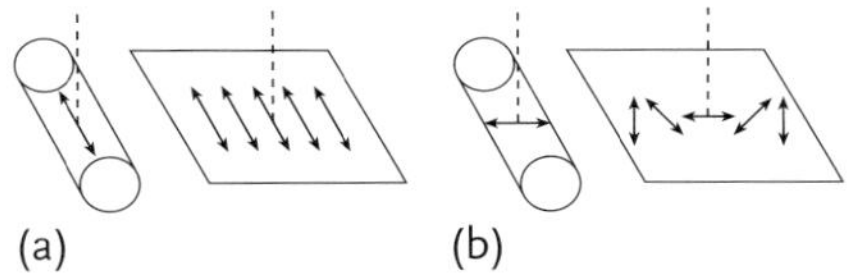

(a) (b)

Figure 6.10 Light polarization (a) parallel and (b) perpendicular to the nanotube axis, shown for both a rolled-up SWNT, and a SWNT unrolled into the graphene layer. The arrows show the light polarization vector, and the dashed lines show the light propagation direction [110].

optical photon in the unrolled graphene layer can be considered as an X-ray photon with respect to spatial considerations, yet the photon energy does not change when the nanotube is unrolled into the graphene layer. Such a "pseudo X-ray" photon is a source of breaking the optical selection rules in the case of perpendicular polarization.

The selection rules for the scattering of electrons by phonons can also be obtained by momentum conservation in 2D graphite. Two cutting lines belonging to the irreducible representations E_μ and $E_{\mu'}$ are separated from each other in the 2D Brillouin zone by $\boldsymbol{k} - \boldsymbol{k}' = (\mu - \mu')2\pi/d_t$, and this is the momentum that the phonon should transfer as a result of the optical transition. As explained in Section 6.4.4, the symmetry of the phonon with such a momentum can be obtained by rolling up the 2D graphene layer and this will yield a phonon with $E_{\mu-\mu'}$ symmetry.

Problems

[6-1] Check the applicability of Eq. (6.2) for group $P(3)$ and for the group of symmetry operations of an equilateral triangle.

[6-2] Propose a unitary matrix and apply the unitary transformation to Γ_R in Eq. (6.7).

[6-3] Show that the traces for the IRs in Eq. (6.5) actually give the characters for the character table in Table 6.2. Show that the Wonderful Orthogonality Theorem works for all possible combinations of IRs. Show that a similar theorem can be made for orthogonality between classes.

[6-4] Consider a CH_4 molecule which has T_d symmetry. Obtain from some textbook the character table for T_d symmetry. Which irreducible representations correspond to the vector (x, y and z)? Show the matrix for a C_3 rotation of this molecule.

[6-5] Which irreducible representations of the T_d group correspond to rotations around the x, y and z axes? Explain that the results do not depend on how you select the axes.

[6-6] Obtain the atom site representation for a CH_4 molecule and decompose your atom site representations into a set of irreducible representations.

[6-7] Calculate the reducible representation for the molecular vibrations for the CH_4 molecule and decompose the reducible representation into irreducible representations. How many vibrational modes are there for a CH_4 molecule?

[6-8] Obtain the symmetries of the Raman and IR-active phonon modes of a CH_4 molecule. In the case of a CH_4 molecule, show that there is no inversion symmetry both by plotting a model of the CH_4 molecule and by showing its atomic coordinates.

[6-9] Now let us consider the single atomic layer of graphite which we call graphene. Consider the character table of the point group for the hexagonal unit cell of monolayer graphene. Obtain the atom site representation of the two carbon atoms of graphene.

[6-10] Obtain the symmetry of the vibrations of graphene at the zone center of the Brillouin zone for graphene.

[6-11] Obtain the symmetry of the tight-binding orbitals of graphene at the zone center of the Brillouin zone.

[6-12] Show that an optical transition from one $2p$ carbon atomic orbital to another $2p$ carbon atomic orbital within the same carbon atom is not allowed. On the other hand, the optical transition from π to π^* energy bands is allowed. Explain what kind of matrix element contributes to such an optically allowed transition? Expand the tight-binding wavefunctions for the optical transition matrix.

[6-13] In the case of graphene, optical transitions occur around the hexagonal corners of the Brillouin zone (K and K' points). When we consider a large unit cell, the K point can be folded into the Γ point (zone center). Plot the extended unit cell and the folded Brillouin zone. What is the point group for the extended unit cell?

[6-14] We can consider a C_2 rotation around the axis at the center of the C–C atomic bond in the direction perpendicular to the bond. However, this rotation is not included in the operations for the hexagonal unit cell of graphene. On the other hand, when we consider the rhombic unit cell of graphene, we can include the C_2 rotation. Discuss the difference in the results of a symmetry description for the two different shapes of the unit cell of graphene.

[6-15] Consider the point group of double layer graphene, in which the two layers have AB stacking. Obtain the irreducible representations for the atom sites and vibrations of bilayer graphene.

[6-16] In the case of double layer graphene with AB stacking, the interlayer interaction can be treated as a perturbation to the unperturbed two graphene layers. What is the irreducible representation describing the interlayer interaction?

[6-17] For double layer graphene, we expect four energy bands derived from the π and π^* bands. Obtain the irreducible representations for the four electronic energy bands at the zone center. Here and in the next graphene related problems, we always consider the AB Bernal stacking.

[6-18] For double layer graphene, show by group theory that the four π bands at the K point consist of one doubly degenerate energy state and two non-degenerate energy states. Discuss the optical selection rules of monolayer graphene and bilayer graphene near the K point.

[6-19] Discuss the Raman selection rules for the vibrations occurring in monolayer graphene and in bilayer graphene.

[6-20] Discuss the symmetry for trilayer graphene and obtain the symmetries of its Raman-active modes. Obtain the optical selection rules for triple layer graphene.

[6-21] Discuss the symmetry for four layer graphene and obtain the symmetries of its Raman-active modes. Obtain the optical selection rules for four layer graphene.

[6-22] When we consider three-dimensional graphite with AB layer stacking, show the point group of the graphite unit cell and discuss the IR and Raman-active modes for graphite.

[6-23] Discuss the symmetry of a unit cell for a graphene ribbon with zigzag edges or one with armchair edges. What is the difference in the symmetry of these unit cells relative to the symmetry of a unit cell for monolayer graphene?

[6-24] The C_{60} molecule has I_h symmetry. Find the atom site irreducible representations and symmetries of the vibrational modes for the C_{60} molecule. How many Raman-active modes are there for the C_{60} molecule and what are their symmetries?

[6-25] Obtain Eqs. (6.22), (6.23) and (6.24).

[6-26] Demonstrate how Eqs. (6.26) and (6.27) are obtained.

[6-27] Consider (n, n) armchair single-wall carbon nanotubes. Obtain Γ^{vec}, $\Gamma^{\text{a.s.}}$, $\Gamma_{\text{lat. mode}}$ and discuss their Raman-active modes.

[6-28] Consider (n, m) chiral single wall carbon nanotubes ($n \neq m$). Obtain Γ^{vec}, $\Gamma^{\text{a.s.}}$, $\Gamma_{\text{lat. mode}}$ and discuss their Raman-active modes.

Part Two Detailed Analysis of Raman Spectroscopy in Graphene Related Systems

7
The G-band and Time-Independent Perturbations

With the materials science and Raman spectroscopy background now in place in Chapters 1–6, we are now ready to start analyzing the Raman spectra of nanocarbons. In the next two chapters we discuss detailed aspects of the Raman G-band (where the notation G comes from graphite). In order to understand the G-band spectra in detail, we have to study the origin of the perturbations to the G-band, caused first by time-independent perturbations, such as strain, which are considered in this chapter, and second by time-dependent perturbations, such as the electron–phonon coupling, which is considered in Chapter 8, where we discuss the effect of temperature and gate voltage on the G-band spectra.

The G-band has two basic properties which make it special for studying sp^2 materials:

- The G-band is present in the Raman spectra of all sp^2 carbon systems, at around 1580 cm^{-1}. It is related to the in-plane C–C bond stretching mode, which gives rise to both the optical in-plane transverse optic (iTO) phonon and the longitudinal optical (LO) phonon branches in graphitic materials.
- Due to the strong C–C bonding and small mass of the C atoms, the G-band in sp^2 carbons has a relatively high Raman frequency in comparison to other materials, and very small perturbations to ω_G can be measured.

Since the carbon atoms in sp^2 carbon materials are neutral (neither positively nor negatively charged), both the iTO phonon and the LO phonon have the same frequency at the zone center of the Brillouin zone.[1] Although the iTO and LO phonon modes are degenerate at the Γ point in both graphite and graphene and are known to comprise the Γ_3^+ doubly degenerate symmetry phonon modes (E_{2g} in point group notation, see Section 6.3.4),[2] only the LO phonon mode has a large Raman intensity. However, in the presence of strain, such as occurs in carbon nanotubes, the LO and iTO phonon modes are mixed with each other, so that both phonon

1) In an ionic crystal, the LO phonon has a higher frequency than the TO phonon mode since the Coulomb interaction acts only on the LO mode.
2) Since there is another E_{2g} symmetry mode in the acoustic phonon branch, we sometimes denote the degenerate iTO and LO optical mode as the $E_{2g}(2)$ or E_{2g2} mode, while the degenerate acoustic mode is denoted by the $E_{2g}(1)$ or E_{2g1} mode.

Raman Spectroscopy in Graphene Related Systems. Ado Jorio, Riichiro Saito, Gene Dresselhaus, and Mildred S. Dresselhaus

ISBN: 978-3-527-40811-5

modes become Raman-active. The iTO and LO phonon frequencies are split into two peaks, and the splitting between the peaks is increased by increasing the strain.

Usually it is hard to observe the splitting of two phonon modes in a material by hydrostatic pressure or by uniaxial strain. However, in the case of the G-band, which has a high-frequency, it becomes possible to observe a clear strain-induced splitting. In fact, while a 1% strain-induced change to a $100\,\mathrm{cm}^{-1}$ Raman feature lies within the $\pm 1\,\mathrm{cm}^{-1}$ precision of most experimental set-ups, a 1% change in the G-band frequency corresponds to $\sim 16\,\mathrm{cm}^{-1}$, which is larger than the natural width of $\sim 10\,\mathrm{cm}^{-1}$ for the G-band features.[3] For this reason, small changes in the physical properties due to strain, as is produced by rolling up the graphene sheet in forming a carbon nanotube, introduce easily measurable strain-induced changes in the G-band feature.

In this chapter we review the detailed G-band properties as a function of strain, which splits the iTO and LO frequencies in graphene (Section 7.1) and carbon nanotubes (Section 7.2). For carbon nanotubes we also consider the effect of quantum confinement (Section 7.3), which is important in both carbon nanotubes and nanoribbons, although the latter will not be treated here.

7.1 G-band in Graphene: Double Degeneracy and Strain

The graphene hexagonal lattice is isotropic in two dimensions. The elastic tensor in graphene is isomorphic to the elastic tensor of the two-dimensional full rotation symmetry group, and this is why nowadays the soccer net is based on hexagonal lattice symmetry.[4] As a result of this symmetry, the LO and iTO phonon modes of graphene are degenerate at the Γ point. This degeneracy is broken when moving away from the Γ point in the Brillouin zone, since the introduction of a phase-directional spatial modulation breaks the rotational symmetry. As a result, the concepts of longitudinal (L) and in-plane transverse (iT) optical (O) modes[5] are expressed with respect to the modulation direction along which the strain is applied.

3) The natural width of a Raman spectral feature is given by the lifetime of the phonons. This will be discussed further in Chapter 8.

4) The hexagonal net was selected for the 1990 world cup in Italy. A hexagonal net produces the shortest length of strings per a unit area and is more flexible and more isotropic than a stiff and anisotropic square net. Since a 120° angle in the hexagonal net is used for the soccer goal, the net can be expanded up to 180°, and the soccer ball can go deeper into the goal net. The shock wave on the net propagates isotropically from the ball, and the ball can be seen to stop clearly at the goal point. The animation can be seen in the following web site: http://www1.gifu-u.ac.jp/~eng/ja/square/2004syou/exciting/exciting.htm.

5) In the transverse mode for two-dimensional materials, we have in-plane and out-of-plane TO modes, while the longitudinal mode is always an in-plane mode. Thus we do not say iLO, but rather we simply say LO for the longitudinal mode.

7.1.1 Strain Dependence of the G-band

When the bond lengths and angles of graphene are modified by strain, the hexagonal symmetry of graphene is broken, and this symmetry-breaking effect splits the LO and iTO mode frequencies [198, 230, 231]. The understanding of this effect comes from elasticity theory [95], which is discussed in many textbooks on the introduction to solid state physics. The dynamic equation for the deformation of the lattice within the linear displacement regime is given by the equations of motion [232]:

$$-M\ddot{u}_i = M\omega_0^2 u_i + \sum_{klm} K_{iklm}\epsilon_{lm} u_k \,, \quad (i, m, k, l = 1, 2)\,, \tag{7.1}$$

where $u_i (i = 1, 2)$ is the in-plane atomic displacement, M is the mass of the carbon atom, and ω_0 is the frequency for the unstrained lattice. Here ϵ_{lm} denotes the strain tensor in the in-plane coordinates which can be obtained by rotating the strain tensor[6] in the phonon propagating direction ℓ (longitudinal) and in its perpendicular direction t (transverse), as $\epsilon_{\ell\ell}$ and ϵ_{tt}. The subscripts ℓ and the t denote, respectively, the in-plane directions of the vibration of the LO and iTO phonon modes [232], so that we can write the strain tensor for u_1 and u_2 as:

$$\begin{pmatrix} \epsilon_{11} & \epsilon_{12} \\ \epsilon_{21} & \epsilon_{22} \end{pmatrix} = \begin{pmatrix} \epsilon_{tt}\cos^2\theta + \epsilon_{\ell\ell}\sin^2\theta & \sin\theta\cos\theta\,(\epsilon_{\ell\ell} - \epsilon_{tt}) \\ \sin\theta\cos\theta\,(\epsilon_{\ell\ell} - \epsilon_{tt}) & \epsilon_{tt}\sin^2\theta + \epsilon_{\ell\ell}\cos^2\theta \end{pmatrix}, \tag{7.2}$$

where the angle θ denotes the angle between u_1 and the iTO (u_2 and LO) phonon direction.

The fourth rank tensor K_{iklm} gives the change in the elastic constant K_{ik} between the displacements u_i and u_k due to ϵ_{lm}, which is defined as:[7]

$$K_{iklm} = \frac{\partial K_{ik}}{\partial \epsilon_{lm}} \,. \tag{7.3}$$

Since both K_{ik} and ϵ_{lm} are second-rank symmetric tensors, they satisfy $K_{ik} = K_{ki}$ and $\epsilon_{lm} = \epsilon_{ml}$ and thus several symmetry relationships for K_{iklm} follow, such as:

$$K_{iklm} = K_{kilm} = K_{kiml} = K_{ikml}\,, \quad \text{and} \quad K_{iklm} = K_{lmik}\,. \tag{7.4}$$

Here the latter condition comes from the fact that this fourth-rank tensor is symmetric for the interchange of two sets of two indices (ik) and (lm).

6) The rotation of a second-rank tensor is given by:
$$\begin{pmatrix} \cos\theta & \sin\theta \\ -\sin\theta & \cos\theta \end{pmatrix} \begin{pmatrix} \epsilon_{tt} & 0 \\ 0 & \epsilon_{\ell\ell} \end{pmatrix} \begin{pmatrix} \cos\theta & -\sin\theta \\ \sin\theta & \cos\theta \end{pmatrix}.$$

7) The reason why the other K_{iklm} components vanish is given by the condition that the tensor K_{iklm} should be invariant under a $2\pi/3$ rotation. In three dimensions, we should add K_{3333}, K_{1133} and K_{3232} and we then get five different force constants. For the combined index (ij) used to define K_{ij}, we use the notation (11) = 1, (22) = 2, (33) = 3, (32) = 4, (13) = 5, (21) = 6. In this notation, the five independent K_{iklm} components are expressed as: $K_{1111} = K_{11}$, $K_{1122} = K_{12}$, $K_{3333} = K_{33}$, $K_{1133} = K_{13}$, and $K_{3232} = K_{44}$.

Further, the hexagonal symmetry restricts the number of independent components of K_{iklm} to a few, namely, K_{1111} and K_{1122} in the two-dimensional motion and there are three different nonzero values that can be expressed in terms of these two components, namely $K_{1111} = K_{2222}$, K_{1122}, and $K_{1212} = (K_{1111} - K_{1122})/2$ in this case [232]. It is noted that K_{1212} is not independent of K_{1111} and K_{1122} and is expressed as $K_{1212} = (K_{1111} - K_{1122})/2$. By defining the following $\tilde{K}$, the motion of the atoms 1 and 2 in the graphene unit cell is well-characterized by the following definitions:

$$\begin{aligned} M\tilde{K}_{11} &\equiv K_{1111} = K_{2222} \\ M\tilde{K}_{12} &\equiv K_{1122} = K_{2211} \\ \frac{1}{2}M\left(\tilde{K}_{11} - \tilde{K}_{12}\right) &\equiv K_{1212} = K_{2112} = K_{1221} = K_{2121}\ . \end{aligned} \tag{7.5}$$

Then Eq. (7.1) becomes as follows:

$$\begin{pmatrix} \Lambda - (\tilde{K}_{11}\epsilon_{11} + \tilde{K}_{12}\epsilon_{22}) & -(\tilde{K}_{11} - \tilde{K}_{12})\epsilon_{12} \\ -(\tilde{K}_{11} - \tilde{K}_{12})\epsilon_{12} & \Lambda - (\tilde{K}_{11}\epsilon_{22} + \tilde{K}_{12}\epsilon_{11}) \end{pmatrix} \begin{pmatrix} u_1 \\ u_2 \end{pmatrix} = \begin{pmatrix} 0 \\ 0 \end{pmatrix}, \tag{7.6}$$

where $\Lambda = \omega^2 - \omega_0^2$ and ϵ_{ij} is expressed by Eq. (7.2). In order to get the solution of $(u_1, u_2)^t \neq (0,0)^t$ (i. e., a nontrivial solution), the determinant of the matrix of Eq. (7.6) should be zero, and this equation is known as the secular equation.

When we put Eq. (7.2) into Eq. (7.6), we get the frequency change $\delta\omega \equiv \omega - \omega_0$ due to strain as follows:[8)]

$$\frac{\delta\omega}{\omega_0} = \frac{\tilde{K}_{11} + \tilde{K}_{12}}{4\omega_0^2}(\epsilon_{\ell\ell} + \epsilon_{tt}) \pm \frac{\tilde{K}_{11} - \tilde{K}_{12}}{4\omega_0^2}(\epsilon_{\ell\ell} - \epsilon_{tt})\ . \tag{7.7}$$

The hydrostatic component of the strain is defined by

$$\epsilon_h = \epsilon_{\ell\ell} + \epsilon_{tt} \tag{7.8}$$

and the shear component by

$$\epsilon_s = \epsilon_{\ell\ell} - \epsilon_{tt}\ . \tag{7.9}$$

The coefficient to ϵ_h in Eq. (7.7) is the Grüneisen parameter λ:

$$\lambda = -\frac{1}{\omega_0}\frac{\partial\omega}{\partial\epsilon_h} = \frac{\tilde{K}_{11} + \tilde{K}_{12}}{4\omega_0^2}\ , \tag{7.10}$$

which describes the shift in frequency for a hydrostatic deformation (strictly speaking, the deformation is hydrostatic when $\epsilon_{\ell\ell} = \epsilon_{tt}$). The coefficient to ϵ_s in Eq. (7.7) is:

$$\beta = \frac{1}{\omega_0}\frac{\partial\omega}{\partial\epsilon_s} = \frac{\tilde{K}_{11} + \tilde{K}_{12}}{4\omega_0^2}\ , \tag{7.11}$$

8) Here we use the fact that $m(\omega^2 - \omega_0^2) = m(\omega + \omega_0)(\omega - \omega_0) \sim 2\omega_0\delta\omega$. All the $\sin\theta$ and $\cos\theta$ terms in Eq. (7.2) disappear after a long calculation. The reason why this disappearance of $\sin\theta$ and $\cos\theta$ terms occurs is that the graphene system is isotropic in plane.

which describes the shift in frequency for shear stress. For a uniaxial strain, $\epsilon_{\ell\ell}$ and ϵ_{tt} are related by the Poisson ratio, defined as:

$$\nu = (\delta w/w)/(\delta l/l)\,, \tag{7.12}$$

where l and w are, respectively, the length and width of a sheet being deformed along length l. First principles calculations give $\epsilon_{tt} = -0.186\epsilon_{\ell\ell}$ for a graphene sheet [231].

7.1.2 Application of Strain to Graphene

Figure 7.1 shows the evolution of the G-band spectra for a graphene sheet subjected to uniaxial strain [231]. Strain (see Figure 7.1a) causes the G-band to split into two peaks, here named G^+ and G^- (see Figure 7.1d). These bands are related to the longitudinal (G^-) and transverse (G^+) atomic motions with respect to the strain direction (where the eigenvectors are defined in Figure 7.1b,c). The labels iTO and LO as used here have no relation to the concept of iLO and iTO in the graphene phonon dispersion relations, where the modes are longitudinal and transverse with respect to a given phonon modulation direction $\boldsymbol{q}$. A clear picture about the phonon eigenvectors is obtained by considering the dependence of the G-band mode intensities as a function of the light polarization direction (see [198, 230, 231]).

In Figure 7.2, the measured values for the strain-dependent shifts are $\partial\omega_{G^+}/\partial\epsilon = -10.8\,\text{cm}^{-1}/\%$ strain and $\partial\omega_{G^-}/\partial\epsilon = -31.7\,\text{cm}^{-1}/\%$ strain. However, these values vary from one research group to another by a factor of ~ 5 [198, 230, 231], mainly due to the difficulty in performing the experiment accurately. Sample preparation and inhomogeneous bending of the sample are examples of experimental difficulties.

Having the G-band frequency shifts from Figure 7.2, one can use Eq. (7.7) by substituting $\epsilon_{tt} = 0$ and $\epsilon_{ll} = \epsilon$ to obtain the coefficients given in Eqs. (7.10) and (7.11) as

$$\lambda = \frac{\delta\omega_{G^+} + \delta\omega_{G^-}}{2\omega_0(1-\nu)\epsilon} \tag{7.13}$$

$$\beta = \frac{\delta\omega_{G^+} - \delta\omega_{G^-}}{\omega_0(1+\nu)\epsilon}\,. \tag{7.14}$$

7.2 The G-band in Nanotubes: Curvature Effects on the Totally Symmetric Phonons

The G-band appears as multiple peaks in a SWNT, while a single peak ($\omega_G \approx 1582\,\text{cm}^{-1}$) is observed for a 2D graphene sheet [112, 233]. Up to six G-band phonons are first-order Raman allowed in chiral SWNTs, although two of them (the totally symmetric A_1 modes, see Figure 6.8) usually dominate the spectra. In this section we discuss the effects of curvature in the A_1 symmetry modes.

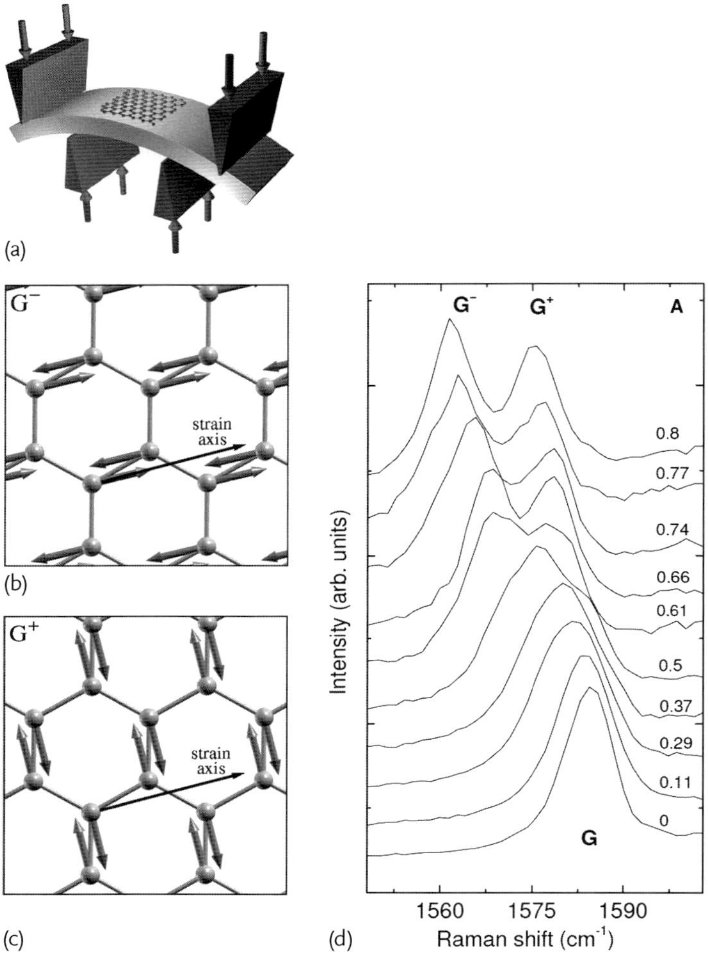

Figure 7.1 The effect of uniaxial strain on graphene. (a) The graphene sheet is deposited on a polymer coated substrate that is bent using the four indicated point supports. (b,c) Eigenfunctions for G^+ (b) and G^- (c) are shown and are determined by density-functional perturbation theory. The direction of the strain axis is indicated for both cases. (d) The G-band spectra thus measured for many values of the applied strain show a splitting into two components, G^+ and G^- that are clearly seen with increasing strain. Note that each spectrum is labeled by its value of the applied strain. Adapted from [231].

7.2.1
The Eigenvectors

In carbon nanotubes, strain exists independent of any external applied force because of nanotube curvature. The system is one-dimensional, so that a longitudinal vibration means atomic motion along the tube axis and a transverse vibration

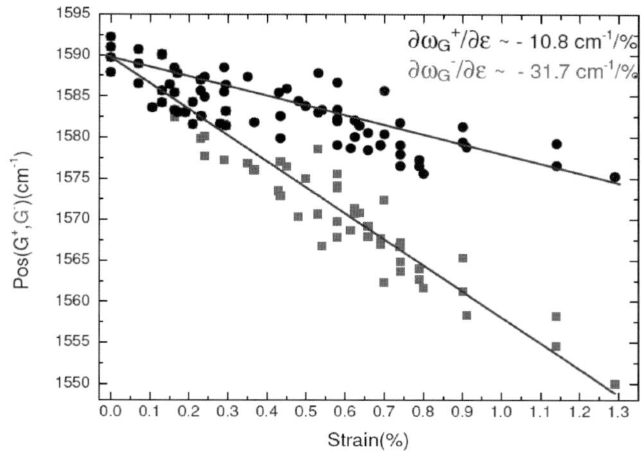

Figure 7.2 G-band frequencies ω_{G^+} and ω_{G^-} for graphene are plotted as a function of applied uniaxial strain. The solid lines are linear fits to the data, and values for the fitted slopes are indicated for both ω_{G^+} and ω_{G^-}. Adapted from [231].

means atomic motion perpendicular to the tube axis [234]. *Ab initio* calculations have been performed for (6,6) and (10,0) achiral and (8,4) and (9,3) chiral nanotubes [234]. The authors found that, for achiral (armchair and zigzag) SWNTs, the strict LO and iTO designations of the G-band phonons remain valid. For chiral nanotubes, however, the phonon eigenvectors lie along different directions relative to the nanotube axis, so that we cannot define strict LO and iTO modes by using the nanotube axis direction. In Figure 7.3a we show the A_1 mode displacements for an (8,4) tube and in Figure 7.3b we show the corresponding displacements for a (9,3) M-SWNT [234]. In Figure 7.3 it can be seen that the displacement of the atoms is along the circumference in the (8,4) S-SWNT, but parallel to the bonds in the (9,3) nanotube, thus showing evidence for sensitive dependence of the atomic displacements on the chiral angle. The smallest angle between the carbon-carbon bonds and the circumference in the (9,3) tube is $30° - \theta = 16.1°$ [234].

Of course the θ result is model-dependent and, therefore, despite the importance and the large number of prior works devoted to Raman scattering in SWNTs [112, 233], there is still controversy about whether the many peaks within this G-band can be assigned to (quasi) LO and iTO type mode behavior. There is also controversy about which features pertain to the three different symmetry types (A_1, E_1 and E_2) related to phonon confinement within the first-order *single resonance* process [112, 227, 233], or if all features belong to a totally symmetric irreducible representation (A_1 symmetry) [235, 236] and originate from a defect-induced *double resonance* Raman scattering process [237]. Phonon confinement will be discussed here in Section 7.3.1, while the double resonance process in this connection will be discussed in Chapters 12 and 13, where we then revisit the double-resonance G-band model. For the moment, we consider only the A_1 symmetry modes with iTO and LO character, for simplicity.

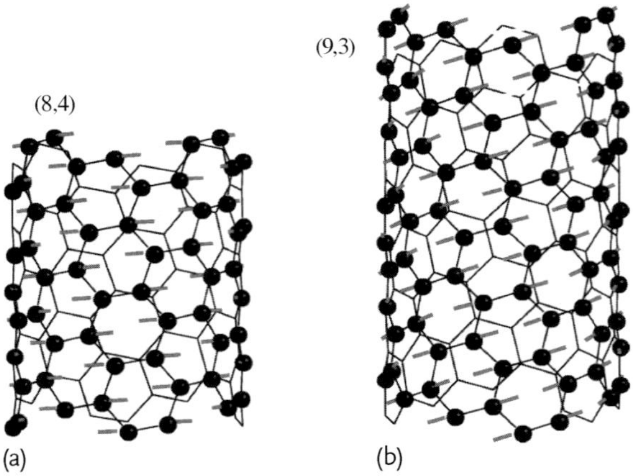

Figure 7.3 (a) The A_1 symmetry high-energy eigenvectors for an (8,4) S-SWNT. The atomic displacements are parallel to the circumference. (b) An A_1 symmetry high-energy eigenvector for a (9,3) M-SWNT. The atomic displacements are parallel to the carbon–carbon bonds. The direction of the helix is indicated by the gray lines [234].

7.2.2
Frequency Dependence on Tube Diameter

The most evident consequence of the strain induced by tube curvature is the ω_G dependence on tube diameter (d_t). Such a dependence has already been introduced in Chapter 4. In Figure 4.12, the filled bullets stand for ω_G from semiconducting SWNTs and the open bullets stand for ω_G from metallic SWNTs. The two most intense A_1 symmetry G peaks, named G^+ and G^- for the higher and lower frequencies, respectively, exhibit the following diameter dependence [179]:

$$\omega_G = 1591 + C/d_t^2 \,, \tag{7.15}$$

where $C_{G^+} = 0$, $C^S_{G^-} = 47.7\,\text{cm}^{-1}\text{nm}^2$, $C^M_{G^-} = 79.5\,\text{cm}^{-1}\,\text{nm}^2$ gives the solid, long dashed and dashed lines in Figure 4.12c, respectively, for semiconducting and metallic SWNTs. Such a dependence is explained as follows. In the time-independent perturbation picture, the ω_G^{LO} mode frequency is expected to be independent of diameter, since the atomic vibrations are along the tube axis. In contrast, the ω_G^{iTO} mode has atomic vibrations along the tube circumference, and increasing the curvature increases the out-of-plane component, thus decreasing the spring constant with a $1/d_t^2$ dependence. This picture holds for S-SWNTs, where G^+ stands for the LO mode, and G^- stands for the iTO mode [179]. However, for M-SWNTs the picture is different: G^+ stands for the iTO mode, and G^- stands for the LO mode [124]. The G-band profile in this case is very different, as shown in the bottom spectrum of Figure 4.12a, and this behavior can only be understood within a time-dependent perturbation picture. This issue is discussed further in Chapter 8.

7.3
The Six G-band Phonons: Confinement Effect

In Section 7.2, we neglected the fact that when rolling up the graphene sheet to form a nanotube, the confinement along the circumferential direction generates a larger number of first-order Raman-active modes. In Section 7.3.1 we discuss mode symmetry and selection rules while in Section 7.3.2 we show how polarization analysis can be used to study the G-band in more depth. For a detailed discussion of the selection rules for the first-order single-resonance Raman scattering process in nanotubes, we direct the reader to Chapter 6.

7.3.1
Mode Symmetries and Selection Rules in Carbon Nanotubes

The A_1, E_1 and E_2 symmetry Raman modes exhibiting zero, two and four nodes along the tube circumference (see Figure 6.8) become Raman-active in SWNTs. These modes are represented using the zone-folding picture in Figure 7.4b. Considering these three symmetries combined with their LO and iTO vibrational nature (displayed in Figure 7.4a), six G-band phonons can be Raman-active in chiral SWNTs (achiral have higher symmetry and only three G-band modes are Raman active). However, their observation depends on the direction of light polarization and on the resonance condition, as discussed below.

Here we select the Z and Y axes as the SWNT axis direction and the photon propagation direction, respectively. Thus we have two independent polarization directions of light, namely parallel (Z) and perpendicular (X) to the nanotube axis. Hereafter we denote a scattering event with incident polarization i and scattered polarization s as (is).[9] Thus we have four different kinds of scattering events: XX,

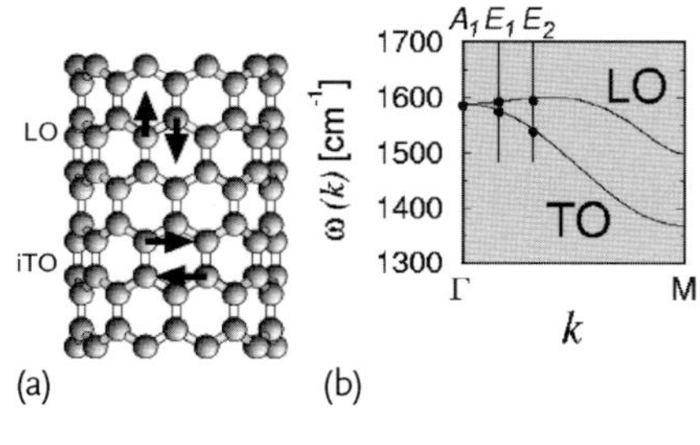

Figure 7.4 (a) Schematic picture of the G-band atomic vibrations along the nanotube circumference and along the nanotube axis of a zigzag nanotube. (b) The Raman-active modes with A_1, E_1, and E_2 symmetries and the corresponding cutting lines $\mu = 0$, $\mu = \pm 1$, and $\mu = \pm 2$ in the unfolded 2D Brillouin zone. The Γ points of the cutting lines are shown by solid dots [80].

9) A more complete notation is $p_i(is)p_s$, named Porto's notation in memory of S.P.S. Porto, where p_i and p_s give the propagation directions for the incident and scattered photons, respectively. Since we only discuss backscattering here, for economy of space we do not use the full notation for the scattering geometry.

Table 7.1 Selection rules for polarization-dependent G-band features and the corresponding resonance conditions. The *Z* and *Y* axes are the SWNT axis direction and the photon propagation direction, respectively. The polarization of the incident and scattered light are given as well as the resonance condition. E_G is the G-band phonon energy [226, 227].

Symmetry of phonon	Scattering event	Resonance
A_1	(ZZ)	$E_{laser} = E_{ii}$, $E_{laser} \pm E_G = E_{ii}$
A_1	(XX)	$E_{laser} = E_{ii\pm 1}$, $E_{laser} \pm E_G = E_{ii\pm 1}$
E_1	(XZ)	$E_{laser} = E_{ii\pm 1}$, $E_{laser} \pm E_G = E_{ii}$
E_1	(ZX)	$E_{laser} = E_{ii}$, $E_{laser} \pm E_G = E_{ii\pm 1}$
E_2	(XX)	$E_{laser} = E_{ii\pm 1}$, $E_{laser} \pm E_G = E_{ii\pm 1}$

XZ, ZX and ZZ. Considering the general case of chiral SWNTs (with C_N symmetry) [226, 227], the first-order Raman signal from isolated SWNTs can only be seen when the excitation laser energy is in resonance with a van Hove singularity (VHS). These selection rules imply that, for isolated SWNTs: (1) A_1 symmetry phonon modes can be observed for the (ZZ) scattering geometry when either the incident or the scattered photon is in resonance with E_{ii}, and for the (XX) scattering geometry when either the incident or the scattered photon is in resonance with $E_{i,i\pm 1}$. (2) E_1 symmetry modes can be observed for the (ZX) scattering geometry for resonance of the incident photon with E_{ii} VHSs, or for resonance of the scattered photon with $E_{i,i\pm 1}$ VHSs, while for the (XZ) scattering geometry for resonance of the incident photon with $E_{i,i\pm 1}$ VHSs, or for resonance of the scattered photon with E_{ii} VHSs. (3) E_2 symmetry phonon modes can only be observed for the (XX) scattering geometry for resonance with $E_{i,i\pm 1}$ VHSs. Therefore, depending on the polarization scattering geometry and resonance conditions, it is possible to observe 2, 4 or 6 G-band peaks. A summary of the polarization dependence and the corresponding resonance conditions is listed in Table 7.1.

7.3.2
Experimental Observation Through Polarization Analysis

First of all, there is a general and simple polarization behavior that one should bear in mind when acquiring the Raman spectra from a sample of aligned SWNTs, which is not accounted for in the selection rules described in Section 7.3.1. Carbon nanotubes behave as antennas, with the absorption/emission of light being highly suppressed for light polarized perpendicular to the nanotube axis, because of the depolarization effect [238, 239]. Here the depolarization effect means that photoexcited carriers screen the electric field of the cross-polarized light [238, 239]. Therefore, if one wants to measure Raman spectra from a sample of aligned carbon nanotubes, the largest Raman intensity will generally be observed for light polarized along the tube axes (ZZ), and almost no signal will be observed for cross polarized light [228, 235, 240], as shown in Figure 7.5a. Furthermore, when acquir-

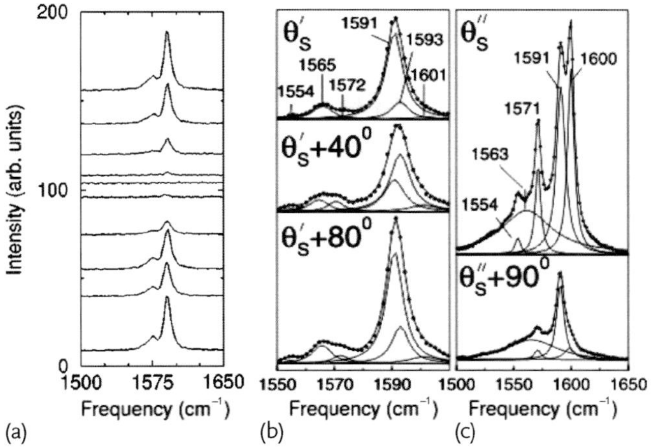

Figure 7.5 (a) Polarization dependence of the G-band from one isolated semiconducting SWNT sitting on a Si/SiO_2 substrate [228]. Both incident and scattered light are polarized parallel to each other and vary from parallel (bottom) to perpendicular (middle) and back to parallel (top) to the tube axis. Parts (b) and (c) show the polarization scattering geometry dependence for the G-band from two isolated SWNTs. The Lorentzian peak frequencies are given in units of cm^{-1}. θ'_S and θ''_S are the incident angles between the light polarization and SWNT axis directions, not known *a priori*. From the relative intensities of the polarization behavior of the G-band modes, $\theta'_S \sim 0°$ and $\theta''_S \sim 90°$ are assigned [226].

ing polarized spectra from a single SWNT with a fixed laser energy, it is not usual to observe Raman signals from both parallel (ZZ) and perpendicular (ZX, XZ, XX) polarization, since the resonance energies for the polarized light in different polarization directions are different from each other (see Table 7.1). This combination of properties make the totally symmetric A_1 mode dominant in the G-band spectra.

However, the most interesting results coming from the polarization analysis are related to the symmetry selection rules for the different phonon/electron symmetries [227, 228, 241], as discussed in Section 7.3.1. Figure 7.5b shows three different G-band Raman spectra from an S-SWNT, but with three different directions for the incident light polarization, that is, θ'_S, $\theta'_S+40°$ and $\theta'_S+80°$, where θ'_S is an angle between the initial polarization directions of the light and the nanotube axis. Six well-defined peaks associated with the G-band features are observed, with different relative intensities for the different polarization geometries, and the symmetries of the various peaks are assigned as follows: 1565 and 1591 $\rightarrow$ A_1 symmetry; 1572 and 1593 $\rightarrow$ E_1 symmetry; 1554 and 1601 $\rightarrow$ E_2 symmetry. Figure 7.5c shows two G-band Raman spectra obtained from another S-SWNT ($\omega_{RBM} = 132\,cm^{-1}$), with θ''_S and $\theta''_S + 90°$. The spectra can be fit using four sharp Lorentzians, and a broad feature at about $1563\,cm^{-1}$. This broad feature (FWHM $\sim$ $50\,cm^{-1}$) is sometimes observed in weakly resonant G-band spectra from S-SWNTs and they are not discussed here.[10] From previous polarization Raman studies [227], the

10) These features are likely generated by the defect-induced double resonance processes, discussed in Section 13.5.

sharp peaks at 1554 and 1600 cm^{-1} should be assigned as E_2 symmetry modes, while the 1571 and 1591 cm^{-1} peaks should be assigned as unresolved ($A_1 + E_1$) symmetry modes, their relative intensities depending on the incident light polarization direction [227].

It is interesting to note the relatively high intensity of the spectra with (XX) polarization observed in Figure 7.5c, indicating resonance with $E_{i,i\pm1}$ optical transitions. For several measured isolated SWNTs, the Raman intensities in Figure 7.5 do not exhibit a substantial reduction for any direction of the incident/scattered light, in contrast to other published results [228, 235, 240], which showed an intensity ratio $I_{ZZ} : I_{XX} \sim 1 : 0$. From our discussion, it is clear that the so-called antenna effect is observed for samples in resonance with only E_{ii} electronic transitions, and that is the case in Figure 7.5a and [228, 235, 240]. However, in general, the intensity ratio ZZ:XX can assume values either larger or smaller than 1, depending on the resonance condition. The samples in [227] exhibit a very large diameter distribution (d_t from 1.3 nm up to 2.5 nm), so that E_{ii} and $E_{i,i\pm1}$ transitions can both occur within the resonance window of the same laser, and an average value of $ZZ : XX = 1.00 : 0.25$ was then observed.

7.3.3 The Diameter Dependence of ω_G

Now that the phonon confinement has been introduced, a more complete picture for the diameter dependence of the G-band than that introduced in Section 7.2.2 can be given. Figure 7.6 shows the G-band phonon frequencies as a function of tube diameter evaluated by zone folding of the graphene phonon dispersion relations (lines) in comparison to *ab initio* calculations (points) [124]. The diameter dependence of the dispersion relations based on zone folding comes from the diameter dependence of the distance between adjacent cutting lines. A single A_1 mode is predicted since the phonon frequencies of the LO and iTO modes are identical to each other at the Γ point in graphite (see Figure 7.4b). Additionally, zone folding shows a relatively small splitting between the longitudinal and transverse E_1 modes, and a larger splitting between the two E_2 modes. A large mode softening is observed for small diameters as the cutting line reaches the maximum of the phonon dispersion for the LO branch in Figure 7.4b.

The full *ab initio* calculations show a similar behavior, but some details are clearly different. Generally, the *ab initio* results are lower in frequency than the zone-folded values. The frequency softening of the *ab initio* points is explained by the fact that curvature weakens the π contribution to the bonds in the circumferential direction, which also explains why the A_1(T) mode[11] is affected most strongly by curvature, whereas the A_1(L) mode is essentially independent of diameter for semiconducting tubes. For diameters of about 1.4 nm, the E_2 modes (squares at approximately 1613 cm^{-1} and 1570 cm^{-1} in Figure 7.6) are symmetrically split by $\pm$22 cm^{-1} with respect to the central graphite frequency $\omega_G = 1592$ cm^{-1} (theo-

11) Here, T and L denote the iTO and LO phonon modes, respectively, as adopted in [124].

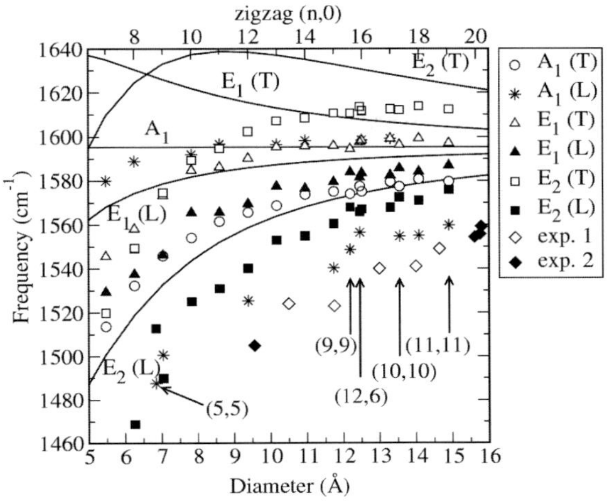

Figure 7.6 Diameter dependence of phonon frequencies of $(n, 0)$ zigzag, (n, n) armchair and (12,6) chiral SWNTs for the G-band calculated by *ab initio* density functional theory (symbols) and by zone folding (lines). Phonons are characterized by their symmetry and whether they are T or L modes. Here L denotes vibration (longitudinal) displacements parallel to the tube axis while T denotes vibrational displacements transverse or perpendicular to the tube axis. The lower axis and upper axis show the diameter and the $(n, 0)$ index scales for zigzag tubes, respectively. [124]

ry). Since the tube curvature shifts the E_1(T) mode to lower values, the frequencies of the E_1(T) and A_1(L) modes (open triangles and asterisks) almost coincide for S-SWNTs and are located at about 1597 cm^{-1}, a little bit higher in frequency than the theoretical central G-band mode of graphite. The E_1(L) and A_1(T) symmetry G-band modes (filled triangles and open circles in Figure 7.6) also have rather similar frequencies and are found at roughly 1580 cm^{-1}, that is, 20 cm^{-1} lower than the E_1(T) and A_1(L) modes. The (n, m) labeled tubes indicate a downshifted $A_1(L)$ for metallic SWNTs, and theseunusual results will be discussed in Chapter 8.

Further confirmation for the G-band mode assignment proposed in Section 7.3.1 comes from comparison of experimental results with *ab initio* calculations. Here we focus on the spectra from semiconducting SWNTs because metallic SWNTs exhibit time-dependent perturbations that will be discussed in Chapter 8. Figure 7.7 plots the G-band mode frequencies for several resonant S-SWNTs vs. the observed ω_{RBM} (bottom axis) and vs. inverse nanotube diameter where the relation $1/d_t = \omega_{\mathrm{RBM}}/248$ was used to label the top axis of Figure 7.7.[12] The spectra are usually fit using 6 peaks, although sometimes only 4 or 2 peaks are used, the spectra beingfit with linewidths approaching the natural linewidth for the G-band modes [242], that is, $\gamma_G \sim 5$ cm^{-1}.

12) The $\omega_{\mathrm{RBM}} = 248/d_t$ relation [176] has been broadly used in the early years of single nanotube spectroscopy (2001–2005), although now we know it represents a special case. How to obtain the tube diameter from the SWNT radial breathing mode frequency (ω_{RBM}) is the subject of Chapter 9.

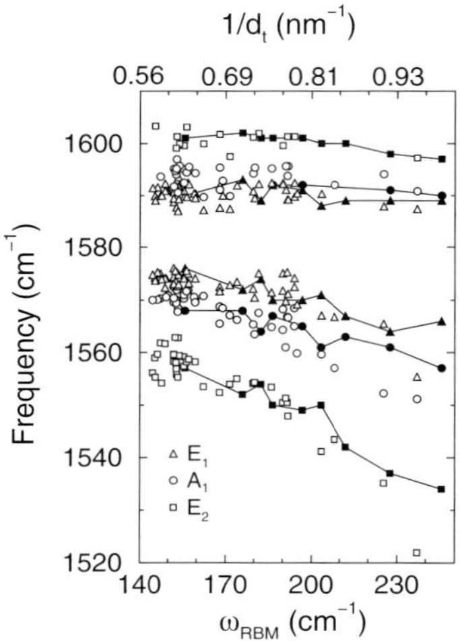

Figure 7.7 Correlation of ω_G and ω_{RBM} shown by plotting ω_G (open symbols) vs. ω_{RBM} (bottom axis) and vs. $1/d_t$ (top axis) for S-SWNTs. Experimental G-band data obtained with $E_{laser} = 1.58$, 2.41 and 2.54 eV. Solid symbols connected by solid lines come from *ab initio* calculations [124] downshifted by 18, 12, 12, 7, 7, 11 cm^{-1} from the bottom to the top of the *ab initio* data, respectively [226].

The solid symbols connected by solid lines in Figure 7.7 come from calculations by Dubay *et al.* (Figure 7.6). The different solid symbols indicate the different mode symmetries: $\bullet \to A_1$, $\blacktriangle \to E_1$, $\blacksquare \to E_2$, in agreement with polarization results (see Figure 7.5). The theoretical points in Figure 7.7 were downshifted by about 1% to fit the experimental data. The observed d_t dependence of the frequencies for each of the 3 higher frequency G$^+$-band modes (A_1, E_1 and E_2) are in very good agreement with theory [124], showing little d_t dependence. For the 3 lower frequency G$^-$-band modes, both *ab initio* calculations and experimental results show a stronger d_t dependence, but *ab initio* calculations seem to slightly underestimate the G$^-$-band mode softening for lower d_t values (mainly for the A_1 symmetry mode). The experimental data from semiconducting SWNTs can be better fit with [226]

$$\omega_G = 1592 - C/d_t^{\beta} , \tag{7.16}$$

with $\beta = 1.4$, $C_{A_1} = 41.4\,\text{cm}^{-1}\,\text{nm}$, $C_{E_1} = 32.6\,\text{cm}^{-1}\,\text{nm}$, and $C_{E_2} = 64.6\,\text{cm}^{-1}\,\text{nm}$.

7.4
Application of Strain to Nanotubes

Different authors have applied externally induced strain to carbon nanotubes, both in bundles [236, 243] and as isolated tubes [244–246]. The elasticity theory presented in Section 7.1.1 is also used to study strain in carbon nanotubes, and it has actually been initially developed for these cases [232]. Measurements on SWNT bundles basically show an overall increase in ω_G by increasing hydrostatic pressure [236, 243]. This result was initially used as evidence for the absence of LO and iTO G-band mode behavior. However, measurements on isolated tubes show a greater richness of effects, which includes both hydrostatic and uniaxial deformation, torsion, bending, etc. Here a large downshift in the E_2 modes was observed and different pressure-induced effects for G^+ vs. G^- were found, depending on (n, m) [247]. For isolated S-SWNTs with uniaxial strain up to 1.65%, shifts in ω_G of up to 40 cm^{-1} were observed [244]. There are still some controversial results regarding strain effects in SWNTs, mostly because of the difficulty to perform the experiments accurately and the need for a large number of measurements to establish and understand the (n, m) dependence.

7.5 Summary

The stretching of the C–C bond in sp^2 graphitic materials gives rise to the so-called G-band. The G-band is highly sensitive to strain effects in sp^2 nanocarbons, and can be used to probe any modification in the flat geometric structure of graphene, such as the strain that is induced by external forces, or even by the curvature when growing a SWNT. This curvature dependence generates a diameter dependence, thus making the G-band a probe also for the tube diameter, while its dependence on externally induced strain is very rich and still controversial. Phonon confinement in SWNTs generate complex selection rules that can be tested using light polarization analysis, although the antenna effect causes the totally symmetric modes to dominate the spectra most of the time. Finally, the G-band of metallic SWNTs is special, and this comes from electron–phonon coupling that can only be treated within time-dependent perturbation theory. This effect generates interesting results in both graphene and carbon nanotubes, related to both temperature and doping. These issues will be discussed in Chapter 8.

Problems

[7-1] Explain that the Coulomb interaction between positive and negative ions in an ionic crystal changes the force constant for the LO phonon mode but not for the iTO phonon mode.

[7-2] When we consider the frequency-dependent dielectric constant $\epsilon(\omega)$, the ratio of the LO to the iTO phonon frequency in an ionic crystals is given by:

$$\frac{\omega_{\text{LO}}}{\omega_{\text{iTO}}} = \frac{\epsilon(0)}{\epsilon(\infty)} ,$$

which was derived by Lyddane, Sachs, and Teller (LST theory). Study the LST theory and obtain the above formula. Check that the formula works for some ionic crystals such as NaCl.

[7-3] The above LST relation is considered for $\boldsymbol{q} = 0$. For a general $\boldsymbol{q}$ vector, we can discuss how the phonon dispersion is modified by the Coulomb interaction by considering the coupling of Maxwell's equations to the equation of motions for the atoms:

$$\ddot{\boldsymbol{x}} + \omega_{\mathrm{iTO}}^2 \boldsymbol{x} - a E_x = 0$$

$$P_x = a\boldsymbol{r} + (\epsilon(\infty) - 1) E_x \ ,$$

where a is given by:

$$a = \omega_{\mathrm{iTO}}[\epsilon(0) - \epsilon(\infty)]^{1/2} \ .$$

Combining these two equations with Maxwell's equations for $D_x = E_x + P_x$ and for H_x, and considering their wave vector $\boldsymbol{q}$, obtain and plot $\omega(q)$.

[7-4] When we pull a hexagonal net in one direction, how much percent of the length of the net can be expanded compared with the length of an undistorted hexagonal net. When we rotate the pulling direction relative to the C–C bond direction, how does the expansion ratio change? Compare these results with the case of a square or triangular net.

[7-5] Consider two carbon atoms which are connected to each other by a spring. Evaluate the force constant in units of eV/Å^2 (use the 1580 cm^{-1} LO phonon mode frequency for sp^2 carbons).

[7-6] Consider a $2\pi/3$ rotation around a carbon atom in the plane of graphene. Make a 3 rotation matrix $D(2\pi/3)$ for the $2\pi/3$ rotation around the z axis.

[7-7] Let us consider a function $f = ax + by + cz + d$. When a vector (x, y, z) is transformed by this $2\pi/3$ rotation into $(x', y', z') = D(2\pi/3)(x, y, z)$ and suppose that f is invariant for the $2\pi/3$ rotation, then show $a = b = c = 0$.

[7-8] Let us consider a function $g = ax^2 + by^2 + cz^2 + dyz + ezx + fxy$. When a vector (x, y, z) is transformed by the $2\pi/3$ rotation as $(x', y', z') = D(2\pi/3)(x, y, z)$ and suppose that g is invariant for the $2\pi/3$ rotation, what are the conditions imposed for the constants a, b, c, d, e, f.

[7-9] A second rank tensor can be defined as a 3×3 matrix a_{ij} in which a is transformed by any invariant operation $(x', y', z') = C(x, y, z)$, where C is a matrix that transforms into $a' = C^{-1}aC$. Because the determinant of a should not be changed by C, show that det(C)=1 and that $a_{ij} = a_{ji}$. Thus a symmetric second rank tensor has six independent components.

[7-10] When we consider a fourth-rank tensor K_{iklm} in three dimension $(i, k, l, m = 1, 2, 3)$, 81 possible variables exist. However, because of the

relationship of Eq. (7.4), K_{iklm} has only 21 independent variables even though there is no symmetry for the system. Explain this statement.

[7-11] What is the relationship between the LO and iTO phonon mode frequencies with the Young's modulus and other force constants?

[7-12] In the D_N symmetry, show that only A_1, E_1, and E_2 phonon modes are Raman-active. If we plot the one-dimensional Brillouin zone (cutting lines), which lines in the 2D Brillouin zone correspond to the Raman-active modes? Answer by constructing a figure.

[7-13] Plot the X, Y, Z axes and show the scattering geometries with the corresponding resonance conditions for carbon nanotubes when we put a carbon nanotube with its axis along the Z axis and when the light is coming from the Y (or Z) direction.

[7-14] The curvature of a nanotube is considered to have a constant strain in one direction. Then estimate the change in the iTO and LO phonon frequencies.

[7-15] Show how to obtain Eqs. (7.13) and (7.14) from the definitions in Section 7.1.1.

[7-16] Apply the theory developed in Section 7.1.1 for carbon nanotubes, and obtain the effect of strain on the transverse and longitudinal modes. You can use [232] as a guide.

[7-17] Study the difference between Eqs. (7.15) and (7.16). Can these two equations be consistent?

[7-18] Make a study of the Poisson ratio and the Grüneisen parameter for graphene, graphite and carbon nanotubes.

8
The G-band and the Time-Dependent Perturbations

In Chapter 7 we learned how to treat strain effects in the G-band of graphene-related systems. This includes pressure effects and other mechanical deformations, such as bending the graphene sheet to "build" a carbon nanotube. The next step would be the study of temperature (T) and doping-dependent effects. However, for an accurate description of such effects in graphene related systems, it is important to understand the dynamics of electron–phonon coupling because the so-called Born–Oppenheimer (or adiabatic) approximation is not valid for graphene.

In this chapter we review the detailed G-band properties connected to dynamic effects of the electron–phonon coupling by introducing the concepts of the Kohn anomaly and by discussing the effect of temperature and doping on the electron–phonon coupling. These effects have to be derived within time-dependent perturbation theory, as explained in Section 8.1, where we discuss the breakdown of the adiabatic approximation. In Section 8.2 we introduce the effect of changing temperature and show how time-dependent perturbation theory applies for interpreting experimental studies of doping by using a gate, as well as by studying temperature-related effects. We address graphene in Section 8.3 and SWNTs in Section 8.4.

8.1
Adiabatic and Nonadiabatic Approximations

To put the time-dependent perturbations in the context of the general vibrational properties, we stress that in most cases, atomic vibrations are treated in the so-called adiabatic approximation. We can use the adiabatic approximation when the electrons move sufficiently fast so that they can follow the small motion of the heavy nuclei. Then the motion of the electrons can be expressed as a function of the *position* of the atom (not as a function of the *momentum* of the atom). However, when the atomic motion is much faster than the time for electron-momentum relaxation by the electron–phonon interaction, the adiabatic approximation is no longer valid. This problem is pictured in Figure 8.1 where the differences in electronic behavior in the adiabatic vs. nonadiabatic approaches are highlighted, respectively, in Figure 8.1b,c [248].

Raman Spectroscopy in Graphene Related Systems. Ado Jorio, Riichiro Saito, Gene Dresselhaus, and Mildred S. Dresselhaus

ISBN: 978-3-527-40811-5

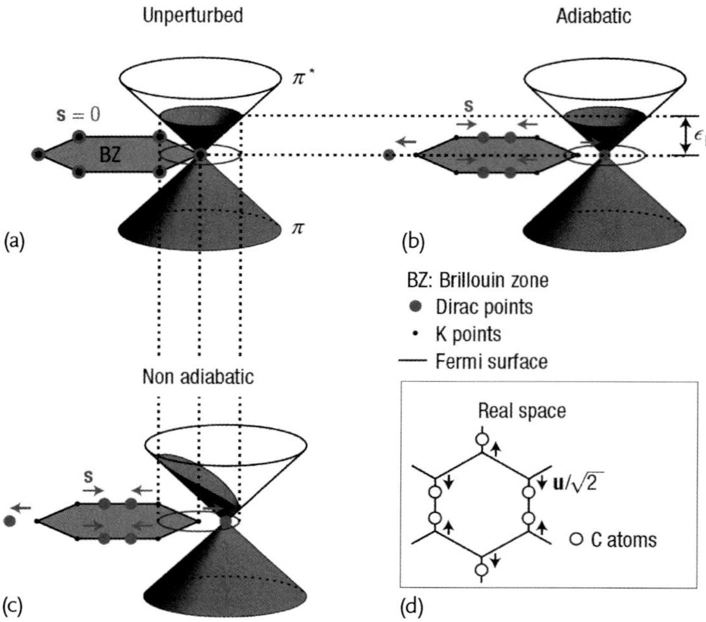

Figure 8.1 Schematic π-band structure of graphene near the high-symmetry K point of the Brillouin zone. Here the graphene is doped and the filled electronic states are shown in gray. (a) Bands of the perfect crystal. The Dirac point is at the K point, the electronic states are filled up to the Fermi energy ϵ_F, and the Fermi surface is a circle centered at K. (b) Bands in the presence of a Γ_6^+ (E_{2g}) lattice distortion. The Dirac points are displaced from K by $\pm s$. Within the adiabatic approximation, the electrons remain in the instantaneous ground state: the bands are filled up to the Fermi energy ϵ_F and the Fermi surface follows the Dirac point displacement. The total electron energy does not depend on s. (c) Bands in the presence of a Γ_6^+ (E_{2g}) lattice distortion. In the nonadiabatic case, the electrons do not have time to relax their momenta to follow the instantaneous ground state. In the absence of scattering, the electron momentum is conserved and a state with momentum k is occupied if the state with the same k was occupied in the unperturbed case. As a consequence, the Fermi surface is the same as in the unperturbed case and does not follow the Dirac cone displacement. The total electron energy increases with s^2, resulting in the observation of a Γ_6^+-phonon softening. (d) Atomic displacement pattern of the Γ_6^+ (E_{2g}) phonon. The atoms are displaced from their equilibrium positions by $\pm u/\sqrt{2}$. Note that the displacement pattern of the Dirac points (in reciprocal space) is identical to the displacement pattern of the carbon atoms (in real space) [248].

The G-band frequency of $\omega_G \sim 1584\,\text{cm}^{-1}$ corresponds to 22 fs as the period for atomic motions. In fact, coherent phonon spectroscopy measurements [43] which can observe oscillations in the transmission probability of light in a material as a function of time at the frequency of G-band phonons, observed a 47 THz oscillation for the G-band which indeed corresponds to 22 fs. The measured electron-momentum relaxation times, due to impurity, electron–electron and electron–phonon scattering processes are all on the order of a few hundred femtoseconds, as deduced from the electron mobility in graphene [249] and from ultrafast spectroscopy measurements in graphite [250, 251]. Thus the virtually excited electrons do not have sufficient time to relax their momenta to reach the instantaneous adiabatic ground state (see the schematics for a nonadiabatic process in Figure 8.1) [248]. Electrons and phonons are thus strongly coupled, and cannot be treated within the usual adiabatic approximation, thus generating a strong G-band frequency dependence on structure, doping (changes in the Fermi level) and temperature.

8.2 Use of Perturbation Theory for the Phonon Frequency Shift

This section starts in Section 8.2.1 by describing the general effect of temperature (T) of phonons, and by highlighting the importance of T on the Fermi distribution. In Section 8.2.2 perturbation theory is used to calculate the phonon frequency shift due to the electron–phonon interaction under nonadiabatic conditions, showing the effect of changes in the Fermi distribution as a function of gate voltage and temperature.

8.2.1 The Effect of Temperature

The change in the phonon frequencies with temperature is a general manifestation of anharmonic terms in the lattice potential energy, which are responsible for the phonon–phonon coupling of the phonon population and of the thermal expansion of the crystal [252]. The effect of temperature on the G-band frequency for different sp^2 nanocarbons has been measured and is represented by:

$$\omega_{\mathrm{G}} = \omega_{\mathrm{G}}^{0} + \chi T \,, \tag{8.1}$$

where ω_{G}^{0} is the G-band frequency in the limit $T \to 0$ and χ is the coefficient for the temperature-dependent correction to ω_{G} (to first order). Table 8.1 gives the values of χ for different sp^2 nanocarbons, found in the literature. Calizo *et al.* [253] found $\omega_{\mathrm{G}}^{0} = 1584\,\mathrm{cm}^{-1}$ and $1582\,\mathrm{cm}^{-1}$ for 1-LG and 2-LG, respectively. They describe the temperature-dependent effects as roughly divided into the *self-energy* shift due to the anharmonic coupling of the phonon modes and to the shift due to the thermal expansion of the crystal,[1] that is [253]:

$$\omega_{\mathrm{G}} - \omega_{\mathrm{G}}^{0} = (\chi_T + \chi_V)\Delta T = \left(\frac{\partial \omega}{\partial T}\right)_V \Delta T + \left(\frac{\partial \omega}{\partial V}\right)_T \Delta T \,. \tag{8.2}$$

In fact, for highly oriented pyrolytic graphite (HOPG) it has been considered that the thermal expansion occurs mainly along the c axis, and the in-plane thermal

1) The thermal expansion of a crystal is also a result of anharmonicity. However, the thermal expansion is also related to changes of the elastic force constants with volume, and these two different physical mechanisms can be considered separately.

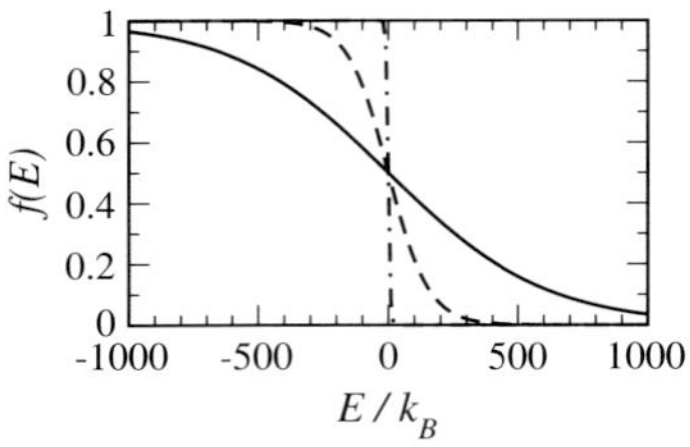

Figure 8.2 The Fermi–Dirac distribution at 300 K (solid), 77 K (dashed), and 4 K (dot-dashed) as a function of energy scaled by k_B, the Boltzmann constant.

Table 8.1 The temperature coefficient in Eq. (8.1) for different sp^2 carbons.

Sample	χ (cm^{-1}/K)	Reference
1-LG	−0.0162	[253]
2-LG	−0.0154	[253]
SWNT	−0.0189	[254]
DWNT	−0.022	[255]
HOPG	−0.011	[256]

expansion is negligible, that is, $\chi = (\chi_T + \chi_V) \approx \chi_T$ for HOPG [256]. The ω_G shifts with temperature have been used to obtain the thermal conductivity of graphene, as explained in [257].

However, for an accurate description of the G-band frequency behavior, we have to consider the electron–phonon coupling, and that the electron population in crystalline structures depends on temperature. This dependence is described by the Fermi–Dirac distribution $f(E)$, which gives the probability that an orbital at energy E will be occupied in an ideal electron gas in thermal equilibrium [95]:

$$f(E) = \frac{1}{\exp[(E - \mu)/k_B T] + 1}, \tag{8.3}$$

where T is the temperature in degrees Kelvin (K), k_B is the Boltzmann constant, and the quantity μ is the chemical potential. At absolute zero temperature, the chemical potential is equal to the Fermi energy ($\mu = E_F$). In general, $f(E) = 1/2$ at $E = \mu$ for Eq. (8.3) and Figure 8.2 shows $f(E)$ for three temperatures of interest. At $T = 0$, the occupation probability is $f(E) = 1$ up to the Fermi level, above which the occupation probability drops rapidly. When the temperature increases, there is a spread in the occupation probability around E_F (see Figure 8.2). These changes in carrier occupation will affect the G-band frequency, as discussed in the following sections.

8.2.2
The Phonon Frequency Renormalization

Within the framework of time-dependent perturbation theory, we consider that when the phonon can excite an electron–hole pair (see Figure 8.3) by the electron–phonon interaction, this virtual process[2] gives rise to a phonon energy renormalization that depends on the electronic structure, the Fermi level and the temperature [141, 196, 258, 259]. The phonons then renormalize the electron energies,[3] while the electrons renormalize the phonon energies.[4] Both of these perturbations occur over a longer time than that observed in Raman spectra, which is on the order of 1 s, and thus we can observe these phenomena in Raman spectra. This electron–phonon coupling gives rise to a controllable modification to the G-band frequency as a function of the gate voltage, and this modification depends strongly on the geometrical structure of the nanomaterial, which applies to both carbon nanotubes and graphene.

The phonon frequency shift due to the electron–phonon (el–ph) interaction of the Γ-point LO- and iTO-phonon modes for graphene (and also for SWNTs) can be calculated by second-order perturbation theory. The phonon energy including the el–ph interaction can be written as:

$$\hbar\omega_\lambda = \hbar\omega_\lambda^{(0)} + \hbar\omega_\lambda^{(2)} , \tag{8.4}$$

(λ = LO, iTO) where $\omega_\lambda^{(0)}$ is the unperturbed phonon frequency without consider-

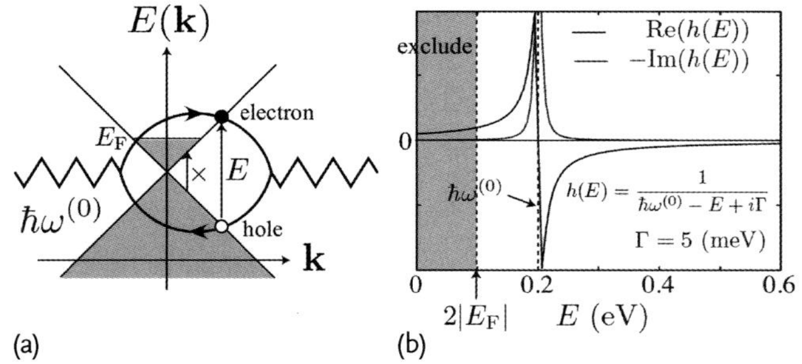

Figure 8.3 (a) An intermediate electron–hole pair state that contributes to the energy shift of the optical phonon modes is depicted. A phonon mode is denoted by a zigzag line and an electron–hole pair is represented by a loop. The low-energy electron–hole pair satisfying $0 \le E \le 2E_F$ is forbidden at zero temperature by the Pauli principle. (b) The energy-dependent real and imaginary parts of the $h(E)$ correction to the phonon energy by an intermediate electron–hole pair state. Especially the sign of the correction depends on the energy of the intermediate state as given by $h(E)$ (see text in Section 8.2) [260].

2) Here a virtual process means the mixing of the wavefunction of the excited states into the ground state wavefunction in perturbation theory. We here consider the electron–phonon interaction as a perturbation.

3) Through the so-called Peierls-like mechanism, that is, the deformation of the electronic structure due to electron–phonon coupling.

4) Resulting in the so-called Kohn anomaly effect, that is, a deformation of the phonon structure due to electron–phonon coupling.

ation of the el–ph interaction, and $\hbar\omega_\lambda^{(2)}$ is the perturbation term given by second-order perturbation theory

$$\hbar\omega_\lambda^{(2)} = 2\sum_{k} \frac{|\langle \mathrm{eh}(\boldsymbol{k})|\mathcal{H}_{\mathrm{int}}|\omega_\lambda\rangle|^2}{\hbar\omega_\lambda^{(0)} - (E_{\mathrm{e}}(\boldsymbol{k}) - E_{\mathrm{h}}(\boldsymbol{k})) + i\Gamma_\lambda} \times [f(E_{\mathrm{h}}(\boldsymbol{k}) - E_{\mathrm{F}}) - f(E_{\mathrm{e}}(\boldsymbol{k}) - E_{\mathrm{F}})] \quad , \tag{8.5}$$

and is the quantum correction to the phonon energy due to electron–hole pair creation as shown in Figure 8.3a. The factor 2 in Eq. (8.5) comes from the spin degeneracy. In Eq. (8.5), $\langle \mathrm{eh}(\boldsymbol{k})|\mathcal{H}_{\mathrm{int}}|\omega_\lambda\rangle$ is the matrix element for creating an electron–hole pair at momentum $\boldsymbol{k}$ by the el–ph interaction with a $q = 0$ phonon, while $E_{\mathrm{e}}(\boldsymbol{k})$ ($E_{\mathrm{h}}(\boldsymbol{k})$) is the electron (hole) energy and Γ_λ is the decay width. In Figure 8.3a, an intermediate electron–hole pair state that has the energy of $E = E_{\mathrm{e}}(\boldsymbol{k}) - E_{\mathrm{h}}(\boldsymbol{k})$ is shown. Thus in Eq. (8.5) we need to sum ($\sum_{\boldsymbol{k}}$) over all possible intermediate electron–hole pair states which can have a much larger energy than the phonon ($E \gg \hbar\omega_\lambda^{(0)}$).

Since $\langle \mathrm{eh}(\boldsymbol{k})|\mathcal{H}_{\mathrm{int}}|\omega_\lambda\rangle$ is a smooth function of $E = E_{\mathrm{e}}(\boldsymbol{k}) - E_{\mathrm{h}}(\boldsymbol{k})$ which appears in the denominator of Eq. (8.5), the contribution to $\hbar\omega_{\lambda^{(2)}}$ from an electron–hole pair depends strongly on its energy. In Figure 8.3b, we plot the real part and imaginary part of the denominator of Eq. (8.5), $h(E) = 1/(\hbar\omega^{(0)} - E + i\Gamma)$ as a function of E in the case of $\hbar\omega^{(0)} = 0.2$ eV and $\Gamma = 5$ meV. Here $\mathrm{Re}(h(E))$ has a positive (negative) value when $E < \hbar\omega^{(0)}$ ($E > \hbar\omega^{(0)}$) and the lower (higher) energy electron–hole pair makes a positive (negative) contribution to $\hbar\omega_\lambda^{(2)}$. Moreover, an electron–hole pair satisfying $E < 2|E_{\mathrm{F}}|$ cannot contribute to the energy shift (shaded region in Figure 8.3a,b) because of the Fermi distribution function $f(E)$ in Eq. (8.5). Thus, the quantum correction to the phonon energy by an intermediate electron–hole pair can be controlled by changing the Fermi energy, E_{F} (see Figure 8.4b). For example, when $|E_{\mathrm{F}}| = \hbar\omega^{(0)}/2$, then $\hbar\omega_\lambda^{(2)}$ takes a minimum value at zero temperature since all positive contributions to $\hbar\omega_\lambda^{(2)}$ are suppressed in Eq. (8.5) (e. g., see dashed line in Figure 8.4b). Since $\mathrm{Re}(h(E)) \approx -1/E$ for $E \gg \hbar\omega^{(0)}$, all high energy intermediate states contribute to phonon softening if we include all the electronic states in the system. Here we introduce a cut-off energy at $E_{\mathrm{c}} = 0.5$ eV as $\sum_{\boldsymbol{k}}^{E_{\mathrm{e}}(\boldsymbol{k})<E_{\mathrm{c}}}$ in order to avoid such a large energy shift in Eq. (8.5). The energy shift due to the high-energy intermediate states ($\sum_{\boldsymbol{k}}^{E_{\mathrm{e}}(\boldsymbol{k})>E_{\mathrm{c}}}$) can be neglected by renormalizing $\hbar\omega^{(0)}$ so as to reproduce the experimental results of the observed Raman spectra [38, 261] since the contribution from $E_{\mathrm{e}}(\boldsymbol{k}) > E_{\mathrm{c}}$ just gives a constant energy shift to $\hbar\omega^{(2)}$. These results do not depend on the selection of the cut-off energy,[5] since E_{c} is much larger than $\hbar\omega^{(0)}$.

The $\mathrm{Im}(h(E))$ in Figure 8.3b is nonzero only very close to $E = \hbar\omega^{(0)}$, which shows that the phonon can resonantly decay into an electron–hole pair with the

5) A cut-off energy is generally taken for setting the upper-limit of the integration in calculating a physical property even when the integration has a contribution above the cut-off energy. In order to avoid the cut-off energy dependence of the results, a smooth function is defined for switching off this contribution. Calculating the phonon frequency in a solid essentially contains the electron–phonon interaction in discussions that were given in the 1950s. See details in [38].

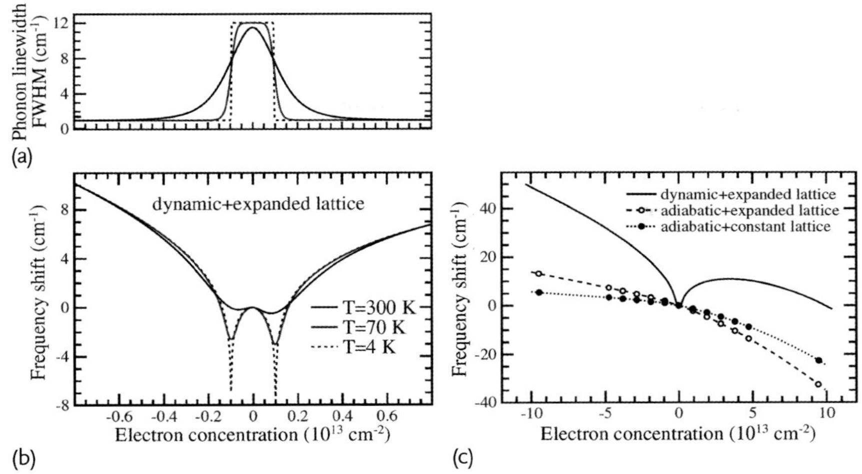

Figure 8.4 (a) Linewidth and (b) frequency of the Raman G-band as a function of the electron concentration. The calculations are made using time-dependent perturbation theory, which considers the dynamic effects (i. e., under the nonadiabatic approximation) and lattice distortion induced by doping. (c) G-band frequency behavior on a larger doping range comparing the expected results for adiabatic vs. nonadiabatic and constant lattice vs. doping induced lattice distortions (extended lattice). Adapted from [258].

same energy as the phonon. It is noted that when $|E_F| > \hbar\omega^{(0)}/2$, then the resonance window width is small, that is $\Gamma_\lambda \approx 0$ at zero temperature, while Γ_λ may take a finite value at a finite temperature (see Figure 8.4a). The plot in Figure 8.3b gives Γ_λ self-consistently[6] calculating $\Gamma_\lambda = -\text{Im}(\hbar\omega_\lambda^{(2)})$ in Eq. (8.5). Figure 8.4a shows the expected behavior of the G-band FWHM by changing the electron concentration (i. e., changing the Fermi level).

The effect of changing temperature on the phonon renormalization is shown in Figure 8.4 by the different line styles, as rationalized by the Fermi distribution related term in Eq. (8.5). As discussed in Section 8.2.1, at $T = 0$ the occupation probability is $f(E) = 1$ up to the Fermi level, above which the occupation probability drops rapidly. This makes a highly singular dependence for the phonon renormalization and linewidth variation at $E_F = \pm\hbar\omega_G/2$ (see Figure 8.4a,b). When the temperature increases, there is a spread in the occupation probability around E_F (see Figure 8.2), thus smoothing out the singularities at $E_F = \pm\hbar\omega_G/2$.[7]

Finally, you may have noticed the asymmetry in the frequency shift in Figure 8.4b for large doping values (e. g., electron concentration $\rightarrow \pm 0.8 \times 10^{13}$ electrons/cm^{-2}). This asymmetry is clearer in Figure 8.4c where the doping range is extended. The Kohn anomaly effect discussed here occurs within a small doping range, where

6) When the initial value of Γ_λ in Eq. (8.5) is the same as $\text{Im}(\hbar\omega_\lambda^{(2)})$, then we can say that the calculation is self-consistent. The value of Γ_λ depends on the electron–phonon interaction (numerator of Eq. (8.5)). This treatment is equivalent to the treatment of Γ_λ when using the uncertainty relation.

7) The Fermi function $f(E)$ in Eq. (8.5) becomes a smooth function of energy at 300 K (see Figure 8.2).

E_F lies near the K point. When higher doping levels take place, lattice distortion induced by doping dictates the ω_G behavior. From Figure 8.4c we see that lattice distortion induced p doping causes a hardening of ω_G, while n doping causes a softening of ω_G. The difference from *weak* and *strong* doping will be discussed further for gate and chemical doping of SWNTs, respectively (Section 8.4.4). In the case of graphene, up to now only gate doping results are available experimentally, as discussed in the next section.

8.3
Experimental Evidence of the Kohn Anomaly on the G-band of Graphene

In this section the effect of doping on the G-band of single-layer graphene (Section 8.3.1) and on the G-band of double-layer graphene (Section 8.3.2) is explicitly considered.

8.3.1
Effect of Gate Doping on the G-band of Single-Layer Graphene

Experimental observation of the effect of doping on the G-band phonon frequency is shown in Figure 8.5 [196]. The G-band is observed to upshift in frequency (Figure 8.5a,b) and to decrease in linewidth (Figure 8.5c) with doping, as predicted by time-dependent perturbation theory. The physics behind this behavior comes from the Pauli exclusion principle. Under increasing doping, the electron–hole interaction for different energies will be forbidden, thereby decreasing the Kohn anomaly effect. At $T = 0$ K, the effect would be abrupt, but for $T \neq 0$ K, there is an energy distribution for the carriers and the Kohn anomaly-induced frequency change tends to saturate when the Fermi level is far from $\hbar\omega_\lambda/2$. The two anomalies at $\pm\hbar\omega_G/2$ are not clearly seen in this experiment due to temperature-induced broadening (see Figure 8.4a). However, a gate voltage dependence for the G-band frequency ω_G was measured at $T = 12$ K, where phonon anomalies at $E_F = \pm\hbar\omega_G/2$ could be clearly distinguished [262]. The 12 K experiment was, however, carried out on bilayer graphene, where another interesting effect occurs, as described in Section 8.3.2.

8.3.2
Effect of Gate Doping on the G-band of Double-Layer Graphene

In bilayer graphene, the unit cell has 4 C atoms rather than 2, and as a result there are two π and two π^*-bands at the K point (see Figure 2.11). In this case, there will be more than two Kohn anomalies in the G-band gate-dependent frequency renormalization (see schema on the right hand side of Figure 8.6) [262]. When the Fermi energy reaches $\pm\hbar\omega_G/2$, the $\pi - \pi^*$ transition from the valence band to the lower conduction band shown in Figure 8.6(I) is no longer allowed, as it is in single-layer graphene. However, the transition from the now filled lowest energy π^*-band

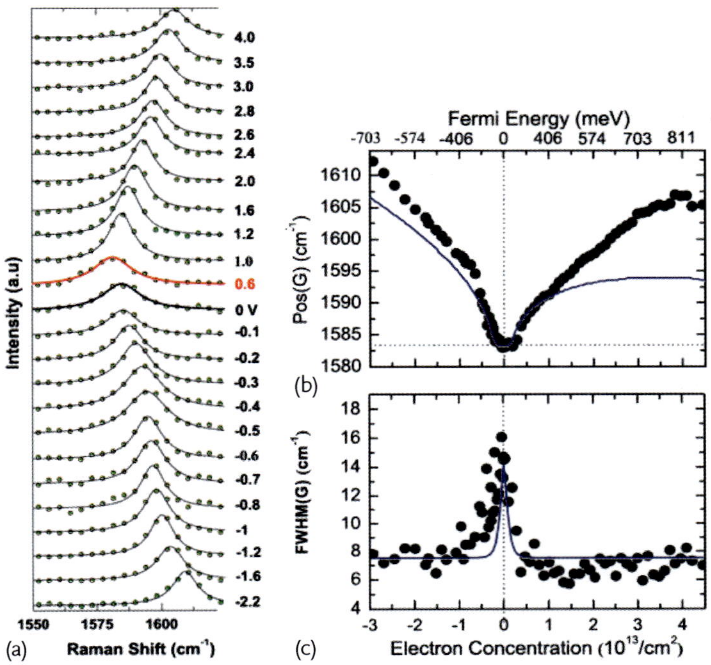

Figure 8.5 The Raman G peak of doped monolayer graphene. (a) The G-band spectra at 295 K for many values of the gate voltage V_g. The red spectrum corresponds to the undoped case, which occurs at $V \neq 0$ due to natural doping of graphene by the environment. (b) The G peak position (frequency) and (c) the linewidth as a function of electron concentration, deduced from the applied gate voltage data. Black circles: measurements; solid line: finite-temperature nonadiabatic calculation. Adapted from [196].

to the higher energy π^*-band, shown by the dashed red arrow in Figure 8.6(II), is possible. When the gate voltage rises further and the Fermi energy reaches the second band, $\pi^* - \pi^*$ transitions are suppressed, as shown in Figure 8.6(III). These effects are seen in the G-band frequency and linewidth of bilayer graphene (see Figure 8.6), where a distinctly different behavior with respect to the monolayer case (see Figure 8.5) is clearly observed for both the G-band frequency and linewidth. Therefore, when discussing graphene systems above, we saw that the renormalization effect changes significantly in going from single-layer to bilayer graphene, and it would change further by increasing the number of layers, although the renormalization effect will become less and less evident with increasing layer number.

8.4 Effect of the Kohn Anomaly on the G-band of M-SWNTs vs. S-SWNTs

The Kohn anomaly is important for systems with an electronic gap smaller than the phonon energy. The Kohn anomaly is, therefore, applicable to the G-band

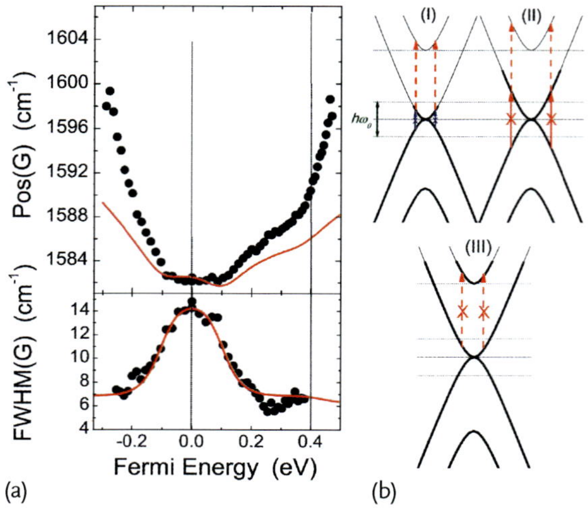

Figure 8.6 (a) Peak frequency (Pos(G)) and linewidth (FWHM(G)) for the Raman G-band feature of doped bilayer graphene vs. Fermi energy. Black circles: measurements; red line: finite-temperature nonadiabatic calculation. (b) Schematics of the electron–phonon coupling at three different doping levels, as indicated by the thicker lines on the electronic bands. Adapted from [262].

in graphene and also in metallic SWNTs. As shown by the "$*$" symbols in Figure 7.6, the LO G-band in metallic SWNTs experiences a large renormalization in frequency, due to the what is called a Peierls-like mechanism [124]. This Peierls-like mechanism is, in fact, due to the electron–phonon interaction, as is also the Kohn anomaly effect, but the effect is much stronger in metallic carbon nanotubes than in graphene, due to the phonon confinement that generates a dynamic gap opening. For semiconducting SWNTs, some renormalization occurs related to virtual transitions, but the effect should be minor for S-SWNTs. Furthermore, in SWNTs, because of their spatial confinement effects, there is a rich behavior depending not only on their metal vs. semiconducting behavior, but also on the SWNT diameter and chirality. We discuss these results here.

8.4.1
The Electron–Phonon Matrix Element: Peierls-Like Distortion

In this section we show that the Kohn anomaly is very different for LO and iTO phonons. The effect of the G-band phonons on the electronic structure is evaluated here first for graphene, within the first-neighbor tight-binding model and the adiabatic approximation. The corresponding effect for nanotubes will then be summarized based on quantum confinement and zone-folding effects. Although the extended tight-binding and nonadiabatic approximations are needed for quantitative studies, the simple pedagogic picture described here can account for the fundamentals of the pertinent physical effect.

Consider the matrix elements H_{AA}, H_{AB}, H_{BA}, and H_{BB} evaluated within the framework of the nearest neighbor π-band orthogonal tight-binding model in the linear in u/a approximation (u is the amplitude of phonon displacements, and $a = \sqrt{3}a_{CC} = 0.246$ nm is the graphene lattice constant):

$$H_{AA} = H_{BB} = E_0 + \epsilon \sum_j^3 (u_{Bj} - u_{A0}) \cdot (r_{Bj} - r_{A0})/a_{CC} \,, \tag{8.6}$$

$$H_{AB} = H_{AB}^* = \sum_j^3 [t + \alpha(u_{Bj} - u_{A0}) \cdot (r_{Bj} - r_{A0})/a_{CC}] \times \exp[ik \cdot (r_{Bj} - r_{A0} + u_{Bj} - u_{A0})] \,. \tag{8.7}$$

Here, E_0 is the atomic orbital energy which is set to zero to define our energy scale, $t = -2.56$ eV is the transfer or hopping integral for graphene, $\epsilon = 39.9$ eV/nm is the on-site electron–phonon coupling (EPC) coefficient, $\alpha = 58.2$ eV/nm is the off-site EPC coefficient [222, 263, 264], r_{Aj} and r_{Bj} are the equilibrium atomic positions shown by the gray and green dots in Figure 8.7a, respectively, u_{Aj} and u_{Bj} are the atomic displacements associated with the Γ_6^+ (E_{2g}, G-band) phonon mode represented by arrows in Figure 8.7a, subscript $j = 0, \ldots, 3$ labels the central atom and its three nearest neighbors, $a_{CC} = 0.142$ nm is the interatomic distance and k is the electron wave vector. Upon substituting u_{Aj} and u_{Bj} from the G-band eigenvectors (Figure 8.7a for the LO phonon) into Eq. (8.6) and setting the graphene determinant equal to zero, we find that $k_F(k'_F)$ oscillates at the phonon frequency with a displacement amplitude $\Delta k_F(\Delta k'_F)$ given by:

$$\begin{aligned} \Delta k_F = -\Delta k'_F &= -\frac{2\sqrt{3}\alpha u}{ta}\hat{y} \quad \text{for LO} \,, \\ \Delta k_F = -\Delta k'_F &= +\frac{2\sqrt{3}\alpha u}{ta}\hat{x} \quad \text{for iTO} \,, \end{aligned} \tag{8.8}$$

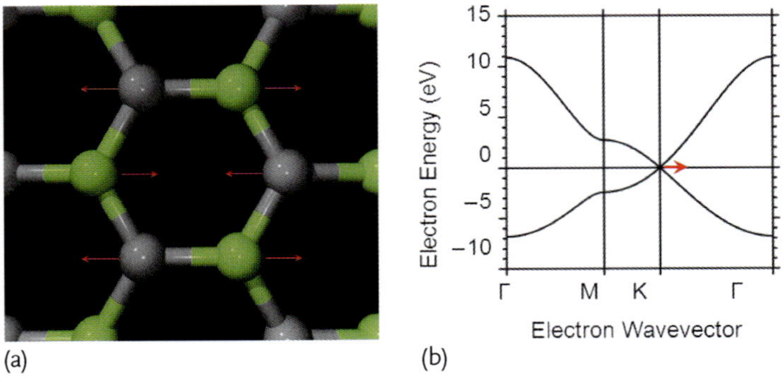

Figure 8.7 (a) Arrows indicate the atomic motions for the G-band mode in graphene. (b) The red arrow indicates the displacement of the $\pi - \pi^*$ crossing point on the $E(\boldsymbol{k})$ diagram when the G-band LO phonon displacement takes place [117].

around the $K(K')$ point in the Brillouin zone [124]. Note that Δk_F and $\Delta k'_F$ are determined by the off-site EPC coefficient α, since the ϵ terms in Eq. (8.6) that are linear in u/a cancel for the u_{Aj} and u_{Bj} vectors [265]. The arrow in Figure 8.7b shows such a Δk_F measurement for the LO phonon. This change in the electronic structure causes a distortion in the electronic matrix element, that is responsible for the EPC.

Now, if we move to metallic nanotubes, the presence of cutting lines in the Brillouin zone due to spatial confinement will play a very important role. For an armchair SWNT in an equilibrium position, a cutting line crosses the K point, and the valence and conduction bands cross (see Figure 8.8a). When a displacement of the G-band iTO phonon takes place (see Figure 8.8b), the $\pi - \pi^*$ crossing point moves along the cutting line direction, and no significant change in the electronic structure occurs. However, when a displacement of the G-band LO phonon takes place (see Figure 8.8c), the $\pi - \pi^*$ crossing point now moves perpendicular to the cutting line direction, thus opening a band gap. This effect changes the total electronic energy of the tube significantly, generating a significant electron–phonon coupling, much stronger than that in graphene. In linear carbon chains, this gap opening decreases the energy enough, so that the phonon softens towards zero frequency and the carbon chain gets distorted (going to the C≡C–C≡C–C bonding configuration from the original C=C=C=C bonding configuration). This distortion is known as the Peierls distortion. In carbon nanotubes, the energy lowering due to the Peierls distortion is not larger than the thermal energy, and therefore

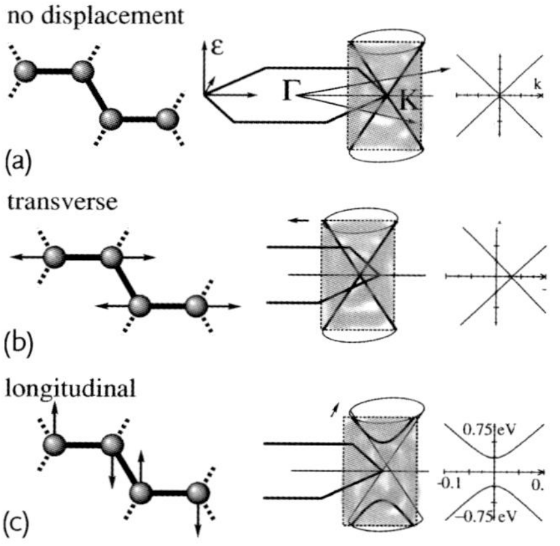

Figure 8.8 (a) Electronic band structure of graphene in the vicinity of the K point. Panels (b) and (c) indicate the changes in the electronic band structure caused by the presence of iTO (transverse) and LO (longitudinal) phonon modes. For the A_1(iTO) mode in armchair SWNTs (and the A_1(LO) mode in zigzag SWNTs), the crossing point shifts away from K towards the Γ point. For the A_1(LO) mode in armchair nanotubes (and the A_1(iTO) mode in zigzag tubes), the crossing point moves perpendicular to the line ΓK, opening a band gap. The thick lines indicate the band structure of an armchair tube obtained by intersecting the gray plane with the two cones. Adapted from [124].

the tube does not distort.[8] However, the LO phonon mode in the nanotube suffers a strong renormalization effect, exhibiting a significant softening in the phonon frequency. The same holds for zigzag and chiral metallic nanotubes but to a lesser degree. Although the cutting line for this case will have a different direction, the Δk_F will change as well, so that the overall picture remains similar. It is similar but not exactly the same, because a small correction will have to be considered when going beyond the first-neighbor tight-binding model, since curvature effects generate a mini-gap opening in non-armchair SWNTs, which can be explained by the extended tight-binding electron–phonon coupling model [117, 222].

8.4.2
Effect of Gate Doping on the G-band of SWNTs: Theory

In Figure 8.9, we show calculated results for $\hbar\omega_\lambda$ LO and TO phonons as a function of E_F for a (10, 10) armchair nanotube generated by gate doping. Here we take values of 1595 cm^{-1} and 1610 cm^{-1} for $\hbar\omega_\lambda^{(0)}$, for the λ = iTO and λ = LO modes, respectively. The energy bars in Figure 8.9 denote Γ_λ values for the decay width, and the extended tight-binding scheme is used to calculate $E_e(\boldsymbol{k})$, $E_h(\boldsymbol{k})$, and the electron wavefunction for $\langle \mathrm{eh}(\boldsymbol{k})|\mathcal{H}_{\mathrm{int}}|\omega_\lambda\rangle$ [267]. Here the el–ph matrix element [203] was obtained using the deformation potential, derived on the basis of density-functional theory by Porezag *et al.* [264]. To obtain the phonon eigenvector, the force constant parameters calculated by Dubay and Kresse [116] were used for the dynamical matrix. The resulting $\hbar\omega_\lambda$ is shown as a function of E_F at room temperature (T = 300 K) and at T = 10 K in Figure 8.9a,b, respectively, where $E_F \neq 0$ is related to gate doping with respect to the equilibrium position at $E_F = 0$, occurring when E_F is at the band crossing point (K point in graphene). It is seen that the iTO mode does not exhibit any energy change, while the LO mode shows both an energy shift and broadening. As we have mentioned above, the minimum energy occurs at $|E_F| = \hbar\omega^{(0)}/2$ ($\approx$ 0.1 eV). There is also a local maximum for the spectral peak at $|E_F| = 0$. The broadening for the LO mode occurs within $|E_F| \leq \hbar\omega^{(0)}/2$ for the lower temperature (10 K), while the broadening has a tail at room temperature for $|E_F| \geq \hbar\omega^{(0)}/2$ in Figure 8.9 [260]. For large $|E_F|$ values, the Kohn anomaly effect is gone and $\omega_G^{\mathrm{LO}} > \omega_G^{\mathrm{iTO}}$, as expected in the time-independent picture (Section 7.3.3) and is the behavior that is generally observed for semiconducting SWNTs.

A continuum model for electrons in a carbon nanotube has been used [260] to explain the lack of an energy shift of the iTO modes for armchair nanotubes. In this work it is shown that the electron–phonon (el–ph) matrix element for electron–hole pair creation by the (A_1) LO and iTO phonon modes is given by:

$$\begin{aligned} \langle \mathrm{eh}(\boldsymbol{k})|\mathcal{H}_{\mathrm{int}}|\omega_{\mathrm{LO}}\rangle &= -igu\sin\theta(\boldsymbol{k}) \,, \\ \langle \mathrm{eh}(\boldsymbol{k})|\mathcal{H}_{\mathrm{int}}|\omega_{\mathrm{iTO}}\rangle &= -igu\cos\theta(\boldsymbol{k}) \,, \end{aligned} \tag{8.9}$$

8) In the case of polyene encapsulated in a SWNT, however, the Peierls distortion is significant [266].

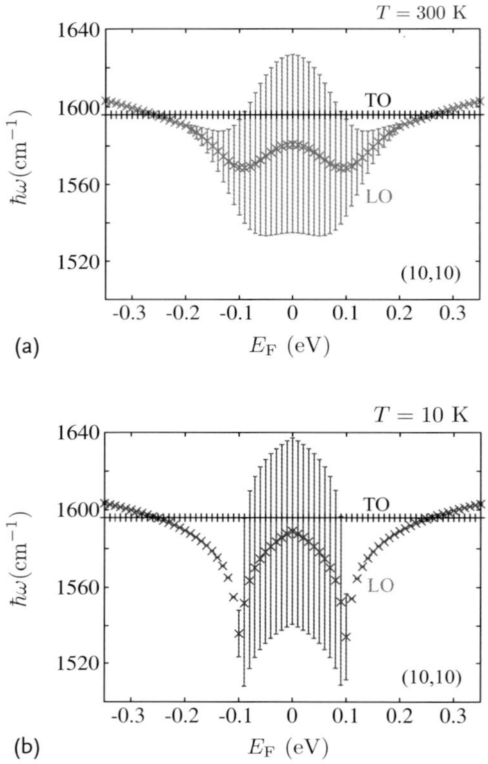

Figure 8.9 The E_F dependence of the LO (gray curve) and iTO (black curve) phonon energy in the case of the (10, 10) armchair nanotube. Part (a) is calculated at room temperature and (b) is at 10 K. Only the energy of the LO mode is shifted, with the iTO mode frequency being independent of E_F. The decay width (Γ_λ) in Eq. (8.5) is plotted as an error-bar [260].

where u is the phonon amplitude, and g is the el–ph coupling constant. Here $\theta(\boldsymbol{k})$ is the angle for the polar coordinate around the K (or K') point in the 2D Brillouin zone, in which the $\boldsymbol{k} = (k_1, k_2)$ point is taken to be on a cutting line for a metallic energy sub-band. The k_1 (k_2) axis is taken in the direction of the nanotube circumferential (axis) direction (see Figure 8.10). Equation (8.9) shows that the matrix element $\langle \text{eh}(\boldsymbol{k})|\mathcal{H}_{\text{int}}|\omega_\lambda\rangle$ depends only on $\theta(\boldsymbol{k})$ but not on $|\boldsymbol{k}|$, which implies that the dependence of this matrix element on E is negligible. For an armchair nanotube, even when considering the curvature-induced distortions, the cutting line for its metallic energy band still lies on the k_2 axis [268]. Thus, we have $\theta(\boldsymbol{k}) = \pi/2$ ($-\pi/2$) for the metallic energy sub-band, which has $k_1 = 0$ and $k_2 > 0$ ($k_2 < 0$). Then, Eq. (8.9) tells us that only the LO mode couples to an electron–hole pair and the iTO mode is not coupled to an electron–hole pair for armchair SWNTs.

In Figure 8.11a, we show calculated results for $\hbar\omega_\lambda$ as a function of E_F for a (15, 0) metallic zigzag nanotube [260]. In the case of zigzag nanotubes, not only the LO mode but also the iTO mode couples to electron–hole pairs. The spectral peak

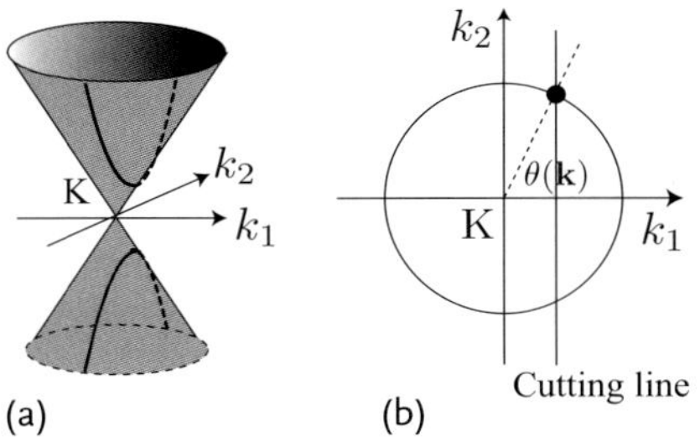

Figure 8.10 (a) Cutting line near the K point for an armchair nanotube. The k_1 (k_2) axis is selected as the nanotube circumferential (in-plane) direction. The amplitude for electron–hole pair creation depends strongly on the relative position of the cutting line from the K point. (b) If the cutting line crosses the K point, then the angle $\theta(\boldsymbol{k})$ ($\equiv \arctan(k_2/k_1)$) takes $\pi/2$ ($-\pi/2$) values for $k_2 > 0$ ($k_2 < 0$). In this case, the LO mode strongly couples to an electron–hole pair, while the iTO mode is decoupled from the electron–hole pair according to Eq. (8.9) [260].

position for the iTO mode is upshifted for $E_F = 0$, since $\text{Re}(\hbar\omega(E))$ for $E < \hbar\omega_{\text{iTO}}$ contributes to a positive frequency shift. It has been shown theoretically [268] and experimentally [269] that, even for "metallic" zigzag nanotubes, a finite curvature opens a small energy gap. When the curvature effect is taken into account, the cutting line does not lie on the K point, but it is then shifted from the k_2 axis. In this case, $\cos\theta(\boldsymbol{k}) = k_1/(k_1^2 + k_2^2)^{1/2}$ is nonzero for the lower energy intermediate electron–hole pair states since $k_1 \neq 0$. Thus, the iTO mode can couple to the low energy electron–hole pair which makes a positive energy contribution to the phonon energy shift. The high energy electron–hole pair is still decoupled from the iTO mode since $\cos\theta(\boldsymbol{k}) \to 0$ for $|k_2| \gg |k_1|$. Therefore, when $|E_F| \leq \hbar\omega_{\text{iTO}}^{(0)}/2$, then $\hbar\omega_{\text{iTO}}$ increases by a larger amount than does $\hbar\omega_{\text{LO}}$. The iTO mode for the small diameter zigzag nanotubes couples strongly to an electron–hole pair because of the stronger curvature effect for small diameter SWNTs. In Figure 8.11b, we show the diameter (d_t) dependence of the $\hbar\omega_\lambda$ for zigzag nanotubes for $E_F = 0$ not only for metallic SWNTs, but also for semiconducting SWNTs. In the case of the S-SWNTs, the LO (iTO) mode appears around 1600 (1560) cm^{-1} without any broadening. Only metallic zigzag nanotubes show an energy shift, and the energy of the LO (iTO) mode decreases (increases) as compared to the semiconducting tubes. In Figure 8.11c, we show a curvature-induced energy gap E_{gap} as a function of d_t for metallic zigzag tubes. The results show that higher (lower) energy electron–hole pairs contribute effectively to the LO (iTO) mode softening (hardening) in metallic nanotubes. In the case of semiconducting nanotubes, we may expect that there is a softening for the LO and iTO modes according to Eq. (8.9). However, the softening is small as compared with that of the metallic nanotubes because the energy of the intermediate electron–hole pair states is much larger than $\hbar\omega_\lambda^{(0)}$ in this case.

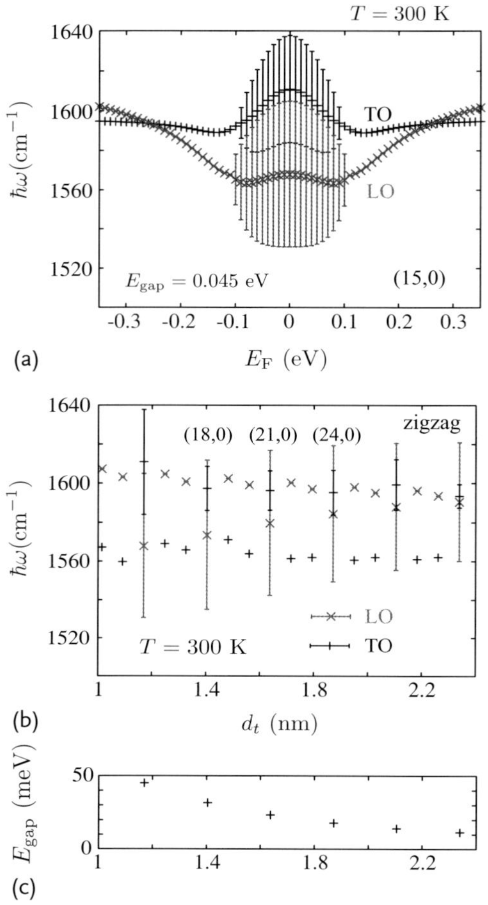

Figure 8.11 (a) The calculated E_{F} dependence of the LO (gray symbols) and iTO (black symbols) phonon frequency for a (15, 0) zigzag nanotube. The frequency of not only the LO mode, but also that of the iTO mode, is shifted due to the curvature effect. The error bars indicate the phonon linewidth. (b) Diameter d_t dependence of the G-band optical phonon frequencies for zigzag nanotubes, including zigzag semiconducting and metallic nanotubes as well as LO and TO modes. (c) The diameter dependence of E_{gap}, where E_{gap} denotes the curvature-induced mini-energy gap [260].

8.4.3
Comparison with Experiments

Before the Kohn anomaly was discussed in the context of graphene and carbon nanotubes, several experimental works had already reported the Fermi level dependence of the LO and iTO modes of bundles and ensembles of carbon nanotubes [48, 49, 270]. These experiments showed a significant change in frequency and linewidth of the broad feature in the G-band associated with metallic nanotubes as E_F was varied. Experiments varying the Fermi level of individual metallic carbon nanotubes [261, 271, 272] later provided more insight into the behavior of the individual modes. In Figure 8.12 we show an experimental intensity map of the G-band spectrum of an individual metallic nanotube as a function of an electrochemical gate voltage. Here there are two distinguishable peaks, a higher frequency mode that does not change appreciably either in frequency or in linewidth as a function of gate voltage, and a broad lower frequency peak, assigned to the LO mode, that upshifts and narrows in linewidth as positive or negative charges are induced on the nanotube.

Unlike the case of graphene, a Kohn anomaly is predicted in metallic carbon nanotubes even in the adiabatic limit. However, use of the adiabatic approximation misses out on key features in the energy range close to the phonon energy. The two signatures of nonadiabatic effects in the Kohn anomaly are: (1) an energy window $E_F < |\hbar\omega_{LO}|$ within which the LO phonon peak is broadened due to the creation of real electron–hole pairs, and (2) a characteristic "W" lineshape of the LO frequency vs. E_F curve caused by the two singularities located at $E_F = \pm\hbar\omega_{LO}$ rather than a single singularity at $E_F = 0$, as predicted within the adiabatic approximation. The data in Figure 8.12 exhibits the characteristic broadening window for metallic SWNTs; however, the "W" shape from the two singularities is not resolved, most likely because of inhomogeneous charging due to trapped charges on the substrate which will result in a smearing of E_F. A more recent experiment [273] using pristine suspended nanotubes that were gated electrostatically was able to resolve this feature in the frequency of the LO mode.

The experiment in Figure 8.12 captures the behavior of the strongly coupled LO mode. The weakly coupled TO mode, if identified at all, is reported to exhibit a flat frequency vs. gate voltage behavior around the Dirac point. This peak has not received much attention, since the TO mode intensity is weak, rendering it difficult

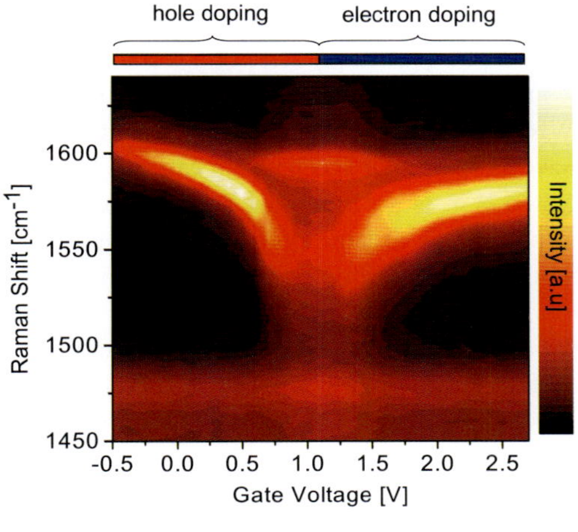

Figure 8.12 Experimental intensity plot of the G-band spectrum of a metallic SWNT as a function of electrochemical gate voltage. The charge neutrality point corresponding to the Dirac point is 1.2 V [261].

to monitor, especially when the G^+ and G^- features lie close to each other in energy. The intensities of the LO and TO modes are known to depend on the chiral angle, with the TO mode being completely silent in the limit of $\theta \rightarrow 0°$ for a zigzag nanotube [272, 274]. Nonetheless, as shown in Figures 8.9 and 8.11, Sasaki *et al.* [275] predict an interesting chiral angle dependence of the TO mode softening. Future experiments on structurally identified and truly isolated individual (n, m) nanotubes, while challenging, will certainly shed light on this topic.

Similar experiments on semiconducting nanotubes reveal that their G-band phonons also experience energy renormalization due to electron–phonon coupling [49, 276]. Since $\hbar\omega_{\mathrm{LO/TO}} < E_{11}^{\mathrm{S}}$ for S-SWNTs, the G-band phonons are unable to create real electron–hole excitations across the semiconducting band gap, and as a result there is no lifetime broadening of the phonon. Nonetheless, virtual electron–hole excitations, which do not conserve energy, do contribute to renormalizing the phonon energy [276]. In the case of semiconducting SWNTs both the TO and LO modes couple to intermediate electron–hole pairs with the TO mode experiencing a greater E_{F}-dependent frequency shift. Interestingly, this effect becomes most significant in larger diameter nanotubes as the band gap energy approaches the phonon energy [276]. Therefore the frequency renormalization in semiconducting nanotubes has the opposite diameter dependence to that of metallic SWNTs, and the effect in S-SWNTs is smaller in magnitude than in M-SWNTs.

8.4.4
Chemical Doping of SWNTs

As discussed in Section 8.2.2, low doping levels suppress the Kohn anomaly, thus causing an upshift in the LO G-band frequency for both p and n doping in SWNTs (see Figure 8.12). However, for higher doping levels the structural distortions are expected to dominate, so that p and n doping should then cause an upshift and downshift, respectively, in the measured phonon frequencies of SWNTs (see Figure 8.4c). Such p or n doping behavior has indeed been observed experimentally for the G-band of MWNTs and SWNTs doped chemically to higher doping levels with different atoms, which behave as donors or acceptors to carbon [34, 277]. The dopant-induced interactions (whether it is an inorganic species such as an alkali-metal donor or a halogen acceptor, or an organic polymer chain or a DNA strand) with the sidewall of a nanotube will perturb the Fermi level of the nanotube through charge-transfer interactions. Since electrons and phonons are strongly coupled to each other, these perturbations will influence the various Raman modes present in carbon nanotubes. For example, the doping with alkali metals like K, Rb, and Cs leads to a softening (or downshift) of 35 cm^{-1} (saturated regime) of the G-band frequencies, and is accompanied by dramatic changes in its lineshape. For SWNT bundles doped with halogens (for example, Br_2), an upshift in the Raman-mode frequencies was observed relative to the corresponding frequencies in the pristine bundles. Details about the Raman characterization of doped SWNTs can be found in [34].

8.5 Summary

In this chapter we addressed the effect of time-dependent perturbations to the G-band spectra of graphene and carbon nanotubes. We showed that, when the adiabatic approximation is not valid, electrons and phonons couple, thus changing both the electron energies (Peierls-like effect) and the phonon energies (Kohn anomaly effect). These effects are strongly dependent on gate voltage and temperature, thus making the G-band work as a probe for nanocarbon doping. The Kohn anomaly is observed in metallic systems, where real electron–hole pair creation can occur by a phonon energy ($\hbar\omega_G$) process, thus strongly influencing the G-band frequency and linewidth of graphene and metallic SWNTs. The effects in metallic SWNTs are stronger than in graphene because of the quantum confinement effect, and this process depends sensitively on diameter and chiral angle. In graphene these effects depend on the number of layers. Semiconducting SWNTs also exhibit a phonon energy renormalization due to electron–phonon coupling, but this renormalization effect is weaker than for the metallic SWNTs because no real anomaly takes place ($E_{gap} > \hbar\omega_G$) for S-SWNTs. Consequently, while the G-band linewidth in graphene and metallic SWNTs is strongly sensitive to whether or not the gate voltage matches the energy of the anomaly, in semiconducting SWNTs the G-band linewidth is basically independent of doping. Finally, putting together the rich behavior of the G-band frequency and linewidth, as discussed in Chapters 7 and 8, we conclude that the Raman G-band provides a highly sensitive probe for studying and characterizing nanocarbons.

Problems

[8-1] Calculate the period of the oscillation for the vibration at $1580\,\text{cm}^{-1}$. Also give the frequency in THz. (Use the fact that $1\,\text{eV} = 8650\,\text{cm}^{-1}$.)

[8-2] Estimate the Raman spectral width in cm^{-1} by using the uncertainty relation when the lifetime of the photoexcited carrier is 500 fs. Repeat the calculation for 50 fs.

[8-3] A typical length of a carbon nanotube is 1 μm. How long does it take for the light to go 1 μm. How many times do carbon atoms oscillate at the G-band frequency during this time?

[8-4] What is the maximum velocity or acceleration of the atomic vibration for a phonon of $1580\,\text{cm}^{-1}$? In this calculation, we should consider the number of phonons to be n.

[8-5] When we use the maximum acceleration in the previous problem, evaluate the force that a π electron feels in this acceleration. Compare this force with the Coulomb force in a carbon atom. You can use $Z = 4$ for the screened

ion core and $r = 0.5$ Å for the radius. Check if the Coulomb potential is sufficiently strong to keep the π electron bound to the carbon atom.

[8-6] (Peierls instability) Consider a linear carbon chain in which the nearest neighbor transfer parameters have alternating values: $t_1, t_2, t_1, t_2, \ldots$. Show that in this case, an energy gap is opened at the zone boundary and the value of the energy gap is proportional to $t_1 - t_2$.

[8-7] In the previous problem, let us consider the electron–phonon parameter α such that $t_i = t_0 - \alpha(x_{i+1} - x_i)$, where x_i denotes the lattice distortion. By an alternative lattice distortion, $x_i = x_0(-1)^i$, the total energy of the electron decreases because of the opening an energy gap at the Fermi energy. On the other hand, because of the lattice distortion, the system loses lattice energy which is proportional to $Kx_0^2/2$ per bond (K is the spring constant). By minimizing the total energy, obtain the optimized x and the energy gap.

[8-8] When the temperature of the nanotubes is either high or very low, how does the phonon softening change as a function of the Fermi energy? Using the Fermi distribution function, and explain your result qualitatively.

[8-9] The G-band phonon becomes soft when the temperature becomes high. Explain the mechanism of phonon softening for high temperature.

[8-10] When the Fermi energy changes, how is the electron–phonon interaction suppressed? Explain and plot qualitatively the phonon frequency as a function of the Fermi energy.

[8-11] Show that, for metallic SWNTs, only the phonon causes an electronic gap opening, independent of the tube chiral angle (this is shown in Figure 8.8 for an armchair SWNT). Here you should show the above result for a zigzag and a chiral SWNT.

[8-12] The RBM (radial breathing mode) frequency can be expected to produce phonon softening by changing the Fermi energy. However, the shift of the RBM frequency is known not to be large ($\sim$ 1–3 cm^{-1}). Consider why the phonon softening of the RBM is small. How about the phonon softening for the D or G$'$-bands?

9
Resonance Raman Scattering: Experimental Observations of the Radial Breathing Mode

In the next three chapters, we present an in depth analysis of the resonance Raman scattering process, that makes possible the observation of the Raman spectra from isolated nanocarbons, such as a single-layer graphene [86], one isolated carbon nanotube [176] or an isolated nanoribbon [83]. Although resonance can occur in any nanocarbon material, in these chapters, we focus on the radial breathing mode (RBM) of nanotubes because the RBM is an especially instructive example of resonance Raman scattering. Because of the low frequency (low energy) of ω_{RBM} and because of the one-dimensional character of carbon nanotubes, the RBM spectra are extremely informative about resonance Raman phenomena. Thus, the study of the RBM spectra can serve to give a clear picture on how Raman spectroscopy can be used to probe the electronic structure of nanotubes. Furthermore, the RBM-related science is so well developed that there are already sufficient experiments and theory in the literature to address most of the information one can generally extract from a Raman feature through its intensity (I_{RBM}), frequency (ω_{RBM}), linewidth (Γ_{RBM}), and the dependence of the three properties on the excitation laser energy (E_{laser}), and also paying attention to environmental effects. Here the environmental effects refer to spectral changes associated with perturbations due to doping or arising from changes in the materials surrounding the SWNT. Since we are dealing with a nanomaterial, any surrounding material will play an important role in the observed optical-related properties, and the RBM spectra can also be used to probe such environmental conditions.

In this chapter, we start in Section 9.1 with the definition of the RBM and a description of its frequency dependence on the tube diameter, which can be simply derived from elasticity theory. In Section 9.2 we review the general optical properties of the RBM spectra in one isolated SWNT, including the resonance Raman effect, the resonance window, Stokes and anti-Stokes phenomena and polarization effects. The final Section 9.3 ends the chapter with an extension of the resonance Raman analysis discussed in Section 9.2 to SWNT samples with a broad (n, m) distribution. These results will serve as a basis for the study of the carbon nanotube electronic structure, theoretically addressed in Chapter 10. Chapter 10 is very interesting from the physics point of view, by departing from the tight-binding description already introduced in Chapter 2, and by discussing the effect of σ–π hybridization and of excitonic effects on the Raman spectra. Chapter 11 addresses

Raman Spectroscopy in Graphene Related Systems. Ado Jorio, Riichiro Saito, Gene Dresselhaus, and Mildred S. Dresselhaus

ISBN: 978-3-527-40811-5

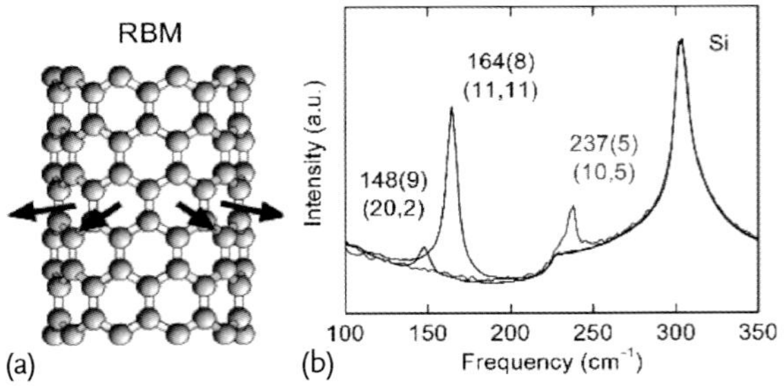

Figure 9.1 (a) Schematic picture of the atomic displacements in the radial breathing mode (RBM). (b) Three superimposed Raman spectra of the RBM of three isolated SWNTs grown by the chemical vapor deposition (CVD) method and contained on a Si/SiO_2 substrate. The spectra are taken at three different spots on the substrate where the RBM Raman signal from resonant SWNTs are found. The RBM frequencies (linewidths) are displayed in cm^{-1}. Also shown are the (n, m) indices assigned from the Raman spectra for each resonant tube. The step in the spectrum at $\sim$ 225 cm^{-1} and the peak at 303 cm^{-1}, common to all spectra, come from the Si/SiO_2 substrate [176].

both the electron–photon and electron–phonon matrix elements and their effect on the observed Raman spectra.

9.1
The Diameter and Chiral Angle Dependence of the RBM Frequency

As suggested by its name, in the radial breathing mode (RBM) all the C atoms are vibrating in the radial direction with the same phase, as if the tube were breathing (see Figure 9.1a). The atomic motion does not break the tube symmetry, that is, the RBM is a totally symmetric (A_1) mode. Since this particular vibrational mode only occurs in carbon nanotubes, it is used to distinguish carbon-based samples containing carbon nanotubes from sp^2 carbon samples that do not contain carbon nanotubes, and to give particular emphasis to samples containing single-wall carbon nanotubes (SWNTs), where the intensity of the RBM is strong compared with other nonresonant spectra coming from the substrate or with other resonance Raman spectra (see Figure 9.1b) [176]. A very important characteristic is the RBM frequency dependence on tube diameter ($\omega_{\mathrm{RBM}} \propto 1/d_t$). Although this dependence was first predicted using force constant calculations [278], an analytical derivation can be made using elasticity theory, that is the subject of Section 9.1.1. Later in Section 9.1.5 shows the small deviations from the simple inverse diameter dependence due to curvature effects and the Kohn anomaly.

9.1.1
Diameter Dependence: Elasticity Theory

Here we show the dependence of the RBM frequency on the SWNT diameter. Elasticity theory describes the energetics of a continuous, homogeneous medium under strain, and follows mostly from Hooke's law (strain being proportional to stress) and Newton's second law. Then the potential energy in an elastic medium is given by [95]:

$$U = \frac{1}{2} \sum_{\lambda=1}^{6} \sum_{\mu=1}^{6} C_{\lambda\mu} e_\lambda e_\mu \,, \tag{9.1}$$

where $C_{\lambda\mu}$ is the stiffness constant which relates strain and stress, and e_λ (or e_μ) is the strain. The sum in Eq. (9.1) is over all possible strain/stress axes ($\lambda, \mu = xx, yy, zz, yz, zx, xy$). Equation (9.1) is a general expansion from a harmonic potential $U = \frac{1}{2} K x^2$. If we consider a uniaxial strain along z, it is common to use the Young's modulus ($Y = C_{zzzz}$), which is defined as the coefficient relating strain/stress to tension/deformation along zz.

The elastic energetics for the RBM can then be described by a one-dimensional-like tension/deformation. The variation in nanotube radius (δR) can be related to a one-dimensional strain e along the radial direction r, which stretches the graphene sheet in the circumferential direction, associated with the nanotube by

$$e = \frac{\delta R}{R} \,, \tag{9.2}$$

and the related elastic energy will be given by

$$U = \frac{1}{2} \int Y e^2 dV = \frac{1}{2} Y V \left(\frac{\delta R}{R} \right)^2 \,, \tag{9.3}$$

from where, by considering a general vibration with a spring constant k $\omega = \sqrt{k/M}$, where k is given by YV/R^2, we get

$$\omega_{\mathrm{RBM}} = \sqrt{\frac{YV}{MR^2}} = \sqrt{\frac{Y}{\rho}} \frac{1}{R} = \frac{A}{d_\mathrm{t}} \,, \tag{9.4}$$

where V is the volume and M is the mass of the cylinder, $\rho = M/V$ is the density and d_t is the tube diameter. The proportionality constant A in Eq. (9.3) can be estimated from the elastic properties of graphite. By describing sound waves in terms of elasticity theory, we see that $\sqrt{Y/\rho}$ is the sound velocity for the longitudinal acoustic mode ($\nu_\mathrm{L} = 21.4$ km/s) [279]. Therefore, A describes the elastic behavior of an isolated SWNT in the large diameter limit, where elasticity theory is expected to be valid, thereby giving $A = 227\,\mathrm{cm}^{-1}\,\mathrm{nm}$ [95, 259, 279, 280].

Figure 9.2 shows a plot of ω_{RBM} vs. d_t for 197 different SWNTs (of which 73 are metallic and 124 semiconducting) [31, 189]. For all the 197 SWNTs, their (n, m) indices were assigned by experiment (extracted from Figure 9.15, discussed in Section 9.3.2) and their diameters were determined by the relation for tube diameter $d_\mathrm{t} = a_{\mathrm{C\text{–}C}} \sqrt{3(n^2 + mn + m^2)}/\pi$, where $a_{\mathrm{C\text{–}C}} = 0.142$ nm is the carbon–carbon distance (see Section 2.3.1). By thus fitting the experimental data shown in Figure 9.2 using the relation $\omega_{\mathrm{RBM}} = \boldsymbol{A}/d_\mathrm{t} + \boldsymbol{B}$, we obtain $\boldsymbol{A} = (227.0 \pm 0.3)\,\mathrm{nm\,cm}^{-1}$ and $\boldsymbol{B} = (0.3 \pm 0.2)\,\mathrm{cm}^{-1}$. This result is in remarkably good agreement with elasticity theory, thus directly connecting one-dimensional carbon nanotubes and their

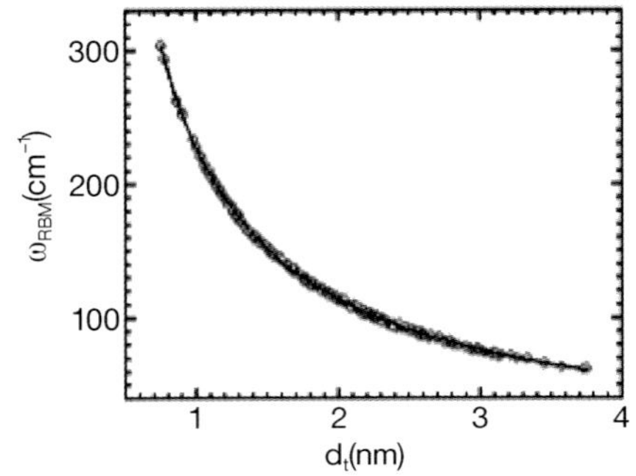

Figure 9.2 Experimental radial breathing mode frequency (ω_{RBM}) as a function of tube diameter (d_t). Open circles represent experimental values and the solid line is given by $\omega_{RBM} = 227.0/d_t + 0.3$ [189].

two-dimensional counterpart graphene from which nanotubes are conceptually derived.

Although experiment and elasticity theory agree perfectly in the experiment analyzed in Figure 9.2, these results have only been obtained, so far, for one specific type of SWNT, that is, SWNTs which are ultra-long, vertically aligned and grown by the water-assisted chemical vapor deposition (CVD) method [281]. Most of the RBM experimental results in the literature have been fitted with the relation $\omega_{RBM} = A/d_t + B$, with values for the parameters A and B varying widely from paper to paper [189, 282], as discussed in the next section.

9.1.2 Environmental Effects on the RBM Frequency

As discussed in the previous section, the RBM resonance Raman scattering (RRS) of SWNTs grown by the water-assisted CVD method [281] follows the simplest linear relation between ω_{RBM} and d_t, namely $\omega_{RBM} = A/d_t$, with the proportionality constant $A = 227.0\,\text{cm}^{-1}\,\text{nm}$, in agreement with the elastic properties of graphene [279], and with a negligible environmental effect ($B \approx 0$) [189]. However, all the other experimental results in the carbon nanotube literature have been fitted with the relation $\omega_{RBM} = A/d_t + B$, with values for A and B varying from one research group to another [172, 176, 180, 183, 184, 283–288]. A nonzero value for the empirical constant factor B prevents the expected limit of a graphene sheet from being achieved, where ω_{RBM} should go to zero when d_t approaches infinity. Therefore, B is supposedly associated with an environmental effect on ω_{RBM}, rather than an intrinsic property of SWNTs. The “environmental effect” here means the effect of the surrounding medium, such as bundling, molecules adsorbed from the air, the surfactant used for the SWNT bundle dispersion, the substrates on which the tubes are sitting, etc. As we will discuss here, all the observed ω_{RBM} values reported in the literature are upshifted from the fundamental relation ($\omega_{RBM} = 227/d_t$, with $B = 0$), the upshift exhibiting a d_t dependence in quantitative agreement with recent predictions which consider the van der Waals interaction between SWNTs and their environment [189].

In Figure 9.3 we compare similar ω_{RBM} Raman spectra taken from two different samples. The gray lines show the ω_{RBM} spectra for the "super-growth" SWNTs which are compared to the black line ω_{RBM} spectra obtained from a SWNT sample grown by the alcohol-assisted CVD method [287]. Comparing the spectra in Figure 9.3a and b it is clear that the ω_{RBM} values for the "alcohol-assisted CVD" sample are upshifted from the "super-growth" ω_{RBM} frequencies.[1]

Figure 9.4a shows the difference between several determinations of $\omega_{\text{RBM}} = A/d_t + B$ found in the literature [172, 176, 180, 183, 184, 286, 287] and the $\omega_{\text{RBM}} = 227.0/d_t$ relation for the "super-growth" samples. All the curves in the literature converge within the 1 to 2 nm d_t range, which is the diameter range for which most of the experimental data were actually obtained. Figure 9.4b shows the difference between the actual experimental values for ω_{RBM} from the literature ($\omega_{\text{RBM}}^{\text{Lit.}}$) [176, 183, 184, 283–288] and for the "super-growth" (S.G) sample ($\omega_{\text{RBM}}^{\text{S.G.}}$), as a function of d_t. All the published results for $\omega_{\text{RBM}}^{\text{Lit.}}$ are grouped in Figure 9.4b on a d_t-dependent upshifted trend for $\Delta\omega_{\text{RBM}} = \omega_{\text{RBM}}^{\text{Lit.}} - \omega_{\text{RBM}}^{\text{S.G.}}$. Therefore, the d_t dependence of the difference between the experimental data in the literature and the fundamental relation $\omega_{\text{RBM}} = 227.0/d_t$ is always of the same sign, as shown in Figure 9.4b.

The problem of addressing the environmental effect on ω_{RBM} is now reduced to solving a simple harmonic oscillator equation for a cylindrical shell subjected to an inwards pressure ($p(x)$) given by [189, 279]:

$$\frac{2x(t)}{d_t} + \frac{\rho}{Y}(1-\nu^2)\frac{\partial^2 x(t)}{\partial t^2} = -\frac{(1-\nu^2)}{Yh}p(x) \,, \qquad (9.5)$$

where $x(t)$ is the displacement of the nanotube in the radial direction, $p(x) = (24\,K/s_0^2)x(t)$, and K (in eV/Å^2) gives the van der Waals interaction strength, s_0 is

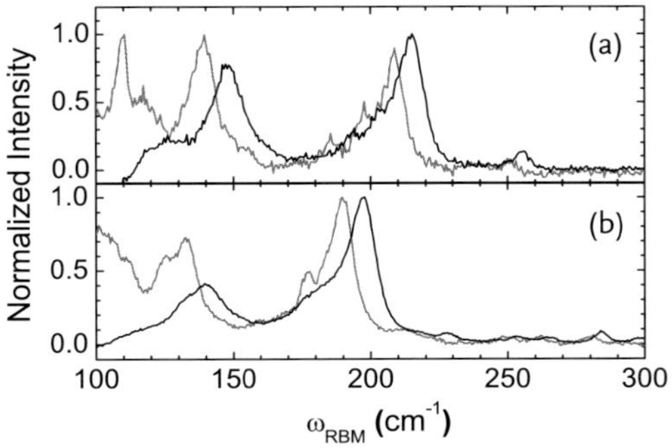

Figure 9.3 The ω_{RBM} spectra for "super-growth" SWNTs (gray) and for "alcohol CVD" SWNTs (black). The spectra are obtained using different laser lines: (a) 590 nm (gray) and 600 nm (black); (b) 636 nm (gray) and 650 nm (black) [189].

1) The differences in the low frequency region (below ~ 120 cm^{-1}) are due to different d_t distributions among various nanotube samples.

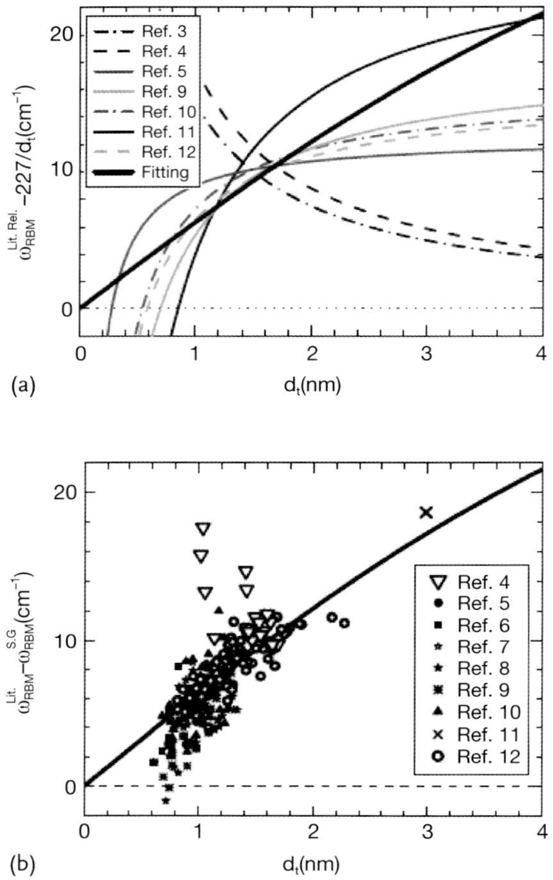

Figure 9.4 (a) Difference between the ω_{RBM} dependence on d_t from the literature ($\omega_{\mathrm{RBM}}^{\mathrm{Lit.\ Rel.}}$) values and the $\omega_{\mathrm{RBM}} = 227.0/d_t$ relation as a function of tube diameter d_t. (b) Difference between the experimental ω_{RBM} data from the literature ($\omega_{\mathrm{RBM}}^{\mathrm{Lit.}}$) and the ω_{RBM} data for the "super-growth" sample ($\omega_{\mathrm{RBM}}^{\mathrm{S.G.}}$) as a function of d_t [189]. Each symbol in (b) represents data from a different reference (see [189] for the references in the legends to (a) and (b) of this figure). The thick solid line is a fit to the data in (b), as discussed in the text, and also shown in (a) [189].

the equilibrium separation between the SWNT wall and the environmental shell, Y is the Young's modulus (69.74×10^{11} g/cm s^2), ρ is the mass density per unit volume (2.31 gm/cm^3), $\nu = 0.5849$ is Poisson's ratio and h represents the thickness of the shell [279]. If $p(x)$ vanishes, Eq. (9.5) gives the fundamental frequency ω_{RBM}^0 for a pristine SWNT in units of cm^{-1},

$$\omega_{\mathrm{RBM}}^0 = \left\{ \frac{1}{\pi c} \left[\frac{Y}{\rho(1-\nu^2)} \right]^{1/2} \right\} \frac{1}{d_t}, \tag{9.6}$$

where the term inside the curly bracket in Eq. (9.6) gives the fundamental value of $A = 227.0\,\text{cm}^{-1}\,\text{nm}$.[2] For a nonvanishing $p(x)$ we have

$$\omega'_{\text{RBM}} = 227.0\left[\frac{1}{d_t^2} + \frac{6(1-\nu^2)}{Yh}\frac{K}{s_0^2}\right]^{1/2}, \tag{9.7}$$

where $[6(1-\nu^2)/Yh] = 26.3\,\text{Å}^2/\text{eV}$. The shift in ω'_{RBM} due to the environment is given by $\Delta\omega_{\text{RBM}} = \omega'_{\text{RBM}} - \omega^0_{\text{RBM}}$. The data in Figure 9.4b is fitted (see thick black solid line) by considering K/s_0^2 in Eq. (9.7) as an adjustable parameter. The best fit is obtained with $K/s_0^2 = (2.2 \pm 0.1)\,\text{meV/Å}^4$. The d_t-dependent behavior of the environmental effect in ω_{RBM} is then established in Figure 9.4 for d_t up to $d_t = 3\,\text{nm}$. A similar environmental effect is obtained for SWNTs in bundles [172, 287], surrounded by different surfactants [180, 183, 184, 283–285], suspended in air by posts [286], or sitting on a SiO_2 substrate [176], but this environmental effect is absent in "super-growth" SWNTs.

For simplicity, all the ω_{RBM} results in the literature which are upshifted from the pristine values due to the van der Waals interaction with the environment can be generally described by:

$$\omega^{\text{Lit.}}_{\text{RBM}} = \frac{227}{d_t}\sqrt{1 + C_e * d_t^2}\,, \tag{9.8}$$

where $C_e = [6(1-\nu^2)/Eh][K/s_0^2]\,(\text{nm}^{-2})$ gives the effect of the environment on ω_{RBM}. Table 9.1 gives the the C_e values fitting the RBM results for several samples in the literature. For $d_t < 1.2\,\text{nm}$, where the curvature effects become important, the environmental effect depends more critically on the specific sample (i. e., C_e for one SWNT sample on SiO_2 may differ from another sample on SiO_2 in the literature), and the observed environmental-induced upshifts range from 1 to $10\,\text{cm}^{-1}$ for small diameter tubes within bundles or wrapped by different surfactants (e. g., SDS (sodium dodecyl sulfate) or single-stranded DNA). This effect gets even stronger when considering the effect of the outer tube on the inner tube in a double-wall carbon nanotube (DWNT), as discussed in the next section.

Table 9.1 Strength of the environmental effect on the RBM frequency as measured by the C_e factor in Eq. (9.8) which fits different SWNT samples in the literature.

C_e	Sample	Reference
0	Water-assisted CVD	Araujo [189]
0.05	HiPCO@SDS	Bachilo [288]
0.059	Alcohol-assisted CVD	Araujo [287]
0.065	SWNT@SiO_2	Jorio [176]
0.067	Free-standing	Pailet [286]

2) Equation (9.6) is different from Eq. (9.3) because in Eq. (9.6) we consider the Poisson ratio $\nu \neq 0$, and the $(1/2\pi c)$ term used to measure frequency in cm^{-1} is given explicitly.

9.1.3
Frequency Shifts in Double-Wall Carbon Nanotubes

The inner and the outer tubes of a DWNT can be either metallic (M) or semiconducting (S).[3] Thus, the following four configurations are possible: M@M, M@S, S@S, and S@M, where S@M denotes an S inner tube inside an M outer tube, following the common notation for fullerenes [289, 290]. Each DWNT configuration is expected to possess distinct electronic properties. In particular, for the S@M configuration, the S inner tube of a DWNT could be regarded as a good approximation for an isolated semiconducting SWNT that is electrostatically shielded and physically protected from the local environment by an outer metallic tube. Therefore, the experimental data from the inner tubes can be used as a standard when compared to SWNTs that are subjected to environmental effects, such as contact with a substrate, water, oxygen, or charged molecular species [290].

Most spectroscopic experiments on DWNTs have been performed on bundles or solution-based samples [291–296], so that it has been inherently difficult to use Raman spectra to investigate which inner (n, m) tubes are actually contained inside the variety of observed outer (n', m') tubes (see Figure 9.5a). In order to quantitatively determine which specific inner and outer tubes actually form each DWNT, one must perform Raman experiments on individual DWNTs (see Figure 9.5c). Techniques that combine the use of E-beam lithography, atomic force microscopy (AFM) and Raman mapping have been developed to measure the Raman spectra from the inner and outer layers of the same individual DWNTs (see Figure 9.5) [289, 290].

An investigation of the Raman spectra of 11 isolated C_{60}-DWNTs, all with (6,5) semiconducting inner tubes and all with the S@M configuration was performed using a single laser excitation energy of $E_{\text{laser}} = 2.10\,\text{eV}$ [290]. The outer tubes of the 11 DWNTs that are formed with a (6,5) inner tube can have different (n, m) designations from one another but some outer tubes will have common (n, m) chiralities. The radial breathing mode (RBM) frequencies $\omega_{\text{RBM,o}}$ for the outer tube for such a DWNT as a function of $\omega_{\text{RBM,i}}$ for the inner tube are shown in Figure 9.6a. In this figure we see that for these 11 individual isolated DWNTs, $\omega_{\text{RBM,o}}$ for the outer tubes varies over a $12\,\text{cm}^{-1}$ range, while $\omega_{\text{RBM,i}}$ for the inner tubes (which all correspond to (6,5) tubes) does not have a constant value, but rather varies over a range of $18\,\text{cm}^{-1}$. This $18\,\text{cm}^{-1}$ variation in the RBM frequency $\omega_{\text{RBM,i}}$ for the inner tube is large, considering that all these inner tubes are (6,5) tubes. These experiments tell us that in forming a DWNT, the inner and outer tubes impose considerable stress on one another. This is suggested by the fact that the nominal wall to wall distances $\Delta d_{t,\text{io}}$ between the inner (i) and outer (o) tubes of the DWNTs are less than the c-axis distance in graphite (0.335 nm). In fact Figure 9.6b shows that $\Delta d_{t,\text{io}}$ values as small as 0.29 nm can be observed, implying a decrease of up

3) In discussing DWNTs, there are two methods for preparing DWNTs, one from heat-treating C_{60} containing SWNTs (called peapods) and denoted by C_{60}-DWNTs, and a second CVD-based method denoted by CVD-DWNTs. Since the two methods lead to DWNTs with different diameter distributions they have somewhat different characteristics.

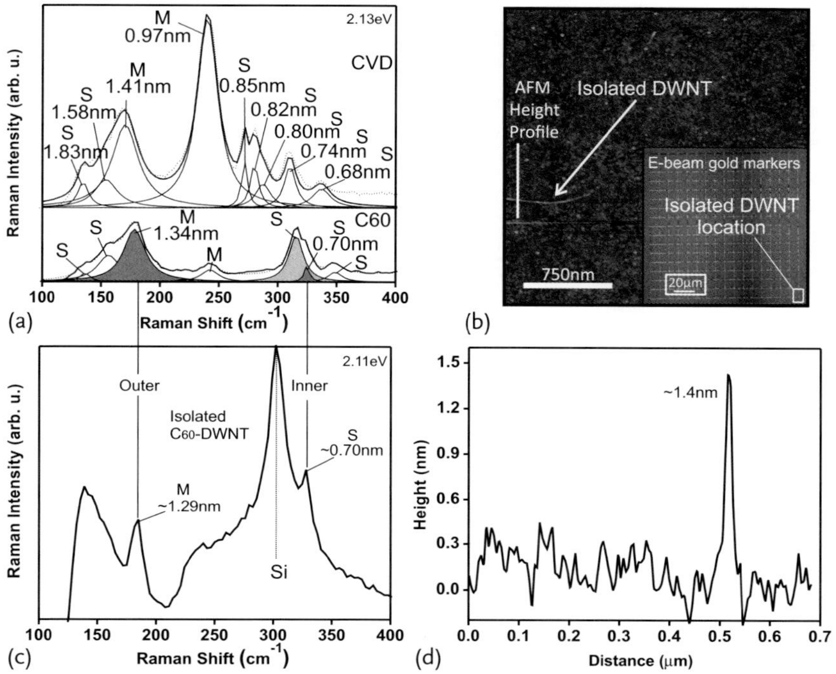

Figure 9.5 (a) Raman spectra for the RBM region for CVD-DWNT and C_{60}-DWNT bundles ($E_{laser} = 2.13$ eV). (b) Atomic force microscope (AFM) image of one individual, isolated DWNT. Inset: Silicon substrate with Au markers showing the location of the DWNT. (c) Raman spectra for the RBM Raman region ($E_{laser} = 2.11$ eV) for an isolated individual C_{60}-DWNT and (d) AFM height profile of the individual, isolated DWNT shown in (b) with the RBM spectrum shown in (c). The vertical lines connecting (a) and (c) show that the ω_{RBM} of the prominent tube diameters observed in the C_{60}-DWNT bundles coincide with the ω_{RBM} of the inner and outer tubes of the isolated C_{60}-DWNTs [290].

to 13% in the wall to wall distance for this set of 11 DWNTs (all of which have (6,5) inner tubes) [290]. In such studies, the tube diameters d_t and wall to wall distances between inner and outer tubes $\Delta d_{t,io}$ were determined from the radial breathing mode frequency-based on the relation between ω_{RBM} and d_t developed for SWNTs (see Section 9.1.2). These estimates for d_t should be considered as nominal values for d_t, and further work is needed to develop a corresponding relation between ω_{RBM} and $1/d_t$ that is valid for DWNTs. Because of the differences in the Coulomb interaction expected for the 4 different DWNT configurations, i. e., S@M, M@S, S@S and M@M, it is expected that even if a linear relation between ω_{RBM} and $1/d_t$ is retained for the inner and outer tubes of each DWNT configuration, the detailed relation will depend on the metallicity configuration of a given DWNT, as given above.

In the case of MWNTs, most of the samples are composed of tubes with diameters too large to exhibit observable RBM features. Although, in a few cases the

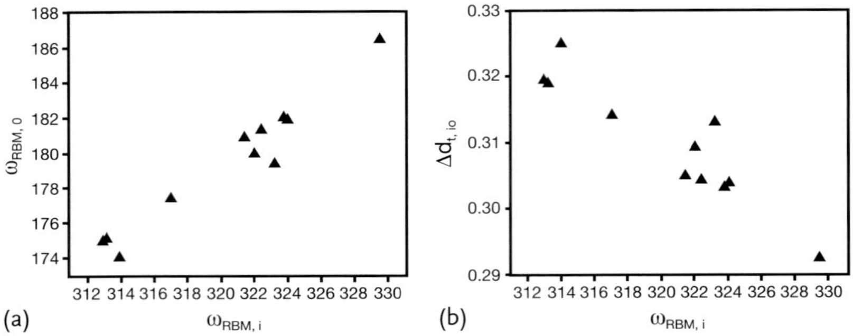

Figure 9.6 All the inner tubes for the 11 peapod-DWNTs in this figure are (6,5) semiconducting tubes. (a) Plot of the $\omega_{RBM,i}$ of the inner tube vs. $\omega_{RBM,o}$ for the outer tubes which pair to form eleven different isolated DWNTs. (b) Plot of the nominal wall to wall distance $\Delta d_{t,io}$ for each of the 11 isolated DWNTs vs. $\omega_{RBM,i}$ shown in (a). An increase in the $\omega_{RBM,i}$ of the inner tubes (all are (6,5) tubes) is accompanied by a decrease in the measured nominal wall to wall $d_{t,io}$ distance [290].

inner tubes have small enough diameters ($d_t \lesssim 2\,\text{nm}$) and their RBM contributions can be seen [297], generally the RBM is not a reliable probe for studying and characterizing MWNTs.

9.1.4 Linewidths

The Raman spectral width is given by the lifetime of phonons. Several mechanisms can be responsible for the linewidth broadening of the resonance Raman features in SWNT spectra, including temperature-dependent effects (anharmonic processes, phonon–phonon and electron–phonon interactions and other effects), tube–tube/tube–substrate interactions, and nanotube defects (vacancies, substitutional and interstitial impurities, 7-5 structural defects, etc.), finite size effects, trigonal warping, as well as the energy separations between the incident or scattered photon and the pertinent van Hove singularity. Linewidth studies are best carried out at the single nanotube level where inhomogeneous broadening effects are minimized and linewidths approaching the natural linewidths for the various processes should be achievable.

Figure 9.7 shows the dependence of the RBM linewidth (Γ_{RBM}) on diameter d_t for 170 SWNTs grown by CVD on a Si/SiO_2 substrate. It is noteworthy that Γ_{RBM} values down to $4\,\text{cm}^{-1}$ are observed in individual SWNTs, since Raman peaks in sp^2 carbons are usually broader [242]. The low Γ_{RBM} values are characteristic of 1D SWNTs. From the 170 data points in Figure 9.7, we clearly observe an increase in the average Γ_{RBM} value (and also in the minimum value) with increasing d_t, that is, with increasing number of atoms along the circumference of the SWNTs. While this result might have a relation to intrinsic confinement effects related to the increase of the tube diameter, a tube flattening due to tube-substrate interaction

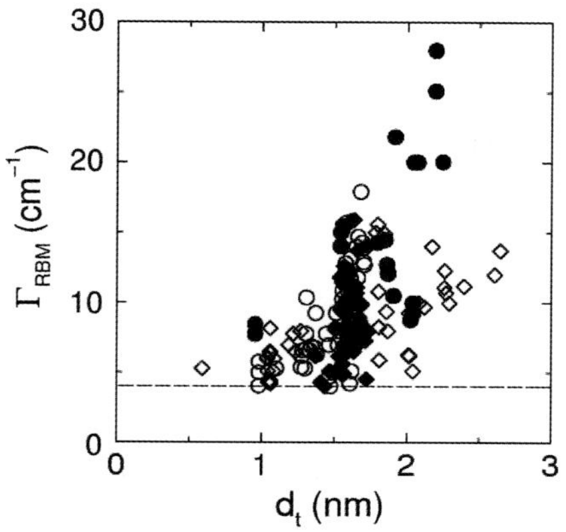

Figure 9.7 Γ_{RBM} vs. d_t for 81 M-SWNTs (filled symbols) and 89 S-SWNTs (open symbols) at 300 K. Circles indicate data obtained with the 2.41 or 2.54 eV laser excitation lines, and diamonds when obtained with a 1.58 eV laser [242].

is also expected to increase with increasing tube diameter, and this flattening may also play an important role in the observed linewidth dependence [298]. In the Raman spectra of SWNTs deposited in Si/SiO_2 substrates, RBM features are not often observed below 90 cm^{-1}, although SWNTs with $d_t > 2.5$ nm are not rare in this type of sample. The limited accuracy in identifying larger diameter tubes is probably caused by too large a broadening of the RBM peak.

9.1.5
Beyond Elasticity Theory: Chiral Angle Dependence

There are two effects that are not considered by elasticity theory. The first is related to the chirality-dependent distortion of the lattice. The second is related to electron–phonon coupling in metallic SWNTs and is associated with the Kohn anomaly (see Section 8.4). These two effects can generate a chiral angle dependence of the RBM frequencies. The first effect should be observable in measurements made on small diameter ($d_t \sim 1$ nm) SWNTs, where the curvature-induced lattice distortion is important. The second is observed only in metallic SWNTs.

Results of ω_{RBM} vs. (d_t, θ) (i. e., (n, m)) were obtained from SWNTs grown by the HiPCO (high pressure CO CVD) method and dispersed in surfactant aqueous solution [299]. The best linear relation fitting the RBM frequency dependence on diameter obtained for this sample was $\omega_{RBM} = 218.3/d_t + 15.9$ (for a discussion of changes to the ω_{RBM} vs. d_t relation see Section 9.1.2). Figure 9.8 shows a plot of the deviations of ω_{RBM} values from the best linear $1/d_t$ dependence that fits all the experimental data ($\Delta\omega_{RBM} = \omega_{RBM} - (218.3/d_t + 15.9)$) as a function of the chiral angle θ. In this figure, one clearly sees deviations of the points from $\Delta\omega_{RBM} = 0$, and these deviations are as large as $\Delta\omega_{RBM} \sim \pm 3$ cm^{-1}, which is much larger than the experimental accuracy ($\simeq 1.0$ cm^{-1}).

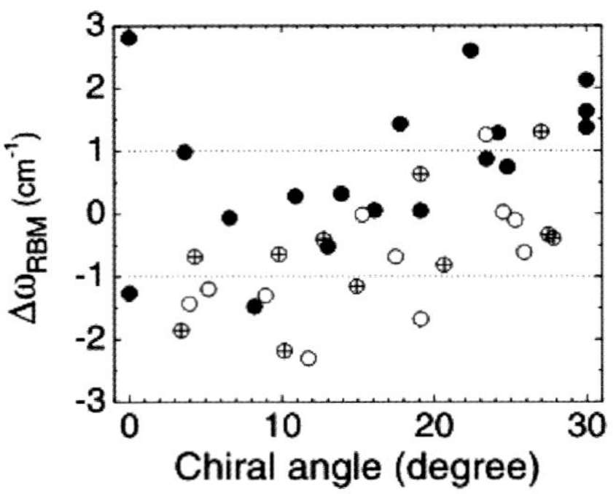

Figure 9.8 Deviation of the experimentally observed RBM frequency (ω_{RBM}) from the linear dependence given by ($218.3/d_t + 15.9$), as a function of θ for a particular HiPCO nanotube sample [299]. Filled, open and crossed circles denote M-SWNTs, type I and type II S-SWNTs, respectively. The dotted lines show an experimental accuracy of $\pm 1\,\text{cm}^{-1}$ [299].

Interesting trends can be seen from the deviations in Figure 9.8. The first is the observation of a systematically larger $\Delta\omega_{\text{RBM}}$ for M-SWNTs (metallic SWNTs, solid bullets) when compared with S-SWNTs (semiconducting SWNTs, open bullets). The second is a $\Delta\omega_{\text{RBM}}$ dependence on the chiral angle θ, showing a clear increase in $\Delta\omega_{\text{RBM}}$ with increasing θ from $0°$ (zigzag) to $30°$ (armchair), and both of these effects are stronger for metallic tubes.

Some of these deviations in ω_{RBM} are due to curvature effects. For small d_t SWNTs, curvature weakens the sp^2 chemical bonds which now have components along the circumferential direction, because of sp^2–sp^3 mixing. As a result, the RBM frequencies decrease with respect to their ideal values as the SWNT diameter decreases. Moreover, curvature destroys the isotropy of the elastic constants in SWNTs and therefore introduces a chirality dependence into ω_{RBM}. All these effects are well documented from a theoretical point of view [182, 300] where, by allowing the atoms to assume relaxed equilibrium positions for each (d_t, θ), the effective diameter changes could be determined. Kürti *et al.* [300] describe in detail the curvature effects on many structural properties of SWNTs. For instance, it is predicted that diameter deviations from the ideal d_t values are roughly the same for zigzag and armchair tubes, but the changes in bond lengths are larger for the two C–C bonds with components along the circumferential direction for zigzag tubes as compared to the three such bonds for armchair tubes with similar diameter. This is a purely geometric effect, related to the directions of the three C–C bonds with respect to the circumferential direction. Therefore, in armchair tubes, the circumferential strain is more evenly distributed between the bonds, leading to smaller bond elongation. Since the RBM softening is directly related to the elongation of bonds along the circumference, a larger softening of ω_{RBM} for zigzag tubes relative to armchair tubes is expected.

Finally, similar to the effect discussed for the G-band in Chapter 8, a phonon frequency shift of the radial breathing mode for M-SWNTs is predicted [275] and observed [301] as a function of Fermi energy, although a much smaller shift

($\sim$ 3 cm^{-1}) due to the Kohn anomaly effect is expected for ω_{RBM} than for ω_{G}. Armchair nanotubes will not show any renormalization-induced frequency shift while zigzag nanotubes will exhibit the maximum phonon softening. This chirality dependence originates from the k-dependent electron–phonon coupling for RBM phonons [275]. In the chiral and zigzag metallic SWNTs, a small energy gap is opened by the curvature of the cylindrical surface. When the curvature-induced gap is larger than $\hbar\omega_{\text{RBM}}$, then the Kohn anomaly effect disappears. Since the gap is proportional to $1/d_t^2$ and $\hbar\omega_{\text{RBM}}$ is proportional to $1/d_t$, there is a lower limit of d_t (1 to 1.8 nm depending on chiral angle) below which we cannot see the Kohn anomaly effect for the RBM phonon [275] .

9.2 Intensity and the Resonance Raman Effect: Isolated SWNTs

The Raman effect as shown in Figures 9.1 and 9.9 for the RBM features is a resonant process. With the physics of ω_{RBM} and Γ_{RBM} in place, next we consider the evolution of the RBM intensity as the laser excitation energy is varied. The range of laser energies over which the resonance Raman spectra is observed is called the resonance window (see Section 4.3.2).

9.2.1 The Resonance Window

Strong resonant effects occur in the Raman scattering from an isolated SWNT when the energy of the incident or scattered light matches an optical transition E_{ii} (see Section 2.3.4), thereby strongly enhancing the Raman signal [112, 136, 171, 176, 282, 303]. Therefore, it is possible to use the resonance Raman effect to study the electronic structure of individual SWNTs, and much effort has therefore been given to measuring the Raman spectra under resonant conditions [176, 302, 304–306]. In this section we review observations of the resonance window for the RBM feature.

Figure 9.9a shows an AFM image of a Si substrate with a thin SiO_2 surface coating [176, 307] and with lithographic markers on an $8 \times 8\ \mu\text{m}^2$ lattice. Isolated SWNTs were grown on top of the substrate by a CVD method (see lines in Figure 9.9b). The light spot ($\sim$ 1 μm diameter) is positioned to be close to a mark ($\sim$ 1 μm size) (see Figure 9.9a) in order to achieve good precision in always returning the light spot to the same position on the sample as E_{laser} is changed. The dashed circles in Figure 9.9a,b display the position where the laser spot is placed, showing the presence of some isolated SWNTs. From the AFM measured SWNT heights, the diameters (d_t) of the 11 SWNTs that lie within the light spot are determined, with d_t ranging from 0.7 nm to 1.9 nm (the AFM precision is about $\pm$0.2 nm). Raman spectra of the sample were measured in the laser excitation wavelength (energy) range 720 nm (1.722 eV) $\leq E_{\text{laser}} \leq$ 785 nm (1.585 eV) with steps of 4 nm ($\sim$ 0.009 eV), as shown in Figure 9.9c,d. All the anti-Stokes

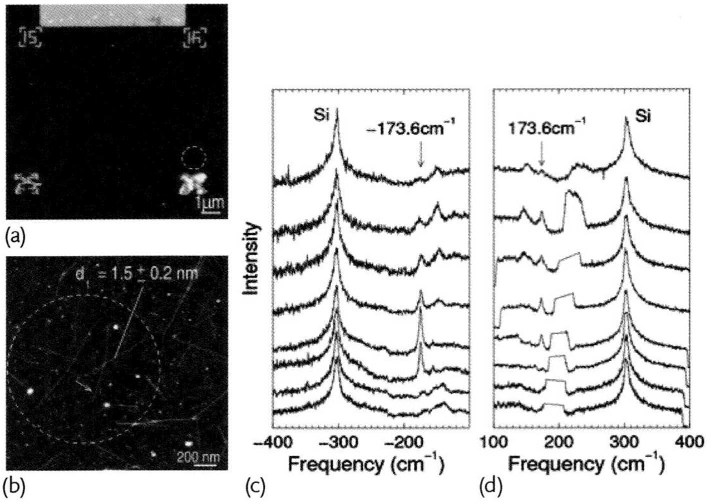

Figure 9.9 (a,b) AFM image of the SWNT sample. Part (a) shows the markers used to localize the spot position (dashed circle) on the substrate during the Raman experiment and for further AFM characterization of the SWNTs that are located within the light spot indicated by the dashed circle in (b). (c) anti-Stokes and (d) Stokes Raman spectra from isolated SWNTs on a Si/SiO_2 substrate for several different laser excitation energies. From bottom to top, the spectra were taken at E_{laser} = 1.623, 1.631, 1.640, 1.649, 1.666, 1.685, 1.703, and 1.722 eV. The excitation was provided by a tunable Ti:Sapphire laser ($P < 10\,\text{mW}$ on the sample) pumped by an Ar ion laser (6 W). The incident light was filtered with a single-monochromator (Macpherson 1200 g/mm), and the scattered light was analyzed with an XY DILOR triple-monochromator, equipped with a N_2 cooled CCD detector. The Stokes signal quality (d) is not as good as that for the anti-Stokes signal (c) due to the frequency-dependent spectrometer efficiency that drops off rapidly with increasing laser wavelength, being worse in the Stokes frequency region. The flat region appearing in all the Stokes spectra in (d) comes from light leakage, and was cut out from the spectra [302].

(Figure 9.9c) and Stokes (Figure 9.9d) spectra were corrected to account for spectrometer efficiency at each laser energy, and the spectra were then normalized by the 303 cm^{-1} Si substrate peak intensities. The anti-Stokes intensities were multiplied by $[n(\omega)+1]/n(\omega)$, where $n(\omega) = 1/[\exp(\hbar\omega/k_B T) - 1]$ is the Bose–Einstein thermal factor, ω is the RBM frequency, k_B is the Boltzmann constant, and T is the temperature (see Section 4.3.2.1). Although high laser power was used to measure the Raman spectra, T was found to be close to room temperature (not higher than 325 K), and this was confirmed by changing the laser power from 1 mW/μm^2 (10 MW/cm^2) to 10 mW/μm^2 (100 MW/cm^2), where the Stokes/anti-Stokes intensity ratio for the 521 cm^{-1} and nonresonant 303 cm^{-1} Si peaks remained constant. Furthermore, the ω_{RBM} peak did not show a temperature-dependent shift, and the intensity ratios between the RBM features and the 303 cm^{-1} Si peaks also remained constant in both the Stokes and anti-Stokes spectra [302].

With the light spot position shown in Figure 9.9a, the Raman spectra were measured with many different laser excitation energies. Figure 9.9 shows the anti-Stokes (c) and Stokes (d) Raman spectra of one light spot for several different laser excitation energies E_{laser}, increasing from the bottom to the top spectra (see caption). In Figure 9.9c,d, we see the RBM feature at 173.6 cm^{-1} appearing and disappearing over the tunable energy range of E_{laser}, thereby allowing us to tune over the whole resonant window of one optical transition energy (E_{ii}) for this resonant SWNT. The linewidth for this $\omega_{\text{RBM}} = 173.6\,\text{cm}^{-1}$ peak is $\Gamma_{\text{RBM}} = 5\,\text{cm}^{-1}$, typical of that for one isolated SWNT (see Section 9.1.4) [176, 242]. The data points in Figure 9.10 show the peak intensity of the 173.6 cm^{-1} RBM feature vs. E_{laser} in the anti-Stokes (a) and Stokes (b) processes, which define the resonance window width Γ for both the anti-Stokes and Stokes processes for the SWNT measured in Figure 9.9.

The RBM peak intensity $I(E_{\text{laser}})$, which is a function of E_{laser}, can be evaluated from Eq. (5.20) (Chapter 5). The first and second factors in the denominator of Eq. (5.20), respectively, describe the resonance effect with the incident and scattered light. Here $+(-)$ applies to the anti-Stokes (Stokes) process for the phonon of energy E_{ph}, while γ_{RBM} gives the inverse lifetime for the resonant scattering pro-

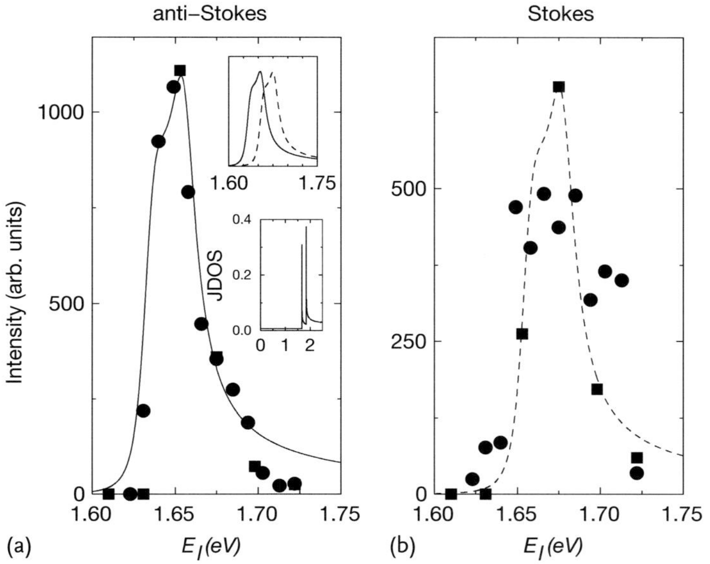

Figure 9.10 Raman intensity vs. laser excitation energy E_l for the $\omega_{\text{RBM}} = 173.6\,\text{cm}^{-1}$ peak (see Figure 9.9) for the (a) Stokes and (b) anti-Stokes Raman processes. Circles and squares indicate two different E_{laser} runs on the same SWNT sample. The line curves indicate the resonant Raman window predicted from Eq. (5.20), with $E_{ii} = 1.655$ eV, $\gamma_r = 8$ meV, but taking the sum over internal states ($\sum_{m,m'}$) outside the square modulus. The upper inset compares the theoretically predicted Stokes and anti-Stokes resonant windows on an energy scale in eV, and the lower insert shows the joint density of states (JDOS) vs. E_{laser} for this SWNT [302].

cess [308]. For simplicity, the matrix elements $M^d M^{ep} M^d$ can be considered to be independent of energy in this small energy range. Here M^d and M^{ep} are, respectively, the matrix elements for the electron-radiation (absorption and emission) and the electron–phonon interactions. Chapter 11 develops the theory for these matrix elements in detail.

The curves in Figure 9.10 show theoretical fits to the experimental data points for the Stokes (dashed line) and anti-Stokes (solid line) resonant windows for an (18,0) metallic zigzag SWNT, using $E_{ph} = 21.5$ meV obtained from $\omega_{RBM} = 173.6\,cm^{-1}$ [302]. Notice the asymmetric lineshape in the resonance windows. These fits were actually obtained in [302] by considering not a coherent Raman scattering process, but an incoherent scattering process, where the sum over the internal states ($\sum_{m,m'}$ in Eq. (5.20)) was taken outside the square modulus. However, this procedure is controversial, since this asymmetry could be generated by other resonance levels lying close in energy.

Disregarding the asymmetry aspect, the width of the resonant windows gives $\gamma_{RBM} = 8$ meV, in good agreement with previous measurements [171, 172, 309] and a transition energy of $E_{ii} = 1.655 \pm 0.003$ eV is also found. The upper inset to Figure 9.10 shows a comparison between the theoretically predicted Stokes and anti-Stokes resonant windows, revealing a shift in these resonant windows due to the resonant condition for the scattered photon, $E_s = E_{ii} \pm E_{ph}$ for the anti-Stokes (+) and the Stokes (−) processes, respectively. Therefore, by using a tunable laser, it is possible to study the resonance window for *one* isolated SWNT, giving its E_{ii} value with a precision better than 5 meV at room temperature. Resonance windows for the RBM mode are, in fact, found to have a dependence on both diameter and chiral angle, as discussed in Section 9.3.2.

9.2.2
Stokes and Anti-Stokes Spectra with One Laser Line

In the nonresonant Raman spectra, the anti-Stokes intensity is always smaller than the Stokes intensity, and the I_{aS}/I_S intensity ratio can be used to measure the sample temperature (see Section 4.3.2.1). However, under sharp resonance conditions the I_{aS}/I_S ratio strongly depends on the difference between the laser excitation energy E_{laser} and the resonance energy E_{ii}. The I_{aS}/I_S intensity ratio for the RBM then depends sensitively on $E_{ii} - E_{laser}$, and the I_{aS}/I_S ratio for the RBM feature at E_{laser} can be used to determine E_{ii} experimentally to within 10 meV and to determine whether the resonance is with the *incident* or *scattered* photon [310].

Figure 9.11 shows both Stokes and anti-Stokes spectra for the RBM for another isolated SWNT sitting on a Si/SiO_2 substrate, which is similar to the SWNTs shown in Figure 9.9. The measured anti-Stokes intensity is already corrected by the Bose–Einstein thermal factor, and a temperature $T = 300$ K was found from the two Si phonon features also present in these spectra. In Figure 9.11a,b the normalized anti-Stokes intensity at $\omega_{RBM} = 253\,cm^{-1}$ is much larger than the Stokes intensity. This asymmetry in intensity between the anti-Stokes and Stokes RBM spectra can be quantitatively analyzed by using resonance Raman theory, and the resulting

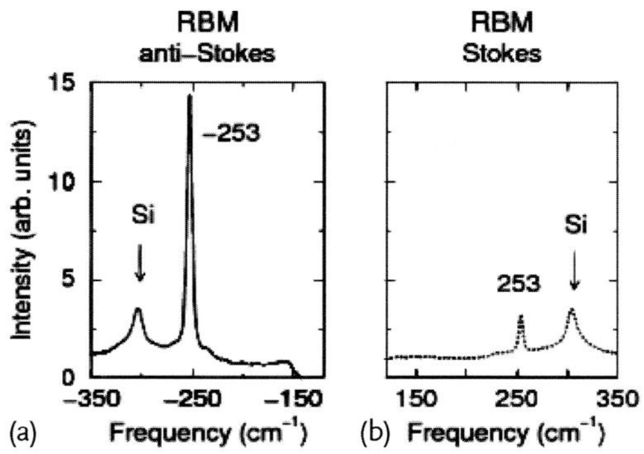

Figure 9.11 (a) Resonant anti-Stokes and (b) Stokes Raman spectra of a (12,1) SWNT (as identified in [310]) on a Si/SiO_2 substrate, using $E_{laser} = 1.579$ eV (758 nm). The peak at $\pm 303\,\mathrm{cm}^{-1}$ comes from the Si substrate. The RBM frequencies are displayed in cm^{-1}.

resonance window $I(E_{laser})$ for both the anti-Stokes and Stokes spectra from one isolated tube can be calculated using Eq. (5.20).

9.2.3
Dependence on Light Polarization

As discussed in Chapter 6, the totally symmetric Raman-active modes (A_1 symmetry) can only be observed when both the incident and scattered light are polarized along the tube (ZZ), or perpendicular to the tube axis (XX). In the (ZZ) scattering configuration, an optical transition is allowed between electronic states with the same angular momenta, that is, $E_{\mu\mu} = \mathbb{E}_\mu^{(v)} \to \mathbb{E}_\mu^{(c)}$ (see Section 6.4.5). Such a transition is equivalent to the usual E_{ii} transitions denoted in the Kataura plot (see Section 2.3.4). In the (XX) scattering configuration, an optical transition is allowed between electronic states with different angular momenta, that is, $E_{\mu\mu} = \mathbb{E}_\mu^{(v)} \to \mathbb{E}_{\mu\pm1}^{(c)}$. Such transitions are usually denoted by $E_{i,i\pm1}$, and they differ in energy from the usual transition energies E_{ii} [311]. There have been strong efforts to characterize such transitions, using polarized photoluminescence spectra [40, 312].

Since the RBM features from isolated SWNTs are seen only under resonance conditions, it is expected that the RBM from a single carbon nanotube will be seen in the (ZZ) and (XX) polarizations for different laser excitation energies. The polarization dependence of the Raman intensity related to the laser excitation energy has been called the *antenna* effect. This antenna effect was first reported by Duesberg *et al.* [235] (see Figure 9.12) and later by others [226–228, 313–315].

In general the intensity of the (XX) polarized spectra should be strongly suppressed by the so-called *depolarization* effect. Ajiki and Ando [238] have calculated the optical conductivity of carbon nanotubes taking into account this depolarization effect, and they found that the absorption of light polarized parallel to the tube axis (Z) is up to 20 times larger than that for perpendicularly polarized light (X).

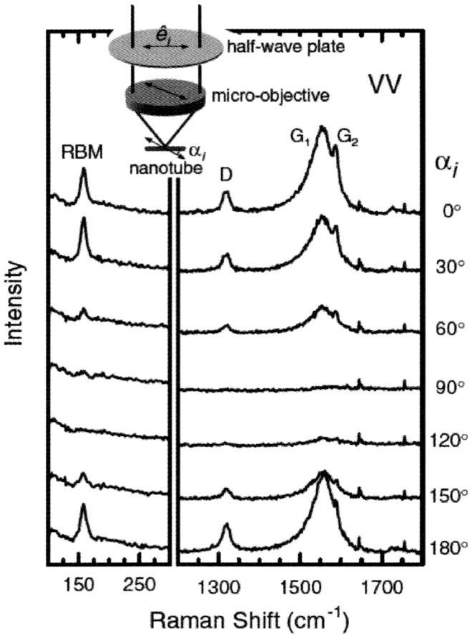

Figure 9.12 Raman spectra of an isolated SWNT (or a thin rope of SWNTs) in the VV configuration for various angles α_i between the tube axis and the polarization of the incident laser beam, as depicted in the inset. For $\alpha_i = 0°$ and $180°$ ($VV = ZZ$), the polarization of the incident radiation is parallel to the axis of the SWNTs determined from scanning force microscopy images with an accuracy of $\pm 10°$ [235].

This leads to a strongly reduced Raman signal when the polarization of the incident radiation is perpendicular to the nanotube axis. This polarization behavior has been demonstrated experimentally in SWNT bundles, where many (n, m) tubes are present and both E_{ii} and $E_{i,i\pm1}$ can be determined for specific tubes using the same sample with the same excitation laser energy [316].

9.3 Intensity and the Resonance Raman Effect: SWNT Bundles

In this section the resonance window analysis introduced in the previous section will be extended to SWNT ensembles. Through the RBM resonance window analysis, we can study the (n, m) dependence of the optical transition energies (E_{ii}). This analysis reveals a great deal of information that goes beyond the simple tight-binding method described in Chapter 2, including $\sigma-\pi$ hybridization and utilizing the science of excitons. The optical transition energies in SWNTs which are sensitive to these excitonic effects have been studied in detail through fluorescence and Raman spectroscopy experiments [80, 183, 185]. Though some aspects of the experiments can be interpreted within the context of a simple, noninteracting electron model [182, 299], it has become increasingly clear that electron–electron interactions also play an important role in determining the optical transition energies. Finally, SWNTs represent one of the best known materials system for the study of exciton photophysics, both from a theoretical and experimental viewpoint. Since SWNTs involve only carbon atoms, theoretical calculations can be carried out by using a relatively simple model Hamiltonian, as discussed in Chapter 10.

9.3.1
The Spectral Fitting Procedure for an Ensemble of Large Diameter Tubes

In isolated SWNTs (Section 9.2.1), or even in ensembles of small d_t SWNTs (briefly discussed in Section 4.4.2, see Figure 4.13a), each radial breathing mode (RBM) in resonance with a given excitation laser line (E_{laser}) is spectrally well-defined in frequency (ω_{RBM}), so that the RBM peaks and resonance profiles can be clearly identified. When larger d_t SWNTs are present and the differences in ω_{RBM} become smaller than the RBM linewidth, the RBM peaks cannot be clearly resolved and the spectra exhibit broad RBM features with contributions from several different (n, m) SWNTs. Therefore, fitting the Raman spectra becomes complex. It is then necessary to establish a systematic procedure to perform the Raman spectral analysis.

Figure 9.13a shows the RBM Raman spectrum obtained from the "alcohol-assisted" SWNTs using $E_{laser} = 1.925$ eV (644 nm) [287]. The bullets show the data points and the solid line shows the fit obtained using 34 Lorentzian curves (the peaks below the spectral curve in Figure 9.13a). Each Lorentzian curve can be related to the RBM from SWNTs with the same (n, m) index. The dark gray Lorentzians represent the RBM from M-SWNTs and the light gray Lorentzians represent the RBM from S-SWNTs. To determine how many Lorentzians should be used to fit

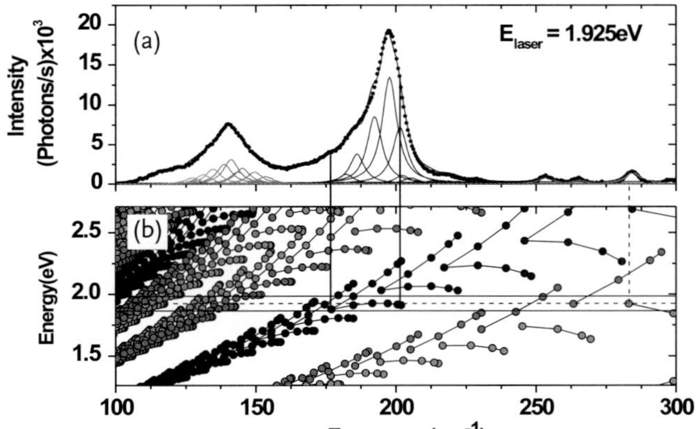

Figure 9.13 (a) Raman spectrum (bullets) obtained with a 644 nm laser line ($E_{laser} = 1.925$ eV). This spectrum was fitted by using 34 Lorentzians (curves under the spectra) and the solid line is the fitting result. (b) The Kataura plot used as a guide for the fitting procedure (from EPAPS material in [287]).

each resonance spectrum, we use the Kataura plot (see Figure 9.13b). The dashed horizontal line in Figure 9.13b represents the excitation energy for the spectrum shown in Figure 9.13a, and the two bold horizontal lines (above and below the dashed line) give the approximate boundary for the RBM resonance profiles (see detailed discussion in Section 9.3.2). To fit the spectrum shown in Figure 9.13a we expect that transitions corresponding to all the bullets inside the rectangle limited by the two bold lines should occur. The vertical bold lines connecting Figure 9.13a,b indicate that the metallic $2n + m = 30$ family is in resonance with the laser excitation energy 1.925 eV, while the dashed vertical line shows the RBM feature from the $(7, 5)$ SWNT.

The difficulty in performing the spectral fitting is due to the large number of Lorentzian curves needed to fit a broad RBM profile [282]. The fitting program tends to broaden and increase some peaks, while eliminating others. If for the same fit one Lorentzian is shifted by a couple of cm^{-1}, the fitting program will return a completely different fitting result. Therefore, some constraints for the peak frequencies and spectral linewidths (full width at half maximum (FWHM)) must be applied. For example, the ω_{RBM} obey the relation $\omega_{\text{RBM}} = (227/d_{\text{t}})\sqrt{1 + C_e/d_{\text{t}}^2}$, which correctly describes environmental effects by changing C_e and this relation is discussed in detail in Section 9.1.2. For lack of information, we may have to require all the Lorentzian peaks in one experimental spectrum to share the same FWHM value.

After analyzing all the spectra such as shown in Figure 9.13a, the Raman intensity at each RBM frequency has to be plotted as a function of E_{laser}. Such a plot gives the resonance profile for the (n, m)-specific SWNTs that have the specified RBM frequency. The RBM peak intensity $I(E_{\text{laser}})$, which is a function of E_{laser}, can be evaluated from Eq. (5.20) or, alternatively, by using a simplification of this equation given by

$$I(E_{\text{laser}}) \propto \left| \frac{1}{(E_{\text{laser}} - E_{ii} - i\gamma_{\text{RBM}})(E_{\text{laser}} - E_{ii} \pm E_{\text{ph}} - i\gamma_{\text{RBM}})} \right|^2 . \tag{9.9}$$

To illustrate the fitting procedure, Figure 9.14 shows three resonance profiles (black bullets), one in the near-infrared range (a), one in the visible range (b), and one in the near-ultraviolet range (c). The three resonance profiles were fitted according to resonance Raman scattering theory (solid line, from Eq. (9.9), and the values obtained for E_{ii} are indicated in Figure 9.14 (as well as for ω_{RBM} and (n, m)) [287, 317]. Notice the resonance window width for SWNTs in bundles (usually within 40–160 meV range) are much broader than for isolated SWNTs (see Figure 9.10).

9.3.2 The Experimental Kataura Plot

In this section, the resonance window analysis is extended to all (n, m) SWNTs, from where we can study the E_{ii} dependence on (d_{t}, θ). Figure 9.15a shows a 2D RBM map for the water-assisted CVD grown (here called the "super-growth", S.G. [281]). SWNT sample. This sample has a very broad diameter distribution, and

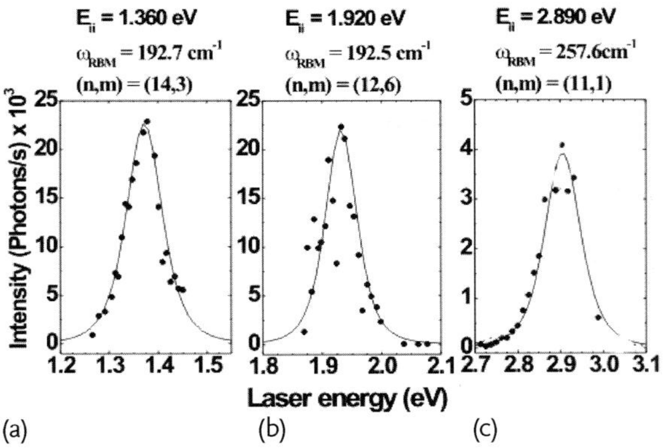

Figure 9.14 Resonance windows for specific (n, m) SWNTs within a bundle. (a) Resonance profile (black dots) in the near-infrared range for $\omega_{\text{RBM}} = 192.7\,\text{cm}^{-1}$. The data for tube (14,3) was fitted (solid line) using Eq. (9.9) with $\gamma_{\text{RBM}} = 0.065\,\text{eV}$ and $E_{ii} = 1.360\,\text{eV}$. (b) Resonance profile in the visible range is shown for $\omega_{\text{RBM}} = 192.5\,\text{cm}^{-1}$ (tube (12,6)), with $\gamma_{\text{RBM}} = 0.045\,\text{eV}$ and $E_{ii} = 1.920\,\text{eV}$. (c) Resonance profile in the near-ultraviolet range is shown for $\omega_{\text{RBM}} = 257.6\,\text{cm}^{-1}$ (tube (11,1)), with $\gamma_{\text{RBM}} = 0.073\,\text{eV}$ and $E_{ii} = 2.890\,\text{eV}$ (from EPAPS material in [287]).

can be used to gain a deep understanding of the SWNT optical properties. For the construction of the experimental Kataura plot in Figure 9.15a, 125 different laser lines were used [189, 317]. By fitting each of the spectra with Lorentzians, (n, m) indices were assigned to 197 different SWNTs.[4)]

Figure 9.15b is a plot of all $E_{ii}^{\text{S.G.}}$ obtained experimentally by fitting the resonance windows extracted from the data in Figure 9.15b, as a function of $\omega_{\text{RBM}}^{\text{S.G.}}$. The observed $E_{ii}^{\text{S.G.}}$ ranges from E_{11}^{S} up to E_{66}^{S} (the superscripts S stand for semiconducting S-SWNTs and M for metallic M-SWNTs). Finally, all the $E_{ii}^{\text{S.G.}}$ data in Figure 9.15b can be fitted using an empirical equation that is discussed below and given by [282, 287, 318]:

$$E_{ii}(p, d_t) = \alpha_p \frac{p}{d_t}\left[1 + 0.467 \log \frac{0.812}{p/d_t}\right] + \beta_p \cos 3\theta / d_t^2\,, \tag{9.10}$$

where p is defined as $1, 2, 3, \ldots, 8$ for E_{11}^{S}, E_{22}^{S}, E_{11}^{M}, E_{33}^{S}, E_{44}^{S}, E_{22}^{M}, E_{55}^{S}, E_{66}^{S}, thus measuring the distance of each cutting line from the K point in the zone-folding procedure. The fitting gave values $\alpha_p = 1.074$ for $p = 1, 2, 3$ and $\alpha_p = 1.133$ for $p \geq 4$. The β_p values for the lower (upper) E_{ii} branches are $-0.07(0.09)$, $-0.18(0.14)$, $-0.19(0.29)$, $-0.33(0.49)$, $-0.43(0.59)$, $-0.6(0.57)$, $-0.6(0.73)$ and -0.65 (unknown) for $p = 1, 2, 3, \ldots, 8$, respectively [317, 318]. The functional form in Eq. (9.10) carries a linear dependence of E_{ii} on p/d_t, as expected from the tight-binding theory plus quantum confinement of the 2D electronic structure of graphene, a logarithmic correction term that comes from many-body interactions, and a θ-dependent term which includes electronic trigonal warping and chirality-dependent curvature

4) The data for Figure 9.2 came from this experiment [189, 317].

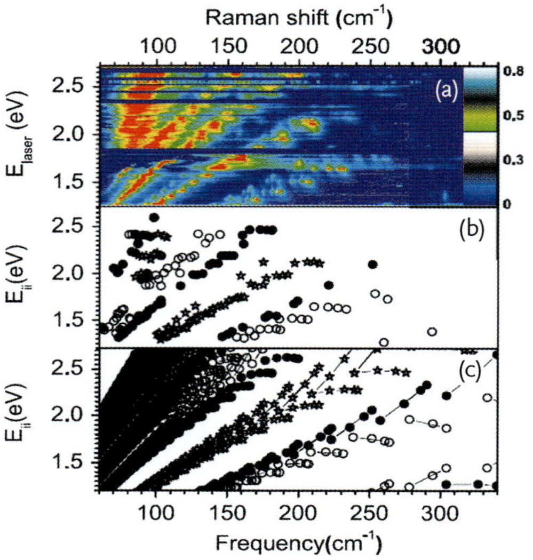

Figure 9.15 (a) RBM resonance Raman map for the "super-growth" (S.G.) SWNT sample [189, 317, 318]. (b) Kataura plot of all transition energies ($E_{ii}^{S.G.}$) that could be experimentally obtained from the resonance windows extracted from (a), as function of ω_{RBM}. (c) Kataura plot obtained from Eq. (9.10) with the parameters that best fit the data in (b). The stars stand for M-SWNTs, the open bullets stand for type I S-SWNTs and the filled bullets stand for type II S-SWNTs [317].

effects ($\sigma - \pi$ hybridization) [287]. The theoretical understanding of all these factors will be discussed in Chapter 10.

9.4 Summary

In this chapter we show how the RBM spectra from single-wall carbon nanotubes can be used to study the concepts of resonance Raman scattering in detail. Although resonance Raman scattering should be observable in every nanocarbon material, the RBM for carbon nanotubes is special because of the one-dimensional physics of carbon nanotubes and the low RBM energy. These two properties, together, generate a very sharp resonance window for RBMs. Furthermore, the RBM frequency depends on tube diameter, as explained here using elasticity theory. Due to this d_t dependence, the RBM from different (n, m) tubes can be identified, and used to study, through the resonance effect, the electronic structure of the carbon nanotubes, as well as environmental effects. This electronic structure is summarized in the empirical Eq. (9.10), which is related to many physical concepts that will be discussed in the next chapter.

Problems

[9-1] Obtain the density of graphite in kg/m^3. Here we can use the in-plane C–C distance which is 1.42 Å and the interlayer distance between two graphene layers which is 3.35 Å and note that about 2% of the carbon atoms in graphite are ^{13}C and the remaining 98% are ^{12}C. For this problem we need at least three digits in numerical accuracy.

[9-2] When the radius of a zigzag SWNT is modified by δR, determine by how much the C–C distance is modified along the circumferential direction.

[9-3] The Young's modulus of graphene is $Y = 1060$ GPa. Obtain the sound velocity of the LA phonon mode by calculating the nanotube density.

[9-4] The Young's modulus of SiC, Fe and diamond are, respectively, $Y = 450$, 200, 1200 GPa. Obtain sound velocities for these materials.

[9-5] Obtain the sound velocities for an ideal air sample in units of m/s and km/h.

[9-6] Obtain the formula for the sound velocity for the TA phonon mode $C_t = \sqrt{G/\rho}$, where G, the shear modulus, is given by $G = Y/(2(1+\nu))$, and where Y and ν are the Young's modulus and Poisson ratio, respectively.

[9-7] Evaluate A in Eq. (9.4) in cm^{-1} units.

[9-8] Check that all terms in Eq. (9.5) are dimensionless.

[9-9] Obtain Eq. (9.6) from Eq. (9.5). Note that a factor $1/2\pi c$ appears when we measure ω in units of cm^{-1}. Using the known values for the various factors, get the value of 227 for ω_{RBM} in units of cm^{-1} for $d_t = 1$ nm.

[9-10] Calculate the shift of ω_{RBM} in Eq. (9.7) for $d_t = 1$ nm and 2nm.

[9-11] Estimate C_e in Eq. (9.8) and the diameter at which the correction term $C_e d_t^2$ becomes 0.21.

[9-12] Explain that the anharmonic term in the vibrational Hamiltonian gives a finite lifetime to the phonon which is responsible for the Raman spectral width.

[9-13] Using the uncertainty relation between energy and time, obtain the phonon lifetime for Raman spectra with a spectral width of 1 cm^{-1} and of 10 cm^{-1}.

[9-14] Evaluate the electric field in V/m for a laser power of 1 $mW/\mu m^2$.

[9-15] For $T = 300$ K and 77 K, what is the intensity ratio of the Stokes to anti-Stokes *nonresonant* Raman signals for a 173.6 cm^{-1} RBM phonon? How about for a 1590 cm^{-1} G-band?

[9-16] Measure the resonance window values from Figure 9.10 and estimate the lifetime of photoexcited carriers. Explain by giving some reasons for which

lifetime is shorter, the lifetime of the photoexcited carriers or the lifetime of phonons.

[9-17] For 785 nm laser light, obtain the wavelength in nm for the Stokes and anti-Stokes scattered light for a 173.6 cm^{-1} RBM phonon.

[9-18] For 785 nm laser light, what is the E_{ii} energy of the scattered light resonance conditions for the Stokes and anti-Stokes Raman spectra for a 173.6 cm^{-1} RBM phonon?

[9-19] Give the expected intensity ratio I_S/I_{aS} between the Stokes and anti-Stokes RBM signals shown in Figure 9.10 for $E_{\mathrm{laser}} = 1.63$ eV, $E_{\mathrm{laser}} = 1.65$ eV and $E_{\mathrm{laser}} = 1.67$ eV. Consider both $T = 0$ K and $T = 300$ K.

[9-20] Explain why the intensity ratio of the Stokes to anti-Stokes resonance Raman intensity for one laser energy might not give the temperature of the sample. For the nonresonance Raman intensity, on the other hand, we may get the information needed to determine the temperature. Why?

[9-21] Consider that the spectra in Figure 9.11 were obtained from a SWNT with $\gamma_{\mathrm{RBM}} = 8$ meV. Using Eq. (9.9), find the value of E_{ii} which gives the observed I_S/I_{aS}.

[9-22] Build your own Kataura plot using Eq. (9.10). Evaluate the E_{22}^S and E_{33}^S energies for the (6,5), (11,1), (10,5) SWNTs by using Eq. (9.10).

[9-23] There are two definitions for the type of semiconducting SWNTs; one is Type I and Type II using $(\mathrm{mod}(2n + m, 3) = 1$ and $= 2)$, and the other is Mod 1 and 2 using $\mathrm{mod}(n - m, 3) = 1, 2$. Show that Type I and Mod 2 (or Type II and Mod 1) are equivalent to each other.

[9-24] In the (n, m) map of SWNTs, show that SWNTs with $2n + m =$ const. have a similar diameter while SWNTs with $n - m =$ const. have a similar chiral angle. Explain that the $2n + m =$ const. family is suitable for studying the chiral angle dependence while that $n - m =$ const. family is suitable for studying the diameter dependence.

10
Theory of Excitons in Carbon Nanotubes

In the resonance Raman spectroscopy of single-wall carbon nanotubes (SWNTs), the optical transition energy from the ith valence band state to the ith conduction band state, E_{ii}, is important for assigning (n, m) values to individual SWNTs. To assign the experimentally observed E_{ii} in single-wall carbon nanotubes, a theoretical development has been carried out with respect to the simple (nearest neighbor) tight-binding (STB) model discussed in Chapter 2. By adjusting the STB (simple tight-binding) parameters, E_{ii} values have been assigned to specific (n, m) SWNTs for a limited region of diameter or energy on the Kataura plot. However, this procedure is not useful for explaining in a systematic way the results obtained from many SWNT samples synthesized by different methods. In this chapter we discuss three aspects going beyond the STB model that are necessary to achieve experimental accuracy:

- Curvature (σ–π hybridization) effects using the extended tight-binding method;
- Excitonic effects using the Bethe–Salpeter equation;
- Dielectric screening effects of excitons.

This chapter begins with a brief description of how curvature effects are introduced into the tight-binding model, to construct what has been called the extended tight-binding (ETB) method (Section 10.1). The curvature effect in SWNTs is responsible for $\sigma - \pi$ hybridization, resulting in a much stronger E_{ii} dependence on the SWNT chiral angle θ than that predicted by the STB picture. In sequence, Section 10.2 gives a broad overview of exciton physics, which is the main part of this chapter. The electron–electron and the electron–hole interactions, generally called many-body effects, change in a significant way the E_{ii} dependence on tube diameter d_t, as well as the relative distance between the different E_{ii} levels. From a theoretical point of view, the importance of excitons to SWNTs was introduced early on by Ando [319], who studied the electronic excitations of nanotubes within a static screened Hartree–Fock approximation. Later on, after experimental results started to show the importance of excitons, detailed first-principles calculations of the effects of many-body interactions on the optical properties were performed for nanotubes with very small diameter [320–323] and some descriptions of excitons in nanotubes based on simpler models [324–327] were also developed.

Raman Spectroscopy in Graphene Related Systems. Ado Jorio, Riichiro Saito, Gene Dresselhaus, and Mildred S. Dresselhaus

ISBN: 978-3-527-40811-5

Following this work, a systematic dependence of exciton effects (including wave function-related phenomena) on the nanotube diameter and chiral angle was developed [186, 328, 329] and this topic is presented in Section 10.4.1. These results are important for providing a quantitative description of the photophysical properties of SWNTs, including the Raman response. Finally, Section 10.5 introduces the importance of the dielectric screening of an exciton by other electrons and by surrounding materials, a topic that is still under development in the science of one-dimensional systems.

10.1 The Extended Tight-Binding Method: σ–π Hybridization

The nearest neighbor (simple) tight-binding (STB) model, developed in Chapter 2, gives the first approximation in constructing a Kataura plot showing the dependence of the transition energies E_{ii} on the tube diameter d_t (Section 2.3.4). The several E_{ii} levels for SWNTs are shown in Figure 2.22 to exhibit a strong $1/d_t$ dependence, that is related to the distance of the cutting line from the K point in the unfolded 2D-graphene Brillouin zone, and there is in addition a small chiral angle θ dependence related to the trigonal warping effect [31]. However, the experimental results of E_{ii} as a function of d_t show from the observed $1/d_t$ dependence of E_{ii} for SWNTs (see Section 9.3.2) that the chirality-dependent pattern (family pattern) which occurs for (n, m) SWNTs with $(2n + m) =$ constant, is actually much larger than that predicted from the STB model. This experimental observation led to the implementation of the ETB model for the explanation of many experimental studies of the photophysics of SWNTs. This $(2n + m)$ spread is mainly attributed to the SWNT curvature effects, which cause a chiral angle dependence in the C–C bond length relaxation in small d_t SWNTs that is missing from the STB approximation.

It had been shown that long-range interactions of the p orbitals are not negligible [330], and the curvature of the SWNT sidewalls results in an important sp^2–sp^3 rehybridization in the small d_t limit. The curvature effect can be included in the tight-binding (TB) model [182, 329] by extending the basis set to the atomic s, p_x, p_y, and p_z orbitals that form the s and p molecular orbitals according to the formalism developed in Chapter 2 (the Slater–Koster formalism [31, 331]). This extended tight-binding (ETB) model utilizes the TB transfer and overlap integrals as a function of the C–C inter-atomic distance calculated within the density functional theory (DFT) framework [263], thus including long-range interactions and bond-length variations within the SWNT sidewall. The atomic p-orbitals are aligned with the cylindrical coordinates of the SWNT sidewall according to a symmetry-adapted scheme [182, 329] in which p_z is orthogonal to the SWNT sidewall, while p_x and p_y are parallel to the SWNT sidewall for each C atom. This choice allows us to consider an 8×8 Hamiltonian for the graphene unit cell of two C atoms (A and B), even for chiral SWNTs with large translational unit cells, thus greatly simplifying the calculations. Further details of the calculational method can be found in [329].

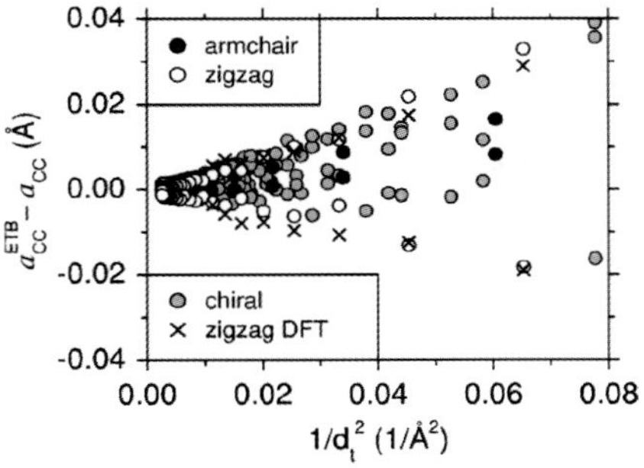

Figure 10.1 Differences between the C–C bond lengths $a_{C\text{–}C}$ in the ETB model (denoted by $a_{C\text{–}C}^{ETB}$) and $a_{C\text{–}C} = 0.142$ nm in the flat graphene layer for many SWNTs as a function of nanotube curvature $1/d_t^2$. Open, closed, and gray dots denote the bond lengths of zigzag, armchair, and chiral SWNTs, respectively, calculated from the ETB model for the optimized SWNT structures. For comparison, crosses show the bond lengths of zigzag SWNTs from DFT calculations [182].

The total energy of the SWNT can be calculated using the short-range repulsive potential obtained from DFT calculations [263], and the geometrical structure optimization can then be performed. To compare the SWNT structures optimized by using the ETB model with the results of other independent geometrical structure optimizations, we plot in Figure 10.1 the change in the C–C bond lengths for each SWNT as a function of nanotube curvature $1/d_t^2$ [182]. For calculating the electronic structure of SWNTs, it is essential to utilize the optimized SWNT structure, since the overlap integrals are very sensitive to the relaxed atomic positions. As a consequence, the θ dependence (i. e., the family pattern in the Kataura plot) increases significantly with decreasing d_t, thereby matching the results observed in the experimental Kataura plot (Section 9.3.2).

10.2
Overview on the Excitonic Effect

The exciton is a bounded electron–hole pair. An exciton in a semiconducting material consists of a photoexcited electron and a hole bound to each other by an attractive Coulomb interaction. In many commonly occurring bulk 3D semiconductors (such as Si, Ge and III–V compounds), the binding energy of an exciton can be calculated by a hydrogenic model with a reduced effective mass and a dielectric constant, giving a binding energy on the order of $\sim$ 10 meV with discrete energy levels lying below the single particle excitation spectra. Thus optical absorption to exciton levels is usually observed only at low temperatures. However, in a single-wall carbon nanotube, because of its 1D properties, the electron–hole binding energy becomes much larger (and can be as large as 1 eV), so that exciton effects can be observed even at room temperature. Thus excitons are essential for explaining optical processes, such as optical absorption, photoluminescence, and resonance Raman spectroscopy in SWNTs.

The following sections presents a broad descriptive picture of the theoretical description of excitons in SWNTs. It starts by briefly addressing the general properties of excitons in general, while also emphasizing the uniqueness of excitons in graphite, SWNTs and C_{60} and the difference in behavior between the excitons in each dimension (2, 1, and 0, as represented by these carbon materials, respectively.) The unusual geometrical structure of sp^2 carbons to which all of these materials relate gives rise to the two special points in the Brillouin zone (K and K'), which are related to one another by time reversal symmetry [80], making these sp^2 carbon systems unique relative to other nanosystems which also have large excitonic effects, but do not have similar symmetry constraints. Differences in symmetry are important and guide electronic structure calculations and the interpretation of experiments. Therefore, an analysis of exciton symmetries in SWNTs is needed to understand in greater detail many aspects of their optical properties, and this is the next topic of this section (Section 10.3). From the group theory analysis, the selection rules for optical phenomena in SWNTs are obtained (and are discussed in Section 10.3.2). Finally, Section 10.4 develops the theory for excitons in carbon nanotubes.

10.2.1
The Hydrogenic Exciton

The simplest treatment for an exciton is given by the Wannier exciton, which can be described by the Schrödinger equation:

$$\left[-\frac{\hbar}{2m_{\mathrm{e}}^{*}}\nabla_{\mathrm{e}}^{2}-\frac{\hbar}{2m_{\mathrm{h}}^{*}}\nabla_{\mathrm{h}}^{2}-\frac{e^{2}}{\epsilon r}\right]\Psi_{\mathrm{ex}}=E_{\mathrm{ex}}\Psi_{\mathrm{ex}}\,, \tag{10.1}$$

where subscripts e, h stand for the electron and the hole of the exciton which are attracted by a Coulomb potential[1] $-e^2/\epsilon r$ (ϵ is the dielectric constant), and m_{e} and m_{h} are the effective mass of the electron and the hole, respectively. By adopting the center of mass coordinate $\boldsymbol{R} = (m_{\mathrm{e}}\boldsymbol{r}_{\mathrm{e}} + m_{\mathrm{h}}\boldsymbol{r}_{\mathrm{h}})/(m_{\mathrm{e}} + m_{\mathrm{h}})$ and the relative distance coordinate $\boldsymbol{r} = (\boldsymbol{r}_{\mathrm{e}} - \boldsymbol{r}_{\mathrm{h}})$, the exciton wavefunction can be given by:

$$\Psi(\boldsymbol{R},\boldsymbol{r}) = g(\boldsymbol{R})\,f(\boldsymbol{r})\,, \tag{10.2}$$

where $g(\boldsymbol{R}) = e^{i\boldsymbol{K}\cdot\boldsymbol{R}}$ describes the movement of an exciton with momentum $\boldsymbol{K}$, and $f(\mathbf{r})$ gives the different exciton levels with solutions obtained by the Schrödinger equation for a hydrogen atom with a reduced mass

$$\frac{1}{\mu}=\left(\frac{1}{m_{\mathrm{h}}^{*}}+\frac{1}{m_{\mathrm{e}}^{*}}\right). \tag{10.3}$$

1) In SI (MKS) units, a Coulomb potential becomes $-e^2/4\pi\epsilon_0\epsilon r$. For a conversion from CGS to MKS units, we include a factor of $1/4\pi\epsilon_0$ along with e^2.

The solution of Eq. (10.1) gives:

$$E(K) = -\frac{\mu e^4}{2\hbar^2\epsilon^2 n^2} + \frac{\hbar^2 K^2}{2(m_e^* + m_h^*)}\,, \quad (n = 1, 2, 3, \ldots)\,. \tag{10.4}$$

The first and second terms in Eq. (10.4) are the excitonic energy levels denoted by the quantum number n and the energy dispersion relation for the center of mass motion of the exciton, respectively. Although the description of excitons in SWNTs is more complex, since the exciton is formed by the mixing of different k states due to the Coulomb interaction (as described below in Section 10.4), the concepts of a dispersion relation, exciton wave vector and excitonic energy levels are closely related to this simplified description.

10.2.2
The Exciton Wave Vector

A single-particle picture of carriers is simple and easy to understand. In a semiconducting material, an electron can be excited from the valence band to the conduction energy band, and the photon energy beyond the band gap goes into the kinetic energy of the excited electron. An excitonic picture, however, cannot be represented by a single-particle model, and we cannot generally use the energy dispersion relations directly to obtain the excitation energy for the exciton. If the electron and hole wavefunctions are localized in the same spatial region, the attractive Coulomb interaction between the electron and hole increases the binding energy, while the kinetic energy and the Coulomb repulsion between the electrons becomes large, too. Thus the optimum localized distances between the e–h (electron–hole) pair determine the exciton binding energy. In the case of a metal, the dielectric screening of the Coulomb interaction by other conduction electrons reduces the Coulomb interaction significantly (where ϵ is infinity) and thus the exciton does not form.[2] The repulsive Coulomb interaction between a photoexcited electron and valence electrons causes the wave vector k for the excited electron to no longer be a good quantum number, since electron–electron scattering occurs and thus the lifetime of an electron becomes finite.

Since the exciton wavefunction is localized in real space, the exciton wavefunction in k space is a linear combination of Bloch wavefunctions with different k states. Thus the definition of k_c and k_v is given by their central values with a width of Δk.[3]

When we consider an optical transition in a crystal, we expect a vertical transition, $k_c = k_v$ (Figure 10.2a), where k_c and k_v are, respectively, the wave vectors of the electron and the hole. The wave vector of the center of mass for the exciton is defined by $K = (k_c - k_v)/2$, while the relative coordinate is defined by $k = k_c + k_v$. Here, we note that the hole (created by exciting an electron) has the opposite sign for its wave vector and effective mass as compared to the electron. The exciton has

2) Metallic SWNTs have shallow exciton bound states relative to those for semiconducting SWNTs.
3) The Fourier transformation of a Gaussian in real space is a Gaussian in k space, too.

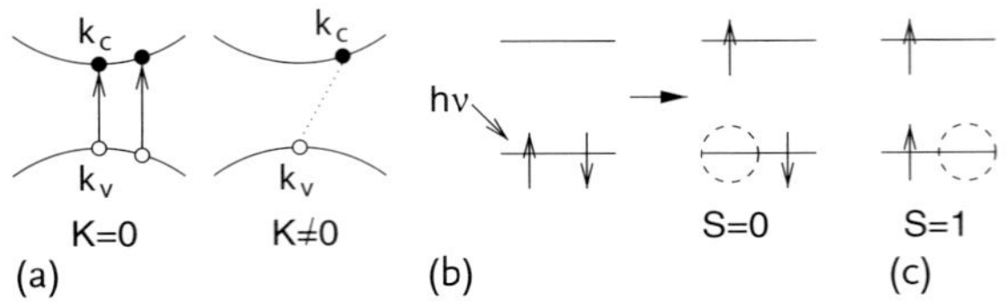

Figure 10.2 (a) A singlet exciton formed at $K = 0$ in a crystal where $k_c = k_v$ (left), at either the band extrema or away from the band extrema if $k_c \neq k_v$, $K \neq 0$, giving rise to a dark exciton (right, see text). (b) When a photon is absorbed by an electron with spin $\uparrow$ (left), we get a singlet exciton ($S = 0$, right). If the spin of the electron is $\uparrow$ we here define the spin of the hole that is left behind as $\downarrow$. (c) A triplet exciton ($S = 1$), that is a dark exciton [187].

an energy dispersion as a function of K, which represents the translational motion of an exciton. Thus only the $K = 0$ ($K < \Delta k$) exciton can recombine by emitting a photon. Correspondingly, a $K \neq 0$ exciton cannot recombine directly to emit a photon and therefore is a dark exciton. Recombination emission for $K \neq 0$ is, however, possible by a phonon-assisted process involving an indirect optical transition.

10.2.3
The Exciton Spin

When we discuss the interaction between an electron and a hole, the definition of the total spin for an exciton is a bit different from the conventional idea of two electrons in a molecule (or a crystal). A hole is a different "particle" from an electron, but, nevertheless, an exchange interaction between the electron and the hole exists, just like for two electrons in a hydrogen molecule.

When an electron absorbs a photon, an electron, for example with spin $\uparrow$ is excited to an excited state as shown in Figure 10.2b, leaving behind a hole at the energy level where the electron with up spin $\uparrow$ had previously been. This hole has not only a wave vector of $-k$ and an effective mass of $-m^*$ as mentioned in Section 10.2.2, but also is defined to be in a spin down $\downarrow$ hole state. The exciton thus obtained (Figure 10.2b) is called a spin singlet exciton, with $S = 0$, since the definition of S for the two-level model shown here is in terms of the two actual electrons that are present,[4] and in this sense the definition for the two actual electrons and for the $S = 0$ exciton are identical. It should also be mentioned that Figure 10.2b does not represent an $S = 0$ eigenstate. To make an eigenstate we must take the antisymmetric combination of the state shown in Figure 10.2b with an electron $\downarrow$ and hole $\uparrow$ [332]. In contrast, a triplet exciton ($S = 1$) can be represented by two electrons, one in the ground state and the other in an excited state to give a total spin of $S = 1$ (Figure 10.2c).[5] For the triplet state in Figure 10.2c, we define the hole to have a spin $\uparrow$ and the resulting state shown here is an eigenstate ($m_s = 1$) for

4) The electric dipole transition does not change the total spin of the ground state which is $S = 0$.
5) The reader should not be confused by having $S = 1$ for the triplet state, since the two spin up electrons are in different energy states.

$S = 1$. We further note that a triplet exciton cannot be recombined by emitting a photon because of the Pauli principle. We call such an exciton, a triplet exciton. A triplet exciton is one type of "dark exciton" (dipole-transition forbidden state).[6] An exchange interaction (> 0) between a hole and an electron works only for $S = 0$ (see Figure 10.2b) and thus the $S = 1$ state in Figure 10.2c has a lower energy than the $S = 0$ state[7] (see also Eq. (10.10) in Section 10.4.1). It should be noted that for the more familiar case of just two electrons, the exchange interaction (< 0) works for the $S = 1$ case and therefore the $S = 1$ state lies lower in energy than the $S = 0$ state.

10.2.4 Localization of Wavefunctions in Real Space

The localization of a wavefunction can be obtained by mixing the Bloch wavefunction labeled by the wave vector k. The equation for determining the mixing of delocalized wavefunctions is called the Bethe–Salpeter equation (Section 10.4.1). The center of mass momentum for an exciton is now a good quantum number in the crystal, while the relative motion of an electron and a hole gives excitonic levels. Thus an exciton is considered to provide a quasi-particle or an elementary excitation with additional freedom, like the plasmon or polariton. By forming an exciton wavefunction, the Hilbert space of the wavefunction for the free particles which describe the electronic states is reduced significantly, and this gives a reduction in the optical absorption for the one-particle spectra. This is known as the oscillator strength sum rule (or f-sum rule). Thus if most of the available oscillator strength for optical absorption is used for the exciton, the spectral intensity for the one-particle transitions is reduced. This situation makes research on excitons for SWNTs more important in the sense that a single-particle excitation has hardly ever been observed in this system in an optical absorption experiment.

The localization length of the exciton in a single-wall carbon nanotube is larger than the diameter of a SWNT but much smaller than the length of a SWNT. This situation gives rise to a predominantly one-dimensional behavior in the optical properties of a SWNT exciton. In a pure 1D exciton, however, the binding energy of the lowest state would be minus infinity. Thus the cylindrical shape of the SWNT is essential for giving a sufficiently large binding energy to the exciton, thus allowing observation of the exciton at room temperature.

6) Spin conversion by a magnetic field could flip a spin and lead to the recombination of the triplet exciton. We will show later another type of dark exciton (E-symmetry exciton).

7) The exchange interaction for the $S = 0$ exciton can be understood as the difference in the interaction energy between two electrons (one at the position of the excited electron and the other at the position of the hole left behind as in Figure 10.2b) and the energy of the $S = 1$ exciton which has no exchange energy (Figure 10.2c).

10.2.5
Uniqueness of the Exciton in Graphite, SWNTs and C_{60}

The electronic structure of a SWNT and of graphite is unique insofar as there are two nonequivalent energy bands near the two hexagonal corners K and K' of the Brillouin zone. We therefore distinguish the regions about K and K' from one another and call them the two valleys of SWNTs and graphite. Although an optical transition occurs vertically in k space, we can consider the electron and the hole in the electron–hole pair to be either in the same valley, or an electron to be in one valley and a hole in the other valley. The latter pair can form an excitonic state in real space, but it never recombines radiatively, since the electron and hole do not have the same k value; we call such a state an E-symmetry exciton (see Section 10.3). An E-exciton is another type of "dark exciton". In addition to the conventional "bright exciton" (an electron–hole pair from the same valley that can recombine radiatively[8)]), the coexistence of many different types of excitons is of importance for understanding the optical properties of SWNTs.

In resonance Raman spectra, photoluminescence or resonance Rayleigh scattering, we can observe a signal even from a single SWNT "molecule". In a one-particle picture of optical processes, a strong enhancement of the optical intensities can be understood in terms of the 1D van Hove singularities (vHSs) in the joint density of states connecting the valence and conduction energy bands. In an excitonic picture, an exciton has an energy dispersion as a function of the center of mass wave vector and we expect 1D vHSs in the excitonic density of states from the ground states, where optical absorption becomes strong and this occurs when the center of mass wave vector vanishes. The assignment of the excitation energy to a SWNT with (n, m) indices works well by interpreting the E_{ii}, which is a one-particle picture concept, in terms of the exciton vHS position. This exciton energy position can be modified by electro-chemical doping or by changing the surrounding materials by use of substrates, solutions or wrapping agents (environmental effects) in the space surrounding a SWNT.

In C_{60}, which is a zero-dimensional molecule [6, 333], excitonic behavior is also observed and the binding energy for C_{60} is estimated to be 0.5 eV, which is of the same order of magnitude in energy as the nanotube exciton. This value for C_{60} is obtained by comparing (i) the optical absorption energy (1.55 eV) and (ii) the energy difference observed by photoelectron emission and inverse photoemission spectroscopy (2.3 eV) [334, 335].

The C_{60} and nanotube excitons exhibit fundamental similarities, both systems being π conjugated, both having similar diameters, and both having singularities in their electronic density of states (molecular levels or a narrow energy band width in the C_{60} crystal). On the other hand, the lowest exciton wavefunction is not homogeneous on the C_{60} ball because the electron and hole have their own molecular orbitals with different symmetries. In contrast, in the nanotube exciton, the

8) We will see in Section 10.3 that even within the same valley, one of the two possible exciton types is a dark exciton because of symmetry requirements.

electron and hole have the same symmetry. The lowest exciton wavefunction is homogeneous around the circumferential direction and is localized only along the tube axis direction, because the range of the Coulomb interaction, U, is larger than the tube diameter and smaller than the length of a SWNT. Furthermore, in the case of the C_{60} crystal, the energy band width of the highest occupied molecular orbital (HOMO) and the lowest unoccupied molecular orbital (LUMO) band is much smaller than the Coulomb interaction, while in the case of the nanotube, the energy band width is larger than U. In a SWNT, the motion of the exciton along the nanotube axis gives an energy dispersion for the exciton while the excitons in C_{60} are localized within a molecule. Though we can use similar experimental and theoretical techniques for considering a molecular exciton for C_{60} and for a SWNT, it is nevertheless important to consider the differences in the physics of a 0D and a 1D system when describing excitons in C_{60} and in SWNTs.

10.3 Exciton Symmetry

Group theory discussions tell us that there are four kinds of spin-singlet excitons corresponding to the symmetries, A_1, A_2, $\mathbb{E}$ and $\mathbb{E}^*$ in a SWNT [135], and that only the excitons with A_2 symmetry are optically allowed. We call A_2 excitons "bright excitons" (dipole-transition allowed states) and all other excitons are dark excitons.

10.3.1 The Symmetry of Excitons

Figure 10.3a shows a schematic diagram of the electronic valence and conduction single-particle bands with a given index μ, for general chiral SWNTs [31]. The ir-

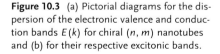

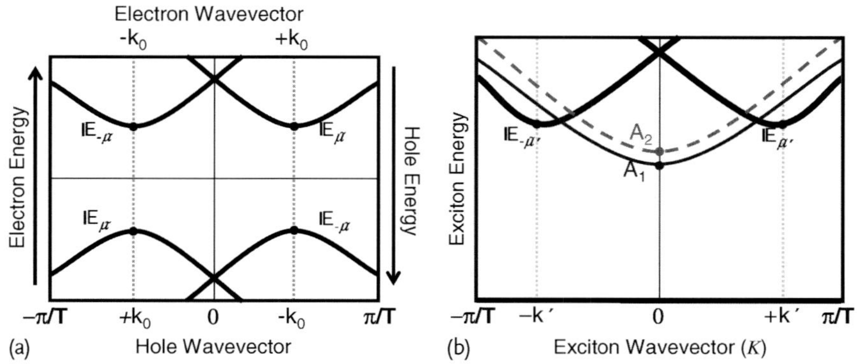

Figure 10.3 (a) Pictorial diagrams for the dispersion of the electronic valence and conduction bands $E(k)$ for chiral (n, m) nanotubes and (b) for their respective excitonic bands. The electron, hole and exciton states at the band edges are indicated by a solid circle and labeled according to the irreducible representation to which they belong [135].

reducible representations of the factor groups of nanotubes are labeled by the angular momentum quantum number μ, which is the index that labels the cutting lines [110]. Here the cutting lines denote the possible k vectors given by the periodic boundary condition around the circumferential direction of a single wall carbon nanotube [110, 336] which can be expressed by $\boldsymbol{k} = \mu \boldsymbol{K}_1 + k\boldsymbol{K}_2/|\boldsymbol{K}_2|$ (where $\mu = 1 - N/2, \ldots, N/2$, and $-\pi/T < k < \pi/T$). Here $\boldsymbol{K}_1$ and $\boldsymbol{K}_2$ are, respectively, the reciprocal lattice vectors along the circumferential and axial directions. N denotes the number of hexagons in the unit cell for SWNTs, and T is the length of the real space unit cell [31]. The electron and hole states at the band-edge are therefore labeled according to their irreducible representations. The exciton wavefunction can be written as a linear combination of products of conduction (electron) and valence (hole) band eigenstates, $\phi_{\mathrm{c}}^{\mu}(\boldsymbol{r}_{\mathrm{e}})$ and $\phi_{\mathrm{v}}^{\mu'*}(\boldsymbol{r}_{\mathrm{h}})$ as [135, 218, 337]

$$\psi(\boldsymbol{r}_{\mathrm{e}}, \boldsymbol{r}_{\mathrm{h}}) = \sum_{\mathrm{v,c},\mu,\mu',k} A_{\mathrm{vc}}\phi_{\mathrm{c}}^{\mu}(\boldsymbol{r}_{\mathrm{e}})\phi_{\mathrm{v}}^{\mu'*}(\boldsymbol{r}_{\mathrm{h}}), \tag{10.5}$$

where v and c stand for valence- and conduction-band states, respectively. $\phi_{\mathrm{c}}^{\mu}(\boldsymbol{r}_{\mathrm{e}})$ and $\phi_{\mathrm{v}}^{\mu'*}(\boldsymbol{r}_{\mathrm{h}})$ are localized functions in real space which are obtained by taking the summation on k. To obtain an accurate solution for the excitonic eigenfunctions (the A_{vc} coefficients in Eq. (10.5)) and eigenenergies, it is necessary to solve the Bethe–Salpeter equation Section 10.4.1, which includes many-body interactions and considers the mixing by the Coulomb interaction of electron and hole states with all the different wave vectors for all the different bands. The Coulomb interaction depends only on the relative distance between the electron and the hole, and thus the many-body Hamiltonian is invariant under the symmetry operations of the nanotube. Each excitonic eigenstate will then transform as one of the irreducible representations of the space group of the nanotube. In general, the electron–hole interaction will mix states with all wave vectors and all bands, but for moderately small-diameter nanotubes ($d_{\mathrm{t}} < 1.5$ nm), the energy separation between singularities in the single-particle JDOS (joint density of states) is fairly large and it is reasonable to consider, as a first approximation, that only the electronic energy sub-bands contributing to a given JDOS singularity E_{ii} will mix to form the excitonic states. Within this approximation, it is possible to employ the usual effective-mass approximation (EMA) and the envelope-function approximation to obtain the exciton eigenfunctions [135, 218, 337]:

$$\psi^{\mathrm{EMA}}(\boldsymbol{r}_{\mathrm{e}}, \boldsymbol{r}_{\mathrm{h}}) = \sum_{\mathrm{v,c}}{}' A_{\mathrm{vc}}\phi_{\mathrm{c}}(\boldsymbol{r}_{\mathrm{e}})\phi_{\mathrm{v}}^{*}(\boldsymbol{r}_{\mathrm{h}})F_{\nu}(z_{\mathrm{e}} - z_{\mathrm{h}}) \,. \tag{10.6}$$

The prime in the summation of Eq. (10.6) indicates that only the electron and hole states associated with the JDOS singularity are included. It is important to emphasize that the approximate wavefunctions Ψ^{EMA} have the same symmetries specified by μ as the full wavefunctions Ψ. The envelope function $F_{\nu}(z_{\mathrm{e}} - z_{\mathrm{h}})$ provides an *ad hoc* localization of the exciton in the relative coordinate $z_{\mathrm{e}} - z_{\mathrm{h}}$ along the axis and ν labels the levels in the 1D hydrogenic series given in Section 10.2.1. The envelope functions will be either even ($\nu = 0, 2, 4\ldots$) or odd ($\nu = 1, 3, 5\ldots$) upon the $z \to -z$ operations.[9] The use of such "hydrogenic" envelope-functions (similar to $f(\mathbf{r})$ in Section 10.2.1) serves merely as a physically grounded guess to the ordering in which the different exciton states might appear. From Eq. (10.6), the irreducible representation of the excitonic state $\mathcal{D}(\psi^{\text{EMA}})$ will then be given by the direct product [135, 218, 337]:

$$\mathcal{D}(\psi^{\text{EMA}}) = \mathcal{D}(\phi_{\text{c}}) \otimes \mathcal{D}(\phi_{\text{v}}) \otimes \mathcal{D}(F_{\nu}), \tag{10.7}$$

where $\mathcal{D}(\phi_{\text{c}})$, $\mathcal{D}(\phi_{\text{v}})$ and $\mathcal{D}(F_{\nu})$ are the irreducible representations of the conduction state, valence state and envelope-function F_{ν}, respectively.

As shown in Figure 10.3a, there are two inequivalent electronic bands in chiral tubes,[10] one with the band edge at $k = k_0$ and the other one at $k = -k_0$. In order to evaluate the symmetry of the excitonic states, it is necessary to consider that the Coulomb interaction will mix the two inequivalent states in the conduction band (electrons) with the two inequivalent states in the valence band (holes). These electron and hole states transform as the 1D representations $\mathbb{E}_{\mu}(k_0)$ and $\mathbb{E}_{-\mu}(-k_0)$ of the C_N point group [135],[11] where the conduction and valence band extrema occur at the same $k = k_0$ (or $-k_0$). This situation gives rise to a van Hove singularity (vHS) in the joint density of states (JDOS) [31, 338]. Taking this into consideration, the symmetries of the exciton states associated with the $\nu = 0$ envelope function, which transform as the $A_1(0)$ representation, can be obtained using the direct product in Eq. (10.7):

$$\begin{aligned} &\left[\mathbb{E}_{\mu}(k_0) + \mathbb{E}_{-\mu}(-k_0)\right] \otimes \left[\mathbb{E}_{-\mu}(-k_0) + \mathbb{E}_{\mu}(k_0)\right] \otimes A_1(0) \\ &\quad = A_1(0) + A_2(0) + \mathbb{E}_{\mu'}(k') + \mathbb{E}_{-\mu'}(-k'), \end{aligned} \tag{10.8}$$

where $k' \sim 2k_0$ and $\mu' = 2\mu$ are the exciton linear momenta and quasi-angular momenta, respectively. Therefore, group theory shows that the set of excitons with the lowest energy is composed of four exciton bands, shown schematically in Figure 10.3. Basically the mixing of two electron and two hole ($\pm\mu$) wave functions generates four exciton states. The mixing of electron and hole states in the same vHSs ($k_{\text{c}} = \pm k_0$, $k_{\text{v}} = \mp k_0$) will give rise to excitonic states, which transform as the A_1 and A_2 irreducible representations of the C_N point group. The excitonic states formed from electrons and holes with $k_{\text{c}} = k_{\text{v}} = \pm k_0$ will transform as the $\mathbb{E}_{\mu'}(k')$ and $\mathbb{E}_{-\mu'}(-k')$ 1D irreducible representations of the C_N point group, with a wave vector k' and an angular momentum quantum number μ'.

The higher-energy exciton states in chiral tubes can be obtained, for instance, by considering the same vHS in the JDOS and higher values of ν. For ν even, the resulting decomposition is the same, since the envelope function also has A_1 symmetry. For odd values of ν, the envelope function will transform as A_2, but that will also leave the decomposition in Eq. (10.8) unchanged. Thus, from the group theory

9) For this symmetry operation, we can use a C_2 axis which is perpendicular to the nanotube axis.
10) The case of achiral nanotubes is given in the problem set for this chapter.
11) Usually E is used to label 2D irreducible representations (IRs) in point groups. In cyclic groups, however, two 1D IRs can be degenerate not by a symmetry in real space, but by time-reversal symmetry. Here these 1D IRs are denoted by $\mathbb{E}$ (see Chapter 6).

point view, both even and odd ν have A_1 and A_2 symmetry excitons. The result is still the same if one now considers higher-energy exciton states derived from higher singularities in the JDOS (for instance, the so-called E_{22} or E_{33} transitions), as long as the angular momentum of the electrons and holes is the same. Therefore, Eq. (10.8) describes the symmetries of all exciton states in chiral nanotubes associated with each E_{ii} transition.

10.3.2
Selection Rules for Optical Absorption

To obtain the selection rules for the optical absorption of the excitonic states, it is necessary to consider that the ground state of the nanotube transforms as a totally symmetric representation (A_1) and that only $K = 0$ excitons can be created (Section 10.2.2). For light polarized parallel to the nanotube axis, the interaction between the electric field and the electric dipole in the nanotube transforms as the A_2 irreducible representation for chiral nanotubes [135]. Therefore, from the four excitons obtained for each envelope function ν, only the A_2 symmetry with $S = 0$ (Section 10.2.3) excitons are optically active for parallel polarized light, and the remaining three states with $S = 1$ are dark states.

Although not related to Raman spectroscopy it is here important to comment that the two-photon absorption experiment [339, 340] represented an important advance for discussing the exciton photophysics of SWNTs. For two-photon excitation experiments, the excitons with A_1 symmetry are accessed, and thus, there will also be one bright exciton for each ν envelope function. The presence of one-photon (two-photon) allowed transitions associated with odd (even) envelope functions result from the presence of two inequivalent vHSs in the first Brillouin zone associated with the two inequivalent carbon atoms in mono-layer graphene. This experiment [339, 340] was considered to prove the excitonic character of the optical levels of SWNTs.

10.4
Exciton Calculations for Carbon Nanotubes

In this section we present some details of the calculations of the excitonic behavior in carbon nanotubes. First a discussion of the Bethe–Salpeter equation is given which is used to calculate the excitonic wavefunctions and their mixing by the Coulomb interaction (Section 10.4.1). Then the energy dispersion of excitons is discussed in Section 10.4.2, while exciton wavefunction calculations are discussed in Section 10.4.3. Finally in Section 10.4.4 the family pattern formation in exciton photophysics is discussed.

10.4.1 Bethe–Salpeter Equation

Here we show how to calculate the exciton energy Ω_n and the wavefunction Ψ^n [186, 311, 319, 321]. Since the exciton wavefunction is localized in real space by a Coulomb interaction, the wave vector of an electron ($\boldsymbol{k}_\mathrm{c}$) or a hole ($\boldsymbol{k}_\mathrm{v}$) is not a good quantum number any more, and thus the exciton wavefunction Ψ_n for the nth exciton energy Ω_n is given by a linear combination of Bloch functions at many $\boldsymbol{k}_\mathrm{c}$ and $\boldsymbol{k}_\mathrm{v}$ wave vectors. The mixing of different wave vectors by the Coulomb interaction is obtained by the so-called Bethe–Salpeter equation [321]

$$\sum_{\boldsymbol{k}_\mathrm{c}\boldsymbol{k}_\mathrm{v}} \left\{ [E(\boldsymbol{k}_\mathrm{c}) - E(\boldsymbol{k}_\mathrm{v})]\delta_{\boldsymbol{k}'_\mathrm{c}\boldsymbol{k}_\mathrm{c}}\delta_{\boldsymbol{k}'_\mathrm{v}\boldsymbol{k}_\mathrm{v}} + K\left(\boldsymbol{k}'_\mathrm{c}\boldsymbol{k}'_\mathrm{v}, \boldsymbol{k}_\mathrm{c}\boldsymbol{k}_\mathrm{v}\right) \right\} \Psi^n(\boldsymbol{k}_\mathrm{c}\boldsymbol{k}_\mathrm{v})$$
$$= \Omega_n \Psi^n\left(\boldsymbol{k}'_\mathrm{c}\boldsymbol{k}'_\mathrm{v}\right) , \qquad (10.9)$$

where $E(\boldsymbol{k}_\mathrm{c})$ and $E(\boldsymbol{k}_\mathrm{v})$ are the quasi-electron and quasi-hole energies, respectively. Here "quasi-particle" means that we add a Coulomb interaction to the one-particle energy and that the particle has a finite lifetime in an excited state. Equation (10.9) represents simultaneous equations for many $\boldsymbol{k}'_\mathrm{c}$ and $\boldsymbol{k}'_\mathrm{v}$ points.

The mixing term of Eq. (10.9) which we call the kernel, $K(\boldsymbol{k}'_\mathrm{c}\boldsymbol{k}'_\mathrm{v}, \boldsymbol{k}_\mathrm{c}\boldsymbol{k}_\mathrm{v})$ is given by:

$$K\left(\boldsymbol{k}'_\mathrm{c}\boldsymbol{k}'_\mathrm{v}, \boldsymbol{k}_\mathrm{c}\boldsymbol{k}_\mathrm{v}\right) = -K^\mathrm{d}\left(\boldsymbol{k}'_\mathrm{c}\boldsymbol{k}'_\mathrm{v}, \boldsymbol{k}_\mathrm{c}\boldsymbol{k}_\mathrm{v}\right) + 2\delta_S K^\mathrm{x}\left(\boldsymbol{k}'_\mathrm{c}\boldsymbol{k}'_\mathrm{v}, \boldsymbol{k}_\mathrm{c}\boldsymbol{k}_\mathrm{v}\right) , \qquad (10.10)$$

with $\delta_S = 1$ for spin singlet states and 0 for spin triplet states (see Section 10.2.3). The direct and exchange interaction kernels K^d and K^x are given by the following integrals [332]:

$$K^\mathrm{d}\left(\boldsymbol{k}'_\mathrm{c}\boldsymbol{k}'_\mathrm{v}, \boldsymbol{k}_\mathrm{c}\boldsymbol{k}_\mathrm{v}\right) \equiv W\left(\boldsymbol{k}'_\mathrm{c}\boldsymbol{k}_\mathrm{c}, \boldsymbol{k}'_\mathrm{v}\boldsymbol{k}_\mathrm{v}\right)$$
$$= \int d\boldsymbol{r}' d\boldsymbol{r} \psi^*_{\boldsymbol{k}'_\mathrm{c}}(\boldsymbol{r}')\psi_{\boldsymbol{k}_\mathrm{c}}(\boldsymbol{r}') w(\boldsymbol{r}', \boldsymbol{r}) \psi_{\boldsymbol{k}'_\mathrm{v}}(\boldsymbol{r}) \psi^*_{\boldsymbol{k}_\mathrm{v}}(\boldsymbol{r}) ,$$
$$K^\mathrm{x}\left(\boldsymbol{k}'_\mathrm{c}\boldsymbol{k}'_\mathrm{v}, \boldsymbol{k}_\mathrm{c}\boldsymbol{k}_\mathrm{v}\right) = \int d\boldsymbol{r}' d\boldsymbol{r} \psi^*_{\boldsymbol{k}'_\mathrm{c}}(\boldsymbol{r}')\psi_{\boldsymbol{k}'_\mathrm{v}}(\boldsymbol{r}') v(\boldsymbol{r}', \boldsymbol{r}) \psi_{\boldsymbol{k}_\mathrm{c}}(\boldsymbol{r}) \psi^*_{\boldsymbol{k}_\mathrm{v}}(\boldsymbol{r}) , \qquad (10.11)$$

where w and v are the screened and bare Coulomb potentials, respectively, and ψ is the quasi-particle wavefunction as discussed below.

The quasi-particle energies are the sum of the single-particle energy ($\epsilon(\boldsymbol{k})$) and self-energy ($\Sigma(\boldsymbol{k})$),

$$E(\boldsymbol{k}_\mathrm{i}) = \epsilon(\boldsymbol{k}_\mathrm{i}) + \Sigma(\boldsymbol{k}_\mathrm{i}) , \quad (i = \mathrm{c}, \mathrm{v}) , \qquad (10.12)$$

where $\Sigma(\boldsymbol{k})$ is expressed by:

$$\Sigma(\boldsymbol{k}_\mathrm{c}) = -\sum_{\boldsymbol{q}} W(\boldsymbol{k}_\mathrm{c}(\boldsymbol{k}+\boldsymbol{q})_\mathrm{v}, (\boldsymbol{k}+\boldsymbol{q})_\mathrm{v}\boldsymbol{k}_\mathrm{c}) ,$$
$$\Sigma(\boldsymbol{k}_\mathrm{v}) = -\sum_{\boldsymbol{q}} W(\boldsymbol{k}_\mathrm{v}(\boldsymbol{k}+\boldsymbol{q})_\mathrm{v}, (\boldsymbol{k}+\boldsymbol{q})_\mathrm{v}\boldsymbol{k}_\mathrm{v}) . \qquad (10.13)$$

In order to obtain the kernel and self-energy, the single-particle Bloch wavefunction $\psi_{\boldsymbol{k}}(\boldsymbol{r})$ and screening potential w are obtained by either a first principles calculation [321] or an extended tight-binding wavefunction and a random phase approximation (RPA) calculation [186]. In the RPA calculation, the static screened Coulomb interaction is expressed by:

$$w = \frac{v}{\kappa \epsilon(\boldsymbol{q})} , \tag{10.14}$$

with a static dielectric constant κ and the dielectric function $\epsilon(\boldsymbol{q}) = 1 + v(\boldsymbol{q})\Pi(\boldsymbol{q})$. By calculating the polarization function $\Pi(\boldsymbol{q})$ and the Fourier transformation of the unscreened Coulomb potential $v(\boldsymbol{q})$, we get information, which is sufficient for describing the exciton energy and wavefunction [186, 319]. For one-dimensional materials, the Ohno potential is commonly used for the unscreened Coulomb potential $v(q)$ for π orbitals [324]

$$v(|\boldsymbol{R}_{u's'} - \boldsymbol{R}_{0s}|) = \frac{U}{\sqrt{((4\pi\epsilon_0/e^2)\, U |\boldsymbol{R}_{us} - \boldsymbol{R}_{0s'}|)^2 + 1}} , \tag{10.15}$$

where U is the energy cost to place two electrons on a single site ($|\boldsymbol{R}_{us} - \boldsymbol{R}_{0s'}| = 0$) and this energy cost is taken as $U \equiv U_{\pi_a \pi_a \pi_a \pi_a} = 11.3\text{eV}$ for π orbitals [324].

10.4.2
Exciton Energy Dispersion

For an electron–hole pair, we introduce wave vectors $\boldsymbol{K}$ for the exciton center of mass and $\boldsymbol{k}$ for the relative motion,

$$\boldsymbol{K} = (\boldsymbol{k}_c - \boldsymbol{k}_v)/2 , \quad \boldsymbol{k} = \boldsymbol{k}_c + \boldsymbol{k}_v . \tag{10.16}$$

The Bethe–Salpeter equation (Eq. (10.9)) is then rewritten in terms of $\boldsymbol{K}$ and $\boldsymbol{k}$. Since the Coulomb interaction is related to the relative coordinate of an electron and a hole, the center-of-mass motion $\boldsymbol{K}$ can be treated as a good quantum number.[12] Thus the exciton energy is given by an energy dispersion as a function of $\boldsymbol{K}$.

In Figure 10.4, we show the two-dimensional Brillouin zone (2D BZ) of graphite and the cutting lines for a (6,5) single-wall carbon nanotube. Since optical transitions occur around the K or K' points in the 2D BZ, we can expect four possible combinations of an electron and hole pair as is discussed in Section 10.3.1 and as is shown in Figure 10.4. The excitons in a SWNT can then be classified according to their $2\boldsymbol{K}$ values. If both the electron ($\boldsymbol{k}_c$) and hole ($\boldsymbol{k}_v$) wave vectors are from the K (or K') region, then $2\boldsymbol{K} = \boldsymbol{k}_c - \boldsymbol{k}_v$ lies in the Γ region and the corresponding exciton is an $A_{1,2}$ symmetry exciton. If an electron is from the K region and a hole is from the K' region, their $2\boldsymbol{K}$ lies in the K region and this exciton is an $\mathbb{E}$ symmetry exciton. If an electron is from the K' region and a hole is from the K region, their $2\boldsymbol{K}$ lies in the K' region and this exciton is an $\mathbb{E}^*$ symmetry exciton.

12) Strictly speaking, when we consider the screening effect of an exciton by other electrons, $\boldsymbol{K}$ is no longer a good quantum number.

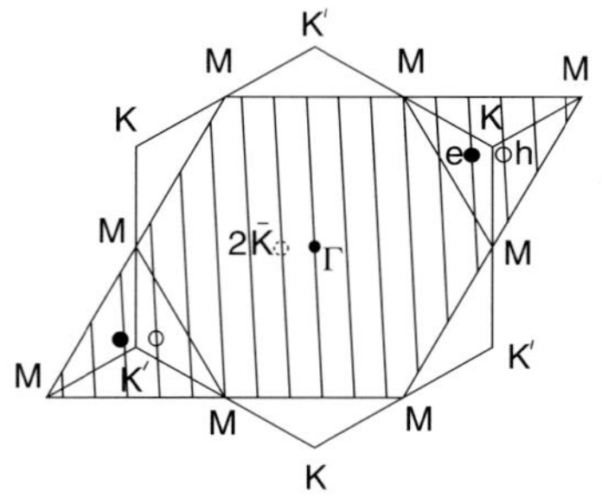

Figure 10.4 The three inequivalent regions in the 2D Brillouin zone of graphite. The cutting lines (Section 10.3.1) for a (6,5) SWNT are shown. The electron–hole pairs and the corresponding center-of-mass momentum $2K = k_c - k_v$ for an $A_{1,2}$ exciton of the (6,5) SWNT are indicated. The electron–hole pair where the electron and hole lie on the second and first cutting lines relative to the K point and the electron–hole pair where the electron and hole lie on the first and second cutting lines relative to the K' point correspond to an E_{12} exciton with the center-of-mass momentum $2K$ on the first cutting line relative to the Γ point [186].

As we discussed in the symmetry section (Section 10.3.1), the exciton wavefunction should be described by an irreducible representation of the group of the wave vector for a SWNT. For A excitons, the electron–hole pair wavefunction $|\boldsymbol{k}_c, \boldsymbol{k}_v\rangle = |\boldsymbol{k}, \boldsymbol{K}\rangle$ with the electron and hole from the K region, and $|-\boldsymbol{k}_v, -\boldsymbol{k}_c\rangle = |-\boldsymbol{k}, \boldsymbol{K}\rangle$ with the electron and hole from the K' region have the same magnitude for $\boldsymbol{K}$. Thus, we can recombine these two electron–hole pairs to get

$$A_{2,1} = |\boldsymbol{k}, \pm, \boldsymbol{K}\rangle = \frac{1}{\sqrt{2}}(|\boldsymbol{k}, \boldsymbol{K}\rangle \pm |-\boldsymbol{k}, \boldsymbol{K}\rangle) . \tag{10.17}$$

Here $|\boldsymbol{k}, +, \boldsymbol{K}\rangle$ and $|\boldsymbol{k}, -, \boldsymbol{K}\rangle$ are antisymmetric (A_2) and symmetric (A_1), respectively, under the C_2 rotation around the axis perpendicular to the nanotube axis.[13)]

10.4.3
Exciton Wavefunctions

In this section we discuss mainly the calculated results relevant to bright excitons [186]. In Figure 10.5, we plot the energy dispersion of $E_{ii}(A_j)$ ($i = 1, 2$; $j = 1, 2$) excitons with spin $S = 0, 1$ for a (6,5) SWNT, where E_{ii} denotes the energy separation of the ith valence band to the ith conduction band. We use the same notation of E_{ii} for the exciton [229], too, for simplicity. The exciton with the largest energy dispersion shows a parabolic energy dispersion relation which reflects the free particle behavior of an exciton with a mass. For the A_1 exciton, $S = 0$ and $S = 1$ are degenerate, since the exchange interaction vanishes by symmetry. Figure 10.5d gives the excitation energy levels for $K = 0$ $E_{11}(A^\nu)$ states. We note that for the spin $S = 0$ states, $E_{11}(A_2^0)$ has a somewhat larger energy than $E_{11}(A_1^0)$. This

13) It might be confusing that $+$ ($-$) corresponds to antisymmetric (symmetric) wavefunctions. But it is a correct statement.

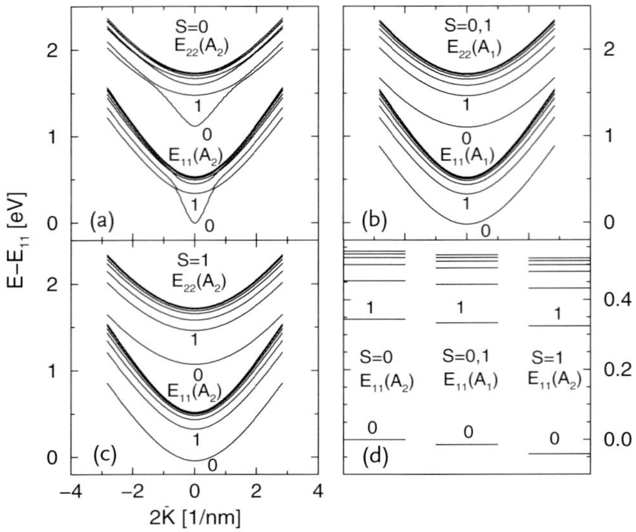

Figure 10.5 The excitation energy dispersions for (a) $E_{11}(A_2)$ ($S = 0$) and $E_{22}(A_2)$ ($S = 0$), (b) $E_{11}(A_1)$ ($S = 0, 1$) and $E_{22}(A_1)$ ($S = 0, 1$), and (c) $E_{11}(A_2)$ ($S = 1$) and $E_{22}(A_2)$ ($S = 1$) excitons for a (6,5) SWNT. The excitation energy levels for $\boldsymbol{K} = 0$ excitons are also shown in (d) [186].

means that the bright A_2 exciton is not the lowest energy state [341]. The Coulomb energy $K^{\rm d}(\boldsymbol{k}', -\boldsymbol{k}; \pm, \boldsymbol{K})$, which is the energy for an inter-valley scattering process, thus has a one order of magnitude smaller energy than the corresponding energy for an intravalley scattering process, $K^d(\boldsymbol{k}', \boldsymbol{k}; \pm, \boldsymbol{K})$. Therefore, the energy difference between $E_{11}(A_2^0)$ and $E_{11}(A_1^0)$ (for $S = 0$) is predicted to be quite small (about 12meV in Figure 10.5d). Moreover, in Figure 10.5d the triplet $E_{11}(A_2^0)$ state lies about 35meV below the singlet $E_{11}(A_2^0)$ state. The energy difference between the triplet and singlet $E_{11}(A_2)$ states is determined by the exchange Coulomb interaction, $K^{\rm x}(\boldsymbol{k}', \boldsymbol{k}; \boldsymbol{K})$ (see Eq. (10.11)), which is about one order of magnitude smaller than the direct Coulomb interaction $K^{\rm d}(\boldsymbol{k}', \boldsymbol{k}; \boldsymbol{K})$ in SWNTs. The energy difference between the singlet $E_{11}(A_2^0)$ state and the $E_{11}(A_1^0)$ state, and the energy difference between the singlet and triplet $E_{11}(A_2^0)$ states are consistent for different calculations [320, 325].

Hereafter, we will mainly discuss the singlet bright exciton $E_{ii}(A_2^0)$ states with $K = 0$. In Figure 10.6 we show the exciton wavefunctions along the nanotube axis of an (8,0) SWNT for several of the $E_{22}(A_2^\nu)$ states with lower excitation energies and with $\nu = 0, 1$, and 2, namely (a) $E_{22}(A_2^0)$, (b) $E_{22}(A_2^1)$ and (c) $E_{22}(A_2^2)$ [186]. Because of the orthogonalization of the wavefunctions, we can see wavefunctions with 0, 1, 2 nodes in Figure 10.6a–c, respectively. The localization of the wavefunction for $E_{22}(A_2^0)$ for the (8,0) SWNT is around 1 nm at full width half maximum intensity. The localization length increases with increasing energy and with increasing nanotube diameter, reflecting the dimensional change from 1D to 2D.

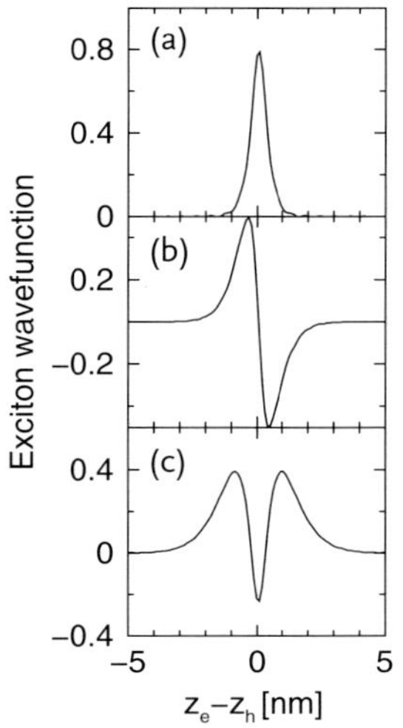

Figure 10.6 The magnitude of the exciton wavefunctions along the nanotube axis of an (8,0) SWNT for the states: (a) $E_{22}(A_2^0)$, (b) $E_{22}(A_2^1)$ and (c) $E_{22}(A_2^2)$ [186].

In a SWNT or in graphite, there are two sublattices, A and B. For $E_{22}(A_2^0)$ and $E_{22}(A_2^2)$, the wavefunctions have a similar amplitude for the A and B sublattices, while for $E_{22}(A_2^1)$, the amplitude of the wavefunction of the electron and hole occupies one of the two sublattice exclusively. The latter behavior of the wavefunction (that the amplitude of the wavefunction can exist only on one sublattice) can be seen for localized edge states. Thus we expect an interesting behavior to occur when the exciton becomes localized at the end of a SWNT.

The $E_{22}(A_2^0)$ and $E_{22}(A_2^2)$ excitons are symmetric and the $E_{22}(A_2^1)$ exciton is antisymmetric upon reflection about the z axis. It then follows that the $E_{22}(A_2^0)$ and $E_{22}(A_2^2)$ excitons are bright and the $E_{22}(A_2^1)$ exciton is dark with respect to linearly polarized light parallel to the z axis.[14] In the two-photon absorption experiments, the $E_{22}(A_2^2)$ exciton becomes bright [339]. For an achiral (armchair or zigzag) SWNT, exciton wavefunctions are either even or odd functions of z because of the inversion center in these SWNTs. Thus, we use A_{2u} or A_{2g} to label $E_{22}(A_2^1)$, or $E_{22}(A_2^0)$ (and $E_{22}(A_2^2)$), respectively, for achiral SWNTs [218].

The localized exciton wavefunction is constructed by mixing many k states in which the mixing coefficients are determined by the Bethe–Salpeter equation (Eq. (10.9)). We found above that the envelope functions for the three wave functions given in Figure 10.6 can, respectively, be fitted to a Gaussian (e^{-Cz^2}, ze^{-Cz^2}, $(Az^2 - B)e^{-Cz^2}$). The mixing coefficients (Fourier transformation) are also localized in k space around one particle k points for a given E_{ii}, and this localization is described by the wavefunction full-width at half maximum magnitude ℓ_k.

14) An important fact in discussing this statement is that the A_2 wavefunction itself has a minus sign under a C_2 rotation or z reflection. Thus an even function of z becomes a dipole-allowed exciton state.

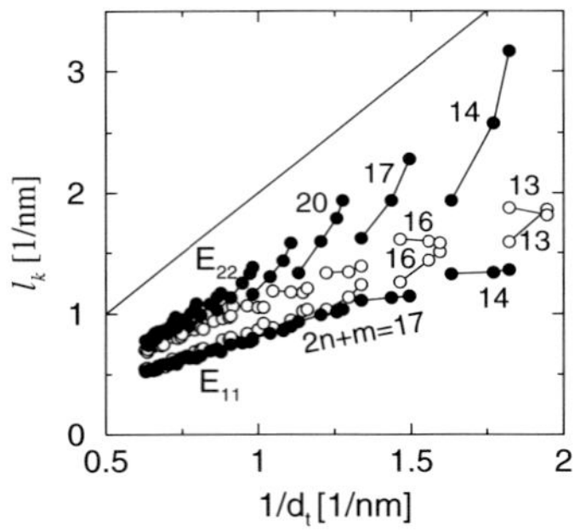

Figure 10.7 The half width ℓ_k of the wave functions in 1D k space for the $E_{11}(A_2^0)$ and $E_{22}(A_2^0)$ states. The cutting line spacing $2/d_t$ is shown by the solid line for comparison. Open and filled circles, respectively, denote SI and SII SWNTs, where SI and SII denote the semiconductor tube type in SWNTs. Integers denote the $2n+m$ values for individual SWNT families [186].

In Figure 10.7, we plot ℓ_k in the 1D k space for the bright exciton states, $E_{11}(A_2^0)$ and $E_{22}(A_2^0)$, and for all SWNTs with diameters (d_t) in the range of $0.5\,\text{nm} < d_t < 1.6\,\text{nm}$. In Figure 10.7, we also plot the cutting line spacing $2/d_t$ by the solid line. An important message here is that ℓ_k is smaller than $2/d_t$ for all SWNTs. This result indicates that one cutting line is sufficient to describe individual $E_{ii}(A)$ states. Consequently, the difficulty in calculating the Bethe–Salpeter equation is reduced significantly for the case of carbon nanotubes. For the higher energy states, $E_{ii}(A_2^\nu)$ states with $\nu = 1, 2, \cdots$, have ℓ_k values that are smaller than that for $E_{ii}(A_2^0)$, since the wavefunctions for $E_{ii}(A_2^\nu)$ are more delocalized in real space. Generally, we can say that the ith cutting line is sufficient to describe the $E_{ii}(A)$, $E_{ii}(\mathbb{E})$ and $E_{ii}(\mathbb{E}^*)$ states[15] and that the ith and $(i+1)$th cutting lines are sufficient to describe the $E_{ii+1}(A)$ and $E_{i+1i}(A)$ states. Since metallic SWNTs (M-SWNTs) have smaller ℓ_k values than semiconducting SWNTs (S-SWNTs), the above conclusion is also valid for M-SWNTs.

The assumption, that we consider only one cutting line, is valid so long as the range of the Coulomb interaction is larger than the diameter d_t of a SWNT. For a typical nanotube diameter ($0.5 < d_t < 2.0\,\text{nm}$), the Coulomb interaction is sufficiently strong for all carbon atoms along the circumferential direction, so that the wavefunction for the E_{ii} exciton becomes constant around the circumferential direction, which is the physical reason why we need only one cutting line. When the diameter is sufficiently large compared to the range of the Coulomb interaction (more than 5 nm), the exciton wavefunction is no longer constant around the circumferential direction (two-dimensional exciton), and then we need to use the kernel from neighboring cutting lines in the Bethe–Salpeter equation.

It is important to mention that for the wavefunction for the $E_{ii+1}(A)$ exciton, which is excited by perpendicularly polarized light (see Section 9.2.3), we must consider two cutting lines (i and $i+1$) for the wavefunction (see Figure 10.4),

15) For the $\mathbb{E}$ exciton, the $\pm i$ states are considered for the electron and the hole.

because of the dipole selection rule. In fact, the calculated exciton has an anisotropy around the circumferential direction in the sense that the electron and hole exist with respect to each other at opposite sides of the nanotube. Since the induced depolarization field [238] cancels the optical field, there is a significant upshift of the energy position of $E_{ii+1}(A)$ relative to the $E_{ii}(A)$ excitonic transition [238, 315]. This upshift in energy has been observed in PL experiments [311, 312] and can also be observed in maps of RRS spectra.

10.4.4 Family Patterns in Exciton Photophysics

Based on resonance Raman spectroscopy studies, it is found that the optical transition energies E_{ii} when plotted against tube diameter exhibit family patterns related to the $2n + m$ = constant families (see Section 9.3). These family patterns are also observed in two-dimensional photoluminescence (PL) plots [180]. The reason why we get family patterns is that (n, m) SWNTs within the same $2n + m$ =constant family have diameters similar to one another and that the E_{ii} values are generally inversely proportional to the nanotube diameter. The small change of the E_{ii} values within the same family is due to the trigonal warping effect of the electronic dispersion around the K (and K') point [229]. The trigonal warping effect and the θ-dependent lattice distorion gives a chirality dependence both for the one-particle energy position at a van Hove singular k point, and the corresponding effective mass. The change of the effective mass for the various SWNTs belonging to the same $(2n+m)$ family is important for determining the exciton binding energy and self-energy for each SWNT.

The energy spread in a family becomes large as the diameter decreases and becomes less than 1 nm. In this case, the simple tight-binding calculation in which we consider only π electrons is not sufficient to reproduce the E_{ii} energy positions. To address this problem, the extended tight-binding (ETB) calculation has been developed (Section 10.1) in which the curvature effect is taken into account by the mixing of the π orbitals with the σ and $2s$ orbitals of carbon. When we then add the density functional form of the many-body effect to the ETB results, we can reproduce nicely the experimental results for the dependence of the E_{ii} on diameter and chiral angle [299, 342].

In Figure 10.8 we plot the exciton Kataura plot for the $E_{11}^S(A_2^0)$ and $E_{22}^S(A_2^0)$ states for S-SWNTs and the $E_{11}^M(A_2^0)$ states for M-SWNTs. Open and filled circles are for Type I and II (SI and SII) SWNTs, respectively, and crossed circles are for M-SWNTs. SI and SII SWNTs are defined by $\mathrm{mod}(2n + m, 3) = 1$ and $\mathrm{mod}(2n + m, 3) = 2$, respectively [336], where mod is the modulus function of integers. The E_{ii} values are the sum of the ETB one-particle energies, the self-energy Σ and the exciton binding energy E_{bd}. We note that a large family spread appears in Figure 10.8, which is consistent with both calculations [182, 299] and experiments [180, 183].

In Figure 10.9, we plot separately each contribution to the ETB (extended tight-binding) transition energy E_{11}, the self-energy Σ of the quasi-particle, and the ex-

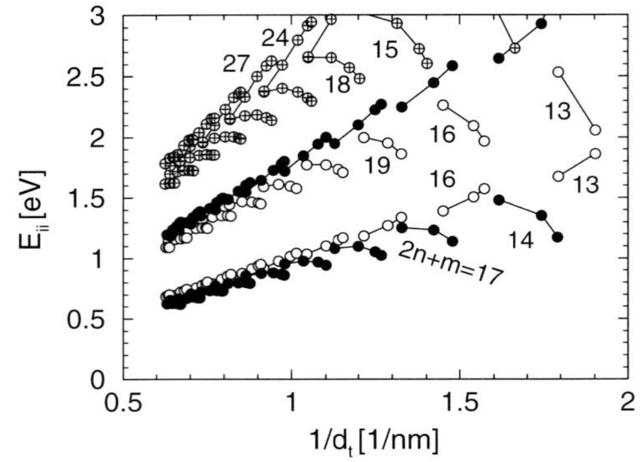

Figure 10.8 The excitation energy Kataura plot based on the extended tight-binding model for $E_{11}^S(A_2^0)$ and $E_{22}^S(A_2^0)$ for S-SWNTs and $E_{11}^M(A_2^0)$ for M-SWNTs. Open and filled circles are for SI and SII SWNTs, respectively, and crossed circles are for M-SWNTs [186].

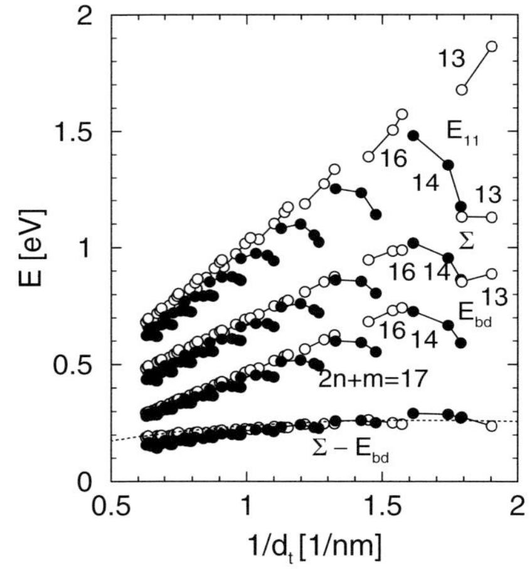

Figure 10.9 The excitation energy E_{11}^S, self-energy Σ, binding energy E_{bd} and energy corrections $\Sigma - E_{\text{bd}}$ based on the ETB model for $E_{11}(A_2^0)$ bright exciton states. Open and filled circles are, respectively, for SI and SII SWNTs. The dashed line is calculated by Eq. (10.18) with $p = 1$ [186].

citon binding energy E_{bd}. We also plot $\Sigma - E_{\text{bd}}$ in the same figure [186]. It is seen that although both Σ and E_{bd} tend to increase the family spread, the two terms almost cancel each other regarding the family spread. This near cancellation leads to a weak chirality dependence, showing that the net energy correction ($\Sigma - E_{\text{bd}}$) to the single-particle energy depends predominantly on the SWNT diameter. Thus, we conclude that the large family spread that is observed in E_{11} originates predominantly from the trigonal warping effect and the θ-dependent lattice distortion in the single-particle spectra. It is known that the logarithmic correction due to the effect of the Coulomb interaction on the dispersion of 2D graphite is not canceled by the exciton binding energy and this effect leads to a logarithmic energy correction $E^{\log}$ given by [181, 299]

$$E^{\log} = 0.55(2p/3d_t)\log[3/(2p/3d_t)] \,, \qquad (10.18)$$

which is the rationale for the logarithmic term is the empirical Eq. (9.10). In Figure 10.9, we plot $E^{\log}$ with $p = 1$ as a dashed line, thus showing that the energy correction $\Sigma - E_{bd}$ follows this logarithmic behavior well. This good agreement for $\Sigma - E_{bd}$ explains why the ETB model works well in considering excitonic and other many-body effects occurring in SWNT photophysics.

10.5 Exciton Size Effect: the Importance of Dielectric Screening

The E_{ii} values are now understood in terms of the bright exciton energy within the framework of a tight-binding calculation which includes curvature optimization [182, 329] and many-body effects [37–39, 186, 343]. The assignments of E_{ii} for SWNTs over a large region of both diameter ($0.7 < d_t < 3.8$ nm) and E_{ii} (1.2–2.7 eV) values and for a variety of surrounding materials are now available [20, 317], thus making it possible to accurately determine the effect of the general dielectric constant κ on E_{ii}. By "general" we mean that κ comprises the screening from both the tube and from the environment. A d_t-dependent effective κ value for the exciton calculation is needed to reproduce the experimental E_{ii} values consistently. This dependence is important for the physics of quasi and truly one-dimensional materials generally and can be used in interpreting optical experiments and environment effects for such materials.

10.5.1 Coulomb Interaction by the 2s and σ Electrons

Figure 10.10 shows a map of experimental E_{ii} values (black dots) [189, 317] from a SWNT sample grown by the water-assisted ("super-growth") chemical vapor deposition method [33, 281]. The resulting data for the E_{ii} transition energies are plotted as a function of the radial breathing mode frequencies ω_{RBM}, as obtained by resonance Raman spectroscopy (RRS) [189, 317, 318]. In Figure 10.10, the experimental values of E_{ii} vs. ω_{RBM} for the "super-growth" sample E_{ii}^{exp} are compared with the calculated bright exciton energies E_{ii}^{cal} (open circles and stars), obtained with the dielectric screening constant $\kappa = 1$. Although E_{ii}^{cal} includes SWNT curvature and many-body effects [186], clearly the E_{ii}^{exp} values are red shifted when compared with theory, and the red shift depends on both ω_{RBM} (i. e., on d_t) and on the optical energy levels (i in E_{ii}).

The E_{ii} values can be renormalized in the calculation by explicitly considering the dielectric constant κ in the Coulomb potential energy given by Eq. (10.14) [344]. Here, κ represents the screening of the e–h (electron–hole) pair by core (1s) and σ electrons (κ_{tube}) and by the surrounding materials (κ_{env}), while $\varepsilon(q)$ explicitly gives

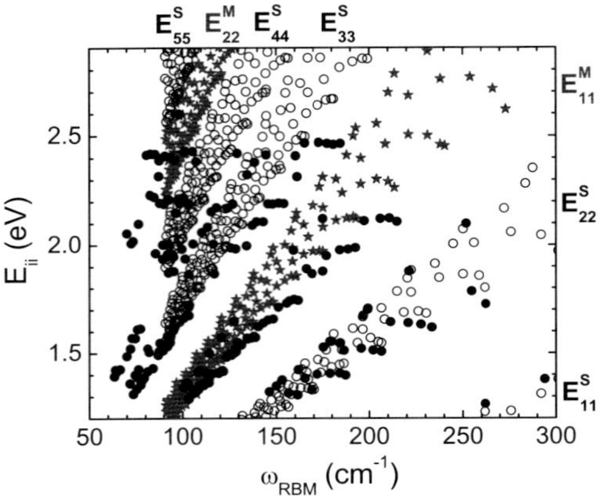

Figure 10.10 Black dots show $E_{ii}^{\exp}$ vs. ω_{RBM} results obtained from resonance Raman spectra taken from a "super-growth" SWNT sample [189, 317]. The black open circles (semiconducting; S-SWNTs) and the dark gray stars (metallic; M-SWNTs) give E_{ii}^{cal} calculated for the bright exciton with dielectric constant $\kappa = 1$ [186]. Along the x axis, E_{ii}^{cal} are translated using the relation $\omega_{\mathrm{RBM}} = 227/d_{\mathrm{t}}$ [189]. Due to limited computer time availability, only E_{ii} for tubes with $d_{\mathrm{t}} < 2.5\,\mathrm{nm}$ (i. e., $\omega_{\mathrm{RBM}} > 91\,\mathrm{cm}^{-1}$) have been calculated. Transition energies E_{ii}^S ($i = 1$ to 5) denote semiconducting SWNTs and E_{ii}^M ($i = 1, 2$) denote metallic SWNTs [190].

the polarization function for π-electrons calculated within the random phase approximation (RPA) [186, 324, 345]. To fully account for the observed energy-dependent E_{ii} redshift, the total κ values are fitted to minimize $E_{ii}^{\exp} - E_{ii}^{\mathrm{cal}}$. The bullets in Figure 10.11 show the fitted κ values as a function of p/d_{t}, which reproduce each experimental E_{ii} value for the assigned (n, m) SWNTs for the "super-growth" SWNT sample. The stars stand for a different SWNT sample, named "alcohol-assisted" SWNTs [346], and these differences are due to different environmental screening (κ_{env}), as discussed later in Section 10.5.2. The integer p corresponds to the distance ratio of the cutting lines from the K point, where $p = 1, 2, 3, 4$ and 5 are for E_{11}^S, E_{22}^S, E_{11}^M, E_{33}^S, and E_{44}^S, respectively [20]. Consideration of the p/d_{t} ratio allows us to compare the κ values of SWNTs with different d_{t} and different E_{ii} using the same plot. As seen in Figure 10.11, the κ values increase with increasing p/d_{t} for different E_{ii} values. The κ values for E_{33}^S and E_{44}^S (Figure 10.11b) appear over a smaller κ region than those for E_{11}^S and E_{22}^S (Figure 10.11a).

The data points in Figure 10.11 can be fit with the empirical relation [190]

$$\kappa = C_\kappa \left(\frac{p}{d_{\mathrm{t}}}\right)^\alpha , \tag{10.19}$$

where the exponent $\alpha = 1.7$ was found to work for all $E_{ii}^{\exp}$, but different C_κ parameters were needed for different samples to reflect the differences in their environmental conditions [190]. For E_{11}^S, E_{22}^S and E_{11}^M, the value $C_\kappa = 0.75$ was obtained for the "super-growth" SWNTs and $C_\kappa = 1.02$ for the "alcohol-assisted" SWNTs (dashed and dotted curves in Figure 10.11a, respectively), and these differences

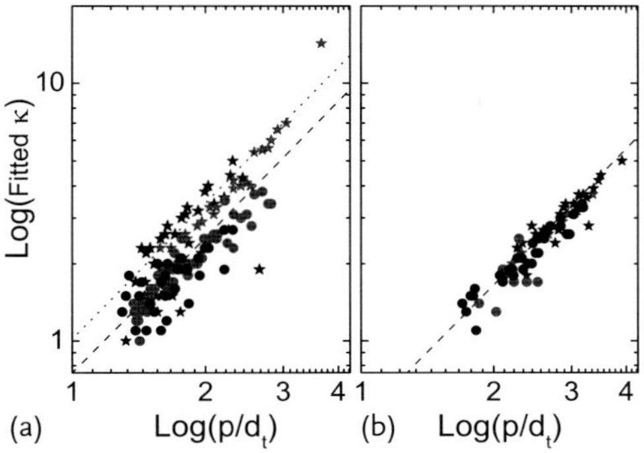

Figure 10.11 The calculated κ values, which are fitted to the experimental E_{ii} values from the "super-growth" (bullets) [317] and "alcohol-assisted" (stars) [287] samples. (a) E_{22}^S (black) and E_{11}^M (dark gray). The dashed and dotted curves are given by Eq. (10.19) with $C_\kappa = 0.75$ and 1.02, respectively. (b) E_{33}^S (black) and E_{44}^S (dark gray). The dashed curves are for Eq. (10.19) with $C_\kappa = 0.49$ [190].

can be understood by κ_{env}. The E_{33}^S and E_{44}^S are fitted using $C_\kappa = 0.49$ for both samples, as shown by the dashed line in Figure 10.11b.

Qualitatively, the origin of the diameter dependence of the dielectric constant presented by Eq. (10.19) consists of: (1) the exciton size and (2) the amount of electric field "feeling" the dielectric constant of the surrounding material. These two factors are connected and the development of an electromagnetic model is needed to fully rationalize this equation. Interestingly, the similarity between the κ values found for E_{22}^S and E_{11}^M shows that the difference between metallic and semiconducting tubes is satisfactorily taken into account by using the RPA in calculating $\varepsilon(q)$ [343]. Also interesting is the different κ behavior that is observed for higher energy levels ($p > 3$), where C_κ is smaller than for E_{ii} with $p \leq 3$, and in this regime κ is independent of the sample environment. Two pictures can be given: (1) the more localized exciton wavefunction (a larger exciton binding energy) for E_{33}^S and E_{44}^S compared with E_{11}^M and E_{22}^S, leads to smaller κ values and a lack of a κ_{env} dependence of the wavefunctions for the E_{33}^S and E_{44}^S excitons; (2) the stronger tube screening (κ_{tube}) leads to an independence regarding κ_{env} and, consequently, leads to a smaller effective κ.

10.5.2
The Effect of the Environmental Dielectric Constant κ_{env} Term

Figure 10.12 shows a comparison between the E_{ii}^{exp} from the "super-growth" SWNT sample (bullets) [317] and from the "alcohol-assisted" SWNT samples (open circles) [287]. From Figure 10.12, we see that besides the changes in ω_{RBM}, as discussed in Section 9.1.2, the E_{ii}^{exp} values from the "alcohol-assisted" SWNTs are generally red shifted with respect to those from the "super-growth" SWNTs. Assuming that κ_{tube} does not change from sample to sample for a particular type of SWNT sample, since the structure of a given (n, m) tube should be the same, these results

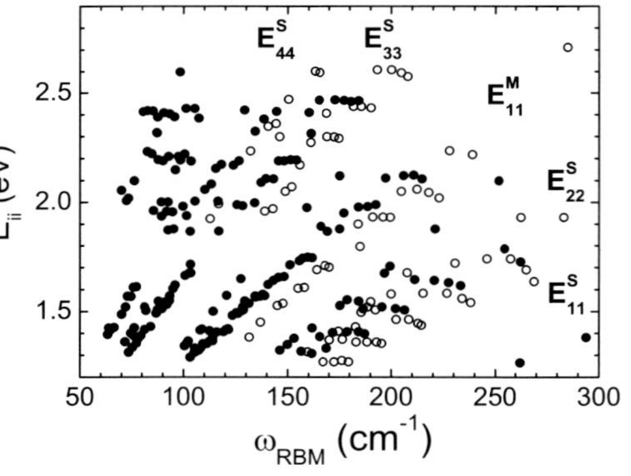

Figure 10.12 E_{ii}^{exp} vs. ω_{RBM} results obtained for the "super-growth" (bullets) and the "alcohol assisted" (open circles) SWNT samples [190].

indicate that the "alcohol-assisted" SWNTs are surrounded by a larger κ_{env} value, than the "super growth" sample, thus increasing the effective κ and decreasing E_{ii} [190].

Looking at Figure 10.11 we can observe the difference in the κ values resulting from fitting the E_{ii}^{exp} to the "super-growth" (bullets) in comparison to "alcohol-assisted" (stars) SWNT samples. For E_{22}^S and E_{11}^M (Figure 10.11a), we see a clear difference for κ up to $p = 3$ when comparing the two samples. However, for E_{33}^S and E_{44}^S (Figure 10.11b), no difference in κ between the two samples can be seen. This means that the electric field of the E_{33}^S and E_{44}^S excitons does not extend much outside the SWNT volume, in contrast to the E_{22}^S and E_{11}^M excitons for which the κ_{env} effect is significant. Since the effect of κ_{env} is relatively small for energies above E_{11}^M, it is still possible to assign the (n, m) values from E_{33}^S and E_{44}^S even if the dielectric constant of the environment is not known, and even though the E_{33}^S and E_{44}^S values are seen within a large density of dots in the Kataura plot.

10.5.3
Further Theoretical Considerations about Screening

The dielectric constant for the materials surrounding the SWNTs cannot be directly used in calculations or in interpreting data, since the electric field exists not only in the surrounding materials but also in the SWNTs themselves. In the calculations shown in Figure 10.13, the dielectric constant κ is treated as a parameter which is used in the Ohno potential and $\Delta E_{ii}^S \equiv \Delta E_{ii}^S(\kappa = 2) - \Delta E_{ii}^S(\kappa = 3) > 0$ is plotted as a function of $1/d_t$. In this figure, we can see the $(2n + m)$ family pattern for type I (S1, $\text{mod}(2n + m, 3) = 1$) and type II (S2, $\text{mod}(2n + m, 3) = 2$) semiconducting SWNTs for ΔE_{11}^S and ΔE_{22}^S. This predicted behavior is consistent with recent experimental results [347, 348].

In Figure 10.14, we plot E_{ii}^S for a (6,5) SWNT as a function of (a) $1/\kappa$ or (b) κ with (solid lines) and without (dashed lines) including the electron screening ef-

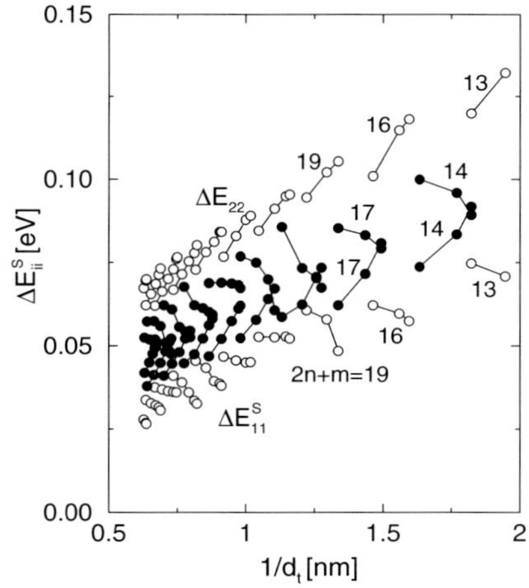

Figure 10.13 Calculated shifts in the E_{11}^S and E_{22}^S transition energies due to changing κ from 3 to 2. Open and filled circles are, respectively, for S1 and S2 type semiconducting SWNTs [186].

fect for the E_{11}^S and E_{22}^S states for a (6,5) SWNT. It is seen that without considering the electron screening effect, E_{ii}^S is approximately linearly dependent on $1/\kappa$. The screening effect will bend the line, reducing the energy shift, especially for the small κ region, for example, $\kappa < 2$. The bending effect arises from the fact that the screening effect by the environment generally provides a dielectric constant, independent of the wave vector $\boldsymbol{q}$, while the effect of the dielectric function $\epsilon(0, q)$ on the E_{ii}^S transition energies resulting from the electron screening effect is a function of q [319]. In Figure 10.14a, we also show the exciton binding energy vs. $1/\kappa$. It is seen that for both E_{11} and E_{22} states, the binding energy approximately scales as

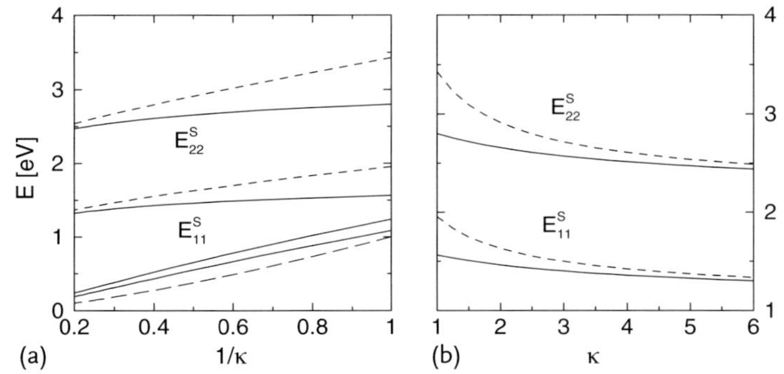

Figure 10.14 The transition energy dependence on κ for states E_{11}^S and E_{22}^S for a (6,5) SWNT. Solid and dashed lines, respectively, do or do not consider the π electron screening effect. (a) Excitation energy vs. $1/\kappa$. The three curves below E_{11} give the $1/\kappa$-dependent exciton binding energies for E_{22}^S and E_{11}^S, and the function $E = (1/\kappa)^{1.4}$ from top to bottom, respectively. (b) Excitation energy vs. κ for E_{11}^S and E_{22}^S [186].

$(1/\kappa)^{1.4}$. This scaling parameter α can be used for estimating the exciton binding energy E_{bd} as:

$$E_{\mathrm{bd}} \propto d_{\mathrm{t}}^{\alpha-2} m^{\alpha-1} \kappa^{-\alpha} , \tag{10.20}$$

where m is the effective mass of the electron or hole [324]. The concept of scaling is useful for explaining the observed family patterns and diameter dependence of the E_{ii}. The diameter dependence on κ is relevant to this scaling rule, which is still not well understood [219].

10.6
Summary

In this chapter we have discussed the exciton science in carbon nanotubes. This discussion starts after introducing the importance of $\sigma - \pi$ hybridization in the curved graphene sheet of carbon nanotubes. Excitons are important in semiconductors generally, but they are especially important in nanomaterials, where the spatial confinement enhances the overlap between optically excited electrons and holes, thus enhancing the exciton binding energy. We here discuss the physics of exciton levels, wave vectors, spin, symmetry aspects, selection rules, energies, wavefunctions, that is, all the aspects important for achieving an accurate description of the optical levels in nanotubes. However, to achieve experimental accuracy, a treatment of the dielectric screening also has to be considered.

The diameter-dependent dielectric constants following Eq. (10.19) reproduce the measured E_{ii} values well for a large region of energy (1.2–2.7 eV) and tube diameter (0.7–3.8 nm). The present treatment for κ is sufficiently accurate for assigning both the $2n + m$ family numbers and the (n, m) SWNTs belonging to each family for different SWNT samples. All the observed E_{ii} vs. (n, m) values are now theoretically described within their experimental precision, considering use of the extended tight-binding model along with many-body corrections plus a diameter-dependent dielectric constant κ (Eq. (10.19)). The empirical Eq. (10.19) is not yet fully understood, and theoretical modeling considering the role of the exciton size is still needed. The results presented here are also consistent with the empirical methodology of Eq. (9.10) [287], and therefore provide justification for this approach.

Problems

[10-1] When the planar sp^2 bond is bent in the circumferential direction by an angle θ which is on the order of $\theta = 0.1$ rad, we expect a large curvature effect. What is the corresponding tube diameter?

[10-2] The curvature effect can be understood by the Slater–Koster method in which the transfer matrix element for π–π orbitals is mixed with σ–σ or-

bitals. Show how the matrix element is modified as a function of θ in Problem 10-1.

[10-3] Applying the wavefunction of Eq. (10.2) to Eq. (10.1), obtain the differential equations for g and f.

[10-4] Solve the one-dimensional hydrogen Schrödinger equation. In particular show that the lowest energy is minus infinity. Obtain the corresponding wavefunction.

[10-5] Solve the two-dimensional hydrogen Schrödinger equation. In this case, we have an angular momentum within a plane. Obtain the energies.

[10-6] When the potential is spherical in three dimensions, the potential is given as a function of r. In this case, show that the angular part of the wavefunction for the Hamiltonian with a spherical potential is given by the spherical harmonics, $Y_{lm}(\theta, \varphi)$.

[10-7] Obtain the differential equation on r for a hydrogen atom with angular momentum ℓ, and solve this equation for the bound states of the hydrogen atom.

[10-8] Explain what are the direct and indirect energy gaps by showing some example in the case of semiconductors. In the case of an indirect energy gap, explain that the exciton cannot only emit a photon.

[10-9] When electron–electron interaction U exists between two electrons, explain that the interaction modifies electronic states mainly near the Fermi energy. In particular, why are the electrons at the bottom of the energy band not affected much by the Coulomb interaction. Consider the two cases that (1) $U \gg W$ (W: energy band width), and (2) $W \gg U$.

[10-10] Explain qualitatively what you expect for the energy dispersion $E(k)$, if electrons have a finite lifetime in the presence of some interaction.

[10-11] When the wavefunction is spatially localized, the wavefunction is expressed by a linear combination of Bloch wavefunctions with many k values. In order to understand this, calculate the Fourier transform of a Gaussian function $f(x) = \exp(-x^2/a^2)$.

[10-12] The spin functions of an electron with $s_z = 1/2$ and $-1/2$ are denoted by α and β, respectively. When we define the spin of a hole by $s_z = -1/2$ and $1/2$, respectively, for the states which are originally occupied by an electron with $s_z = 1/2$ and $-1/2$ and denoted by $\beta_{\rm h}$ and $\alpha_{\rm h}$, obtain the total spin function of a singlet ($S = 0$) and a triplet ($S = 1$) exciton.

[10-13] When we consider the electric-dipole transition which does not depend on spin, explain that the $S = 1$, $S_z = -1, 0, 1$ exciton states are all dark exciton states.

[10-14] In the presence of a Coulomb interaction, an energy difference between the $S = 1$ and $S = 0$ states of two electrons appears, which is known as the exchange energy. Explain the physical meaning of the exchange energy in terms of the Pauli exclusion principle, the Hartree–Fock approximation, and the spin of two electrons. Why are the $S = 1$ states lower in energy than the $S = 0$ states?

[10-15] Explain the reason why the $S = 0$ exciton does not relax quickly to the lower energy $S = 1$ states.

[10-16] Derive Eq. (10.8) for achiral SWNTs and draw schematics similar to those of Figure 10.3 for both zigzag and armchair tubes.

[10-17] Illustrate figures similar to Figure 10.4 for the E_{11} exciton levels for A_1, A_2 and E symmetries. What happens for E_{22} or E_{12} excitons? Note that two configurations are coupled when explaining A_1 and A_2 excitons (see Eq. (10.17)).

[10-18] Why does the plus sign in Eq. (10.17) correspond to the antisymmetric A_2 exciton? Discuss how the wavefunctions of the valence and conduction bands change by a C_2 rotation.

[10-19] Compare Eq. (10.18) with the logarithmic part of Eq. (9.10) and discuss the results.

[10-20] Show that the effective mass for the E_{33}^S or E_{44}^S exciton is larger than that for E_{11}^S or E_{22}^S.

[10-21] By considering the $1s$ energy of a hydrogen atom, show that the exciton binding energy becomes large when the effective mass of the electron and hole is large. From this, show that the exciton binding energy of the E_{33}^S or E_{44}^S exciton is larger than that for the E_{11}^S or E_{22}^S exciton.

[10-22] When the dielectric constant of the surrounding material κ_{env} is large, which exciton feels κ_{env} more, E_{11}^S or E_{33}^S? Explain by illustrating the electric field for these excitons.

[10-23] In a $2n + m =$ const. family, which SWNTs have a larger effective mass, near armchair SWNTs or near zigzag SWNTs?

[10-24] Using the fact that the distance of the K point from the Γ point in the direction perpendicular to the cutting line is $K\Gamma = (2n + m)/3$, illustrate the cutting lines for the E_{11}^S to E_{44}^S transitions for S1 and S2 SWNTs in the two-dimensional Brillouin zone.

[10-25] For E_{11}^S, which semiconducting type has a larger effective mass, S1 or S2? How about for E_{22}^S?

11
Tight-Binding Method for Calculating Raman Spectra

This chapter focuses on the physics involved in calculating Raman spectra, building on the experimental observation of resonance Raman spectra in Chapter 9, and the excitonic nature of the optical transitions discussed in Chapter 10. Besides the strong E_{laser} dependence due to resonance effects, as described in the previous chapters, the Raman intensity also depends on the strength of the matrix elements in the numerator of Eq. (5.20) (see Chapter 5). In the RBM intensity (I_{RBM}) analysis developed in Chapter 9, the matrix elements were assumed to be constant, and this procedure was found to be sufficiently accurate for extracting the resonance transition energies (E_{ii}). However, I_{RBM} strongly depends on the SWNT chirality, that is, on the (n, m) indices, and this dependence requires calculation of the matrix elements. These calculations can only be developed after obtaining the basic electronic structure developed in Chapter 10, and these calculations are not only interesting for explaining the (n, m) dependence of I_{RBM}, but also are needed for developing a fundamental understanding of the physics behind the electron–photon and electron–phonon coupling which strongly influence the observed Raman spectra. Furthermore, the Raman intensity also depends on the resonance window width (γ_r), which broadens the resonance condition while avoiding a singularity in the denominator of Eq. (5.20). The science behind this γ_r parameter can also be calculated and these effects are described in this chapter.

Much of the physics comes from the anisotropic optical absorption in graphene, and these topics will be described here. Although this chapter focuses on both the (n, m) dependence of I_{RBM} and the theoretical description for explaining such a dependence, we only consider the RBM feature as a case study because of its rich behavior and because both experiment and theory are well developed for the RBM feature. The theory discussed here is also important for describing the Raman response for graphene systems in general.

In Section 11.1 we present a few general considerations involved in the calculation of the Raman spectra for graphene and carbon nanotubes, while Section 11.2 summarizes experimental results for the (n, m) dependence of the Raman intensity for carbon nanotubes. Results for a theoretical calculation of the electronic structure are presented in Section 11.3 based on a simple tight-binding calculation. Section 11.4 to Section 11.7 presents an overview of the principles involved in calculating the electron and phonon states and the various pertinent matrix el-

Raman Spectroscopy in Graphene Related Systems. Ado Jorio, Riichiro Saito, Gene Dresselhaus, and Mildred S. Dresselhaus

ISBN: 978-3-527-40811-5

ements and in calculating the Raman intensity. In Section 11.8 we present the parameters for extending the calculations to excitons. In Section 11.9 we combine all the results needed to formulate a calculation of the first-order Raman intensity for carbon nanotubes. In Section 11.10 we introduce the physics of γ_r. We conclude this chapter in Section 11.11 with a brief summary.

11.1
General Considerations for Calculating Raman Spectra

When we calculate the Raman intensity for sp^2 carbons, we need many computational programs in order to calculate physical properties, such as (1) the electronic energy band structure and (2) the phonon dispersion relations. Using the phonon eigenvectors, we can obtain the non-resonant Raman intensity using the so-called bond-polarization theory, in which the polarization induced by the phonon amplitude is proportional to the amplitude of the phonon vibrations [31, 278]. As we discussed in the previous chapters, resonance Raman scattering is essential to describe results for carbon nanotubes, since non-resonance Raman theory cannot account for the interesting phenomena that are observed for sp^2 carbons but can only serve as a basis for symmetry arguments. For calculating the resonance Raman spectra, further physical properties, such as (3) the optical dipole transition matrix elements and (4) the electron–phonon matrix elements are needed. Especially when we consider the general structure of sp^2 carbons (nanotubes, deformed graphene, etc.), the interaction of not only the π electrons, but also the $2s$ and σ electrons should be included. Here, we start with simple tight-binding methods to establish general principles and we use extended tight-binding methods for more detailed calculations, when we consider only π electrons or when we consider a combination of $2s$, σ and π electrons, respectively. Thus we discuss below (1) and (2) within the simple tight-bonding model in Section 11.3 and within the extended tight-binding model in Section 11.4, and then we discuss the more detailed calculations for (3) and (4) within the extended tight-binding model for the remainder of the chapter. Further, in the case of SWNTs, we must consider (5) exciton states, as well as (6) the exciton–photon interaction and (7) the exciton–phonon interaction and their respective matrix elements. Using either electron–phonon or exciton–phonon matrix elements, (8) the resonance window can be determined by evaluating the lifetime of the photoexcited carriers. (9) The phonon life time which is relevant to the Kohn anomaly effect is also calculated by the electron–phonon and exciton–phonon interactions. Within the tight-binding method, computer programs have been developed for calculating the Raman spectra for sp^2 carbons by the authors and their co-workers.

First principles approaches can do the same types of calculations once a set of programs for determining the matrix elements are ready for use. However, it seems that most of the packaged software do not yet contain the programs needed for calculating the electron–photon or electron–phonon matrix elements. Furthermore, it would take a large amount of computational time to obtain the resonance Raman spectra or the Raman excitation profile if we obtained such results by first principle calculations. The tight-binding method introduced in Section 11.3 is therefore useful to apply to sp^2 carbon materials. In this chapter we show the basic description for topics (1)–(8). For topic (9), we refer the reader to Chapter 8.

This chapter does not describe the calculational methods in detail, but rather overviews the principles of the calculations by introducing the relevant references and the relationship between the calculational programs. This analysis will be useful for the readers in order to understand what is needed for analyzing experimental Raman data and what are the essential points for understanding the observed phenomena.

11.2
The (n, m) Dependence of the RBM Intensity: Experiment

For analyzing the importance of the matrix elements to the Raman cross-section, we compare the RBM intensity for different (n, m) SWNTs satisfying the full resonance condition (i. e., $E_{\text{laser}} = E_{ii}$). The RBM spectra from the "super-growth" water-assisted SWNTs have already been extensively analyzed with respect to their resonance energies E_{ii} and their adherence to the $\omega_{\text{RBM}} = 227/d_t$ cm^{-1}nm relation, important for the (n, m) assignment (see Chapter 9). These Raman spectra are now ready for an accurate intensity analysis, which is discussed in the present chapter.

High-resolution transmission electron microscopy (HRTEM) imaging was applied to experimentally obtain the d_t distribution of a given super-growth sample, which was then used for RRS characterization [349]. It was here assumed that SWNTs of different chiral angles are equally abundant in the growth process. The relative (n, m) population of the SWNTs must scale as the d_t distribution times $1/d_t$, since the number of different (n, m) species of a given diameter scales linearly with d_t. Also, chiral SWNTs are twice as populous as achiral ones, since both right-handed and left-handed isomers are present in a typical sample. Figure 11.1a shows the intensity calibrated experimental RRS map. The strong features are related to RRS RBM features with E_{ii} transitions.[1] Notice the intensity variation from peak to peak in Figure 11.1a. Especially noticeable is the change in intensity within a given $(2n + m)$ = constant branch. The RBM signal gets stronger when going from larger to smaller chiral angles. Since each spectrum ($S_{(\omega, E_{\text{laser}})}$) is the sum of the individual contributions to the observed intensity of all SWNTs, the spectral intensity can be written as:

$$S_{(\omega, E_{\text{laser}})} = \sum_{n,m} \left[\text{Pop}_{(n,m)} I_{(n,m)}^{E_{\text{laser}}} \frac{\Gamma/2}{(\omega - \omega_{\text{RBM}})^2 + (\Gamma/2)^2} \right], \tag{11.1}$$

1) A close inspection of the experimental resonant Raman spectral (RRS) map shows some low-intensity features associated with cross-polarized transitions (E_{12}^S) and RBM overtones. For our purposes, it is safe to ignore cross-polarized and overtone features, since their total contribution to the RRS map is less than 4%.

where $\mathrm{Pop}_{(n,m)}$ is the population of the (n, m) nanotube species, $\Gamma = 3\,\mathrm{cm}^{-1}$ is the experimental average value for the full width at half maximum intensity of the tube's RBM Lorentzian lineshape, ω_{RBM} is the frequency of its RBM and ω is its Raman shift. Each nanotube in the sample contributes to the RBM RRS spectra with one Lorentzian, whose total integrated area $(I_{(n,m)}^{E_{\mathrm{laser}}})$ for the Stokes process at a given excitation laser energy (E_{laser}) is given by:

$$I_{(n,m)}^{E_{\mathrm{laser}}} = \left| \frac{\mathcal{M}}{(E_{\mathrm{laser}} - E_{ii} + i\gamma_r)(E_{\mathrm{laser}} - E_{\mathrm{ph}} - E_{ii} + i\gamma_r)} \right|^2 , \tag{11.2}$$

where $E_{\mathrm{ph}} = \hbar\omega_{\mathrm{RBM}}$ is the energy of the RBM phonon, E_{ii} is the energy corresponding to the ith excitonic transition, γ_r is a damping factor and $\mathcal{M}$ represents the matrix elements for the Raman scattering by one RBM phonon of the (n, m) nanotube. The values for E_{ii} and ω_{RBM} were determined experimentally (see Chapter 9). $\mathcal{M}$ and γ_r were found by fitting the experimental RBM RRS map with Eq. (11.1) using the functions:

$$\mathcal{M} = \left[\mathcal{M}_{\mathrm{A}} + \frac{\mathcal{M}_{\mathrm{B}}}{d_{\mathrm{t}}} + \frac{\mathcal{M}_{\mathrm{C}}\cos(3\theta)}{d_{\mathrm{t}}^2}\right]^2$$
$$\gamma_r = \gamma_{\mathrm{A}} + \frac{\gamma_{\mathrm{B}}}{d_{\mathrm{t}}} + \frac{\gamma_{\mathrm{C}}\cos(3\theta)}{d_{\mathrm{t}}^2} , \tag{11.3}$$

where $\mathcal{M}_i$ and γ_i (i = A, B, C) are adjustable parameters that also account for their dependence on (n, m) or equivalently on (d_{t}, θ). The best values for $\mathcal{M}_i$ and γ_i, considering the excitonic transitions E_{22}^S and the lower branch of E_{11}^M are listed in Table 11.1 for d_{t} in nm, γ_r in meV and $\mathcal{M}$ in arbitrary units since the Raman intensity is usually given by arbitrary units.

Using Eqs. (11.1)–(11.3) the values in Eq. (11.1), we obtain the modeled RRS map shown in Figure 11.1b, which accounts for the experimentally observed results shown in Figure 11.1a very well. To understand the (n, m) dependence of

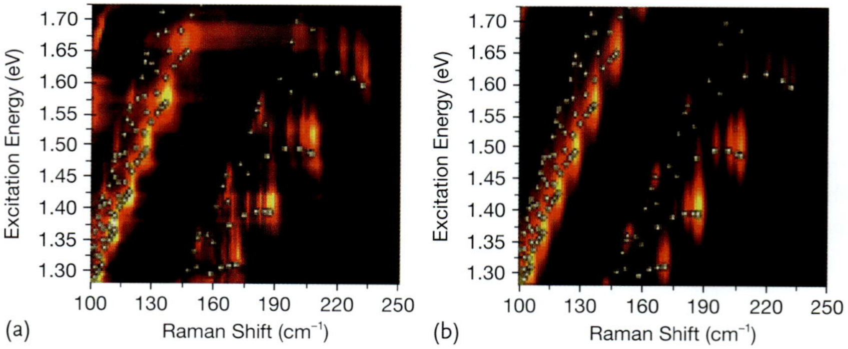

Figure 11.1 (a) Experimental RRS map for the radial breathing mode feature. The intensity calibration was made by measuring a standard tylenol sample. (b) Modeled map obtained by using Eq. (11.1) and the same laser excitation energies range as in (a) [349].

Table 11.1 Fitted parameters $\mathcal{M}_i$ and γ_i for metallic, semiconductor type 1 and type 2 tubes. These parameters are to be used in Eq. (11.3) with d_t in nm, yielding $\mathcal{M}$ in arbitrary units and γ_i in meV.

Type	$\mathcal{M}_A$	$\mathcal{M}_B$	$\mathcal{M}_C$	γ_A	γ_B	γ_C
M	1.68	0.52	5.54	23.03	28.84	1.03
S_1	−19.62	29.35	4.23	−3.45	65.10	7.22
S_2	−1.83	3.72	1.61	−10.12	42.56	−6.84

both $\mathcal{M}$ and γ_r, described by Eq. (11.3), the next sections address their (n, m) dependence theoretically. Before discussing the direct calculation of the matrix elements and resonance window widths, we first describe the calculations for electron and phonon states.

11.3 Simple Tight-Binding Calculation for the Electronic Structure

In this section first, we review the tight-binding method for the electronic energy bands for sp^2 carbons and then apply this method to the electronic structure of graphite. In a perfectly periodic system, any wavefunction for an energy band satisfies the Bloch theorem [118]. A Bloch function is a basis wavefunction which satisfies the Bloch theorem. Typical examples of Bloch functions are plane wave and tight-binding wavefunctions. A large (complete) set of the plane wave functions can express the energy band precisely by solving the Hamiltonian as a function of the wave vector k. The computational accuracy of the plane wave expansion is determined by how many plane waves are adopted in the computation for a given unit cell which is specified by a cut-off energy. The plane wave expansion for energy band calculations is mainly used for first principle calculations which does not explicitly specify the atom species contained in the structure of the solid or of the molecule. As far as we consider only the sp^2 carbon system, the tight-binding method is useful for understanding the physics and for saving computational time.

As described in Section 2.2, the tight-binding wavefunction $\Psi_j(\boldsymbol{k})$, where j denotes the energy band index, is given by a linear combination of a small number of tight-binding Bloch wave functions $\Phi_{j'}$

$$\Psi_j(\boldsymbol{k}, \boldsymbol{r}) = \sum_{j'=1}^{N} C_{jj'}(\boldsymbol{k}) \Phi_{j'}(\boldsymbol{k}, \boldsymbol{r}) \,, \quad (j = 1, \cdots, N) \,, \tag{11.4}$$

where $C_{jj'}(\boldsymbol{k})$ are coefficients to be determined and N is the number of atomic orbitals in the unit cell. From an atomic orbital φ_j in the unit cell, we can construct the tight-binding Bloch function Φ_j as

$$\Phi_j(\boldsymbol{k}, \boldsymbol{r}) = \frac{1}{\sqrt{N_u}} \sum_{\boldsymbol{R}}^{N_u} e^{i\boldsymbol{k}\cdot\boldsymbol{R}} \varphi_j(\boldsymbol{r} - \boldsymbol{R}) \,, \quad (j = 1, \cdots, N) \,, \tag{11.5}$$

where the summation takes place over N_u lattice vectors $\boldsymbol{R}$ in the crystals. When we define the Hamiltonian and the overlap matrices, $\mathcal{H}_{jj'}(\boldsymbol{k})$ and $\mathcal{S}_{jj'}(\boldsymbol{k})$, respectively,

$$\mathcal{H}_{jj'}(\boldsymbol{k}) = \langle \Phi_j | \mathcal{H} | \Phi_{j'} \rangle \,, \quad \mathcal{S}_{jj'}(\boldsymbol{k}) = \langle \Phi_j | \Phi_{j'} \rangle \,, \quad (j, j' = 1, \cdots, N) \,. \tag{11.6}$$

The Schrödinger equation is given in terms of the simultaneous equations

$$\sum_{j'=1}^{N} \mathcal{H}_{jj'}(\boldsymbol{k}) C_{ij'} = E_i(\boldsymbol{k}) \sum_{j'=1}^{N} \mathcal{S}_{jj'}(\boldsymbol{k}) C_{ij'} \quad (i = 1, \cdots, N) \,. \tag{11.7}$$

Defining a column vector by

$$\boldsymbol{C}_i = \begin{pmatrix} C_{i1} \\ \vdots \\ C_{iN} \end{pmatrix} . \tag{11.8}$$

Then Eq. (11.7) is expressed by

$$\mathcal{H}\boldsymbol{C}_i = E_i(\boldsymbol{k}) \mathcal{S} \boldsymbol{C}_i \,. \tag{11.9}$$

By using a numerical calculation for the diagonalization of a given $\mathcal{H}$ and $\mathcal{S}$ for each $\boldsymbol{k}$, we get the energy eigenvalues $E_i(\boldsymbol{k})$ and eigenfunctions $\boldsymbol{C}_i(\boldsymbol{k})$.[2)]

The ij matrix element of $\mathcal{H}$ is expressed by

$$\begin{aligned} \mathcal{H}_{ij}(\boldsymbol{k}) &= \frac{1}{N_u} \sum_{R,R'} e^{ik(R-R')} \langle \varphi_i(r - R') | \mathcal{H} | \varphi_j(r - R) \rangle \\ &= \sum_{\Delta R} e^{ik(\Delta R)} \langle \varphi_i(r - \Delta R) | \mathcal{H} | \varphi_j(r) \rangle \,, \end{aligned} \tag{11.10}$$

where $\Delta R \equiv R - R'$ is the distance between carbon atoms in graphite as illustrated in Figure 11.2 and in the second line of Eq. (11.10), we use the fact that $\langle \varphi_i(r - R') | \mathcal{H} | \varphi_j(r - R) \rangle$ only depends on ΔR. Similarly, $\mathcal{S}$ is given by:

$$\mathcal{S}_{ij}(\boldsymbol{k}) = \sum_{\Delta R} e^{ik(\Delta R)} \langle \varphi_i(r - \Delta R) | \varphi_j(r) \rangle \,. \tag{11.11}$$

The tight-binding parameters for $\mathcal{H}$ and $\mathcal{S}$ in Eqs. (11.10) and (11.11) are defined, respectively, by $\langle \varphi_i(r - \Delta R) | \mathcal{H} | \varphi_j(r) \rangle$ and $\langle \varphi_i(r - \Delta R) | \varphi_j(r) \rangle$ for some nearest neighbor ΔR from knowledge of the $\varphi_j(r)$ atomic orbitals. The tight-binding parameters are given so as to reproduce the experimental atomic energy dispersion data obtained by angle-resolved photoemission spectroscopy (ARPES) or first principles calculations. A typical parameter set (TBP) is listed in Table 11.2 for the 3NN (3rd nearest neighbor tight-binding parameters coupling three graphene layers) as shown in Figure 11.2 [350].

2) We use the LAPAC library (zhegv) for the diagonalization of a Hermitian matrix $\mathcal{A}$ in terms of a positive definite Hermitian matrix $\mathcal{B}$, that obeys $\mathcal{A}\boldsymbol{C} = E\mathcal{B}\boldsymbol{C}$. The output of the "zhegv" subprogram is $E_i(\boldsymbol{k})$ and $\boldsymbol{C}_i(\boldsymbol{k})$.

Table 11.2 Third nearest neighbor tight-binding (3NN TB) parameters for few-layer graphene and graphite. All values are in electron volt except for the parameters s_0–s_2 which are dimensionless. The parameters coming from fits to LDA and GW calculations are shown. The 3NN Hamiltonian is valid over the whole two (three)-dimensional Brillouin zone (BZ) of graphite (graphene layers) [350].

TBP	3NN TB-GW[a]	3NN TB-LDA[a]	EXP[b]	3NN TB-LDA[c]	ΔR, pair[d]
γ_0^1	–3.4416	–3.0121	–5.13	–2.79	$a/\sqrt{3}$, AB
γ_0^2	–0.7544	–0.6346	1.70	–0.68	a, AA and BB
γ_0^3	–0.4246	–0.3628	–0.418	–0.30	$2a/\sqrt{3}$, AB
s_0	0.2671	0.2499	–0.148	0.30	$a/\sqrt{3}$, AB
s_1	0.0494	0.0390	–0.0948	0.046	a, AA and BB
s_2	0.0345	0.0322	0.0743	0.039	$2a/\sqrt{3}$, AB
γ_1	0.3513	0.3077	–	–	c, AA
γ_2	–0.0105	–0.0077	–	–	$2c$, BB
γ_3	0.2973	0.2583	–	–	$(a/\sqrt{3}, c)$, BB
γ_4	0.1954	0.1735	–	–	$(a/\sqrt{3}, c)$, AA
γ_5	0.0187	0.0147	–	–	$2c$, AA
E_0[e]	–2.2624	–1.9037	–	–2.03	
Δ[f]	0.0540[g]	0.0214	–	–	

a fits to LDA and GW calculations[350]
b fit to ARPES experiments by Bostwick *et al.* [351]
c fit to LDA calculations by Reich *et al.* [115]
d in-plane and out-of-plane distances between a pair of A and B atoms.
e the energy position of π orbitals relative to the vacuum level.
f difference of the diagonal term between A and B atoms for multi-layer graphene.
g the impurity doping level due to unintended dopants is adjusted in order to reproduce the experimental value of Δ in graphite.

As seen in Table 11.2, many research groups have so far obtained a set of tight-binding parameters for graphite, carbon nanotubes and graphene. In Figure 11.2,

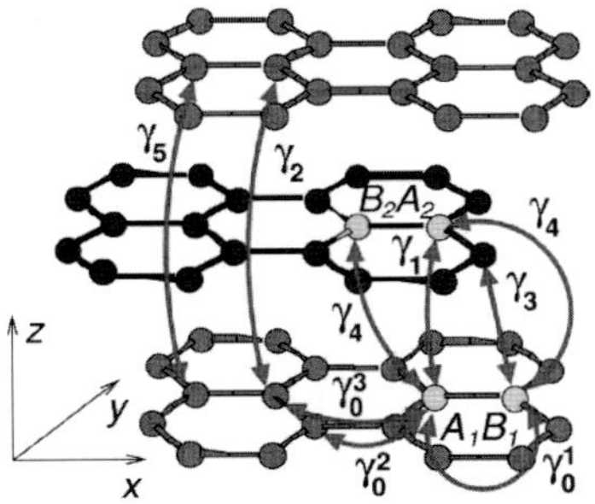

Figure 11.2 Identification of the various Slonczewski–Weiss parameters which describe the tight-binding parameters for a pair of carbon atoms separated by a distance ΔR in graphite [350].

we show a definition of the tight-binding parameters for the Hamiltonian in Eq. (11.10) for pairs of carbon atoms separated by their corresponding distances ΔR [350] in Table 11.2. The notation γ_i for the tight-binding parameters follows that of Slonczewski and Weiss [97] in which $\gamma_0^j, (j = 1, 2, 3)$ denotes the in-plane parameters with the jth nearest neighbors up to the 3rd nearest neighbor (3NN). As far as we consider transport properties near the K point of the first Brillouin zone, the in-plane nearest neighbor parameter γ_0^1 is sufficient. However, when we consider optical transition phenomena around the K point, the further neighbor terms (see Figure 11.2) γ_0^2 and γ_0^3 are also necessary [115]. The parameters γ_1, γ_3, and γ_4 denote interactions between carbon atoms in adjacent layers (see Figure 11.2), while the parameters γ_2 and γ_5 couple carbon atoms between the second nearest neighbor layers. The parameters γ_3 and γ_4 introduce a $\boldsymbol{k}$-dependent interlayer interaction and γ_2 determines a small energy dispersion along the KH direction in the three-dimensional Brillouin zone which gives rise to an electron and a hole carrier pocket, such as is observed for the semimetal graphite.

As for the overlap tight-binding parameters, s_0, s_1, and s_2, in Eq. (11.11) we only consider the in-plane parameters. These parameters are essential for describing the asymmetry between the valence and conduction energy bands of graphite relative to the Fermi energy. The values of these parameters for graphite provide a larger energy band width for the conduction band than that for the valence band consistent with detailed optical and transport measurements in graphite [16].

11.4
Extended Tight-Binding Calculation for Electronic Structures

The simple tight-binding parameters obtained in Section 11.3 are for π orbitals for graphite or few layer graphene. When we consider SWNTs with different diameters, we must consider new tight-binding parameters for each SWNT diameter since the curvature of the cylindrical tube surface mixes π orbitals with σ orbitals (or equivalently ($2p_x$ and $2p_y$) and $2s$ orbitals). Furthermore, when we consider an optimization of the bond length or of the geometrical structure, we need to calculate tight-binding parameters as a function of the various C–C bonds.

The extended tight-binding (ETB) calculation is a tight-binding calculation for π, σ, and $2s$ orbitals, in which the tight-binding parameters for a pair of orbitals and for a particular ΔR are given as a function of the pertinent C–C bond lengths. The functions of the tight-binding parameters are given by first principles calculations for several sp^2 molecules or solids so as to reproduce the optimized C–C bond lengths and bond angles of many sp^2 materials. For carbon systems, the tight-binding parameters as a function of the C–C bond lengths have been calculated by Porezag [263], and this calculation is adopted for optimized calculations of SWNTs with diameters smaller than 1 nm [182]. The ETB calculation of smaller diameter SWNTs reproduces well the observed family pattern[3)] of the optical transition energies. The observed family pattern is considered to be due to a curvature effect.

In quantum chemistry calculations, great effort has been put into calculating these tight-binding parameters as a function of the bond lengths for many different elements of the periodic table using so-called semi-empirical methods. MNDO, MINDO, AM3, and PM5 are names of the parameter sets for popular semi-empirical methods, which are used in many chemistry molecular level calculations packages such as MOPAC and Gaussian, etc. An advantage of the ETB or the semi-empirical method is that the calculation of the optimized structure is easy and thus the speed of the calculation is fast. Thus even when first principle calculations are employed, semi-empirical calculations are in fact used for a preliminary determination of the initial values.

11.5 Tight-Binding Calculation for Phonons

As described in Section 3.2, the phonon energy dispersion can be calculated by a set of springs which connect some nearest neighbor atoms and these theoretical calculations are fitted to experimentally derived phonon dispersion relations, such as are obtained from neutron or X-ray inelastic scattering measurements. The equations of motion used to describe such vibrations are given by:

$$M_i \ddot{\boldsymbol{u}}_i = \sum_j K^{(ij)}(\boldsymbol{u}_j - \boldsymbol{u}_i) , \quad (i = 1, \ldots, N) , \tag{11.12}$$

where M_i is the mass of the ith atom and $K^{(ij)}$ represents the 3×3 force constant tensor. The summation on j is taken over jth nearest neighbor atoms so as to reproduce the phonon energy dispersion relation. When we use the Bloch theorem for deriving an expression for $\boldsymbol{u}_i$ to obtain the amplitude of the vibration with the phonon wave vector $\boldsymbol{q}$ and frequency ω, we get,

$$\boldsymbol{u}_{\boldsymbol{q}}^{(i)} = \frac{1}{\sqrt{N_u}} \sum_{\boldsymbol{R}_i} e^{i(\boldsymbol{q}\cdot\boldsymbol{R}_i - \omega t)} \boldsymbol{u}_i , \tag{11.13}$$

where the sum is taken over all N_u $\boldsymbol{R}_i$ vectors in the crystal for the ith atoms in the unit cell. The equation for $\boldsymbol{u}_{\boldsymbol{q}}^{(i)}$ $(i = 1, \ldots, N)$, where N is the number of atoms in the unit cell, is given by [31]:

$$\left[\sum_j K^{(ij)} - M_i \omega^2(\boldsymbol{q}) I \right] \boldsymbol{u}_{\boldsymbol{q}}^{(i)} - \sum_j K^{(ij)} e^{i\boldsymbol{q}\cdot\Delta\boldsymbol{R}_{ij}} \boldsymbol{u}_{\boldsymbol{q}}^{(j)} = 0 , \tag{11.14}$$

3) Optical transition energies E_{ii} of an (n, m) SWNT with $2n + m$ =constant show a similar energy to one another. When we plot E_{ii} as a function of the diameter, we can see a pattern for SWNTs with $2n + m$ =constant. The observed pattern is called the family pattern and the values of $2n + m$ are called the family number. (See Section 10.4.4.)

where I is a 3×3 unit matrix and $\Delta \boldsymbol{R}_{ij} = \boldsymbol{R}_i - \boldsymbol{R}_j$ is the relative coordinate of the ith atom with respect to the jth atom. The simultaneous equations implied by Eq. (11.14) with $3N$ unknown variables $\boldsymbol{u}_q \equiv (\boldsymbol{u}_q^{(1)}, \boldsymbol{u}_q^{(2)}, \cdots, \boldsymbol{u}_q^{(N)})^t$, for a given $\boldsymbol{k}$ vector, can be solved by diagonalization of the $3N \times 3N$ matrix in brackets, which we call the dynamical matrix, and here we look for a non-trivial solution for $\boldsymbol{u}_q \neq \boldsymbol{0}$ for each $\boldsymbol{q}$ point.

11.5.1
Bond Polarization Theory for the Raman Spectra

Once the phonon eigenvectors $\boldsymbol{u}_k$ are obtained, we can estimate the Raman intensity by use of bond polarization theory. The Raman intensity $I_{\eta'\eta}(\omega)$ for $\mathcal{N}$ atoms in the unit cell is then calculated by the empirical bond polarizability model [352, 353]

$$I_{\eta'\eta}(\omega) \propto \omega_L \omega_S^3 \sum_{f=1}^{3\mathcal{N}} \frac{\langle n(\omega_f)\rangle + 1}{\omega_f} \left| \sum_{\alpha\beta} \eta'_\alpha \eta_\beta P_{\alpha\beta,f} \right|^2 \delta(\omega - \omega_f) \,, \quad (11.15)$$

where ω_L, ω_S, and ω_f are, respectively, the frequencies for the incident and scattered light[4] and the fth phonon frequency at the Γ point, and η and η' are the corresponding unit vectors for the incident and scattered light polarization, respectively. Here $\langle n(\omega_f)\rangle = 1/(\exp(\hbar\omega_f/k_B T) - 1)$ is the Boltzmann occupation number for the phonon, while $P_{\alpha\beta,f}$ is the derivative of the electronic polarization tensor with respect to a change in the C–C bond, in which α and β denote the Cartesian components. Bond polarization theory can be applied to graphite, graphene and nanotubes.

The polarization tensor is modified by the phonon vibration in which the electronic polarization $P_{\alpha\beta,f}$ is proportional to the amplitude of the phonon vibration. After some calculation, we get [31]

$$\begin{aligned} P_{\alpha\beta,f} = -\sum_{\ell B} \Bigg[& \frac{\boldsymbol{R}_0(\ell,B)\cdot\boldsymbol{\chi}(\ell|f)}{R_0(\ell,B)} \times \left\{ \left(\frac{\alpha'_\parallel(B) + 2\alpha'_\perp(B)}{3} \right) \delta_{\alpha\beta} \right. \\ & \left. + \left(\alpha'_\parallel(B) - \alpha'_\perp(B)\right) \left(\frac{R_{0\alpha}(\ell,B) R_{0\beta}(\ell,B)}{R_0(\ell,B)^2} - \frac{1}{3}\delta_{\alpha\beta} \right) \right\} \\ & + \left(\frac{\alpha_\parallel(B) - \alpha_\perp(B)}{R_0(\ell,B)} \right) \left\{ \frac{R_{0\alpha}(\ell,B)\chi_\beta(\ell|f) - R_{0\beta}(\ell,B)\chi_\alpha(\ell|f)}{R_0(\ell,B)} \right. \\ & \left. - \frac{\boldsymbol{R}_0(\ell,B)\cdot\boldsymbol{\chi}(\ell|f)}{R_0(\ell,B)} \times \frac{2R_{0\alpha}(\ell,B) R_{0\beta}(\ell,B)}{R_0(\ell,B)^2} \right\} \Bigg] \,, \end{aligned} \quad (11.16)$$

where $\boldsymbol{\chi}(\ell|f)$ denotes the unit vectors of the fth normal modes of the ℓth atom, B denotes a bond which is connected to the ℓth atom in the unit cell, and $\mathbf{R}(\ell, \mathbf{B})$ is

4) Since $\omega_L \sim \omega_S$ for the incident and scattered light, we can say that the intensity is proportional to ω_S^4 which is discussed in Section 5.5.

Table 11.3 Bond lengths and Raman polarizability parameters for single-wall carbon nanotubes and for various carbon-related molecules.

Molecule	Bond lengths [Å]	$\alpha_{\parallel} + 2\alpha_{\perp}$ [Å^3]	$\alpha_{\parallel} - \alpha_{\perp}$ [Å^3]	$\alpha'_{\parallel} + 2\alpha'_{\perp}$ [Å^2]	$\alpha'_{\parallel} - \alpha'_{\perp}$ [Å^2]
CH_4[a]	C–H (1.09)	1.944			
C_2H_6[a]	C–C (1.50)	2.016	1.28	3.13	2.31
C_2H_4[a]	C=C (1.32)	4.890	1.65	6.50	2.60
C_{60}[b]	C–C (1.46)		1.28	2.30 ± 0.01	2.30 ± 0.30
	C=C (1.40)		0.32 ± 0.09	7.55 ± 0.40	2.60 ± 0.36
C_{60}[a]	C–C (1.46)		1.28 ± 0.20	1.28 ± 0.30	1.35 ± 0.20
	C=C (1.40)		0.00 ± 0.20	5.40 ± 0.70	4.50 ± 0.50
SWNT[c]	C=C (1.42)		0.07	5.96	5.47
SWNT[d]	C=C (1.42)		0.04	4.7	4.0

a S. Guha *et al.* [352].
b D. W. Snoke *et al.* [354].
c A. M. Rao *et al.* (unpublished data which is used in their work [136]).
d R. Saito *et al.* [278].

the corresponding vector from the ℓth atom to the neighboring atom ℓ' specified by B. Here $\alpha'_{\parallel}(B)$ and $\alpha'_{\perp}(B)$ are the radial derivatives of $\alpha_{\parallel}(B)$ and $\alpha_{\perp}(B)$, that is

$$\alpha'_{\parallel}(B) \equiv \frac{\partial \alpha_{\parallel}(B)}{\partial R(\ell, B)}, \quad \text{and} \quad \alpha'_{\perp}(B) \equiv \frac{\partial \alpha_{\perp}(B)}{\partial R(\ell, B)}, \tag{11.17}$$

respectively. The values of $\alpha_{\parallel}(B)$, $\alpha_{\perp}(B)$, $\alpha'_{\parallel}(B)$ and $\alpha'_{\perp}(B)$ have been reported by some groups empirically as a function of the bond lengths between two carbon atoms or between carbon-hydrogen atoms, and these values are listed for carbon nanotubes in Table 11.3.

Thus once we obtain the phonon eigenvectors, the Raman intensity can be calculated by using the empirical values for the appropriate bond polarizability parameters (see Table 11.3). The Raman intensity thus obtained is for a non-resonance Raman signal. However, one can use these results to specify qualitatively which Raman-active modes give a relatively strong signal or how the Raman signal changes by changing the polarization direction of the light.

11.5.2 Non-Linear Fitting of Force Constant Sets

The force constant matrix is obtained by minimizing the least square values of $\mathcal{F}$,

$$\mathcal{F} \equiv \sum_k \frac{\left(f_k^{\text{obs.}} - f_k^{\text{cal.}}\right)^2}{\sigma_k^2}, \tag{11.18}$$

Table 11.4 Force constant parameters of graphite in units of 10^4 dyn/cm. Here the subscripts *r*, *ti*, and *to* refer to radial, transverse in-plane, and transverse out-of-plane force constants, respectively.

Fitted force constants	Jishi *et al.* neutron [123]	Grüneis *et al.* theory[133]	Dubay *et al.* *ab initio* [124]	Maultzsch *et al.* X-ray [129]	Zimmerman *et al.* theory [355]
$\phi_r^{(1)} =$	36.50	40.37	44.58	39.28	41.80
$\phi_r^{(2)} =$	8.80	2.76	7.31	6.34	7.60
$\phi_r^{(3)} =$	3.00	0.05	–5.70	–6.14	–0.15
$\phi_r^{(4)} =$	–1.92	1.31	1.82	2.53	–0.69
$\phi_{ti}^{(1)} =$	24.50	25.18	11.68	11.36	15.20
$\phi_{ti}^{(2)} =$	–3.23	2.22	–3.74	–3.18	–4.35
$\phi_{ti}^{(3)} =$	–5.25	–8.99	6.67	9.27	3.39
$\phi_{ti}^{(4)} =$	2.29	0.22	0.52	–0.40	–0.19
$\phi_{to}^{(1)} =$	9.82	9.40	10.00	10.18	10.20
$\phi_{to}^{(2)} =$	–0.40	–0.08	–0.83	–0.36	–1.08
$\phi_{to}^{(3)} =$	0.15	–0.06	0.51	–0.46	1.00
$\phi_{to}^{(4)} =$	–0.58	–0.63	–0.54	–0.44	–0.55

where $f_{\boldsymbol{k}}^{\text{obs.}}$, $f_{\boldsymbol{k}}^{\text{cal.}}$, and $\sigma_{\boldsymbol{k}}$ denote, respectively, the observed phonon frequency, the calculate phonon frequency and the error bar (or the weight) of the observed phonon frequency at the $\boldsymbol{k}$ point. The force constant parameter sets are determined by a non-linear fitting procedure so as to minimize $\mathcal{F}$. The fitting procedure is not so easy when using a large number of the force constants, since there are many local minima of $\mathcal{F}$ in the parameter space. In order to avoid this problem by getting out of a local minimum to a global minimum, some computer techniques are required.[5)]

In Table 11.4, we list the force constants labeled by ϕ for the earliest result of a fit to inelastic neutron scattering data by Jishi *et al.* [123], theoretical data by Grüneis *et al.* [133] and by Dubay *et al.* [124], inelastic X-ray scattering data by Maultzsch *et al.* [129] and theoretical work by Zimmerman *et al.* [355]. The force constant matrix tensor K_{ij} for a given C–C bond is calculated by rotating the diagonal matrix for a pair of two carbon atoms on the x axis whose diagonal elements are ϕ_r, ϕ_{ti} and ϕ_{to}, which are clearly defined relative to the C–C bond direction [31].

An important quantity that we should consider is the so-called *force constant sum rules* [356] which are used to impose the condition $\omega = 0$ for acoustic phonons. Since all forces K_{ij} are internal forces between the ith and jth atoms, there should be no total force on the system. Otherwise, the materials would automatically move or rotate around their center of mass. Since the sum of internal forces vanishes, the

5) The non-linear fitting procedure depends on the systems considered and there is not a unique way to get to the global minimum. A recommended way to approximate the global minimum is to increase the parameters one by one using the optimized values of the previous calculation as the initial values when adding one additional parameter.

translational invariance is automatically satisfied. However, the rotation around the center of the mass is not always satisfied by a finite number of force constant sets. When we consider the rotation around the ith atom, if the following condition for the jth tangential force constants $\phi_{\text{ti}}^{(j)}$ and $\phi_{\text{to}}^{(j)}$ [355] is satisfied

$$\sum_j^n n_j \phi_t^{(j)} \Delta R_{ij}^2 = 0 \,, \quad (t = ti \quad \text{or} \quad to) \,, \tag{11.19}$$

then the corresponding increase of the potential energy around the ith atom will disappear. Here n_j and ΔR_{ij} are, respectively the number and the distance of the jth atom from the original ith atom. In the case of graphene, there are four tangential phonon modes, that is, iTA, iTO, oTA, and oTO modes, for which Eq. (11.19) must be satisfied for $\phi_{\text{ti}}^{(j)}$ and $\phi_{\text{to}}^{(j)}$. For a force constant set for radial motion, since the force and ΔR_{ij} are parallel to each other, no torques occur. This force constant sum rule is essential for finding the zero values at the Γ point in the Brillouin zone for the acoustic phonon branches. At other $\boldsymbol{k}$ points, the phonon eigenvalues should never be negative (or imaginary), since the dynamical matrix is a positive Hermitian matrix.

The force constant rules given by (Eq. (11.19)) can be used in non-linear fitting procedures (such as for a Lagrange multiplier method) or we can simply define the outermost force constant parameter so as to satisfy Eq. (11.19).

11.6 Calculation of the Electron–Photon Matrix Element

The electron–photon matrix element is calculated in terms of an electron dipole-transition of π electrons. The optical dipole transition from a π (2p) state to an unoccupied π state of an electron within an atom is forbidden. Thus the optical transition between a π and a π^* energy band is possible for the nearest neighbor electron–photon matrix elements as shown below.

The perturbation Hamiltonian of the dipole transition is given by:

$$H_{\text{opt}} = \frac{ie\hbar}{m} \boldsymbol{A}(t) \cdot \nabla \,, \tag{11.20}$$

where $\boldsymbol{A}$ is the vector potential. Here we adopt the Coulomb gauge $\nabla \cdot \boldsymbol{A}(t) = 0$. In this case, the electric field of the light is given by $\boldsymbol{E} = i\omega \boldsymbol{A}$. Hereafter we consider only a linear polarization of the light and thus the vector potential $\boldsymbol{A}$ is given by:

$$\boldsymbol{A} = \frac{-i}{\omega} \sqrt{\frac{I}{c\epsilon_0}} \exp(\pm i\omega t) \boldsymbol{P} \,, \tag{11.21}$$

where $\boldsymbol{P}$ is the unit vector (polarization vector) which specifies the direction of $\boldsymbol{E}$, while I is the intensity of the light in W/m^2 and ϵ_0 is the dielectric constant for vacuum using SI units. The "±" sign corresponds to the emission ("+") or absorption ("−") of a photon with frequency ω.

The matrix element for optical transitions from an initial state $\imath$ at $\boldsymbol{k} = \boldsymbol{k}_\imath$, denoted by $\Psi^\imath(\boldsymbol{k}_\imath)$ to a final state f denoted by $\Psi^f(\boldsymbol{k}_f)$ at $\boldsymbol{k} = \boldsymbol{k}_f$, is defined by

$$\boldsymbol{M}_{\text{opt}}^{f\imath}(\boldsymbol{k}_f, \boldsymbol{k}_\imath) = \left\langle \Psi^f(\boldsymbol{k}_f) \right| H_{\text{opt}} \left| \Psi^\imath(\boldsymbol{k}_\imath) \right\rangle . \tag{11.22}$$

The electron–photon matrix element in Eq. (11.22) is calculated by

$$\boldsymbol{M}_{\text{opt}}^{f\imath}(\boldsymbol{k}_f, \boldsymbol{k}_\imath) = \frac{e\hbar}{m\omega_\rho}\sqrt{\frac{I_\rho}{c\epsilon_0}}\, e^{i(\omega_f - \omega_\imath \pm \omega)t}\, \boldsymbol{D}^{f\imath}(\boldsymbol{k}_f, \boldsymbol{k}_\imath) \cdot \boldsymbol{P} , \tag{11.23}$$

where the electric dipole vector between initial states $\imath$ and final states f denoted by $\boldsymbol{D}^{f\imath}(\boldsymbol{k}_f, \boldsymbol{k}_\imath)$ is defined by

$$\boldsymbol{D}^{f\imath}(\boldsymbol{k}_f, \boldsymbol{k}_\imath) = \langle \Psi^f(\boldsymbol{k}_f) | \nabla | \Psi^\imath(\boldsymbol{k}_\imath) \rangle . \tag{11.24}$$

For a given polarization, $\boldsymbol{P}$, when $\boldsymbol{D}$ is parallel to $\boldsymbol{P}$, the optical absorption (or stimulated emission) is the largest, while when $\boldsymbol{D}$ is perpendicular to $\boldsymbol{P}$, the optical absorption is absent.

11.6.1
Electric Dipole Vector for Graphene

Let us now consider the electric dipole vector for graphene [220]. The wavefunction in Eq. (11.4) with $n = 2$ is given by $\Psi(\boldsymbol{k}) = C_{\text{A}}\Phi_{\text{A}}(\boldsymbol{k}, \boldsymbol{r}) + C_{\text{B}}\Phi_{\text{B}}(\boldsymbol{k}, \boldsymbol{r})$, in which Φ is the Bloch wavefunction for $2p_z$ atomic orbitals for the A and B sites of graphene. Then the electric dipole vector $\boldsymbol{D}^{f\imath}(\boldsymbol{k}_f, \boldsymbol{k}_\imath)$ for graphene is given by:

$$\begin{aligned}\boldsymbol{D}^{f\imath}(\boldsymbol{k}_f, \boldsymbol{k}_\imath) = {} & C_{\text{B}}^{f*}(\boldsymbol{k}_f)C_{\text{A}}^{\imath}(\boldsymbol{k}_\imath)\langle \Phi_{\text{B}}(\boldsymbol{k}_f, \boldsymbol{r})|\nabla|\Phi_{\text{A}}(\boldsymbol{k}_\imath, \boldsymbol{r})\rangle \\ & + C_{\text{A}}^{f*}(\boldsymbol{k}_f)C_{\text{B}}^{\imath}(\boldsymbol{k}_\imath)\langle \Phi_{\text{A}}(\boldsymbol{k}_f, \boldsymbol{r})|\nabla|\Phi_{\text{B}}(\boldsymbol{k}_\imath, \boldsymbol{r})\rangle .\end{aligned} \tag{11.25}$$

Since the $2p_z$ orbital and the $\partial/\partial z$ component of ∇ all have odd symmetry with respect to the z mirror plane, the z component of $\boldsymbol{D}$ becomes zero. When we expand the Bloch function into an atomic orbital, the leading term of $\langle \Phi_{\text{A}}(\boldsymbol{k}_f, \boldsymbol{r})|\nabla|\Phi_{\text{B}}(\boldsymbol{k}_\imath, \boldsymbol{r})\rangle$ is the atomic matrix element between nearest neighbor atoms

$$m_{\text{opt}} = \langle \phi(\boldsymbol{r} - \boldsymbol{R}_{\text{nn}})|\frac{\partial}{\partial x}|\phi(\boldsymbol{r})\rangle , \tag{11.26}$$

where $\boldsymbol{R}_{\text{nn}}$ is the vector between nearest neighbor atoms along the x axis.

When we use approximate coefficients $C_{\rm A}$ and $C_{\rm B}$ for a $\boldsymbol{k}$ point around $\boldsymbol{K} = (0, -4\pi/(3a))$,

$$C^{\rm c}_{\rm A}(\boldsymbol{K}+\boldsymbol{k}) = \frac{1}{\sqrt{2}}, \quad C^{\rm c}_{\rm B}(\boldsymbol{K}+\boldsymbol{k}) = \frac{-k_y + ik_x}{\sqrt{2}k},$$
$$C^{\rm v}_{\rm A}(\boldsymbol{K}+\boldsymbol{k}) = \frac{1}{\sqrt{2}}, \quad C^{\rm v}_{\rm B}(\boldsymbol{K}+\boldsymbol{k}) = \frac{k_y - ik_x}{\sqrt{2}k}, \qquad (11.27)$$

where c and v denote the conduction and valence energy bands, respectively. The electric dipole vector is then given by:

$$\boldsymbol{D}^{\rm cv}(\boldsymbol{K}+\boldsymbol{k}) = \frac{3m_{\rm opt}}{2k}(k_y, -k_x, 0)\,. \qquad (11.28)$$

In Figure 11.3a we plot the normalized directions of $\boldsymbol{D}^{\rm cv}(\boldsymbol{k})$ as arrows over the 2D Brillouin zone of graphene. Around the K points, the arrows show a vortex behavior. Note also that the rotational directions of $\boldsymbol{D}^{\rm cv}(\boldsymbol{k})$ around the K and K' points are opposite to each other in Figure 11.3.

In Figure 11.3b we plot the values of the oscillator strength $O(\boldsymbol{k})$ in units of $m_{\rm opt}$ (see Eq. (11.26)) on a contour plot. Here $O^{\rm cv}(\boldsymbol{k})$ is defined by

$$O^{\rm cv}(\boldsymbol{k}) = \sqrt{\boldsymbol{D}^{cv*}(\boldsymbol{k}) \cdot \boldsymbol{D}^{\rm cv}(\boldsymbol{k})}\,. \qquad (11.29)$$

From Figure 11.3b it is clear that the oscillator strength $O^{\rm cv}(\boldsymbol{k})$ has a maximum at the M points and a minimum at the Γ point in the Brillouin zone. The $\boldsymbol{k}$ dependent $O^{\rm cv}(\boldsymbol{k})$ will be relevant to the type-dependent photoluminescence intensity of a single-wall carbon nanotube [357] though we need to consider the electric dipole vector for each carbon nanotube individually in terms of its diameter and chiral angle [220, 358, 359].

For getting the optical absorption intensity, we take the inner dot product $\boldsymbol{D}^{\rm cv}(\boldsymbol{k}) \cdot \boldsymbol{P}$ up to terms linear in k_x and k_y for a given polarization vector $\boldsymbol{P} = (p_x, p_y, p_z)$

$$\boldsymbol{P} \cdot \langle \Psi^{\rm c}(\boldsymbol{k})|\nabla|\Psi^{\rm v}(\boldsymbol{k})\rangle = \pm\frac{3m_{\rm opt}}{2k}(p_y k_x - p_x k_y)\,. \qquad (11.30)$$

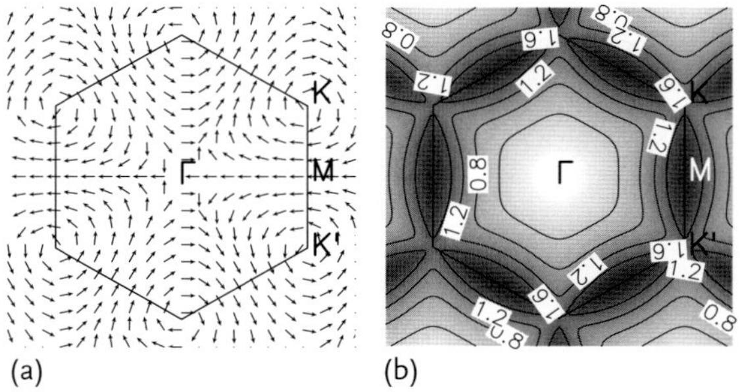

Figure 11.3 (a) The normalized electric dipole vector $\boldsymbol{D}^{\rm cv}(\boldsymbol{k})$ is plotted as a function of $\boldsymbol{k}$ over the 2D BZ. (b) The oscillator strength in units of $m_{\rm opt}$ as a function of $\boldsymbol{k}$ is plotted over the 2D BZ. The separation of two adjacent contour lines is $0.4m_{\rm opt}$. The darker areas have a larger value for the oscillator strength [220].

This result shows that the line $p_y k_x - p_x k_y = 0$ in the 2D BZ becomes a node in the optical absorption for a given $\boldsymbol{P} = (p_x, p_y)$. In the case of graphene, however, the optical transition events take place along equi-energy contours around the K points, and we cannot see the node in Figure 11.3b. If $\boldsymbol{k}$-dependent optical absorption measurements could be carried out, this node could then be determined experimentally.

11.7 Calculation of the Electron–Phonon Interaction

The electron–phonon interaction is expressed by a modification to the tight-binding parameters by the lattice vibrations. In this section, we rewrite the wavefunction appearing in Eqs. (11.4) and (11.5) using a different notation which is suitable for calculating the electron–phonon matrix elements [222]:

$$\Psi_{a,\boldsymbol{k}}(\boldsymbol{r}) = \frac{1}{\sqrt{N_u}} \sum_{s,o} C_{s,o}(a,\boldsymbol{k}) \sum_{\boldsymbol{R}_t} e^{i\boldsymbol{k}\cdot\boldsymbol{R}_t} \phi_{t,o}(\boldsymbol{r} - \boldsymbol{R}_t) , \tag{11.31}$$

where $s = A$ and B is an index denoting each of the distinct carbon atoms of graphene, and $\boldsymbol{R}_t$ is the equilibrium atom positions relative to the origin. Here $\phi_{t,o}$ denotes the atomic wave functions for the orbitals $o = 2s, 2p_x, 2p_y$, and $2p_z$ at $\boldsymbol{R}_t$, respectively. The atomic wave functions are selected as real functions.[6)]

When we consider the potential energy term, V, then V can be expressed by the atomic potentials,

$$V = \sum_{\boldsymbol{R}_t} v(\boldsymbol{r} - \boldsymbol{R}_t) , \tag{11.32}$$

where v in Eq. (11.32) is the Kohn–Sham potential of a neutral pseudo-atom [263]. The matrix element for the potential energy between the initial and final states $\Psi_{\mathrm{i}} = \Psi_{a,\boldsymbol{k}}$ and $\Psi_{\mathrm{f}} = \Psi_{a',\boldsymbol{k}'}$ is

$$\begin{aligned} \langle \Psi_{a',\boldsymbol{k}'}(\boldsymbol{r}) | V | \Psi_{a,\boldsymbol{k}}(\boldsymbol{r}) \rangle &= \frac{1}{N_u} \sum_{s',o'} \sum_{s,o} C^*_{s',o'}(a',\boldsymbol{k}') C_{s,o}(a,\boldsymbol{k}) \\ &\times \sum_{u'} \sum_{u} e^{i(-\boldsymbol{k}'\cdot\boldsymbol{R}_{u',s'} + \boldsymbol{k}\cdot\boldsymbol{R}_{u,s})} m(t',o',t,o) , \end{aligned} \tag{11.33}$$

with the matrix element $m(t', o', t, o)$ for the atomic potential given by:

$$m(t',o',t,o) = \int \phi_{s',o'}(\boldsymbol{r} - \boldsymbol{R}_{t'}) \left\{ \sum_{\boldsymbol{R}_{t''}} v(\boldsymbol{r} - \boldsymbol{R}_{t''}) \right\} \phi_{s,o}(\boldsymbol{r} - \boldsymbol{R}_t) d\boldsymbol{r} . \tag{11.34}$$

6) Combining m and $-m$ states of spherical harmonic function Y_{lm}, we can get a real atomic wave function.

The atomic matrix element m comes from an integration over three centers of atoms, $\boldsymbol{R}_t$, $\boldsymbol{R}_{t'}$, and $\boldsymbol{R}_{t''}$. We neglect m for the cases for which all centers are different from one another. When we consider only two center integrals, m consists, respectively, of off-site and on-site matrix elements m_α and m_λ as follows:

$$m_\alpha = \int \phi_{s',o'}(\boldsymbol{r} - \boldsymbol{R}_{t'}) \{v(\boldsymbol{r} - \boldsymbol{R}_{t'}) + v(\boldsymbol{r} - \boldsymbol{R}_t)\} \phi_{s,o}(\boldsymbol{r} - \boldsymbol{R}_t) d\boldsymbol{r} ,$$

$$m_\lambda = \int \phi_{s',o'}(\boldsymbol{r} - \boldsymbol{R}_t) \left\{ \sum_{\boldsymbol{R}_{t'} \neq \boldsymbol{R}_t} v(\boldsymbol{r} - \boldsymbol{R}_{t'}) \right\} \phi_{s',o}(\boldsymbol{r} - \boldsymbol{R}_t) d\boldsymbol{r} . \tag{11.35}$$

When we consider the amplitude of a phonon mode $\boldsymbol{S}(\boldsymbol{R}_t)$, the potential variation due to a lattice vibration is given by:

$$\begin{aligned} \delta V &= \sum_{\boldsymbol{R}_t} v[\boldsymbol{r} - \boldsymbol{R}_t - \boldsymbol{S}(\boldsymbol{R}_t)] - v(\boldsymbol{r} - \boldsymbol{R}_t) \\ &\approx - \sum_{\boldsymbol{R}_t} \nabla v(\boldsymbol{r} - \boldsymbol{R}_t) \cdot \boldsymbol{S}(\boldsymbol{R}_t) . \end{aligned} \tag{11.36}$$

Under first-order perturbation theory, the electron–phonon matrix element is defined as [203, 222, 360–362]

$$\begin{aligned} M_{a,\boldsymbol{k} \to a',\boldsymbol{k}'} &= \langle \Psi_{a',\boldsymbol{k}'}(\boldsymbol{r}) | \delta V | \Psi_{a,\boldsymbol{k}}(\boldsymbol{r}) \rangle \\ &= -\frac{1}{N_u} \sum_{s',o'} \sum_{s,o} C^*_{s',o'}(a', \boldsymbol{k}') C_{s,o}(a, \boldsymbol{k}) \\ &\quad \times \sum_{u',u} e^{i(-\boldsymbol{k}' \cdot \boldsymbol{R}_{u',s'} + \boldsymbol{k} \cdot \boldsymbol{R}_{u,s})} \delta m(t', o', t, o) , \end{aligned} \tag{11.37}$$

where $\delta m(t', o', t, o)$ is the atomic deformation potential. Then δm can also be separated into two parts,

$$\delta m = \delta m_\alpha + \delta m_\lambda , \tag{11.38}$$

with the off-site and on-site deformation potentials δm_α and δm_λ, respectively, given by:

$$\begin{aligned} \delta m_\alpha &= \int \phi_{s',o'}(\boldsymbol{r} - \boldsymbol{R}_{t'}) \{\nabla v(\boldsymbol{r} - \boldsymbol{R}_{t'}) \cdot \boldsymbol{S}(\boldsymbol{R}_{t'}) \\ &\qquad + \nabla v(\boldsymbol{r} - \boldsymbol{R}_t) \cdot \boldsymbol{S}(\boldsymbol{R}_t)\} \phi_{s,o}(\boldsymbol{r} - \boldsymbol{R}_t) d\boldsymbol{r} , \\ \delta m_\lambda &= \delta_{\boldsymbol{R}_t, \boldsymbol{R}_{t'}} \int \phi_{s',o'}(\boldsymbol{r} - \boldsymbol{R}_{t'}) \\ &\qquad \times \left\{ \sum_{\boldsymbol{R}_{t''} \neq \boldsymbol{R}_{t'}} \nabla v(\boldsymbol{r} - \boldsymbol{R}_{t''}) \cdot \boldsymbol{S}(\boldsymbol{R}_{t''}) \right\} \phi_{s',o}(\boldsymbol{r} - \boldsymbol{R}_{t'}) d\boldsymbol{r} , \end{aligned} \tag{11.39}$$

where the off-site and on-site atomic deformation potentials are, respectively, the corrections to off-diagonal and diagonal Hamiltonian matrix elements and both terms have the same order of magnitude [363].

When using the Slater–Koster scheme to construct tight-binding Hamiltonian matrix elements between two carbon atoms [263], the carbon $2p$ orbitals are chosen to be along or perpendicular to the bond connecting the two atoms. The four fundamental hopping and overlap integrals are (ss), $(s\sigma)$, $(\sigma\sigma)$, and $(\pi\pi)$. We follow the same procedure as was used to construct the deformation potential matrix elements $\langle\phi|\nabla v|\phi\rangle$. We introduce the matrix elements,

$$\begin{aligned}\boldsymbol{\alpha}_{\mathrm{p}}(\tau) &= \int \phi_\mu(\boldsymbol{r})\nabla v(\boldsymbol{r})\phi_\nu(\boldsymbol{r}-\boldsymbol{\tau})d\boldsymbol{r} = \alpha_{\mathrm{p}}(\tau)\hat{I}(\alpha_{\mathrm{p}})\,,\\ \boldsymbol{\lambda}_{\mathrm{p}}(\tau) &= \int \phi_\mu(\boldsymbol{r})\nabla v(\boldsymbol{r}-\boldsymbol{\tau})\phi_\nu(\boldsymbol{r})d\boldsymbol{r} = \lambda_{\mathrm{p}}(\tau)\hat{I}(\lambda_{\mathrm{p}})\,,\end{aligned} \tag{11.40}$$

where $\hat{I}(\alpha_{\mathrm{p}})$ and $\hat{I}(\lambda_{\mathrm{p}})$ are unit vectors in Eq. (11.40) describing the direction of the off-site and on-site deformation potential vectors $\boldsymbol{\alpha}_{\mathrm{p}}$ and $\boldsymbol{\lambda}_{\mathrm{p}}$, respectively [363], and $p=\mu\nu$. The $2p$ orbital ϕ_μ (ϕ_ν) is along or perpendicular to the bond connecting the two carbon atoms and τ is the distance between the two atoms.[7] In Figure 11.4, we show the non-zero matrix elements for the (a) off-site $\boldsymbol{\alpha}_{\mathrm{p}}$ and (b) on-site $\boldsymbol{\lambda}_{\mathrm{p}}$ atomic deformation potentials for 2s, σ and π atomic orbitals.

In Figure 11.5, the calculated values of $\boldsymbol{\alpha}_{\mathrm{p}}$ and $\boldsymbol{\lambda}_{\mathrm{p}}$ are plotted as a function of interatomic distance between two carbon atoms [222]. At $r=1.42$ Å, the bond length between a carbon atom and one of its nearest neighbors, we have $\alpha_{\pi\pi}\approx 3.2$eV/Å and $|\lambda_{\pi\pi}|\approx 7.8$eV/Å, and $|\alpha_{\pi\sigma}|\approx 24.9$eV/Å. In order to calculate the electron–phonon matrix element of Eq. (11.37) for each phonon mode, the amplitude of the

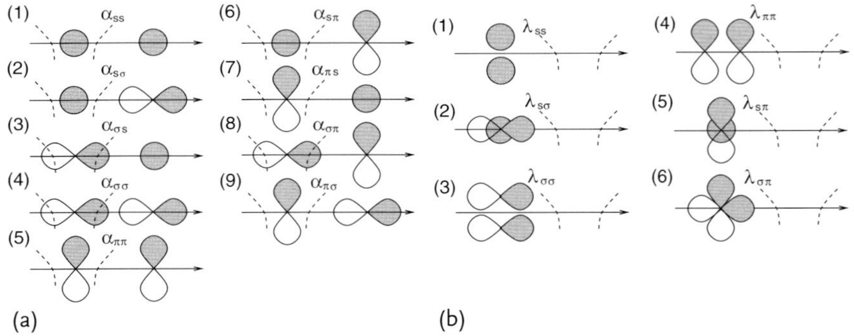

Figure 11.4 (a) The nine non-zero off-site deformation potential vectors $\boldsymbol{\alpha}_{\mathrm{p}}$. The dashed curves represent the atomic potentials. (b) The six non-zero on-site deformation potential vectors $\boldsymbol{\lambda}_{\mathrm{p}}$. The dashed curves represent the atomic potentials. For λ_{ss}, $\lambda_{\sigma\sigma}$, $\lambda_{\pi\pi}$ the two same orbitals are illustrated by shifting them with respect to each other [222].

7) From the matrix element $\boldsymbol{\alpha}_{\mathrm{p}}$, we can deduce another matrix element,

$$\begin{aligned}\boldsymbol{\beta}_{\mathrm{p}}(\tau) &= \int \phi_\mu(\boldsymbol{r})\nabla v(\boldsymbol{r}-\boldsymbol{\tau})\phi_\nu(\boldsymbol{r}-\boldsymbol{\tau})d\boldsymbol{r}\\ &= \int \phi_\nu(\boldsymbol{r})\nabla v(\boldsymbol{r})\phi_\mu(\boldsymbol{r}+\boldsymbol{\tau})d\boldsymbol{r} = \beta_{\mathrm{p}}(\tau)\hat{I}(\beta_{\mathrm{p}}).\end{aligned} \tag{11.41}$$

However, the integral in Eq. (11.41) can be expressed by α terms [222].

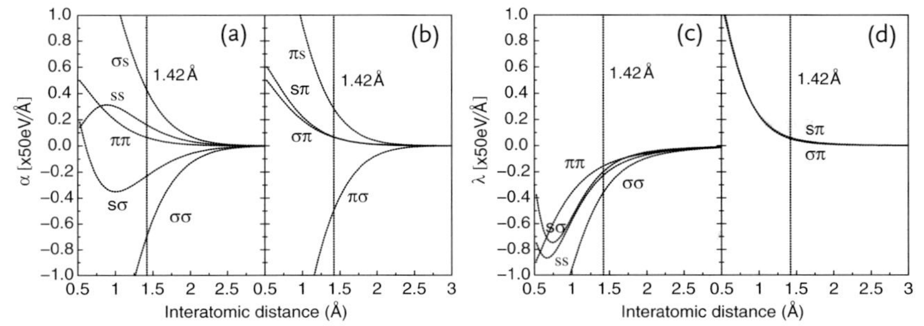

Figure 11.5 (a,b) α_{p} and (c,d) λ_{p} as a function of interatomic distance. The vertical line corresponds to 1.42 Å which is the C–C distance in graphite [222].

atomic vibration $S(R_t)$ for the phonon mode (ν, q) is calculated by

$$S(R_t) = A_\nu(q)\sqrt{\bar{n}_\nu(q)}e^\nu(R_t)e^{\pm i\omega_\nu(q)t} . \tag{11.42}$$

In Eq. (11.42) $\pm$ is for phonon emission (+) and absorption (−), respectively, and A, $\bar{n}$, e, and ω are the phonon amplitude, number, eigenvector, and frequency, respectively. At equilibrium, the phonon number in Eq. (11.42) is determined by the Bose–Einstein distribution function $n_\nu(q)$ for phonon ν,

$$n_\nu(q) = \frac{1}{e^{\hbar\omega/k_{\mathrm{B}}T} - 1} . \tag{11.43}$$

Here $T = 300$ K is the lattice temperature at room temperature and k_{B} is the Boltzmann constant. For phonon emission, the phonon number $\bar{n} = n + 1$ while for phonon absorption, $\bar{n} = n$. The amplitude of the zero-point phonon vibration is

$$A_\nu(q) = \sqrt{\frac{\hbar}{N_u m_{\mathrm{C}}\omega_\nu(q)}} , \tag{11.44}$$

and the phonon eigenvector $e^\nu(R_t)$ is given by diagonalizing the dynamical matrix in Eq. (11.14).

11.8
Extension to Exciton States

Here we mention the procedure used to the extend the electron–photon and electron–phonon interactions into exciton–photon and exciton–phonon interactions, respectively. The detailed description is described in the original papers [186, 188] and in the exciton chapter in this book (see Chapter 10).

In the case of carbon nanotubes, a pair of photoexcited electrons and holes forms an exciton. It should be mentioned that graphene has no excitonic states. Since the exciton binding energy for nanotubes is large (on the order of 1 eV), the exciton can exist even at room temperature. Because of the Coulomb interaction between photoexcited electrons and the holes that are left behind, the exciton wavefunction is localized in real space. When we do not consider the Coulomb interaction explicitly in the electronic structure calculation, the Bloch theorem holds for the electronic structure and thus the wavenumber $\boldsymbol{k}$ is a good quantum number in the sense that all wavefunctions and their energies are given as a function of $\boldsymbol{k}$. The wavefunction for $\boldsymbol{k}$ is delocalized over the lattice. In order to get spatially localized wavefunctions, we mix Bloch wavefunctions with many wave vectors $\boldsymbol{k}$ with one another by the Coulomb interaction so as to minimize the excitonic energy. A mathematical picture of the Fourier transformation from real space to $\boldsymbol{k}$ space is useful. That is, when the exciton wavefunction is more localized in real space, the exciton wavefunction is more delocalized in the $\boldsymbol{k}$ space. The equation for how to mix the different $\boldsymbol{k}$ states is the so-called Bethe–Salpeter equation [186].

In Section 10.4.1 we discussed the Bethe–Salpeter equation. After solving the Bethe–Salpeter equation, we get the exciton wavefunction $|\Psi_{\boldsymbol{q}}^{n}\rangle$ with a center-of-mass momentum $\boldsymbol{q}$ as

$$|\Psi_{\boldsymbol{q}}^{n}\rangle = \sum_{\boldsymbol{k}} Z_{\boldsymbol{k}c,(\boldsymbol{k}-\boldsymbol{q})v}^{n} c_{\boldsymbol{k}c}^{+} c_{(\boldsymbol{k}-\boldsymbol{q})v}|0\rangle , \tag{11.45}$$

where $Z_{\boldsymbol{k}c,(\boldsymbol{k}-\boldsymbol{q})v}^{n}$ is the eigenvector of the nth ($n = 0, 1, 2, \cdots$) state of the Bethe–Salpeter equation, $c^{+}(c)$ is the creation (annihilation) operator for the electron in the conduction (valence) band with momentum $\boldsymbol{k}(\boldsymbol{k} - \boldsymbol{q})$, and $|0\rangle$ is the ground state. The summation on $\boldsymbol{k}$ is taken over the two-dimensional Brillouin zone (2D BZ) and $Z_{\boldsymbol{k}c,(\boldsymbol{k}-\boldsymbol{q})v}^{n}$ is localized in $\boldsymbol{k}$ space near the $\boldsymbol{k}$ point where an electron–hole pair is created.

11.8.1
Exciton–Photon Matrix Element

The exciton–photon matrix element is given by a linear combination of the electron–photon matrix element $D_{\boldsymbol{k}}$ at $\boldsymbol{k}$, weighted by $Z_{\boldsymbol{k}c,\boldsymbol{k}v}^{n*}$

$$M_{\text{ex-op}} = \langle \Psi_0^{n} | H_{\text{el-op}} | 0 \rangle = \sum_{\boldsymbol{k}} D_{\boldsymbol{k}} Z_{\boldsymbol{k}c,\boldsymbol{k}v}^{n*} , \tag{11.46}$$

where Ψ_0^{n} is the exciton wavefunction with $\boldsymbol{q} = \boldsymbol{0}$ center-of-mass momentum. Since the center-of-mass momentum should be conserved before and after an optical transition, it is sufficient to consider the case of $\boldsymbol{q} = \boldsymbol{0}$.

In the case of a single-wall carbon nanotube, since the lattice structure is symmetric under a C_2 rotation around an axis which is perpendicular to the nanotube axis and goes over the center of a C–C bond, the C_2 exchange operation between A and B carbon atoms in the hexagonal lattice is equivalent to the exchange of $\boldsymbol{k}$ and $-\boldsymbol{k}$ states. Since the exciton wavefunction of a carbon nanotube should transform as an irreducible representation of the C_2 symmetry operation, we get A_1, A_2, E, and E^* symmetry excitons (Section 10.3.1). For example, the A_1 and A_2 exciton wavefunctions which are symmetric and antisymmetric under a C_2 rotation, respectively, are given by:

$$|\Psi_0^n(A_{1,2})\rangle = \frac{1}{\sqrt{2}}\sum_k Z^n_{kc,kv}\left(c^+_{kc}c_{kv} \mp c^+_{-kc}c_{-kv}\right)|0\rangle\ , \tag{11.47}$$

where $\boldsymbol{k}$ and $-\boldsymbol{k}$ are located around the K and K′ points, respectively, and $-$ $(+)$ in $\mp$ corresponds to an A_1 (A_2) exciton.[8]

When we use the relation $D_{\boldsymbol{k}} = D_{-\boldsymbol{k}}$, the excitonic-optical (ex-op) matrix elements for the A_1 and A_2 excitons are given by:

$$M_{\text{ex–op}}(A_1^n) = 0\ ,$$
$$M_{\text{ex–op}}(A_2^n) = \sqrt{2}\sum_k D_{\boldsymbol{k}} Z^{n*}_{kc,kv}\ . \tag{11.48}$$

Equation (11.48) directly indicates that A_1 excitons are dark and only A_2 excitons are bright, which is consistent with the predictions by group theory (Section 10.3.2) [218]. Because of the spatially localized exciton wavefunction, the exciton–photon matrix elements are greatly enhanced (on the order of 100 times) compared with the corresponding electron–photon matrix elements.

11.8.2
The Exciton–Phonon Interaction

Using the creation and annihilation operators, the electron–phonon (el–ph) coupling for a phonon mode $(\boldsymbol{q}, \nu)$ has the form

$$H_{\text{el–ph}} = \sum_{kq\nu}\left[M^{\nu}_{k,k+q}(c)c^+_{(k+q)c}c_{kc} - M^{\nu}_{k,k+q}(v)c^+_{(k+q)v}c_{kv}\right]\left(b_{q\nu} + b^+_{q\nu}\right)\ , \tag{11.49}$$

where $M(c)$ $(M(v))$ is the el–ph matrix element for the conduction (valence) band, and $b^+_{q\nu}$ $(b_{q\nu})$ is a phonon creation (annihilation) operator for the νth phonon mode at $\boldsymbol{q}$.

From Eq. (11.49), we obtain the exciton–phonon matrix element between the initial state $|\Psi^{n1}_{q1}\rangle$ and a final state $|\Psi^{n2}_{q2}\rangle$,

$$\begin{aligned} M_{\text{ex–ph}} &= \left\langle \Psi^{n2}_{q2}\right| H_{\text{el–ph}} \left|\Psi^{n1}_{q1}\right\rangle \\ &= \sum_k\left[M^{\nu}_{k,k+q}(c)Z^{n2*}_{k+q,k-q1}Z^{n1}_{k,k-q1} - M^{\nu}_{k,k+q}(v)Z^{n2*}_{k+q2,k}Z^{n1}_{k+q2,k+q}\right]\ , \end{aligned} \tag{11.50}$$

with $\boldsymbol{q} = \boldsymbol{q}_2 - \boldsymbol{q}_1$ accounting for momentum conservation. We can see that the exciton–phonon interaction is obtained by taking the average of the electron–phonon matrix element $M^{\nu}_{k,k+q}$ weighted by the exciton wavefunction.

8) What is confusing but correct is that "−" corresponds to a symmetric wavefunction and that "+" corresponds to an anti-symmetric wavefunction in Eq. (11.47).

Compared with the exciton–photon matrix elements, the exciton–phonon matrix elements are not enhanced significantly since the interaction area between the exciton and the phonon is decreased for the exciton wavefunction relative to that for band states.

11.9 Matrix Elements for the Resonance Raman Process

Combining all the matrix elements discussed above, we can formulate the first-order Stokes Raman intensity due to the electron–phonon interaction as

$$I_{\mathrm{el}} = \left| \frac{1}{L} \sum_{\boldsymbol{k}} \frac{D_{\boldsymbol{k}}^2 \left[M_{\mathrm{el-ph}}(\boldsymbol{k} \to \boldsymbol{k}, c) - M_{\mathrm{el-ph}}(\boldsymbol{k} \to \boldsymbol{k}, v) \right]}{\left[E - E_{\mathrm{cv}}(\boldsymbol{k}) + i\gamma_r \right] \left[E - E_{\mathrm{cv}}(\boldsymbol{k}) - E_{\mathrm{ph}} + i\gamma_r \right]} \right|^2 , \tag{11.51}$$

where γ_r is a broadening factor defining the resonance Raman window. The γ_r value is calculated by the life time of the photoexcited carriers [308, 364] by using the uncertainty relationship. The lifetime of the photoexcited carriers is calculated by the transition probability for emitting a phonon through the electron–phonon (or exciton–phonon) interaction. For more details see Section 11.10 on the resonance Raman window.

When we use the exciton–photon and exciton–phonon interactions which apply to single-wall carbon nanotubes, the Raman intensity for the exciton I_{ex} is given by:

$$\begin{aligned} I_{\mathrm{ex}} &= \left| \frac{1}{L} \sum_a \frac{M_{\mathrm{ex-op}}(a) M_{\mathrm{ex-ph}}(a \to b) M_{\mathrm{ex-op}}(b)}{(E - E_a + i\gamma_r)(E - E_a - E_{\mathrm{ph}} + i\gamma_r)} \right|^2 \\ &= \left| \frac{1}{L} \sum_a \frac{M_{\mathrm{ex-op}}(a)^2 M_{\mathrm{ex-ph}}(a \to a)}{(E - E_a + i\gamma_r)(E - E_a - E_{\mathrm{ph}} + i\gamma_r)} \right|^2 . \end{aligned} \tag{11.52}$$

In the second line of Eq. (11.52), we assume that the virtual state b can be approximated by the real state a. In the case of a first-order Raman process, since $\boldsymbol{q} = \boldsymbol{0}$, the matrix element of Eq. (11.50) is simplified as

$$M_{\mathrm{ex-ph}} = \sum_{\boldsymbol{k}} \left[M_{\boldsymbol{k},\boldsymbol{k}}^{\nu}(c) - M_{\boldsymbol{k},\boldsymbol{k}}^{\nu}(v) \right] |Z_{\boldsymbol{k},\boldsymbol{k}}|^2 . \tag{11.53}$$

When we consider the second-order Raman intensity, we should consider $\boldsymbol{q} \neq \boldsymbol{0}$ phonon scattering. In this case, the exciton–phonon interaction between an A_2 exciton state and an E exciton state is important, where the E exciton state consists of an electron near the K point and a hole near the K' point and vice versa.

In this chapter, we show that we can calculate the resonance Raman spectra and the intensity combined with many calculation programs. Further, using the electron–phonon (or exciton–phonon) interaction, we can calculate the resonance window γ_r by calculating the lifetime of the photoexcited carriers (Section 9.2.1). Further, using the electron–phonon interaction for metallic energy bands, we can estimate the Raman spectral width in the case of the Kohn anomaly (Section 8.4). There are many possibilities for using these interactions such as for coherent-phonon spectroscopy [365].

11.10 Calculating the Resonance Window Width

The width of the resonance window or the γ_r value in Eq. (11.52), of the Raman excitation profile in quantum mechanics is related by the uncertainty relation to the lifetime of the carriers. Usually, the dominant contribution to the lifetime of the carriers in the Raman spectra is in an inelastic scattering process by the emitting and absorbing phonons. In this section, we show how to calculate the carrier lifetime [360] by considering electron–phonon matrix elements [203, 222] and the Fermi Golden rule. For metallic systems (M-SWNTs), the electron–plasmon coupling contributions can shorten the lifetime (broaden the γ_r values), as discussed in [308].

The transition rate for an excited electron scattered to another electronic state can be evaluated for phonon scattering. The inverse of this transition rate is called the relaxation time τ [357, 361], which is inversely proportional to the resonance window, that is, to the γ_r value, and γ_r satisfies the uncertainty principle:

$$\gamma_r = \frac{\hbar}{\tau} \,. \tag{11.54}$$

The transition rate for the scattering per unit time of an excited electron from an initial state $\boldsymbol{k}$ to all possible final states $\boldsymbol{k}'$ by the νth phonon mode can be obtained by the Fermi Golden Rule [357],

$$\begin{aligned}\frac{1}{\tau_\nu} &= W_{\boldsymbol{k}}^{\nu} \\ &= \frac{S}{8\pi M d_\mathrm{t}} \sum_{\mu',k'} \frac{|D_\nu(\boldsymbol{k},\boldsymbol{k}')|^2}{\omega_\nu(\boldsymbol{k}'-\boldsymbol{k})} \left[\frac{dE(\mu',k')}{dk'}\right]^{-1} \\ &\quad \times \left\{ \frac{\delta(\omega(\boldsymbol{k}')-\omega(\boldsymbol{k})-\omega_\nu(\boldsymbol{k}'-\boldsymbol{k}))}{e^{\beta\hbar\omega_\nu(\boldsymbol{k}'-\boldsymbol{k})}-1} \right. \\ &\quad \left. + \frac{\delta(\omega(\boldsymbol{k}')-\omega(\boldsymbol{k})+\omega_\nu(\boldsymbol{k}'-\boldsymbol{k}))}{1-e^{-\beta\hbar\omega_\nu(\boldsymbol{k}'-\boldsymbol{k})}} \right\} ,\end{aligned} \tag{11.55}$$

where S, M, d_t, β, and μ' denote the area of the 2D graphite unit cell, the mass of a carbon atom, the diameter of a SWNT, $1/k_\mathrm{B}T$, and the cutting line indices of the final state, respectively. Here $D_\nu(\boldsymbol{k},\boldsymbol{k}')$ is a matrix for scattering an electron from $\boldsymbol{k}$ to $\boldsymbol{k}'$ by the νth phonon mode. The relaxation process is restricted to satisfying energy-momentum conservation. The two terms in brackets in Eq. (11.55) represent the absorption and emission processes, respectively, of the νth phonon mode with energy $\hbar\omega_\nu(\boldsymbol{k}'-\boldsymbol{k})$.

With this result for the case of S-SWNTs, we can get calculated γ_r values in agreement with experiment, by just considering the electron–phonon coupling model [308]. We can see that the γ_r value shows a strong dependence on chirality and diameter for S-SWNTs. However, the γ_r value calculation for M-SWNTs needs an additional contribution, such as might come from the electron–plasmon interaction, because the calculated γ_r value that considers only the electron–phonon interaction is not consistent with experimental results, that is, calculations give a considerably underestimated γ_r value compared with experiment. Here new physics, such as the interaction between the excited electron in the conduction band and the plasmon associated with two linear energy bands have to be considered. In order to apply a detailed electron–plasmon effect to the γ_r value calculation, more in-depth research has to be done.

11.11 Summary

The diameter distribution of a pristine SWNT sample can be determined by HRTEM and this result was compared with the RBM RRS map for the same sample. Under the assumption of an equal distribution of chiral angles, the RRS RBM cross-section of the SWNTs was determined and it was seen that it can be well represented by a simple empirical formula. The RBM intensity can then be used in the inverse process to yield the sample's diameter distribution.

The observed (n, m) dependent matrix elements and resonance window widths were then addressed theoretically, based on a tight-binding model for electrons and phonons in graphene and carbon nanotubes. Excitons are also addressed. The calculational methods are important not only for explaining the RBM results, but also for interpreting the Raman spectra of sp^2 carbon systems more generally.

Problems

[11-1] Consider two π orbitals on two carbon atoms. When we assume that the nearest neighbor tight-binding parameters for the Hamiltonian and overlap matrices are γ_0 and s_0, obtain the eigenenergies for the bonding and anti-bonding orbitals.

[11-2] Obtain a tight-binding Hamiltonian matrix for single-layer graphene. Consider two cases: (1) consider only the nearest-neighbor interaction and (2) consider up to the third nearest-neighbor interaction.

[11-3] In the previous problem, also consider the overlap matrix elements.

[11-4] Obtain a tight-binding Hamiltonian matrix for bilayer graphene. Consider the two cases: (1) consider only the γ_1 band parameter and (2) consider the γ_1, γ_3 and γ_4 band parameters. In the case of bilayer graphene, we also need to consider the energy E_0^e.

[11-5] Obtain a tight-binding Hamiltonian matrix for trilayer graphene in which we consider band parameters from γ_0 to γ_5. In the case of trilayer graphene, we also need to consider the energy Δ which differentiates A and B carbon sites from one another.

[11-6] For the case of bilayer graphene, obtain the electronic energy dispersion numerically. What is the role of γ_3 and γ_4 in the energy dispersion near the K point?

[11-7] In the case of single-layer graphene, show explicitly the relationship between the force constant sum rule for ϕ_{ti} and that for ϕ_{to}.

[11-8] Show that the in-plane and out-of-plane phonon modes are decoupled in the dynamical matrix when we use the force constant models.

[11-9] In the case of phonons around the K point, we can consider zone folding in which the K point phonon modes are folded into the Γ point if we consider a super-cell structure. What kind of super-cell structure is needed for obtaining the zone folding for the K point phonon mode?

[11-10] Discuss the selection rule between the atomic orbitals which are specified by ℓ, m, n atomic quantum numbers. Show the optically allowed states for 2p orbitals as an initial state.

[11-11] Show that the relationship between the electric field $\boldsymbol{E}$ for light in vacuum and the vector potential $\boldsymbol{A}$ is given by $\boldsymbol{E} = i\omega \boldsymbol{A}$.

[11-12] Using the Poynting vector, $I = EB/\mu_0 = E^2/(\mu_0 c)$, obtain Eq. (11.21).

[11-13] Using the simplest tight-binding Hamiltonian, obtain the wavefunction coefficients in Eq. (11.27) around $\boldsymbol{K} = (0, -4\pi/(3a))$.

[11-14] Obtain the wavefunction coefficient around $\boldsymbol{K}' = (0, 4\pi/(3a))$, and the dipole vector $\boldsymbol{D}^{\text{cv}}(\boldsymbol{K}' + \boldsymbol{k}) = [3m_{\text{opt}}/2k](-k_y, k_x, 0)$. Explain that the rotational directions of $\boldsymbol{D}^{\text{cv}}(\boldsymbol{k})$ around the K and K' points are opposite to each other.

[11-15] When the crystal potential (Eq. (11.32)) is periodic, show that the matrix elements of Eq. (11.33) have non-zero values only for $\boldsymbol{k} = \boldsymbol{k}'$.

[11-16] In Figure 11.6, another set of off-site deformation potential vectors $\boldsymbol{\beta}_{\text{p}}$ are shown. By comparing $\boldsymbol{\beta}_{\text{p}}$ with $\boldsymbol{\alpha}_{\text{p}}$ in Figure 11.4, show that the following relationships between $\boldsymbol{\beta}_{\text{p}}$ and $\boldsymbol{\alpha}_{\text{p}}$ hold:

$$\begin{aligned}
&\boldsymbol{\beta}_{ss} = -\boldsymbol{\alpha}_{ss}\,, \quad \boldsymbol{\beta}_{s\sigma} = \boldsymbol{\alpha}_{\sigma s}\,,\\
&\boldsymbol{\beta}_{\sigma s} = \boldsymbol{\alpha}_{s\sigma}\,, \quad \boldsymbol{\beta}_{\sigma\sigma} = -\boldsymbol{\alpha}_{\sigma\sigma}\,,\\
&\boldsymbol{\beta}_{\pi\pi} = -\boldsymbol{\alpha}_{\pi\pi}\,, \quad \boldsymbol{\beta}_{s\pi} = \boldsymbol{\alpha}_{\pi s}\,,\\
&\boldsymbol{\beta}_{\pi s} = \boldsymbol{\alpha}_{s\pi}\,, \quad \boldsymbol{\beta}_{\sigma\pi} = -\boldsymbol{\alpha}_{\pi\sigma}\,,\\
&\boldsymbol{\beta}_{\pi\sigma} = -\boldsymbol{\alpha}_{\sigma\pi}\,.
\end{aligned} \tag{11.56}$$

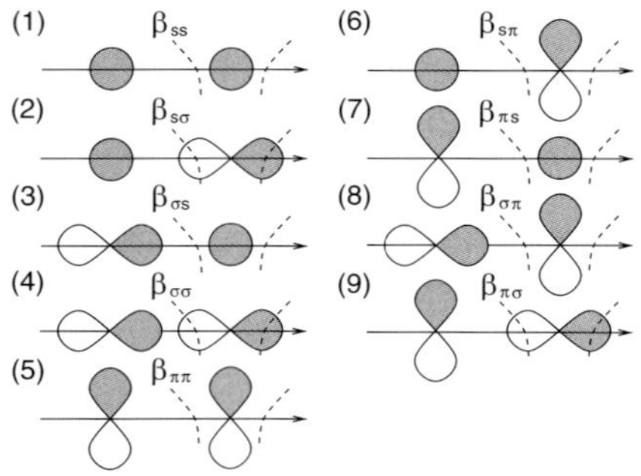

Figure 11.6 The nine non-zero off-site deformation potential vectors $\boldsymbol{\beta}_{\rm p}$. The dashed curves represent the atomic potential.

[11-17] Consider the E exciton wavefunction and show that E exciton states do not contribute to an optical absorption (or emission).

[11-18] When we consider the phonon emission from a photoexcited carrier by the electron–phonon interaction, show by illustrating in a figure that there are 24 possible final states in graphene which satisfy energy-momentum conservation.

[11-19] In Eq. (11.49), we consider the scattering of a hole in the second term. Explain why the minus sign in $-M^{\nu}_{k,k+q}(\nu)$ appears.

[11-20] When the excitation energy is not resonant with the energy separation between an occupied and an unoccupied state, time-dependent perturbation theory tells us that the optical transition probabilities to possible real states are non-zero. Then we can consider a virtual state whose energy is determined by the energy of the light. Show that the virtual state can be expressed by a linear combination of the real states. Especially when the virtual state position is close to a real state, show that the coefficient of the real state is then close to unity and thus we can approximate the virtual state by the real state.

12
Dispersive G′-band and Higher-Order Processes: the Double Resonance Process

All kinds of sp^2 carbon materials exhibit a strong Raman feature which appears in the frequency range 2500–2800 cm^{-1}, as shown in Figure 1.5. Together with the G-band (1585 cm^{-1}), this higher-frequency feature in the spectrum is also a Raman signature of sp^2 carbon materials, and is called the G′-band[1] to emphasize that it is a Raman-allowed mode for graphitic sp^2 carbons. Interestingly, the G′-band is a second-order two-phonon process and, intriguingly, it exhibits a strong frequency dependence on the laser excitation energy. This *dispersive behavior* ($\omega_{G'} = \omega_{G'}(E_{\text{laser}})$) is unusual in Raman scattering, since Raman-active mode frequencies usually do not depend on laser excitation energy. Together with many other Raman peaks (e. g., features around 2450 and 3240 cm^{-1} in Figure 4.14), the G′-band in sp^2 carbons pertains to the class of higher-order Raman spectra, which include overtone and combination modes (see Section 4.3.2.7). The G′-band in particular is a second-order process related to a phonon near the K point in graphene, activated by a double resonance process, which is responsible for its dispersive nature and causes a strong dependence on any perturbation to the electronic and/or phonon structure of graphene. For this reason, the G′ feature provides a very sensitive probe for characterizing sp^2 nanocarbons. For example, the G′-band can be used for differentiating between single and double-layer graphene with Bernal interlayer stacking order, as discussed in Section 4.4.3 and for probing aspects of the electronic structures of SWNTs. The present chapter discusses the science behind these higher-order Raman scattering processes and how they can be used to characterize sp^2 nanocarbons in general. In particular, we review in this chapter a number of the characteristics and properties of the most intense second-order feature in the Raman spectrum for sp^2 carbons, namely, the G′-band. In Section 12.1 the general aspects of higher-order Raman processes are considered, while in Section 12.2 double resonance phenomena are reviewed for graphene and generalized to other spectral features in Section 12.3. Strictly speaking, we should call this effect the "multiple-resonance process", since more than two resonances may take place, as discussed

1) The G′-band is also called the 2D band in the graphene literature. Here we use the conventional G′-band notation for the following reasons: (1) The G′-band is not a defect-induced process while the D and D′-bands denote defect-induced Raman features (see Chapter 13). (2) 2D is conventionally used to denote two-dimensional systems, and is so used also in the sp^2 carbon literature.

Raman Spectroscopy in Graphene Related Systems. Ado Jorio, Riichiro Saito, Gene Dresselhaus, and Mildred S. Dresselhaus

ISBN: 978-3-527-40811-5

in Section 12.2.1. However, most of the experimental results can be explained with the double resonance and, since it is largely treated in the literature with this name, we keep it here. The double resonance process in carbon nanotubes is discussed in Section 12.4, while Section 12.5 provides a brief summary of Chapter 12.

12.1 General Aspects of Higher-Order Raman Processes

A two-phonon emission process is a second-order Raman process, described classically by considering anharmonic terms in the polarizability tensor (see Section 4.3.1):

$$\alpha = \alpha_0 + \alpha_1 \sin \omega_q t + \frac{\partial^2 \alpha}{\partial Q_1 \partial Q_2} Q_1 Q_2 \,, \tag{12.1}$$

where Q_1 and Q_2 are amplitudes for the two phonons. The last term in Eq. (12.1) gives a Raman shift of $\pm\omega_1 \pm \omega_2$ for the two phonons.

In quantum mechanics, the two-phonon process is described by using fourth-order perturbation theory and the scattering intensity can be calculated using Eqs. (5.22) and (5.23), which are reproduced here:

$$I(\omega, E_{\text{laser}}) \propto \sum_i \left| \sum_{m',m'',\omega_1,\omega_2} J_{m',m''}(\omega_1, \omega_2) \right|^2 , \tag{12.2}$$

where the summation is taken over two intermediate electronic states m and m' and the corresponding phonon frequencies ω_1 and ω_2, and also for the initial states j, after taking the square of the scattering amplitude, $J_{m,m'}$ that is given by:

$$\begin{aligned} &J_{m',m''}(\omega_1, \omega_2) \\ &= \frac{M^{\text{d}}(\boldsymbol{k}, im'') M^{\text{ep}}(-\boldsymbol{q}, m''m') M^{\text{ep}}(\boldsymbol{q}, m'm) M^{\text{d}}(\boldsymbol{k}, mi)}{(E_{\text{laser}} - \Delta E_{mi} - i\gamma_{\text{r}})(E_{\text{laser}} - \Delta E_{m'i} - \hbar\omega_1 - i\gamma_{\text{r}})} \\ &\times \frac{1}{(E_{\text{laser}} - \Delta E_{m''i} - \hbar\omega_1 - \hbar\omega_2 - i\gamma_{\text{r}})} \,. \end{aligned} \tag{12.3}$$

Equation (12.3) differs slightly from Eq. (5.23), by the addition of the broadening factor γ_{r},[2)] which gives the width of the resonance window, that is, the energy uncertainty related to the lifetime of the excited state, and avoids having J going to infinity when the system is in resonance. Figure 12.1 shows a Feynman diagram for a two-phonon Stokes Raman process. Of course, like for first-order processes in general (see Section 5.3), different scattering orderings are possible for this two-phonon process and each distinct process should be considered separately, using similar Feynman diagrams when making a very accurate intensity analysis of

2) In principle, the three γ_{r} factors in Eq. (12.3) can be different. However, there is no experimental evidence presently available to provide separate values for each of these γ_{r} factors.

Figure 12.1 Feynman diagram for a two-phonon Stokes Raman process.

higher-order processes. In general, energy and momentum conservation for the incident (i) and scattered (s) electrons requires:

$$E_s = E_i \pm E_{q1} \pm E_{q2} \tag{12.4}$$

$$k_s = k_i \mp q_1 \mp q_2, \tag{12.5}$$

where + (−) in Eq. (12.4) and − (+) in Eq. (12.5) correspond to phonon absorption and emission with the wave vectors q_1 and q_2. By considering $k_s \approx k_i$ (see Section 4.3.2.6), momentum conservation requires $q_2 \approx -q_1$.

Any two-phonon scattering process involving the same Raman-active ($M^{ep} \neq 0$) phonon mode twice, and just obeying $q_2 = -q_1$ gives a Raman spectrum with a double frequency. Such a process can be observed for any solid. However, any $q \neq 0$ wave vector can be involved in two-phonon processes, and thus the corresponding Raman signal without any resonance effect is broad. Moreover, the anharmonic (or fourth-order) perturbation is relatively weak compared with the α_1 term in Eq. (12.1).[3] Interestingly, in sp^2 carbon materials the special electronic and phonon structures are such that resonance processes can take place for well-selected internal electron–phonon scattering processes. These resonance processes strongly enhance the second-order Raman spectra from specific phonons, through the so-called double resonance (DR) Raman process [159] described in Section 12.2. This then generates a unique Raman spectra for sp^2 carbon materials. There are two factors that govern the Raman intensity, thereby enhancing the probability of a scattering event for specific phonons:

1. The double resonance condition, which enhances J in Eq. (12.3) by minimizing two of the three factors in the denominator of Eq. (12.3) at the same time;
2. A strong electron–phonon matrix element, which enhances J by enhancing the numerator in Eq. (12.3).

The next sections discuss how these two factors occur in graphene (Section 12.2), and in carbon nanotubes (Section 12.4), focusing mostly on the G′-band, which is the most intense DR feature and is also one of the most dispersive features. Generalization of the DR process to many features in the Raman spectra of sp^2 nanocarbons is presented in Section 12.3.

3) This picture is different from higher-order Raman spectra for molecules, where the phonon amplitudes are large enough to enhance anharmonic effects and the electronic energy levels are discrete (and do not exhibit q-dependent effects).

12.2
The Double Resonance Process in Graphene

In this section the double resonance (DR) process introduced in Section 12.1 is reviewed for graphene, starting with a more detailed description of the double resonance process in Section 12.2.1. This is followed by a discussion of the dependence of this phenomenon on E_{laser} in Section 12.2.2 and on the number of layers of graphene in Section 12.2.3. Emphasis is also given in Section 12.2.4 to the use of the G′-band in characterizing graphene samples with regard to their stacking order along the c crystallographic axis.

12.2.1
The Double Resonance Process

When a photon with a given energy is incident on monolayer graphene, it will excite an electron from the valence band to the conduction band vertically in momentum space (gray arrow in Figure 12.2a). Since the graphene energy band does not have an energy gap, we always have an electron with wave vector k for any E_{laser} which satisfies $E_{\text{laser}} = E^{\text{c}}(k) - E^{\text{v}}(k)$. The photoexcited electron at k is then scattered by emitting a phonon with wave vector q to a state at $k - q$, as shown by the black arrow in Figure 12.2a. The phonon emission in Figure 12.2a corresponds to intervalley scattering in which the phonon q vector connects two energy bands at the K and K' points of the Brillouin zone. If there is a phonon in the vibrational structure of graphene with the wave vector q and phonon energy E_q so that this photon can connect the two conduction electronic states, this phonon scattering process will be resonant. A double resonance process (electron–photon and electron–phonon scattering, shown by the two solid dots in Figure 12.2a) will then take place.

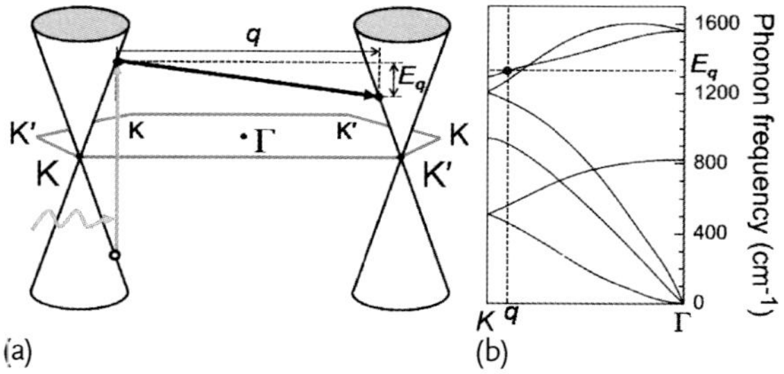

Figure 12.2 (a) This schematic diagram shows cones representing the electronic dispersion (energy vs. momentum) near the Fermi level at the K and K' points in the hexagonal Brillouin zone of graphene. The light-induced electron–hole formation and the one-electron–one-phonon scattering process taking place in the double resonance process are indicated by gray and black arrows, respectively. (b) The phonon dispersion in graphene is displayed [366], in which the phonon wave vector q measured from the Γ point and energy E_q are given by a dot.

Using a quantum mechanical description of this phenomenon, the two processes described above, and shown in Figure 12.2, give a large scattering amplitude J, since they minimize the first two terms in the denominator of Eq. (12.3). The full Raman process requires two phonons with wave vectors q and $-q$ for momentum conservation, and both the electron–hole creation by the incident photon and the subsequent electron–hole recombination by emitting scattered light are shown in Figure 12.3. In Figure 12.3, we show the corresponding intravalley two-phonon scattering processes in which the phonon q vector connects two conduction band states with the same K (or K') points. In graphene, both intravalley and intervalley (from/to K to/from K', see Figure 12.2) scattering processes can occur.

The processes shown in Figure 12.3a–c represent Stokes processes, where resonance with (a) the incident and (b) the scattered photons take place, in addition to the internal electron–phonon scattering process. While the electron–phonon resonant scattering shown in Figure 12.3b minimizes the second term in the denominator of Eq. (12.3), resonance with the incident/scattered photons minimizes the first/third term in the denominator of Eq. (12.3). Panels (d–f) of Figure 12.3 show the respective anti-Stokes processes. Note that, for the same laser excitation line, a different q vector will give rise to the double resonance process for Stokes and anti-Stokes processes ($q_S \neq q_{aS}$). A more detailed discussion can be found in [367].

The intervalley and intravalley double resonance two-phonon processes are, respectively, relevant near 2700 cm^{-1} (the G′-band), and near 3240 cm^{-1}. The 3240 cm^{-1} peak arises from a second-order process for a $q \neq 0$ phonon near the maximum frequency in the phonon dispersion in Figure 12.2b near the Γ point,

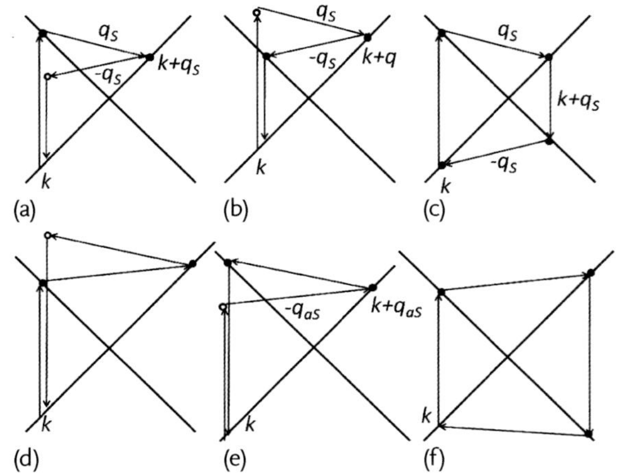

Figure 12.3 The full multiple resonance Raman process, with two electron–phonon scattering events, requiring both q and $-q$ phonons for momentum conservation, and the electron–hole recombination for scattered light emission. Double resonance with the incident light is shown in (a), while double resonance with the scattered light is shown in (b). Part (c) shows a fully resonant process, where the $-q$ scattering also takes place on the hole. Parts (a–c) are for Stokes (S) processes, while (d–f) show the corresponding anti-Stokes (AS) processes [160, 178, 367, 368].

for which the momentum transfer q is small in the intravalley process. The reason why the G′ feature is much more intense than the 3240 cm^{-1} feature has nothing to do with intravalley vs. intervalley scattering processes, but rather with the electron–phonon matrix elements, that are much stronger for phonons near the K point than for phonons near the Γ point.

Intriguing is the fact that the G′-band in mono-layer graphene is more intense than the first-order Raman-allowed G-band. Some argue that this is an indication that the dominating process is not the double-resonance but actually the fully-resonant scattering process shown in Figure 12.3c [369, 370]. However, this very strong G′-band intensity could also be related to different electron–phonon matrix elements for near K and near Γ point phonons. It is true that the fully resonant process in Figure 12.2c,f should, in principle, be much more probable than the others which exhibit a virtual (nonresonant) state. However, this is only valid if the electron and hole electronic dispersion relations are symmetric. Since the electron wave function overlap in graphene results in a different normalization for the valence and conduction bands, an electron–hole dispersion asymmetry is introduced, and for this reason the two processes could select DR phonons with somewhat different q vectors. This asymmetry is relatively small and is generally neglected in common descriptions of the electronic structure of graphene in terms of mirror band cones. More theoretical and experimental work is required to fully understand the differences in electron–phonon vs. hole-phonon scattering, including differences in the matrix elements that have not yet been addressed theoretically.

The slope of the energy dispersion $\partial E/\partial k$ is called the *group velocity*. When we consider only the direction of the group velocity for the initial k, there are two possibilities for scattered $k - q$ states as shown in Figure 12.4 where each of the intervalley (Figure 12.4a,c) and intravalley (Figure 12.4b,d) scattering processes correspond to backward (Figure 12.4a,b) and forward (Figure 12.4c,d) scattering. Here the backward (forward) scattering means that the direction of the group velocity

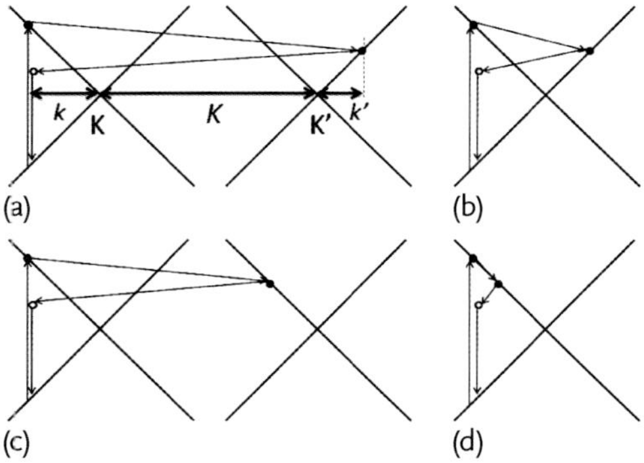

Figure 12.4 The double-resonance Stokes Raman processes for intervalley (a,c) and intravalley (b,d) scattering. Here (a,b) relates to the backward scattering process with $q_{\text{DR}} = k + k'$ and (c,d) relates to the forward scattering process with $q_{\text{DR}} = k - k'$, with k and k' measured from the K point. K is the distance between the K and K' points, k (k') is the distance of the resonant states from K (K'), as defined in (a) [160, 178, 367, 368].

does (does not) change after scattering. The corresponding q vectors for intervalley scattering are given by:

$$q = K + q_{\text{DR}} = K + k + k' \approx K + 2k \quad \text{(backward scattering)} \tag{12.6}$$

$$q = K + q_{\text{DR}} = K + k - k' \approx K \quad \text{(forward scattering)} , \tag{12.7}$$

where K is the distance between K and K', and k (k') here is measured from the K (K') point, which means q_{DR} is the phonon wave vector distance from the K (K') point. In the case of intravalley scattering, we just put $K = 0$ in Eq. (12.6) and (12.7). Since the phonon energy is usually small compared to the excited energy levels, $k \approx k'$, these two double resonance conditions approach $q_{\text{DR}} = 2k$ and $q_{\text{DR}} = 0$ (commonly used in the literature [158–160]). As already stated, the $q_{\text{DR}} \approx 2k$ wave vector gives rise to the G′-band, while the $q_{\text{DR}} \approx 0$ wave vector gives rise to a DR feature from the iTO phonon very close to the K point, consistent with the Raman peak observed around 2450 cm^{-1} in Figure 4.14 (see peak assignment summary in Chapter 14). However, there are some controversies about the origin of this 2450 cm^{-1} feature (denoted by G*), since ω_{G^*} is also consistent with the frequency of another combination mode, which is further discussed in Section 12.3. The $q_{\text{DR}} \approx 0$ processes are expected to be less intense than the $q_{\text{DR}} \approx 2k$ because the destructive interference condition is exactly satisfied for $q_{\text{DR}} = 0$ [371]. However, the asymmetric (density of states-like) lineshape of the 2450 cm^{-1} feature in Figure 4.14 seems to be representative of the density of $\boldsymbol{q}$ vectors fulfilling the double resonance process, discussed in the next paragraph.

While all the important double resonance conditions have already been introduced, the picture discussed up to now is not the full story because graphene is a two-dimensional material. For a given laser energy, not only is the electron–hole excitation process shown in Figure 12.2 resonant, but any similar process within a circle in these cones defined by E_{laser} (see Figure 12.5) is also resonant. Furthermore, the mechanism of double resonance (DR) is actually satisfied by any phonon whose wave vector connects two points on two circles around K and K', as shown in Figure 12.5 [367]. (Here we neglect the trigonal warping effect of the constant energy surface of graphene near $K(K')$ for simplicity.) A phonon with wave vector q connects two points along the circles with radii k and k' around the K and K' points, respectively, where the difference between k and k' (for $k \neq k'$) comes from the energy loss from the electron to the phonon.[4)] By translating the vector q to the Γ point, and considering all possible initial and final states around the K and K' points, the doughnut-like figure shown in Figure 12.5 is generated. Therefore, there is a large set of q vectors fulfilling the double resonance condition. However, there is also a high density of phonon wave vectors $\boldsymbol{q}$ satisfying the DR mechanism for which the end of the wave vectors measured from the Γ point are on the inner and outer circles of the "doughnut" in Figure 12.5. Therefore, the radii of the inner and outer circles around K'' (see Figure 12.5) are, respectively, $k - k'$ and $k + k'$.

4) Here q is the real phonon wave vector, measured from the Γ point, while in defining q_{DR}, k and k' are measured from K.

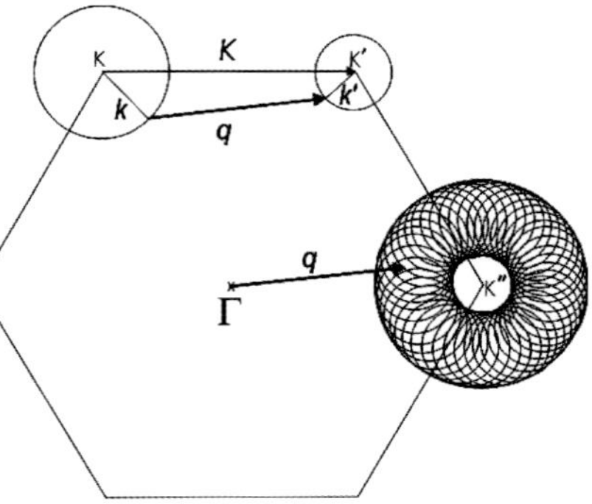

Figure 12.5 One of the possible double resonance (DR) Stokes Raman processes involving the emission of a phonon with wave vector $-\boldsymbol{q}$. The set of all phonon wave vectors $\boldsymbol{q}$ which are related to transitions from points on the two circles around K and K' gives rise to the collection of small circles around the K'' point obeying the vector sum rule $\boldsymbol{q} = \boldsymbol{K} - \boldsymbol{k} + \boldsymbol{k}'$ (here we neglect the trigonal warping effect). Note that this collection of circles is confined to a region between the two circles with radii $q_{\mathrm{DR}} = k + k' \approx 2k$ and $q_{\mathrm{DR}} = k - k' \approx 0$. The differences between the radii of the circles around K and K' and thus the radius of the inner circle around K'' are actually small in magnitude and are here artificially enlarged for clarity in presenting the concepts of the double resonance process [367].

Exactly as given by the 1D model (Eqs. (12.6) and (12.7)) these are the phonons associated with the singularities in the density of $\boldsymbol{q}$ vectors that fulfill the double resonance requirements, and they are expected to make a significant contribution to the second-order Raman scattering process. However, for a full description and lineshape analysis, considering the 2D model generates an asymmetric density of states that is consistent with the asymmetric lineshape that is observed for the 2450 cm^{-1} feature in Figure 4.14.

12.2.2
The Dependence of the $\omega_{\mathrm{G}'}$ Frequency on the Excitation Laser Energy

As described in Section 12.2.1, when a photon with a given energy (E_{laser}) is incident on graphene, it will excite an electron from the valence to the conduction band. This electron can be resonantly scattered by a phonon with the correct wave vector q and phonon energy E_q to satisfy the double resonance conditions. Figure 12.6 shows that, if E_{laser} is changed, the correct wave vector q and phonon energy E_q that will fulfill the double resonance conditions will also change. This effect gives rise to the dispersive nature of the G′-band, which comes from an *intervalley* double resonance (DR) Raman process involving an electron with wave vector k in the vicinity of the K point and two iTO phonons with wave vectors $q_{\mathrm{DR}} \approx 2k$, where both k and q_{DR} are measured from the K point (see Section 12.2.1).

Figure 12.7a shows the Raman spectra in the region of both the G* (∼2450 cm^{-1}) and G′ (∼ 2700 cm^{-1}) bands for different laser excitation energies. Also the G*-band can be either explained by the $q \approx 0$ DR relation, or by the $q \approx 2k$ rela-

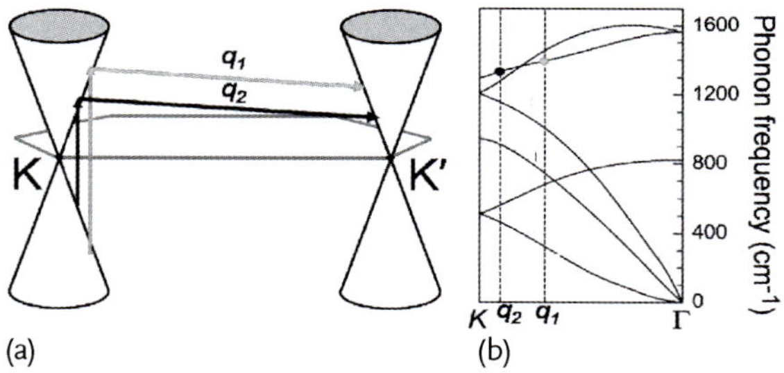

Figure 12.6 (a) The schematic diagram shows the light-induced electron–hole formation and one electron-one phonon scattering event taking place in the double resonance process with two different excitation laser energies (associated with phonon wave vectors q_1 and q_2, respectively), which are indicated by the gray and black arrows. (b) The phonon dispersion in graphene is shown where the phonon wave vector q that fulfills the double resonance requirements for each E_{laser} is also indicated in terms of the wave vectors q_1 and q_2.

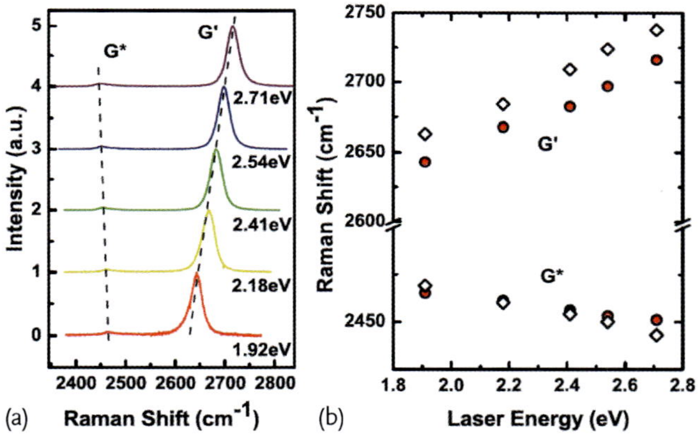

Figure 12.7 (a) Raman spectra of the G′ and the G*-bands of monolayer graphene for 1.92, 2.18, 2.41, 2.54 and 2.71 eV laser excitation energy. (b) Dependence of $\omega_{G'}$ and ω_{G^*} on E_{laser}. The red circles correspond to the graphene data and the lozenges correspond to data for turbostratic graphite. From [372].

tion applied to an *intervalley* process, but involving one iTO phonon and one LA phonon [371, 382]. Figure 12.7b shows the G′ and G* frequencies $\omega_{G'}$ and ω_{G^*} as a function of E_{laser} for graphene and turbostratic graphite (for which the stacking between graphene layers is random). The G′-band exhibits a highly dispersive behavior with $(\partial\omega_{G'}/\partial E_{\text{laser}}) \simeq 88\,\text{cm}^{-1}/\text{eV}$ for monolayer graphene, $95\,\text{cm}^{-1}/\text{eV}$ for turbostratic graphite [372], and $106\,\text{cm}^{-1}/\text{eV}$ for carbon nanotubes (see Section 12.4 and [173]). The G*-band exhibits a much less pronounced dispersion, with $(\partial\omega_{G^*}/\partial E_{\text{laser}}) \simeq -10$ to $-20\,\text{cm}^{-1}/\text{eV}$ and of opposite sign for both monolayer and turbostratic graphite [371, 372], and null for carbon nanotubes [373]. The G* feature is further discussed in Section 12.3.

To analyze the experimental data from Figure 12.7b, one has to consider the electron and phonon dispersion of a graphene monolayer, discussed in Chapters 2 and 3. Near the K point, the electron and phonon dispersions can be approximated by the linear relations $E(k) = \hbar v_F k$ and $E(q_{DR}) = \hbar v_{ph} q_{DR}$, respectively, where $v_F = \partial E(k)/\partial k$ and $v_{ph} = \partial E(q)/\partial q$ are the electron and phonon velocities near the K point, respectively (usually v_F is called the Fermi velocity, $v_F \sim 10^6$ m/s). The k (q_{DR}) is the electron (phonon) wave vector measured with respect to the K point, so that the general and approximated conditions for the double resonance Raman are given by:

$$\begin{aligned} E_{laser} &= 2 v_F k \\ E_{ph} &= v_{ph} q_{DR} \\ q_{DR} &= k \pm k' \,, \end{aligned} \tag{12.8}$$

where E_{laser} and E_{ph} are, respectively, the laser and phonon energies, and k' is the scattered electron wave vector near the K' point in the graphene Brillouin zone. It is important to remember that we are dealing here with combination modes, so that the observed E_{ph} has to reflect this combination. For example, for the G′-band, the observed G′-band energy is given by $E_{G'} = 2E_{ph}$, where E_{ph} is the energy for the iTO phonon mode at q_{DR}.[5)] Making another commonly used approximation in Eqs. (12.8), that is, $q_{DR} = k + k' \approx 2k$, then $E_{G'}$ can be written as:

$$E_{G'} = 2\frac{v_{ph}}{v_F} E_{laser} \,. \tag{12.9}$$

A drawback to the use of the DR Raman features to define the electron and phonon dispersion relations is that the measured values depend on both v_{ph} and v_F, and one has to be known to obtain the other. Adding to this problem, the physics of the phonon dispersion for graphene near the K point is rather complex due to the Kohn anomaly, which was discussed for the G-band ($q \to 0$) in Section 8.2, and the Kohn anomaly also occurs for phonons at $q \to K$. The high frequency of the iTO phonon when combined with the Kohn anomaly near the K point are together responsible for the strong dispersive behavior observed for $\omega_{G'}$. The exact values for v_{ph} and v_F are still under debate since they depend on the complex physics of many-body effects [86, 355, 366].

12.2.3
The Dependence of the G′-band on the Number of Graphene Layers

Because of the dispersive behavior of the G′-band, it can be used to characterize graphene layers in terms of their dispersive behavior and to distinguish between different types of graphene in terms of the number of layers and the stacking of these layers (see Figure 12.8). To explain this behavior, we first turn to the electronic properties of bilayer graphene with AB Bernal layer stacking (as also occurs in graphite), since this bilayer graphene structure is probed by resonance Raman

5) It is only when crystalline disorder is present that the first-order $q \neq 0$ phonons can be observed, as discussed in Chapter 13.

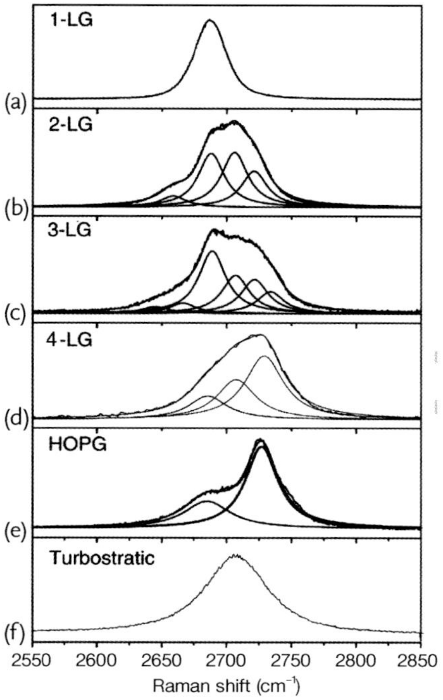

Figure 12.8 The G′ Raman band measured with $E_{\text{laser}} = 2.41\,\text{eV}$ for (a) 1-LG, (b) 2-LG, (c) 3-LG, (d) 4-LG, (e) HOPG and (f) turbostratic graphite. The splitting of the G′ Raman band opens up in going from mono- to three-layer graphene and then closes up in going from 4-LG to HOPG (see text) [84].

scattering. Since the electronic structure of graphene changes with layer stacking (see Section 2.2.4), such changes in layer stacking can be probed by the double resonance features, and most sensitively by the G′-band. Bilayer graphene has a richer G′-band spectrum (Figure 12.8b) than its monolayer counterpart (Figure 12.8a), because of its special electronic structure, consisting of two conduction bands and two valence bands (see Figure 12.9). From the double resonance (DR) Raman process in bilayer graphene with AB stacking, it is possible to distinguish four Lorentzians in the experimental Raman spectra for each laser line [86, 374]. The DR Raman model can then be used to relate the electronic and phonon dispersion of bilayer graphene with the experimental dependence of $\omega_{G'}$ on E_{laser} [192].

Figure 12.9a shows the dispersion of each one of the four peaks that comprise the G′-band as a function of E_{laser} for bilayer graphene, as shown in Figure 12.9b–e. Each one of the DR Raman processes obeying the selection rules (see Chapter 6) gives rise to one of the G′ peaks and is labeled as P_{ij} $(i, j = 1, 2)$ in Figure 12.9 [192], which connects two energy band E_i and E_j. Since the iTO phonon along the KM direction increases its frequency with increasing wave vector q, the highest frequency of the G′ peak for a given E_{laser} energy is associated with the P_{11} process, which also has the largest wave vector (q_{11}). The smallest wave vector q_{22}

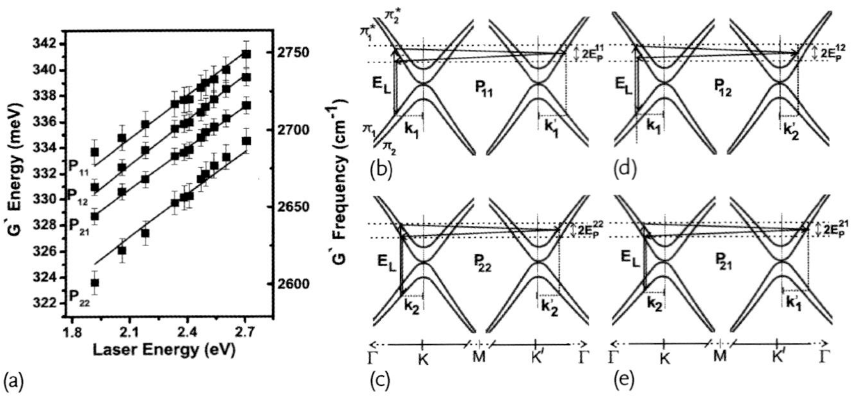

Figure 12.9 (a) Plot of the frequency of four peaks arising from the four G′-band peaks vs. E_{laser} observed in bilayer graphene. These four peaks arise from the four process shown in (b–e) to comprise the G′-band of bilayer graphene plotted in (a) as a function of laser energy [192].

is associated with the process P_{22}, which gives rise to the lowest frequency peak of the G′-band. The two intermediate frequency peaks of the G′-band are associated with processes P_{12} and P_{21} [192]. Increasing the number of layers increases the number of possible G′-band scattering processes. Trilayer graphene already has 15 possibilities [98, 217], but the frequency spacing between these peaks is not large enough to allow identification of each scattering event (Figure 12.8c). The situation gets even more complex for N-layer graphene ($N > 3$), though the G′-band spectra at a typical E_{laser} such as 2.41 eV starts to get simpler in appearance (Figure 12.8d for 4-LG), converging to a two-peak structure in highly oriented pyrolytic graphite (HOPG, $N \to \infty$, Figure 12.8e). The two-peak structure of the G′-band in HOPG (Figure 12.8e) is the result of a convolution of an infinite number of allowed DR processes in what turns out to be a three-dimensional electron and phonon dispersion. A geometrical approach for the understanding of the evolution of the G′-band in the Raman spectrum from monolayer graphene to bulk graphite (HOPG) has been discussed in [375].

12.2.4
Characterization of the Graphene Stacking Order by the G′ Spectra

A long time before graphene had been isolated, Raman spectroscopy had been used to quantify the structural ordering along the c axis in graphite, since the G′-band is very sensitive to the stacking order [155–157]. Nemanich and Solin were the first to show the change from one peak to two peaks in the profile of the G′-band in the Raman spectra obtained from polycrystalline graphite and crystalline graphite, respectively [150, 151]. Lespade *et al.* [155, 156] performed a Raman spectroscopy study on carbon materials heat treated at different temperatures T_{htt} and observed that, by increasing T_{htt}, the G′-band changes from a one peak to a two peak feature (see Figure 12.8e,f). They associated this evolution with the degree of graphitization of the samples and suggested that the origin of the two-peak structure of the G′-band in crystalline graphite was related to the stacking order occurring along the c axis. Recently, the evolution of the G′-band from a single to a few graphene layers [86, 197, 374], and its complete evolution from the 2D to 3D aspect (from one to two peaks) has been quantitatively systematized (see previous section and [376]). Furthermore, Barros *et al.* have used the G′-band to identify three G′-band peaks due to the coexistence of 2D and 3D graphite phases in pitch-based graphitic foams [82].

Finally, it is important to mention that in CVD-grown graphene the stacking of the layers is often not AB Bernal stacking, and this lowering in symmetry results in a broadened single G′ peak for sample regions containing monolayer or bilayer graphene [71, 377, 378]. Thus the use of G′-band Raman spectroscopy to assign the number of layers needs to be viewed with caution since the G′-band lineshape is also strongly related to the stacking order of these layers.

12.3 Generalizing the Double Resonance Process to Other Raman Modes

The sp^2 carbons exhibit several combination modes and overtones, as shown in Figure 12.10 for graphite whiskers [379]. Basically all the branches in the phonon dispersion can be observed in such Raman features which obey the double resonance condition [160]. Many of the peaks observed in the spectra of Figure 12.10 below 1650 cm^{-1} are actually one-phonon bands activated by defects, as discussed in Chapter 13. Above 1650 cm^{-1} the observed Raman features are all multiple-order combination modes and overtones, some of which are also activated by defects.

As shown in Figure 12.10, the double resonance peaks change frequency with changing E_{laser}, and they can be fitted onto the phonon dispersion diagram shown in Figure 12.11 using DR theory (Section 12.2.2). The data points displayed in Figure 12.11 all stand for the $q_{\mathrm{DR}} \approx 2k$ DR backward resonance condition, the ones near Γ and K coming from intravalley and intervalley scattering processes, respectively. Actually, in the Raman spectra there are no characteristic features distinguishing peaks associated with the intravalley from the intervalley scattering processes, or even from the $q_{\mathrm{DR}} \approx 2k$ or $q_{\mathrm{DR}} \approx 0$ resonance conditions. All we have in hand is the E_{laser} dependence of each peak that has to fulfill one of the DR processes and to fit the predicted phonon dispersion relations. For example, the data points near-K, assigned as the iTO+LA combination mode (TO+LA in Figure 12.11) could alternatively be assigned to a $q_{\mathrm{DR}} \approx 0$ process, since this combination mode is very weakly (or non) dispersive [373]. Supporting this assignment is the asymmetric DR phonon-density of states-like shape observed for this peak, and against this identification is the destructive interference working towards DR Raman processes at exactly $q = K$ [129]. The debate about the iTO+LA combination mode assignment near the K point remains for future clarification. Near Γ the dispersive behavior is clear and the assignment is unquestionable [380].

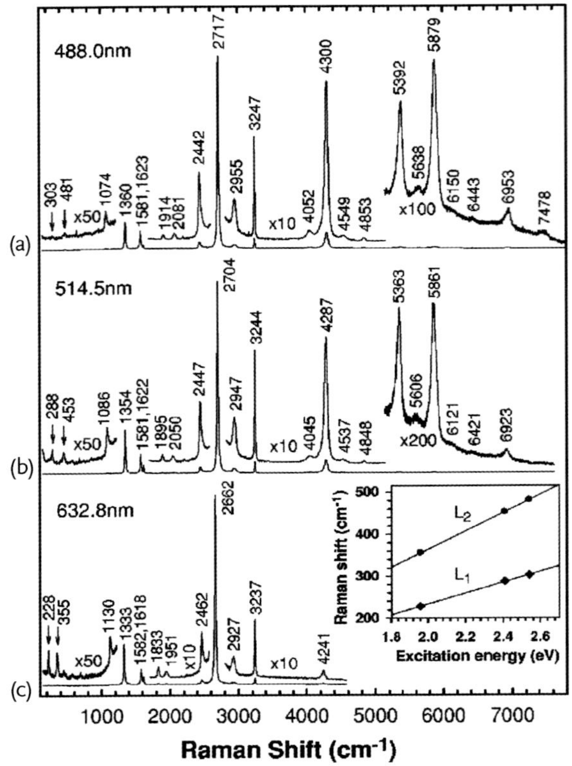

Figure 12.10 (a–c) Raman spectra of graphite whiskers obtained at three different laser wavelengths (excitation energies) [379]. Note that some phonon frequencies vary with E_{laser} and some do not. Above 1650 cm^{-1} the observed Raman features are all multiple-order combination modes and overtones, though some of the peaks observed below 1650 cm^{-1} are actually one-phonon bands activated by defects, as discussed in Chapter 13. The inset to (c) shows details of the peaks labeled by L_1 and L_2. The L_1 and L_2 peaks, which are dispersive, are explained theoretically by defect activation of double resonance one-phonon processes (see Chapter 13) involving the acoustic iTA and LA branches, respectively [160].

12.4
The Double Resonance Process in Carbon Nanotubes

For one isolated SWNT, the signal is limited to a fixed optical transition E_{ii} by the van Hove singularities, and we can say that the "dispersive" behavior will be "quantized". The effect of multiple bands of resonances can be seen in SWNT bundles where the vertical stripes in Figure 12.12a indicate resonance windows for a given resonance band. The modes appearing in Figure 12.12b for the spectral region between 400–1200 cm^{-1} are called intermediate frequency modes (IFMs), since their frequencies lie between the common RBM and G modes. The IFM features are attributed to combination modes (oTO $\pm$ LA [381, 382]), but it is not yet clear whether

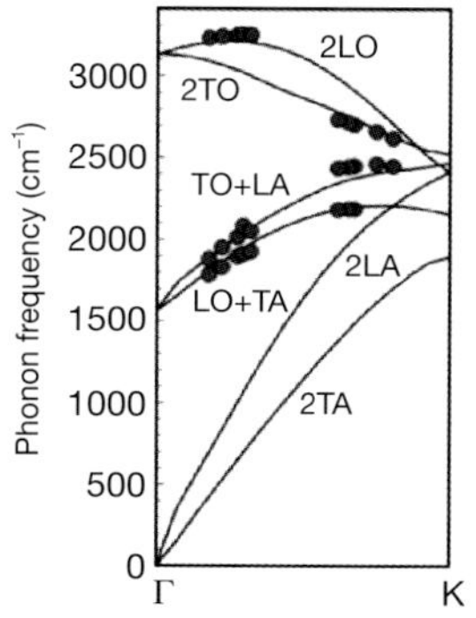

Figure 12.11 Two-phonon dispersion of graphite based on second-order double resonance peaks in the Raman spectra (circles). Solid lines are dispersion curves from *ab initio* calculations considering combination modes and overtones with totally symmetric irreducible representations. Adapted from [371].

these modes are Raman-active or induced by disorder (see Chapter 13). Theory relates their observation to quantum confinement along the tube length [383], and some supporting experimental evidence has been found for such an effect [384]. The IFM picture is not yet fully understood, but it represents a generalization of the double resonance effect. It is interesting to comment here that the DR theory for the dispersive features was actually developed for a one phonon inelastic scatter-

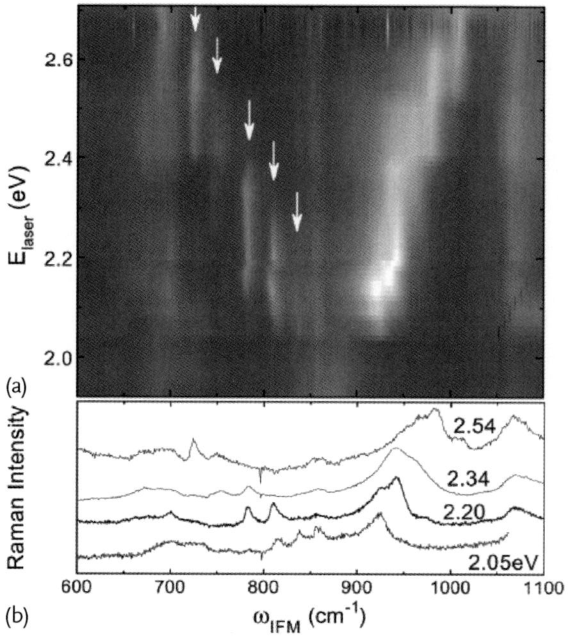

Figure 12.12 (a) Two-dimensional plot of the E_{laser} dependence for the Raman spectra of SWNT bundles in the intermediate frequency mode (IFM) range. The light areas indicate high Raman scattering intensity. Arrows point to five well-defined ω_{IFM} features. (b) Raman spectra in the IFM range taken with $E_{\text{laser}} = 2.05, 2.20, 2.34$, and 2.54 eV [381].

ing event plus one elastic scattering event activated by defects, such as the D-band, and this process is discussed in Chapter 13.

Like for the other sp^2 carbons, the G′ Raman feature in the carbon nanotube spectra provides unique information about the electronic structure of both semiconducting and metallic SWNTs. In analogy to graphene, SWNTs show dispersive behavior, although some unique characteristics are observed due to the one-dimensional structure of SWNTs. In Section 12.4.1 we show the G′ behavior in SWNT bundles, where most of the (n, m)-dependent uniqueness is averaged out, but there remains a close relation between SWNTs and graphene and some anomalous results can still be observed experimentally. In Section 12.4.2 we discuss the G′-band in isolated tubes, where anomalous effects related to their 1D structure are discussed. Regarding the analogy with multilayer graphene, carbon nanotubes have double-wall nanotubes (DWNTs), three-wall, MWNTs, etc. However, the literature on DWNTs is advancing rapidly and the complexity and richness associated with the differing properties of DWNTs having semiconducting vs. metallic outer and inner tubes make a detailed discussion of this large topic [289, 385, 386] outside the scope of this book. Many layers MWNTs exhibit large diameter nanotubes, thus approaching graphite [387].

12.4.1 The G′-band in SWNTs Bundles

SWNTs bundles exhibit behaviors not found at the individual tube level. Of particular interest is the novel dispersion of the G′-band observed in SWNT bundles. The inset to Figure 12.13a shows the dispersion of the $\omega_{G'}$ frequency in SWNT bundles. Fitting the observed linear dispersion with E_{laser} for SWNTs [173] gives:

$$\omega_{G'} = 2040 - 106 E_{\text{laser}} . \tag{12.10}$$

However, different from graphene and graphite, the G′-band dispersion in SWNTs exhibits a superimposed oscillatory behavior as a function of E_{laser}, as shown in Figure 12.13a when the linear dispersion is subtracted from the data points. Such a behavior is not directly related to the uniqueness of the electronic structure, but rather is due to the $\omega_{G'}$ dependence on tube diameter.

The frequency of the measured G′-band feature depends on tube diameter because of a force constant softening, associated with the curvature of the nanotube wall that is dependent on d_t. Experiments on isolated tubes show the d_t dependence of $\omega_{G'}$ to be [389]

$$\omega_{G'} = \omega_{G'_0} - 35.4/d_t , \tag{12.11}$$

where $\omega_{G'_0}$ is the laser energy dependent value observed in graphene (the limit of an infinite diameter tube). This diameter dependence of $\omega_{G'}$ is responsible for the oscillatory behavior observed in Figure 12.13a. The vertical lines in Figure 12.13b denote the diameter range of the SWNT bundle used in the G′-band dispersion experiment. When moving along an arrow in Figure 12.13b by increasing the excitation laser energy, (for example, within the E_{22}^S sub-band by changing E_{laser}, above

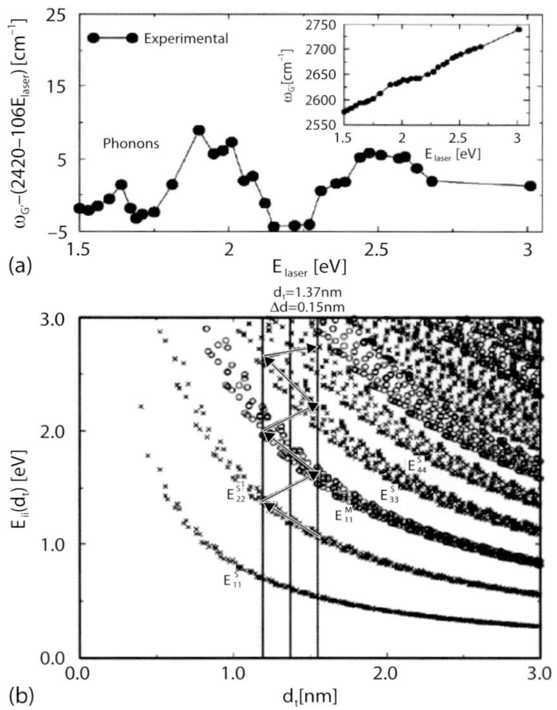

Figure 12.13 (a) Oscillatory dispersion G′-band data for $\omega_{G'}$ for a SWNT bundle sample taken from [173] after subtracting the linear dispersion 2420 + 106E_{laser} from the $\omega_{G'}$ vs. E_{laser} data shown in the inset. (b) Optical transition energies E_{ii} as a function of diameter for SWNTs (the Kataura plot). The vertical lines denote the diameter range 1.37+ or −0.15 nm of the SWNT bundle used in the G′-band dispersion experiment shown in (a) [388].

1 eV), different SWNTs with different diameters enter and leave resonance with a given E_{ii} optical transition. By increasing the laser energy, the diameter decreases, thus increasing the expected energy due to the double resonance process. When the resonance condition with E_{laser} jumps from, for example, E^S_{22} to E^M_{11} (around E_{laser} = 1.5 eV), the diameter jumps to higher values. This process modulates the $\omega_{G'}$ dispersion, as observed by the oscillatory behavior in $\omega_{G'}$ vs. E_{laser} seen in Figure 12.13a.

It is important to make clear that the "continuous" G′-band frequency dispersion observed in Figure 12.13 is a result observed in the bundles where different tubes enter and leave resonance, thus probing all the unfolded two-dimensional Brillouin zone. For one isolated SWNT, the signal is limited to a fixed optical transition E_{ii} by the van Hove singularities, and we can say that the "dispersive" behavior is quantized (see the results on the IFMs in Section 12.4, Figure 12.12). To fully appreciate the 1D-confinement effects on the G′ spectra, experiments on the isolated SWNT level have to be discussed, and this is the topic of Section 12.4.2.

12.4.2
The (*n*, *m*) Dependence of the G′-band

This section gives an appreciation of the effect of 1D confinement on the G′ feature in SWNTs. In the case of SWNTs, the resonance condition is restricted to $E_{laser} \approx E_{ii}$ (the transition energy between van Hove singular energies). This fact gives rise to a $\omega_{G'}$ dependence on the SWNT diameter (see Section 12.4.1) *and* chiral angle.

Figure 12.14 shows the cutting lines for two metallic SWNTs, one zigzag and one armchair, in the unfolded 2D Brillouin zone of graphene. The van Hove singularities occur where a cutting line is tangent to an equi-energy contour,[6] thus causing a chiral angle dependence on the excited states k_i, which are the states responsible for the dominant optical spectra observed for SWNTs, including the double resonance features. The presence of cutting lines in carbon nanotubes will, therefore, affect the dispersive Raman features [388] arising from the double resonance process [158–160]. The effects are general for many DR features (see Section 12.3), but we discuss here only the most intense double resonance Raman feature, the G′-band, since the G′-band dispersion is very large and, therefore, provides the most accurate experimental results for this effect.

The two-peak G′-band Raman features observed from semiconducting and metallic isolated nanotubes are shown in Figure 12.15a,b, respectively, where the (n, m) indices for these nanotubes are assigned as (15,7) and (27,3) [176]. The presence of two peaks in the G′-band Raman feature indicates the resonance with both the incident E_{laser} and scattered $E_{laser} - E_{G'}$ photons, respectively, with two different van Hove singularities (VHSs) for the same nanotube. E_{laser} and $E_{laser} - E_{G'}$ are defined in Figure 12.15a,b below the G′-band profiles, by the outer and inner equi-energy contours near the 2D Brillouin zone of the graphene layer. The wave vectors corresponding to the resonance VHSs are also shown. For the double resonance process in graphite, the momentum conservation for the electron–phonon interaction couples the electronic $\boldsymbol{k}$ and phonon $\boldsymbol{q}$ wave vectors by the relation $\boldsymbol{q} \approx -2\boldsymbol{k}$,[7] where both the electronic and phonon wave vectors are measured from

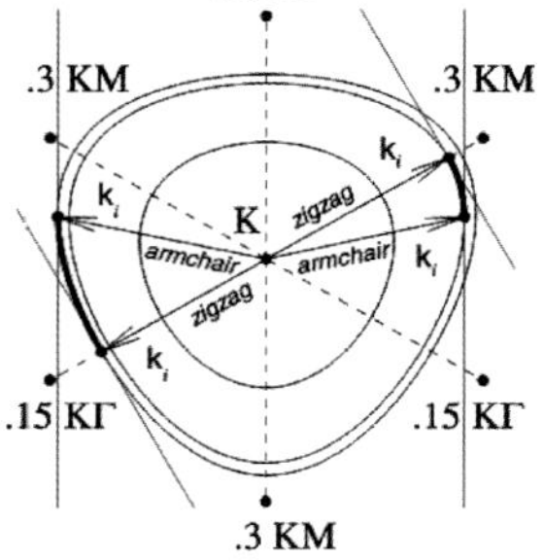

Figure 12.14 Cutting lines for two metallic SWNTs, one zigzag and one armchair, in the unfolded 2D Brillouin zone of graphene. The wave vectors k_i point with arrows to the locations where the van Hove singularities occur [390].

6) The equi-energies in Figure 12.14 have to include the trigonal warping effect to capture the full chirality dependence that is observed experimentally.

7) Equation (12.6) provides a consideration of the modulus only. The correct vectorial correlation between k and q would have a minus sign.

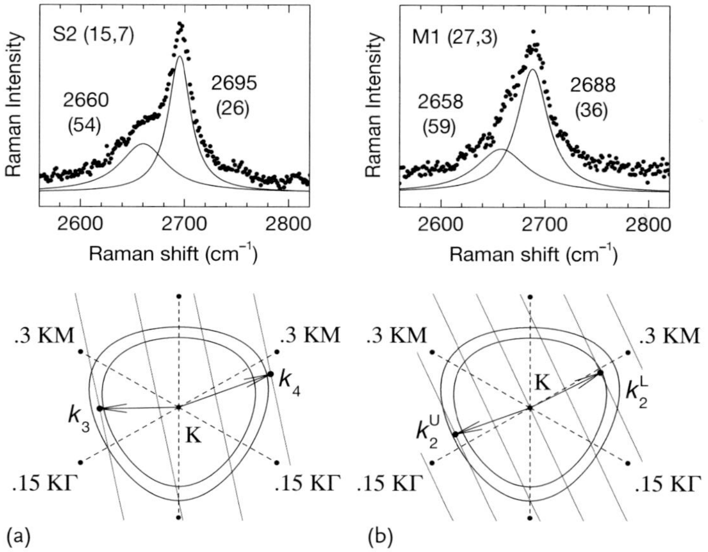

Figure 12.15 The G′-band Raman spectra for (a) a semiconducting (15,7) and (b) a metallic (27,3) SWNT, showing two-peak structures [390–392], respectively. The *vicinity* of the *K* point in the unfolded Brillouin zone is shown in the lower part of the figure, where the equi-energy contours for the incident $E_{\text{laser}} = 2.41\,\text{eV}$ and the scattered $E_{\text{laser}} - E_{G'} = 2.08\,\text{eV}$ photons, together with the cutting lines and wave vectors for the resonant van Hove singularities ($E_{33}^S = 2.19\,\text{eV}$, $E_{44}^S = 2.51\,\text{eV}$, $E_{22}^{M(L)} = 2.04\,\text{eV}$, $E_{22}^{M(U)} = 2.31\,\text{eV}$), are shown.

the nearest *K* point in the Brillouin zone [393]. For the double resonance process in carbon nanotubes, the introduction of cutting lines superimposed on the 2D Brillouin zone changes the equality relation $\boldsymbol{q} = -2\boldsymbol{k}$, which is slightly different in nanotubes relative to graphene or graphite because the allowed k vectors for carbon nanotubes are no longer continuous [390].

The two peaks in Figure 12.15a and b can be associated with the phonon modes of the wave vectors $\boldsymbol{q}_i = -2\boldsymbol{k}_i$, where $i = 3, 4, 2\text{L}, 2\text{U}$ for E_{33}^S, E_{44}^S, E_{22}^{ML}, and E_{22}^{MH}, respectively,[8] and the electronic wave vectors $\boldsymbol{k}_i$ are shown in the lower part of Figure 12.15. For the S-SWNT shown in Figure 12.15a, the resonant wave vectors $\boldsymbol{k}_3$ and $\boldsymbol{k}_4$ have different magnitudes, $k_4 - k_3 \simeq K_1/3$, resulting in twice the difference for the phonon wave vectors, $q_4 - q_3 \simeq 2K_1/3 = 4d_t/3$, so that the splitting of the G′-band Raman feature arises from the *phonon dispersion* $\omega_{\text{ph}}(q)$ around the *K* point. In contrast, for the metallic nanotube (M-SWNT) shown in Figure 12.15b, the resonant wave vectors $\boldsymbol{k}_2^{\text{L}}$ and $\boldsymbol{k}_2^{\text{H}}$ have roughly equal magnitudes and opposite directions away from the *K* point, so that the splitting of the G′-band Raman feature for metallic nanotubes arises from the *anisotropy* of the phonon dispersion $\omega_{\text{ph}}(q)$ around the *K* point [390], which we call the phonon trigonal warping effect. Overall,

8) The density of states of E_{22}^{M} for an M-SWNT splits into the two peaks (higher (H) and lower (L) energy peaks) except for armchair nanotubes, where a degeneracy in their frequency occurs.

the presence of two peaks in the double resonance Raman features of isolated carbon nanotubes is associated with quantum confinement effects expressed in terms of cutting lines. Correspondingly, the two-peak structure of the double resonance Raman features is not observed in 2D graphitic materials, such as in a monolayer graphene sheet. As stated above, the G′-band doublet structure observed in 3D graphitic materials is attributed to the interlayer coupling [157].

Finally, the G′-band is not the only feature to exhibit an (n, m) dependence for SWNTs. Actually, all the double resonance features may exhibit such a dependence. The stronger the dispersive behavior, the larger the (n, m) dependence. The smaller the tube diameter, the larger are the frequency shift effects. The (n, m) dependence of other combination modes, such as the iTO+LA combination mode near Γ, have been studied in detail (see [380]).

12.5 Summary

In this chapter we introduced the double resonance effect, important for explaining the observation of combination modes and overtones in sp^2 carbon materials. Many strong and weak Raman peaks can be assigned to two-phonon, second-order double resonance processes, while others are one-phonon disorder-activated processes, as discussed in Chapter 13. The peak frequencies usually exhibit small deviations depending on their sp^2 structure (single vs. multi-layer graphene, ribbons, tubes with different diameters and chiral angles, etc.), but in general the peak frequencies reflect the phonon dispersion relation of graphene subjected to the double resonance selection rules. This phonon dispersion effect allows us to probe the interior of the Brillouin zone with light scattering. Usually this is not possible using light because of the small momentum of photons as compared to the momentum range within the Brillouin zones of typical materials, so that neutron or electron scattering is generally applied to study such phenomena. The sensitivity to the detailed sp^2 structure makes these features, and especially the strong G′-band, a powerful tool for quantifying the number of graphene layers and the stacking order in graphite, and for studying various chirality-dependent effects in carbon nanotubes. The double resonance bands are strongly sensitive to changes in the electronic and vibrational structure in general, and serve as a sensitive probe for such effects.

Problems

[12-1] Polarization $\boldsymbol{P}$ is expressed by $\boldsymbol{P} = \alpha \boldsymbol{E}$, where is α is a polarization tensor, which is given by Eq. (12.1). When $\boldsymbol{E} = \boldsymbol{E}_0 \exp(i\omega t)$, $Q_i = Q_i^0 \exp(i\omega_i t)$, $(i = 1, 2)$, show that the last term of Eq. (12.1) gives the frequency terms of $\omega \pm \omega_1 \pm \omega_2$.

[12-2] In an optical process such as photoabsorption or phonon emission, the probability for the occurrence of an optical process is proportional to the square of the matrix elements of the electromagnetic perturbation in time-dependent perturbation theory. Explain how we get energy conservation and momentum conservation from perturbation theory. Especially show that these conservation rules are not always necessary because of the uncertainty principle, which becomes important for short times or short lengths.

[12-3] The $\pm$ symbols in Eq. (12.4) correspond to phonon absorption and emission with the wave vectors q_1 and q_2, respectively. Then explain, in the corresponding momentum conservation rule of Eq. (12.5), why we should not use $\pm$ but rather use $\mp$ by illustrating the scattering process in the two-dimensional Brillouin zone.

[12-4] Using Eqs. (12.2) and (12.3), analyze how the intensity of a second-order Raman process involving two phonons with q and $-q$ wave vectors changes when one, two or three denominators are minimized (select a constant value for γ_r and draw the resonance window). Make a comparative analysis of the intensities for the six processes depicted in Figure 12.3.

[12-5] Discuss how the triple-resonance process in Figure 12.3c,f depends on the symmetry between the valence and conduction band. For this discussion, show how the total intensity would change when the slopes for the electron and hole dispersions are different. Analyze the results as a function of γ_r and the phonon linewidth Γ_q.

[12-6] Since the electronic dispersion in graphene near the K point is characterized by a Fermi velocity of $v_F = 1 \times 10^6$ m/s and the G′-band dispersion is given by $(\partial\omega_{G'}/\partial E_{laser}) \simeq 88\,\text{cm}^{-1}/\text{eV}$, find the phonon dispersion for the iTO phonon branch in graphene. Compare the result you find with the result obtained by the force constant model described in Section 3.1.3. You will see why the addition of electron–phonon coupling, not accounted for in the force constant model, is very important for describing the phonon dispersion near the K point and, consequently, for explaining the dispersive behavior of the G′-band.

[12-7] Explain qualitatively how to obtain the circles around the K and K' points in Figure 12.5 by using the electron and phonon energy dispersion. Estimate the diameters of these circles for $E_{laser} = 2.41$ eV, and a phonon frequency of $1350\,\text{cm}^{-1}$ and compare these lengths with the ΓK distance.

[12-8] Plot the phonon density of states as a function of energy for possible DR q vectors, as shown in Figure 12.5, by assuming that all circles are regular (not deformed) circles. Here assume that the phonon dispersion is proportional to the distance of q from the K point in the two-dimensional Brillouin zone.

[12-9] Considering the electron and phonon dispersions near the K point, calculate the difference in frequency between the Stokes and anti-Stokes process-

es for the G′-band (intervalley process), equivalent to the Stokes and anti-Stokes processes depicted in Figure 12.3a,d for the intravalley scattering process.

[12-10] Considering the double resonance condition $q \approx 0$, do you expect a dispersive (E_{laser}-dependent) behavior? Explain why.

[12-11] Draw a picture similar to Figure 12.5, but considering the trigonal warping effect. What differences do we expect to find in the double resonance features when the trigonal warping effect is included?

[12-12] Derive Eq. (12.9).

[12-13] Consider: (a) the average G′ dispersion $(\partial\omega_{G'}/\partial E_{\text{laser}}) = 106\,\text{cm}^{-1}/\text{eV}$; (b) the G′ diameter dependence given by Eq. (12.11); (c) the Kataura plot. Describe quantitatively the oscillatory behavior expected for the G′-band in SWNT bundles, considering a diameter distribution $1.2 \leq d_t \leq 1.6\,\text{nm}$ for the above three cases. For which diameter distribution would the oscillatory behavior become an averaged linear dispersion? Analyze your result as a function of the G′-band Raman peak width for one given SWNT.

[12-14] Explain how the inclusion of trigonal warping in the phonon dispersion relation would generate a chiral angle (θ) dependence for the G′-band of a hypothetical SWNT with the same d_t and as θ is changed from 0 to 30°.

[12-15] There is a mirror symmetry for n layer graphene (n is an odd number) in which a mirror is parallel to the graphene layer. The π energy band and phonon modes are either symmetric (S) or anti-symmetric (AS) with regard to the mirror operation. Discuss the selection rules for optical transitions and for the electron–phonon interaction for four possible combinations of S and AS energy bands. How about the case when n is an even number?

[12-16] When you look at Figure 12.3a,b, it is easy to see that the incident and scattered double resonance processes in graphene occur with the same $(q, -q)$ wave vector pair. Explain why two peaks are observed in the G′-band of SWNTs when both the incident and scattered light are in resonance with two different optical transition energies.

[12-17] Explain why there is no possibility of a second-order Raman process with the combination modes of two phonons, one causing an intravalley and one causing an intervalley scattering event.

[12-18] Explain that the density of states (DOS) of E_{ii}^M is split into two peaks from the two cutting lines near the K point except for armchair nanotubes. Please specify the cutting line in the two-dimensional Brillouin zone which corresponds to the DOS peak at the higher energy for each i of E_{ii}^M.

13
Disorder Effects in the Raman Spectra of sp^2 Carbons

In general, disorder-induced symmetry-breaking plays a very important role in the determination of several materials properties such as transport properties and the relaxation of photoexcited carriers. In particular, sp^2 carbons which have high symmetry are sensitive to symmetry-breaking defects. Disorder and symmetry-breaking are observed sensitively by spectroscopy which depends strongly on crystal symmetry [95, 118]. The presence of disorder in sp^2 hybridized carbon systems, leads to rich and intriguing phenomena in their resonance Raman spectra, thus making Raman spectroscopy one of the most sensitive and informative techniques to characterize disorder in sp^2 carbon materials. Raman spectroscopy has thus become a key tool and is widely used to identify disorder in the sp^2-network of different carbon structures, such as diamond-like carbon, amorphous carbon, nanostructured carbon, as well as carbon nanofibers, nanotubes and nanohorns [20, 168, 394].

Figure 13.1a shows the Raman spectra of crystalline graphene, exhibiting the first-order Raman-allowed G-band. When graphene is bombarded by a low dose of Ar^+ ions (10^{11} Ar^+/cm^2), point defects are formed and the Raman spectra of the disordered graphene exhibit two new sharp features appearing at 1345 cm^{-1} and 1626 cm^{-1} for $E_{laser} = 2.41$ eV, as seen in Figure 13.1b. These two features have been called the D and D′-bands, respectively, to denote disorder. These bands are dispersive, and they are observed at these special frequencies when excited with a 514 nm wavelength (2.41 eV) laser. Finally, when the periodic system is strongly disordered by a large ion dose (10^{15} Ar^+/cm^2), the Raman spectrum resembles the profile of the density of states for the higher-energy optical phonon branch (Figure 13.1c) [194, 195].

The basic description of disorder-induced peaks in the Raman spectra of sp^2 carbons comes from the double resonance model discussed in Chapter 12, and is associated with the following considerations. The requirement that only phonons at the center of the Brillouin zone are first-order Raman-allowed[1] ($q = 0$) by symmetry comes from momentum conservation, and momentum conservation is associated with translational symmetry. The effect of breaking the translational symmetry of

1) Not all zone-center phonons are Raman-allowed. Only phonons which behave like a symmetric second-order rank tensor, such as xy, $x^2 - y^2$, can be Raman-active modes (see Chapter 6).

Raman Spectroscopy in Graphene Related Systems. Ado Jorio, Riichiro Saito, Gene Dresselhaus, and Mildred S. Dresselhaus

ISBN: 978-3-527-40811-5

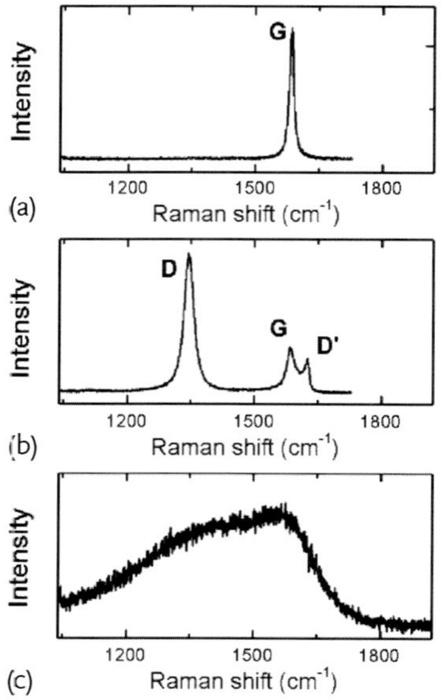

Figure 13.1 The first-order Raman spectrum of (a) crystalline graphene, (b) defective graphene, (c) and fully disordered single-layer graphene deposited on a SiO_2 substrate. These spectra are all obtained with $E_{\text{laser}} = 2.41\,\text{eV}$ [195].

crystals by introducing disorder into the lattice is the breakdown of momentum conservation, through the activation of phonons at interior k points of the Brillouin zone. Disorder-induced lattice distortions could also lead to a break down of other symmetry-based selection rules, thus also activating $q = 0$ phonons that are forbidden by the symmetry of the unperturbed crystal structure. For example, the out-of-plane TO phonon mode at $850\,\text{cm}^{-1}$, which is not Raman-active but is instead IR (infrared)-active, can be observed as a weak feature in Raman spectra in the presence of defects. Usually the effect of disorder on the Raman spectra of crystalline materials is a broadening of the Raman-allowed peaks (e. g., G and G′), the observation of new features (e. g., D and D′) related to symmetry forbidden scattering processes and, at high disorder levels, the observation of a phonon-density-of-states like spectra. These changes are all related to the turning on of new scattering processes by the progressive break-down of symmetry and by the introduction of new wave vectors to conserve momentum.

This chapter starts, in Section 13.1, with a brief taste of what would be a general quantum mechanical description of a defect-induced Raman effect. In Section 13.2 the defect-induced double resonance scattering processes are considered in detail, which describes, based on the electron and phonon dispersion relations, the frequencies of the disorder-induced Raman peaks and their dependence on E_{laser}. In Section 13.3, the D and D′ peak intensities are discussed, that is, how they evolve with increasing amounts of disorder. We address two systems, ion-bombarded graphene and nanographite, where disorder is represented by point defects and by boundaries, respectively. In Section 13.4 we show that the zigzag edge has a special symmetry so that the D-band double resonance process is forbidden, although the D′ feature can be seen in the zigzag edge spectra. This is the first attempt [161] to use Raman spectroscopy to study the "atomic structure" of the defect and these results can be used to differentiate between zigzag and armchair graphene edges and grain boundaries. In sequence, Section 13.5 discusses specific details regarding the disorder found in one-dimensional carbon nanotubes, which includes multiple features in the G-band arising from disorder. Finally, in Section 13.6 a different concept is discussed, that is the effect that defects cause on Raman-allowed peaks, due to local electron and phonon energy renormalization. This effect has been observed by near-field optical measurements on the G′ feature in doped SWNTs and represents a new route for future research [191]. Section 13.7 summarizes the main points discussed in this chapter.

13.1 Quantum Modeling of the Elastic Scattering Event

For a full quantum description of the disorder-induced effect in the Raman spectra of SWNTs, it is necessary to calculate the Raman intensity $I(\omega, E_{\text{laser}})$ of a disorder-induced band, which is due to a double resonance scattering process [80], given by:

$$I(\omega, E_{\text{laser}}) = \sum_i \left| \sum_{a,b,c,\omega_{\text{ph}}} \frac{M_{\text{op}}(k, ic) M_{\text{def}}(-q, cb) M_{\text{ep}}(q, ba) M_{\text{op}}(k, ai)}{\Delta E_{ai}(\Delta E_{bi} - \hbar\omega_{\text{ph}})(\Delta E_{ci} - \hbar\omega_{\text{ph}})} \right|^2 \tag{13.1}$$

where $\Delta E_{ai} = (E_{\text{laser}} - (E_a - E_i) - i\gamma_{\text{r}})$ and γ_{r} denotes a broadening factor. Here subscripts i, a, b, and c, respectively, denote the initial state, the excited state, the first scattered state of an electron by a phonon, and the second scattered state of an electron by a defect. M_{op}, M_{ep} and M_{def} denote the electron–photon, electron–phonon and electron-defect scattering matrix elements, respectively. We use the fact that $E_b = E_c$ since the scattering from b to c is an elastic scattering process induced by the defect. Therefore, the new feature of Eq. (13.1) is the presence of the M_{def} matrix element, which describes the elastic scattering by defects. As previously discussed, in quantum mechanics the scattering processes can occur in different orders and, for example, the process in which elastic scattering occurs first in the double resonance process is also possible and has to be considered. For a given initial and final state i, all intermediate states are added before taking the square in Eq. (13.1).

An elastic electron scattering from electron state k to k' can be expressed by the matrix element [395]

$$M_{k'k} = \langle \Psi^{\text{c}}(k') | V | \Psi^{\text{c}}(k) \rangle \,, \tag{13.2}$$

in which $\Psi^c(k)$ is the conduction-band wavefunction of two-dimensional (2D) graphite (or monolayer graphene) at wave vector k, and $V = V_0 + V_{def}$ is the potential term of the Hamiltonian with crystal potential (V_0) and a defect perturbation potential (V_{def}). Here $\Psi(k)$ is expanded by the Bloch wavefunction Φ_s, and Φ_s is expressed in terms of the atomic wavefunction, $\phi(r - R_s)$ (see Chapter 2), thus giving:

$$M_{k'k} = \frac{1}{N_u} \sum_{s,s'} C^*_{s'}(k') C_s(k) \sum_{R_s, R_{s'}} \exp\left(-ik' R_{s'} + ik R_s\right) V_{s's} , \tag{13.3}$$

where $V_{s's}$ is the atomic matrix element of V defined by $\langle \phi(R_{s'}) | V | \phi(R_s) \rangle$. When $V = V_0$, then $V_{s's}$ depends only on $R_{s'} - R_s$, and also $M_{k'k}$ has a nonvanishing value only for $k' = k$[2] which implies crystal momentum conservation. If we remove a carbon atom from the site s', the atomic tight-binding matrix elements $\langle \phi(R_{s'}) | V | \phi(R_s) \rangle$ containing s' become zero and then elastic scattering from k to k' occurs. When we consider the tight-binding method within the nearest neighbor interaction, then the tight-binding γ_0 parameter becomes zero for the atoms that are nearest neighbors to the site s'. This means that we add $-\gamma_0$ parameters for the three nearest s sites as tight-binding parameters for V_{def}, and Eq. (13.3) for $k' \neq k$ then becomes:

$$M_{k'k} = -\frac{\gamma_0}{N_u} C^*_{s'}(k') C_s(k) \sum_{R_s} \exp\left(-ik' R_{s'} + ik R_s\right) , \tag{13.4}$$

where a summation on R_s is taken only for the three nearest neighbor atoms to the s' atom.

However, this expression for $M_{k'k}$ is too simple and a fundamental correction to this expression is necessary. Since the impurity potential for the missing atom ($-\gamma_0$) is of the same order as for the tight-binding parameter of the electronic structure (γ_0), the matrix element cannot be expressed within the lowest order of perturbation theory and we must consider higher-order terms. In order to obtain such higher-order terms, we consider the correction to the wavefunction $\Psi^c(k)$ due to the presence of such defects. In fact, the perturbation to $\Psi^c(k)$ as an electron is scattered around the defect mixes many $\Psi^c(k')$ wave functions for different wave vectors with $\Psi^c(k)$ for $k' \neq k$. The perturbed wavefunction Φ in the presence of an impurity potential V_{def} which is defined by the difference between the unperturbed potential and the potential after adding the defect, is given by:

$$\Phi(k) = \Psi^c(k) + \sum_{k'} \frac{\langle \Psi^c(k') | V_{def} | \Psi^c(k) \rangle}{E(k') - E(k) + i\gamma} \Psi^c(k') , \tag{13.5}$$

where γ is a broadening factor due to the finite lifetime of carriers due to the defect scattering (introduced by the uncertainty relation). It is noted that k in $\Phi(k)$ no longer has the meaning of "as a function of k" but rather is modified by the

2) The umklapp scattering process $k' = k + G$ (where G is a reciprocal lattice vector) also occurs.

correction to $\Psi^{c}(k)$. Thus, when we put Eq. (13.5) into Eq. (13.2) to redefine $M_{k'k}$, we now obtain:

$$\begin{aligned} M_{k'k} &\equiv \langle \Phi(k')| V_{\text{def}} |\Phi(k)\rangle \\ &= \langle \Psi^{c}(k')| V_{\text{def}} |\Psi^{c}(k)\rangle \\ &\quad + \sum_{k''} \frac{\langle \Psi^{c}(k')| V_{\text{def}} |\Psi^{c}(k'')\rangle \langle \Psi^{c}(k'')| V_{\text{def}} |\Psi^{c}(k)\rangle}{E(k'') - E(k) + i\gamma} \,. \end{aligned} \tag{13.6}$$

Equation (13.6) then gives the next order correction to the matrix element for elastic scattering. Since we consider the elastic scattering $E(k'') = E(k)$, in Eq. (13.6), the value of γ can not be neglected. Using the Fermi Golden rule for a second-order time-dependent perturbation, γ is proportional to the sum of $|M_{k'k}|^2$ which is inversely proportional to the lifetime of the k state, and the γ values are determined self-consistently. When we substitute $\Phi(k')$ for $\Psi^{c}(k')$ in the last term of Eq. (13.5), we can iteratively obtain the expansion of the perturbation series. The corresponding $M_{k'k}$ is defined iteratively. An infinite series for this expansion of the scattering matrix elements is called the *T-matrix* [396, 397]. Here we do not go into detail regarding the quantum theory of the T-matrix, but rather we point out that the calculation of the T-matrix is necessary for discussing elastic scattering.

An important fact is that the Fourier transform of $V_{\text{def}}(r)$ to q space to obtain $V_{\text{def}}(q)$ determines the range of the defect potential. When V_{def} is a short-range potential such as a point defect, a dominant contribution to $V_{\text{def}}(q)$ comes from a large range of q values since intervalley scattering from the K to K' valley (or vice versa) is important. On the other hand, when V_{def} is a long-range potential, then intravalley scattering is dominant. In the case of intravalley scattering, then backward scattering is absent for the quantum interference effect, which is significant in the case of single-wall carbon nanotubes [396, 397].

Figure 13.2 shows the calculated spectral D-band Raman intensity for a nanoribbon with an armchair edge, for three different laser energies of 1.90 eV (solid line), 2.30 eV (dashed line) and 2.70 eV (dotted line), as described in [395]. The defects here are the edges with an armchair atomic structure, and the sp^2 periodicity is broken by the missing atoms at the edge bonds. The elastic matrix elements are taken in the calculation at the lowest order of $M_{k'k}$ as in Eq. (13.2), which is here given analytically [395]. When the armchair edge exists in the direction of x, then the k_x component of k conserves momentum, while k_y changes its sign (by reflection), which corresponds to intervalley scattering. The $M_{k'k}$ matrix element for the zigzag edge does not contribute to intervalley scattering,[3] which means the D-band should be absent when measuring zigzag edges, as observed experimentally (see Section 13.4).

When comparing the results in Figure 13.2 with experiment, the frequency dependence of the D-band feature on E_{laser} is well described (see inset to Figure 13.2). The D-band peak come from the iTO phonon dispersion branch near the K point for which $q \approx 2k$, as dictated by the double resonance process [159, 160], and this

3) In the case of the zigzag edge, k_x changes its sign and this corresponds to intravalley scattering (K to K, or K' to K').

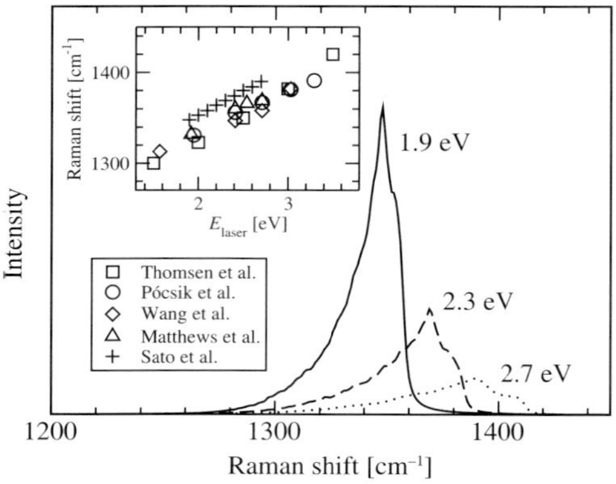

Figure 13.2 The calculated Raman spectra of the D-band for $E_{\text{laser}} = 1.90$ eV (solid line), 2.30 eV (dashed line), 2.70 eV (dotted line) considering a nanoribbon with an armchair edge [395]. The inset shows a comparison of ω_D vs. E_{laser} between various works [159, 395, 398–400], where crosses represent the calculated results discussed here [395].

topic is discussed more fully in Section 13.2. However, an accurate description of the scattering intensity behavior (e. g., its dependence on both E_{laser} and ribbon width, and the matrix element dependence on q generating an asymmetric D-band lineshape) has still not been fully achieved. This is not only because the T-matrix was not considered correctly, but because other aspects such as the phonon coherence length and the resonance window width γ have not been considered self-consistently. Therefore, while the present section gives a taste on how to fully treat the process quantum mechanically, and indicates an accurate description of disorder-induced Raman features, for an accurate quantum mechanical description of the Raman intensities, more work is still needed.

13.2
The Frequency of the Defect-Induced Peaks: the Double Resonance Process

Defects break the momentum conservation requirement $q = 0$ for the first-order Raman-allowed phonons, so that, in principle, any scattering event involving phonons in the interior of the Brillouin zone ($q \neq 0$), would then be allowed. However, as discussed in Chapter 12, in sp^2 carbon materials the resonant electron–phonon scattering processes connecting real electronic states (see Figure 13.3a) minimize the denominators in Eq. (13.1), that is, these resonant processes are privileged and have much higher probabilities, so that the spectra are dominated by the double resonance scattering processes. While momentum conservation in a perfect lattice can only be fulfilled by $q = 0$ phonons or two-phonon scattering processes with $q - q = 0$, as discussed in Chapter 12, in the presence of disorder

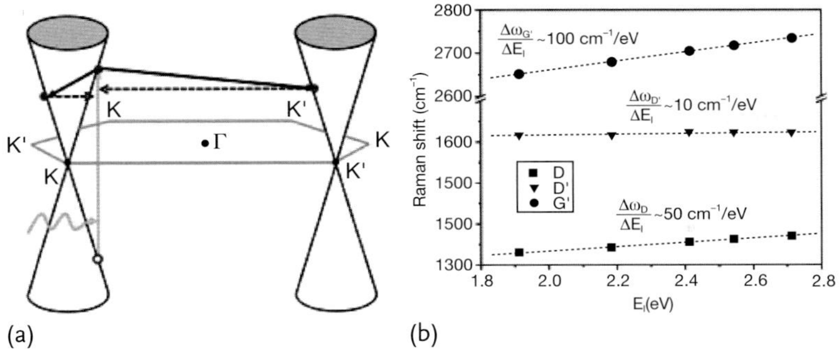

Figure 13.3 (a) Schematics showing the electronic dispersion near the Fermi level at the K and K' points in the hexagonal Brillouin zone of graphene. The light-induced electron–hole formation is indicated by a gray arrow. The two resonant electron–phonon scattering processes associated with the D (intervalley) band and the D′ (intravalley) band are indicated by the black arrows. The dashed arrows indicate elastic scattering induced by defects. (b) Laser energy dependence or dispersion of the frequency of the D, D′ and G′-bands [394].

the momentum conservation can be satisfied through an elastic scattering process by a defect, as represented by dashed arrows shown in Figure 13.3a.

Therefore, the frequencies of the defect-induced peaks are well explained by the double resonance processes discussed in Chapter 12, although some special details have to be taken into account. For example, from a frequency analysis, we see that both the D and D′-band scattering events shown in Figure 13.3 are the one-phonon processes that are related to the two-phonon processes observed at 2700 cm^{-1} (G′ $\sim$ 2D) and at 3240 cm^{-1} (G″ $\sim$ 2D′) in Figure 4.14. The dispersions of the frequencies of the D, D′ and G′-bands are shown in Figure 13.3b by plotting their frequency dependence on E_{laser}. The slope associated with the G′-band is about 100 cm^{-1}/eV and is two times larger than the slope of the D-band (50 cm^{-1}/eV). The D′-band also exhibits a weak dispersive behavior, the slope being about 10 cm^{-1}/eV [394]. However, there is no exact matching between the D and G′-bands (i. e., $\omega_{G'} \neq 2\omega_D$) because their physical processes have some differences. As discussed in Section 12.2.1, different q vectors give rise to different double resonance processes in the Stokes (S) and anti-Stokes (aS) processes ($q_S \neq q_{aS}$). The same applies to the disorder-induced bands and more: within a one-phonon Stokes double resonance process, different wave vectors q will be introduced if we consider the elastic scattering taking place either before or after the inelastic phonon scattering (see Figure 13.4) [80, 367].

The D and D′-bands are not the only disorder-induced one-phonon peaks in the Raman spectra for disordered sp^2 materials (see Figure 12.10). Similar to the two-phonon processes discussed in Section 12.3, which can occur through any combination or overtone of the six dispersive phonon energy branches in graphene, the disorder-induced Raman frequencies can be related to any of the six phonon branches of 2D graphite with the appropriate wave vector which fulfills the double resonance condition. The intravalley and intervalley double resonance processes

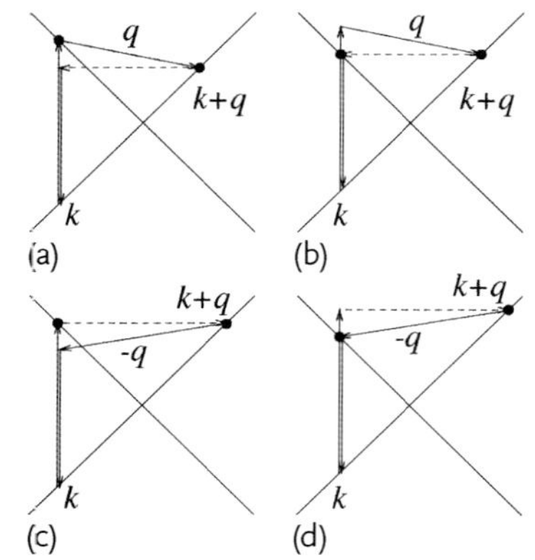

Figure 13.4 Four different one-phonon intravalley second-order double resonance Stokes processes. For each process, the dashed lines denote an elastic scattering process and black dots are shown for the resonant points. For the two-phonon second-order processes, the dashed line of each figure is changed to be an inelastic phonon emission process and thus only the (a) and (d) processes are possible for two-phonon scattering processes [160] (see the (a) and (d) panels in Figure 12.3).

are mediated by phonons near the Γ and K (or K') points, respectively, and we can change both the resonant k and q values by changing E_{laser}, as determined by Eq. (12.8). Thus by using electronic structure information, we can determine the phonon dispersion relations around the K and the Γ points, by considering intervalley and intravalley processes, respectively. For both intervalley and intravalley processes, the fitting between the observed Raman frequencies and the dispersion relations depend upon the different possibilities for second-order double resonance processes. Four of them are exemplified in Figure 13.4 for the intravalley Stokes Raman scattering process, as is for example pertinent to the D′ band.

Figure 13.5a plots the E_{laser} dependence of the disorder-induced double resonance peaks, as obtained considering a linear electronic dispersion (see Chapter 2) and the phonon dispersion in Figure 13.5b. The lower horizontal axis of Figure 13.5a correlates the E_{laser} values with the phonon wave vectors q indicated in the upper horizontal axis for Γ to $K/4$ and for $3K/4$ to K. These q values are also shown in Figure 13.5b and are related to intravalley and intervalley processes obeying $q \approx 2k$ (see Eq. (12.8)) [160]. It is noted that the linear relationship between E_{laser} and k, and consequently between E_{laser} and q is valid for $E_{\text{laser}} < 3.0$ eV (see Figure 2.10 in Chapter 2), and the vertical dotted lines in Figure 13.5b show limits for the q wave vectors imposed by $E_{\text{laser}} < 3.0$ eV. By comparing Figure 13.5a,b, it is easy to correlate the double resonance peaks with the six different phonon branches in graphite. Solid and open circles correspond to the phonon modes around the K and the Γ points, respectively. Nondispersive features are also seen in Figure 13.5a, and they come from the $q \approx 0$ double resonance condition (see Eqs. (12.6) and (12.7), and related text).

Finally, by using the E_{laser} vs. q relations given in Figure 13.5a, we can take experimental values from several published papers giving Raman frequencies observed

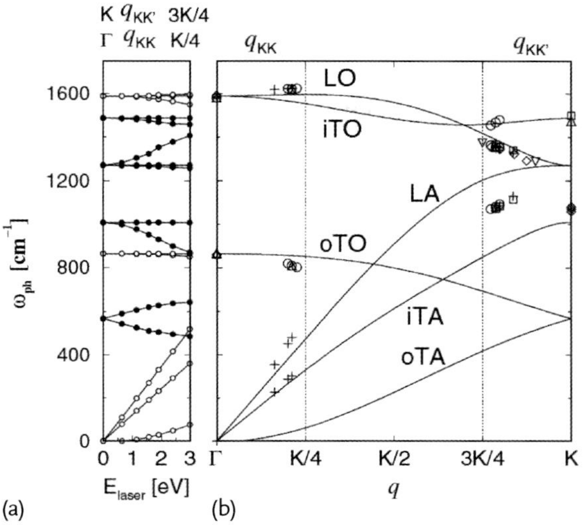

Figure 13.5 (a) Calculated Raman frequencies for the double resonance condition in 2D graphite as a function of E_{laser} (bottom horizontal axis) and the corresponding q vector along Γ-K (top horizontal axis). Solid and open circles correspond to phonon modes around the K and the Γ points, respectively. The q_{KK} vectors from Γ to $K/4$ are shown by open circles and the $q_{KK'}$ vectors from $3K/4$ to K are shown by solid circles. (b) The six graphite phonon dispersion curves (lines) and experimental Raman observations (symbols) consistent with double resonance theory [160].

using different laser lines for various sp^2 carbons (e. g., Figure 12.10) and plot all these data points in Figure 13.5b. Notice that in the Raman spectra there is no information on whether the peak comes from an intravalley or intervalley process, and on obeying the $q \approx 0$ or $q \approx 2k$ resonance conditions. We choose among these possible scattering assignments by considering where the experimental data would best fit in the phonon dispersion in Figure 13.5b. The discrepancies can be due to wrong scattering assignments or inaccurate phonon dispersion relations. In fact, these double resonance peak assignments have been largely used to improve the theoretical modeling of phonons in sp^2 nanocarbons [366].

13.3 Quantifying Disorder in Graphene and Nanographite from Raman Intensity Analysis

As discussed in Section 13.2, the frequencies of the disorder-induced features in the Raman spectra for sp^2 carbons are well-explained by the double resonance model. What remains to be established for making Raman spectroscopy a powerful tool to characterize disorder in sp^2 materials is how to relate specific defects to their corresponding disordering processes, and how to obtain quantitative information about the amount of such defects in the lattice. As briefly discussed in Section 13.1, more experimental and theoretical work is still needed for an accurate quantum mechanical description of disorder-induced Raman processes. However, some work has already been done and phenomenological models have been developed, as described below.

Extensive work in the field of amorphous carbon leads to a description of the amorphization trajectory for carbon materials (see Figure 4.7c in Section 4.4.1). However, to achieve a quantitative description of these phenomena, the effects of disorder on the electron and phonon properties have to be probed in both momentum space (*k*-space) and real space (*r*-space), which means Raman spectroscopy has to be combined with microscopy experiments. For example, transmission electron microscope (TEM) or scanning tunneling microscopy (STM) can characterize disorder in the crystal *r*-space by probing the local surface density of electronic states, with atomic-level resolution. Simultaneous in-situ TEM and Raman measurements are, in principle, possible. However, a special experimental set up and special sample preparation methods would be needed. Usually, STM and Raman spectroscopy cannot be easily correlated with each other, since optical spectroscopy probes a volume that is limited by the light penetration depth, while STM is mostly sensitive to surfaces. In this context, the possibility of exfoliating graphite to pull out a single graphene sheet, provides an ideal situation in which microscopy and spectroscopy can be correlated to probe disorder effects in both *r*-space and *k*-space. The initial efforts in this research direction are now discussed.

13.3.1 Zero-Dimensional Defects Induced by Ion Bombardment

The controlled use of ion implantation to study defects in sp^2 carbons is a well established technique [401]. These experiments are normally carried out as a function of ion dose and for different ion species and different ion energies. Low mass ions at low ion fluence introduce point defects. Increasing the ion dose causes an increasing density of point defects and eventually causes the damaged regions to overlap. In this section we discuss effects from Ar^+ implantation as a function of ion dose and the resulting damage to HOPG [402] and graphene [194, 195].

Consecutive Ar^+ ion bombardment and Raman spectroscopy experiments were performed on monolayer graphene samples [194]. Low energy ions (90 eV), experimentally confirmed to barely exceed the threshold value for the displacement of surface C atoms, were used to produce structural defects in the graphene layer, thereby avoiding cascade effects.[4] The bombardment ion doses span the typical values that are used for ion implantation studies, starting with 10^{11} Ar^+ impacts per cm^2, which corresponds to one defect per 4×10^4 C atoms, and going up to 10^{15} Ar^+/cm^2, denoting the onset of full disorder in graphene. STM images show that for up to 10^{12} Ar^+/cm^2, the ion bombardment-induced defects are isolated from each other, with each defect causing a rather large disordered area in the STM images ($\sim$ 1 nm radius). Near a 10^{13} Ar^+/cm^2 dose, the disordered areas start to

4) Cascade effects are effects whereby a scattered C atom with a large energy hits another C atom iteratively. Similar phenomena can be seen in the chain reaction of dominos.

coalesce and the surface exhibits a mixture of ordered and disordered regions. At a 10^{14} Ar^+/cm^2 dose and above, the hexagonal crystalline pattern can no longer be observed by probing the local density of electronic states by STM. Analysis of the STM images at each ion dose gives the defect concentration from which we can extract the average distance between defects, $L_D = \sigma^{-1/2}$, where σ is the density of defects. Therefore, the σ and L_D values can be obtained from the STM images by the direct counting of defects [194]. For the highest ion doses of 10^{15} Ar^+/cm^2, when the effect of defects start to coalesce, we consider that the defect density increases linearly with bombardment time.

Figure 13.6 shows the Raman spectra of a graphene monolayer subjected to the ion bombardment procedure as described above. From the pristine sample (bottom spectrum) to the lowest bombardment dose in Figure 13.6 (10^{11} Ar^+/cm^2), the D-band process is activated, showing a very small intensity relative to the G peak. Within the bombardment dose range 10^{11}–10^{13} Ar^+/cm^2, the intensities of the disorder peaks increase. A second disorder-induced peak around $\sim 1620\,cm^{-1}$ (the D′-band) also becomes evident, but we do not focus on this feature here. Above 10^{13} Ar^+/cm^2, the Raman spectra start to broaden significantly and end up exhibiting a profile similar to the graphene phonon density of states (PDOS). From the 10^{14} (top spectrum) to 10^{15} Ar^+/cm^2 (not shown) dose, the Raman scattering response develops its PDOS-like profile, showing a lineshape broadening with no change in peak frequencies.

Quantifying the development of disorder in a graphene monolayer can be achieved by plotting the I_D/I_G data as a function of the average distance between defects L_D, as shown in Figure 13.7. The I_D/I_G ratios are here obtained

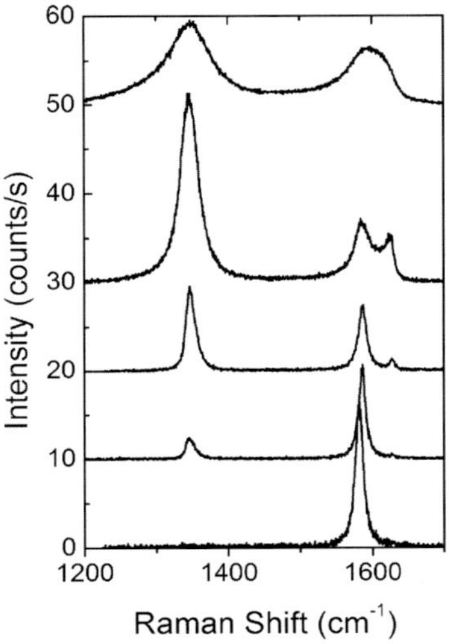

Figure 13.6 Evolution of the first-order Raman spectra using a $\lambda = 514\,nm$ laser of a graphene monolayer sample deposited on an SiO_2 substrate, subjected to Ar^+ ion bombardment. The ion doses are from the bottom to the top, 0, 10^{11}, 10^{12}, 10^{13} and 10^{14} Ar^+/cm^2. The spectra in this figure are also displaced vertically for clarity [194].

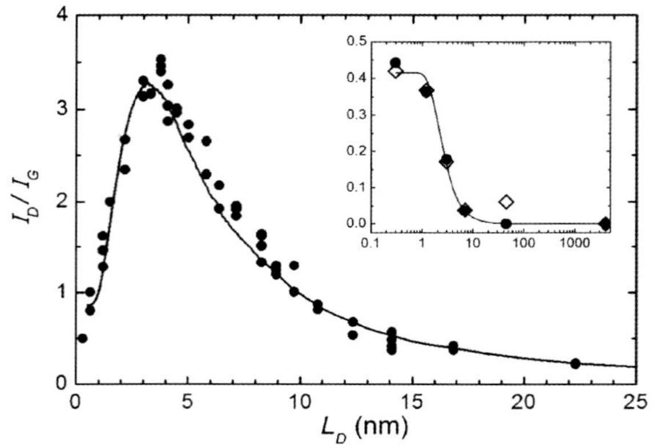

Figure 13.7 The I_D/I_G data points from three different monolayer graphene samples as a function of the average distance L_D between defects, induced by the ion bombardment procedure. The solid line is a modeling of the experimental data with Eq. (13.7). The inset shows a plot of I_D/I_G vs. L_D on a log scale for two samples: (i) a ~ 50-layer graphene sample; (ii) a 2 mm-thick HOPG sample, whose measured values are here scaled by $(I_D/I_G) \times 3.5$ [194].

by considering the peak intensity at the fixed D-band (1345 cm^{-1}) and G-band (1585 cm^{-1}) frequencies. The I_D/I_G ratio has a nonmonotonic dependence on L_D, increasing initially with increasing L_D up to $L_D \sim 3.5$ nm where I_D/I_G has a peak value, and then decreasing for $L_D > 3.5$ nm. This result is similar to the proposed amorphization trajectory for graphitic nanocrystallites (see Section 4.4.1), and such a behavior suggests the existence of two disorder-induced competing mechanisms contributing to the Raman D-band. These competing mechanisms are the basis for a phenomenological model for the L_D dependence of I_D/I_G that is now described (Section 13.3.2).

13.3.2
The Local Activation Model

The results in Figure 13.7 are modeled by assuming that a single impact of an ion on the graphene sheet causes modifications on two length scales, here denoted by r_A and r_S (with $r_A > r_S$), which are the radii of two circular areas measured from the impact point (see Figure 13.8). Within the shorter radius r_S, structural disorder from the impact occurs. We call this the structurally-disordered or *S*-region. For distances larger than r_S but shorter than r_A, the lattice structure is preserved, but the proximity to a defect causes a mixing of Bloch states near the *K* and *K'* valleys of the graphene Brillouin zone, thus causing a break-down of the selection rules, and leading to an enhancement of the D-band. We call this the defect activated or *A*-region. In qualitative terms, an electron–hole excitation will only be able to "see" the structural defect if the electron–hole pair is created sufficiently close to the defect and if the excited electron (or hole) lives long enough for the defective

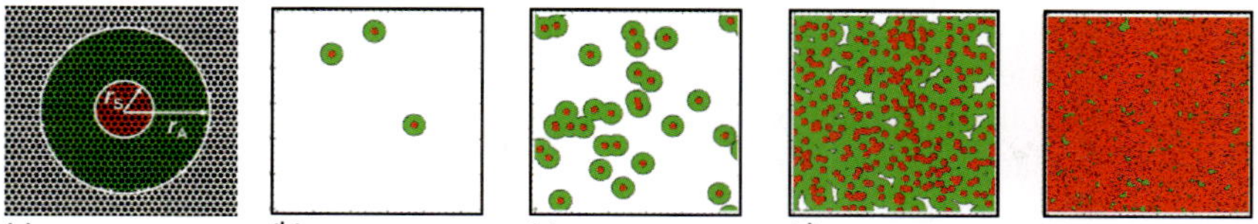

(a) (b) (c) (d) (e)

Figure 13.8 (a) Definition of the "activated" A-region (darkest gray) and "structurally disordered" S-region (dark gray). The radii are measured from the impact point which is chosen randomly in our simulation. Parts (b–e) shows $55\,\mathrm{nm} \times 55\,\mathrm{nm}$ portions of the graphene simulation cell, with snapshots of the structural evolution of the graphene sheet for different defect concentrations: (b) $10^{11}\,\mathrm{Ar}^+/\mathrm{cm}^2$; (c) $10^{12}\,\mathrm{Ar}^+/\mathrm{cm}^2$; (d) $10^{13}\,\mathrm{Ar}^+/\mathrm{cm}^2$; (e) $10^{14}\,\mathrm{Ar}^+/\mathrm{cm}^2$, like in Figure 13.6 [194].

region to be probed by Raman spectroscopy. If the Raman scattering process occurs at distances larger than $\ell = r_A - r_S$ from the defective region, the wave vector k is a good quantum number for analyzing the scattering selection rules and those regions that remain well ordered will only contribute significantly to the G-band. Our phenomenological model for the I_D/I_G ratio is given as a function of the average distance between two defects, L_D:

$$\frac{I_D}{I_G}(L_D) \propto I_D(L_D) = C_A f_A(L_D) + C_S f_S(L_D)\,, \tag{13.7}$$

where I_G is considered as constant (independent on L_D), and f_A and f_S are simply the fractions of the A and S areas in the sheet, respectively, with respect to the total area. Although both the A and S regions can break momentum conservation, giving rise to a D-band, the A-regions will contribute most strongly to the D-band, while the S-regions will make less contribution to the D-band due to the breakdown of the lattice structure itself.

We now describe stochastic simulations (see Figure 13.8b–e) used to implement the phenomenological model for the I_D/I_G ratio. The structurally-disordered (S) region is shown in light gray in Figure 13.8a and the activated (A) region is shown in dark gray in Figure 13.8a. The structural evolution of a graphene sheet under ion bombardment was simulated by randomly choosing a sequence of impact positions on a ($50\,\mathrm{nm} \times 50\,\mathrm{nm}$) sheet. The following set of rules for each event was defined: (1) pristine regions (white area in Figure 13.8b–e) may turn into S (light gray) or A (dark gray) regions, depending on the proximity to the impact point; (2) similarly, A-regions may turn into S (light gray); (3) S-regions always remain S regions. Then, the initially pristine sheet evolves, as the number of impacts increase, to become mostly activated, leading to an increase of the D-band, and later the mostly structurally-disordered regions become increasingly widespread, leading to a decrease of the D-band. Snapshots of this evolution are shown in Figure 13.8b–e for the same argon ion concentrations as in Figure 13.6. The stochastic simulations of the bombardment process, with the impact points for the ions chosen at random, combined with Eq. (13.7) with parameters $C_A = 4.56$, $C_S = 0.86$, $r_A = 3\,\mathrm{nm}$ and $r_S = 1\,\mathrm{nm}$, give the full line curve in Figure 13.7, which is in excellent agreement with the experimental results (points) in this figure [194].

The nonmonotonic behavior in Figure 13.7 can be understood by considering that, for low defect concentrations (large L_D), the total area contributing to scattering is proportional to the number of defects, giving rise to a $I_D/I_G = (102 \pm 2)/L_D^2$ dependence that works well for $L_D > 2r_A$. Upon increasing the defect concentration, the activated regions start to overlap and these regions eventually saturate. The D-band intensity then reaches a maximum and a further increase in the defect concentration decreases the D-band intensity because the graphene sheet starts to be dominated by the structurally-disordered areas.

The length scale $r_S = 1$ nm, which defines the structurally-disordered area, is in perfect agreement with the average size of the disordered structures seen in the STM images. This parameter should not be universal, but it is specific to the bombardment process. The Raman relaxation length ℓ for the defect-induced resonant Raman scattering in graphene for a laser energy of 2.41 eV, is found to be $\ell = r_A - r_S = 2$ nm.[5] The C_A parameter in Eq. (13.7) is a measure of the maximum possible value of the I_D/I_G ratio in graphene, which would occur in a hypothetical situation in which $K - K'$ wave vector mixing would be allowed everywhere, but no damage would be made to the hexagonal network of carbon atoms. C_A should then be defined by the electron–phonon matrix elements, and the value $C_A = 4.56$ is then in rough agreement with the ratio between the electron–phonon coupling for the iTO phonons evaluated between the Γ and the K points [366]. The C_S parameter is the value of the I_D/I_G ratio in the highly disordered limit, which has not yet been addressed theoretically.

Finally, for practical use, it is important to have an equation relating I_D/I_G to L_D, and such an equation can be obtained by solving rate equations for the bombardment process (see problem set). The entire regime ($0 \to L_D \to \infty$) can be fitted using:

$$\frac{I_D}{I_G} = C_A \frac{r_A^2 - r_S^2}{r_A^2 - 2r_S^2} \left[\exp\left(\frac{-\pi r_S^2}{L_D^2} \right) - \exp\left(\frac{-\pi (r_A^2 - r_S^2)}{L_D^2} \right) \right] + C_S \left[1 - \left(\frac{-\pi r_S^2}{L_D^2} \right) \right] . \tag{13.8}$$

Fitting the data in Figure 13.7 with Eq. (13.8) gives $C_A = (4.2 \pm 0.1)$, $C_S = (0.87 \pm 0.05)$, $r_A = (3.00 \pm 0.03)$ nm and $r_S = (1.00 \pm 0.04)$ nm, also in excellent agreement with experiment and consistent with the parameters obtained within the computational modeling [194].

The present model provides a method to accurately quantify the density of defects σ or, equivalently, the average distance between defects ($L_D = \sigma^{-1/2}$) in graphene. Before the defects start to coalesce ($L_D > 6$ nm in the present case), the expected behavior occurs, that is, $I_D/I_G = A/L_D^2$, where $A = (102 \pm 2)$ nm^2 was found. When the defects start to coalesce there is a competition between two disorder mechanisms, and Eq. (13.8) can be used for a quantitative analysis to determine the relative importance of each mechanism. The present results discussed

5) Here we discuss the relaxation length for the excited electron, and this should not be confused with the relaxation length for the phonons.

for graphene are similar to what has been observed in ion bombarded HOPG [402], although some details are different. First, for HOPG a larger G-band is always observed due to the contribution from the undisturbed under-layers. Second, for graphene, above 10^{15} Ar^+/cm^2, the spectra show a decreased intensity, indicating full amorphization or partial sputtering of the graphene layer. For HOPG, above 10^{15} Ar^+/cm^2, I_D/I_G saturates and no further change is observed in the Raman spectra because of the large number of layers that have been amorphitized and/or sputtered. This behavior is seen in the inset to Figure 13.7, which shows the I_D/I_G evolution for two HOPG samples of different thicknesses. Despite differences in absolute values, which depend on the number of undisturbed under-layers (the diamond data in the inset to Figure 13.7 coming from a thicker HOPG sample was scaled by $\times 3.5$), the I_D/I_G values increase and saturate when increasing the ion dose.

A study of the I_D/I_G evolution as a function of L_D depending on the number of layers N has also been developed [195]. The I_D/I_G behavior was observed to scale with N, clearly demonstrating the lower energy ions (90 eV) used in the experiment were not able to do cascade effects, but the process is limited generally to one defect per bombarding ion. For few-layer graphene samples ($N = 1, 2, 3$), the normalized evolution of I_D/I_G increases with increasing the number of defects (increasing the "activated area"[194]), and further saturates and decreases. This decrease is due to the take-over of the activated area by the "disordered area", as introduced previously for 1-LG [194]. However, the decrease in I_D/I_G for larger ion doses is less evident the larger the N. For many-layers graphene ($\sim$ 50 and higher), the normalized evolution of I_D/I_G with increasing number of defects is a monotonic increase, since there are always more graphene layers to be bombarded.

In summary, this section gives a clear picture of the basic mechanisms behind the evolution of the disorder-induced D-band for point defects, which is given by a competition between the structurally-damaged area, and the D-band-activated area relative to the total area. Since this is basically a geometric interplay, different results will be obtained when moving from the "zero-dimensional" defects caused by ion bombardment in graphene, and the "one-dimensional" defects represented by the boundaries of a nanocrystalline graphene sample or a graphite crystal. These nanocrystalline systems actually represent an extensively studied system, usually formed by the annealing of diamond-like carbon films formed by sputtering [168], and are discussed in the next section.

13.3.3 One-Dimensional Defects Represented by the Boundaries of Nanocrystallites

In 1970, Tuinstra and Koenig [148, 149] performed systematic Raman and X-ray diffraction studies of many graphitic samples with different in-plane crystallite sizes L_a. These authors concluded that the ratio of the D and G-band intensities (I_D/I_G) is inversely proportional to L_a, which was determined from the width of the X-ray diffraction peaks. After this pioneering work, the ratio I_D/I_G was used for many years to estimate L_a in disordered carbon materials. Knight and White [403]

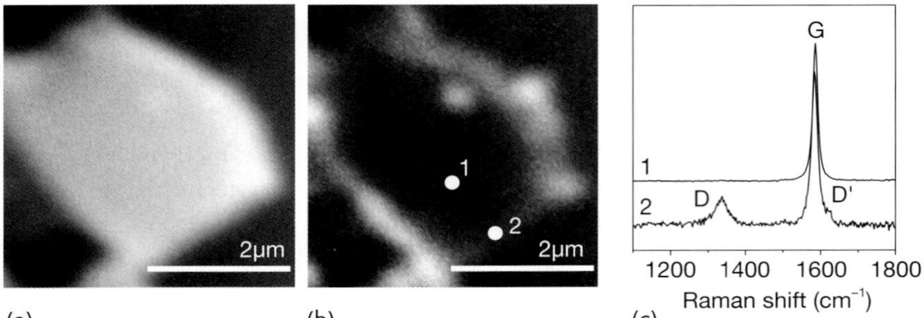

(a) (b) (c)

Figure 13.9 (a) G-band and (b) D-band confocal (300 nm resolution) Raman images of a graphite crystallite deposited on a glass substrate. In (c) the Raman spectra obtained in regions 1 and 2 (white circles depicted in panel (b)) are shown. The laser excitation comes from a HeNe ($\lambda = 633$ nm) laser using an experimental setup described in [394].

later summarized the Raman spectra of various graphitic systems measured using the $\lambda = 514.5$ nm ($E_{\text{laser}} = 2.41$ eV) laser line, and they derived an empirical expression which allows the determination of L_a from the (I_D/I_G) ratio [403]. Later, a general formula was developed giving I_D/I_G vs. L_a for nanographite systems for any excitation laser energy in the visible range, as presented below [404].

Figure 13.9a,b show two confocal Raman images of a 6 nm-high HOPG crystallite deposited on a glass substrate. Figure 13.9a shows a Raman image of the crystallite, obtained by plotting the spatial dependence of the G-band intensity, while in Figure 13.9b the spatial dependence of the intensity of the disorder-induced D-band is shown and here the boundary of the crystallite is highlighted. Figure 13.9c shows two Raman spectra, one taken at an interior point of the crystallite, and the other at the edge. It is clear from Figure 13.9a–c that the G-band intensity is uniform over the whole graphite surface, while the D-band intensity is localized where the crystalline structure is not perfect, mostly at the edges of the crystallite. Notice also that the D-band intensity varies from edge to edge, and this D-band intensity is dependent on the light polarization direction and the atomic structure at the edge, as is discussed in Section 13.4.

For evaluation of the I_D/I_G dependence on the crystallite dimensions, one can consider a square of side L_a, for which the intensity of the G-band will vary as $I_G \propto L_a^2$. The intensity of the D-band will, however, depend on the width δ of the "border" where the D-band is activated, given by $I_D \propto L_a^2 - (L_a - 2\delta)^2$. The intensity ratio will then be given by:

$$\frac{I_D}{I_G} = \alpha \left[4 \left(\frac{\delta}{L_a} - \frac{\delta^2}{L_a^2} \right) \right] , \tag{13.9}$$

where α is dependent on the appropriate matrix elements [394].

In the limit $L_a \gg \delta$, Eq. (13.9) can be simplified to yield the famous Tuinstra–Koenig relation

$$\frac{I_D}{I_G} = C(E_{\text{laser}})/L_a , \tag{13.10}$$

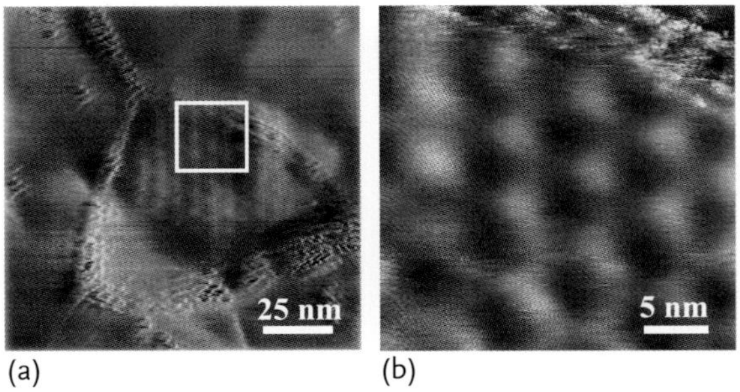

Figure 13.10 Scanning tunneling microscopy (STM) images with atomic resolution obtained from the surface of a nanographite crystallite of a sample with $L_a = 65\,\text{nm}$. (a) A Moire pattern at the crystallite surface is observed. (b) Magnification of the region delineated by the white square in part (a) [405].

where the values of the empirical constant $C(E_{\text{laser}})$ change from paper to paper in the literature. One could then expect that, once the relaxation length and matrix element ratio were measured for the D-band scattering in ion-bombarded graphene (Section 13.3.2), these values could just be transferred here to obtain α and δ. However, these factors depend on the structurally-disordered area (S_S), which is not well defined for the nanographite. Figure 13.10 shows two scanning tunneling microscopy (STM) images with atomic resolution obtained from the surface of a crystallite in a sample with $L_a = 65\,\text{nm}$. The atomic arrangement of the carbon atoms observed in these pictures indicates that the samples are formed by nanographitic crystallites, but with a clearly disordered grain boundary between crystallites [405]. Variability associated with these grain boundaries may be responsible for the different I_D/I_G vs. L_a results observed in the literature.

Furthermore, the empirical constant $C(E_{\text{laser}})$ has been known to depend on E_{laser} since 1984 [152], but $C(E_{\text{laser}})$ has been quantitatively developed only more recently [404], using experimental results from nanographites with different L_a values prepared from diamond-like carbon (DLC) films heat treated at different temperatures T_{htt} [404]. Before heat treatment, the sp^3 and sp^2 carbon phases coexist in the samples, but the sp^3 phases completely disappear for $T_{\text{htt}} > 1600\,°\text{C}$ [406]. STM images of the samples obtained at different heat treatment temperatures $T_{\text{htt}} \geq 1800\,°\text{C}$ show that these samples correspond to aggregates of nanographite crystallites, and show increasing L_a with increasing T_{htt}. The evolution of the (100) X-ray diffraction peak obtained using synchrotron radiation, for the samples heat treated at different T_{htt} also give a measure of the crystallite sizes, by evaluating L_a from the Scherrer relation $L_a = 1.84\lambda/\beta \cos\theta$, where λ is the synchrotron radiation wavelength (0.120 nm), θ is the position of the (100) diffraction peak, and β is the half-height width of the (100) peak of graphite in 2θ (rad) units [404]. The mean crystallite sizes obtained by X-ray diffraction range from 20 to 500 nm in size, and the X-ray values are in good agreement with the L_a values obtained directly from the STM images [376, 404, 405].

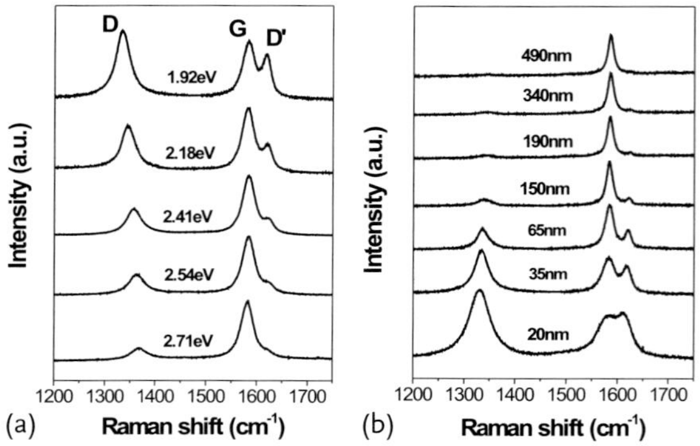

Figure 13.11 The first-order Raman spectra of (a) the nanographite sample heat treated at 2000°C ($L_a = 35$ nm), for five different laser energy values (1.92 eV, 2.18 eV, 2.41 eV, 2.54 eV, and 2.71 eV). (b) Nanographite samples with different crystallite sizes L_a (in mm) using 1.92 eV laser excitation energy [404].

Shown in Figure 13.11a are results from Raman scattering experiments performed at room temperature with different E_laser values, showing spectra of the D, G, and D′-bands for the $T_\mathrm{htt} = 2000°$C sample ($L_a = 35$ nm) for five different E_laser values (1.92 eV, 2.18 eV, 2.41 eV, 2.54 eV, and 2.71 eV). The spectra are normalized to the G-band intensity, and clearly the $(I_\mathrm{D}/I_\mathrm{G})$ ratio is strongly dependent on E_laser. Figure 13.11b shows the Raman spectra using $E_\mathrm{laser} = 1.92$ eV for samples with different T_htt values, thereby giving rise to samples with different crystallite sizes L_a [404].

Figure 13.12a shows a plot of $(I_\mathrm{D}/I_\mathrm{G})$ vs. $1/L_\mathrm{a}$ for all samples and using the five different E_laser values from Figure 13.11. Noting that $I_\mathrm{D}/I_\mathrm{G}$ for a given sample is strongly dependent on E_laser, we see that all these curves collapse on to the same curve in the $(I_\mathrm{D}/I_\mathrm{G})E_\mathrm{laser}^4$ versus L_a plot shown in Figure 13.12b, demonstrating that the ratio $I_\mathrm{D}/I_\mathrm{G}$ is inversely proportional to the fourth power of E_laser. Thus, a general equation is obtained for the determination of the nanographite crystallite

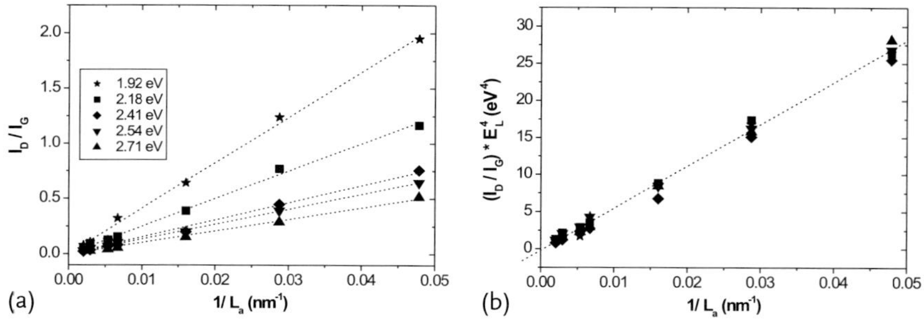

Figure 13.12 (a) The intensity ratio $I_\mathrm{D}/I_\mathrm{G}$ for nanographite samples is plotted versus $1/L_\mathrm{a}$ using five different laser excitation energies. (b) All curves shown in part (a) collapse onto the same curve in the $(I_\mathrm{D}/I_\mathrm{G})E_L^4$ vs. $(1/L_\mathrm{a})$ plot where $E_L \equiv E_\mathrm{laser}$ [404].

size L_{a} using any laser line in the visible range [404]

$$L_{\mathrm{a}}(nm) = \frac{560}{E_{\mathrm{laser}}^4}\left(\frac{I_{\mathrm{D}}}{I_{\mathrm{G}}}\right)^{-1} = (2.4 \times 10^{-10})\lambda_{\mathrm{laser}}^4\left(\frac{I_{\mathrm{D}}}{I_{\mathrm{G}}}\right)^{-1}, \tag{13.11}$$

where the laser excitation is given in terms of both E_{laser} (eV) or wavelength (nm).

13.3.4
Absolute Raman Cross-Section

Measuring the absolute cross-section for the Raman scattering processes is not trivial, since the Raman signal depends strongly on the specific setup (specific optics), on alignment, excitation wavelength. This is the reason why the intensity ratio $I_{\mathrm{D}}/I_{\mathrm{G}}$ has been used systematically for quantifying disorder. Some argue that the intensity ratio $I_{\mathrm{D}}/I_{\mathrm{G}'}$ should be chosen because both the D-band and G′-band involve a very similar phonon (intervalley iTO, $q \approx 2k$). However, the D and G′-bands differ strongly in energy, and different Raman setups would give different responses.

However, Cançado *et al.* [405] have done all the calibration procedures for measuring the absolute Raman cross-section of the D, G, D′ and G′-bands (see Section 5.5). In this work, the dependence of $I_{\mathrm{D}}/I_{\mathrm{G}}$ on L_a was shown to come from I_{D}, while I_{G} was found to be independent of L_a within the measured L_a range (from 20 to 500 nm). For the E_{laser} dependence, the double resonance features were shown to be E_{laser} independent, while the I_{G} shows the E_{laser}^4 dependence expected from scattering theory (see Section 5.5 and [405]). It is not yet known if the E_{laser}^4 dependence will be also observed for zero-dimensional (e. g., ion-induced) defects. Actually, one of the open fields in the Raman spectroscopy of sp^2 carbons is what rules determines intensity of the double resonance features. Sato *et al.* (see Section 13.1 and [395]) and Basko [370] have done some theoretical work on this topic, but the results are still not at the level of explaining experimental observations.

13.4
Defect-Induced Selection Rules: Dependence on Edge Atomic Structure

Besides defect quantification, it is important to discuss how disorder depends on the specific defect. An example of a result that was successful in distinguishing different defects from one another is the study of the edge of a graphite sample, analyzing the orientation of the carbon hexagons with respect to the edge axis, thereby distinguishing the so-called zigzag edge arrangements from the armchair or random atomic edge structures [161]. As discussed here, the armchair (zigzag) edge structure can be identified spectroscopically by the presence (absence) of the D-band. This effect can be understood by applying double resonance theory to a semi-infinite graphite crystal and by considering the one-dimensional character of the defect, as discussed below.

The most common case of disorder-induced features in the Raman spectra of graphite-related materials occurs in samples formed by aggregates of small crystallites. In this case, the crystallite boundaries form defects in real space. Since the crystallites have different sizes and their boundaries are randomly oriented, the defect wave vectors exhibit all possible directions and values. Therefore, the existence of a defect with momentum exactly opposite to the phonon momentum is always possible, giving rise to double resonance processes connecting any pair of points on the circles around the K and K' points. In this case, the intensity of the D-band is isotropic and does not depend on the light polarization direction. However, in the case of edges, the D-band intensity is anisotropic because the double resonance process cannot then occur for any arbitrary pair of points. Since, the edge defect in real space is well localized in the direction perpendicular to the edge, it is completely delocalized in this direction in reciprocal space and, therefore, the wave vector of such a defect assumes all possible values perpendicular to the step edge. Hence, the defect associated with a step edge has a one-dimensional character and it is only able to transfer momentum in the direction perpendicular to the edge.

Figure 13.13a shows the edges with zigzag (top) and armchair (bottom) atomic structure, separated from each other by 150°. The wave vectors of the defects associated with these edges are represented by $\boldsymbol{d}_a$ for the armchair edge and $\boldsymbol{d}_z$ for the zigzag edge. Figure 13.13b shows the first Brillouin zone of 2D graphite oriented according to the lattice in real space, as shown in Figure 13.13a. Note that for intervalley scattering, only the armchair $\boldsymbol{d}_a$ vector is able to connect points belonging to circles centered at two nonequivalent K and K' points. Considering usual laser energies (< 3 eV), the radius of the circles around the K' and K points are not large

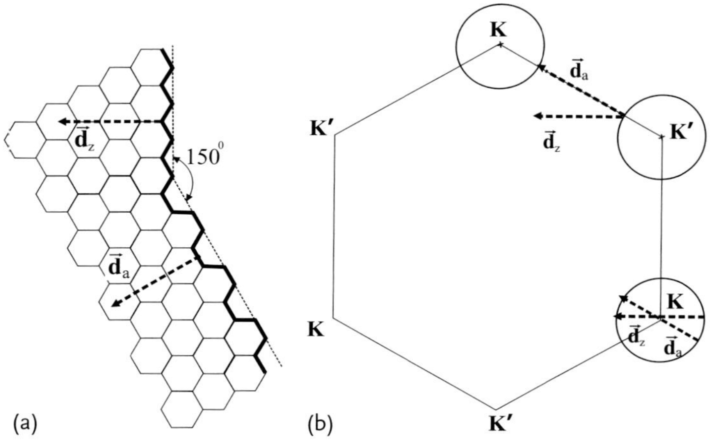

Figure 13.13 (a) Schematic illustration of the atomic structure of edges with the zigzag and armchair orientations. The boundaries can scatter electrons with momentum transfer along $\boldsymbol{d}_z$ for the zigzag edge, and $\boldsymbol{d}_a$ for the armchair edge. (b) First Brillouin zone of 2D graphite, showing defect-induced intervalley and intravalley scattering processes. Since $\boldsymbol{d}_z$ cannot connect the K and K' points, the defect-induced double resonance intervalley process is forbidden at zigzag edges [161].

enough to allow the connection of any $\boldsymbol{k}'$ and $\boldsymbol{k}$ states by a zigzag $\boldsymbol{d}_z$ vector. Therefore, the intervalley double resonance process associated with this defect cannot occur for a perfect zigzag edge. The mechanism depicted in Figure 13.13b thus shows that the D-band scattering process is forbidden in a zigzag edge [161].

On the other hand, the D′-band, around 1620 cm^{-1}, is given by an intravalley process, which connects points belonging to the same circle around the K (or K') point (see K point in Figure 13.13b). Therefore, the intravalley process can be satisfied by both $\boldsymbol{d}_a$ and $\boldsymbol{d}_z$ vectors and, for this reason, the observation of the D′-band should be independent of the edge structure.

Finally, when measuring an armchair edge, the D-band intensity will depend strongly on the light polarization direction with respect to the edge direction. The D-band intensity will be a maximum when the light is polarized along the edge, and zero when the light polarization is crossed with respect to the edge direction. This result is related to the theoretical calculations that predict an anisotropy in the optical absorption (emission) coefficient of 2D graphite (see Eq. (11.30) in Section 11.6.1), given by [220]:

$$W_{\text{abs,ems}} \propto |\boldsymbol{P} \times \boldsymbol{k}|^2 , \tag{13.12}$$

where $\boldsymbol{P}$ is the polarization of the incident (scattered) light for the absorption (emission) process, and $\boldsymbol{k}$ is the wave vector of the electron measured from the K point. Correlating Eq. (13.12) with the D-band intensity dependence on the light polarization direction with respect to the edge is one of the problems at the end of this chapter.

In summary, the one-dimensional edge defect selects the direction of the electron and phonon wave vectors associated with the disorder-induced Raman process, and causes a dependence of the Raman D-band intensity on the atomic structure of the edge (strong for an armchair edge and weak for a zigzag edge). This discussion, therefore, represents an effort to improve our understanding of the influence of the defect structure on the Raman spectra of sp^2 carbon systems, which may be very useful to characterize defects in nanographite-based devices. The first experimental evidence for the selection rules discussed here was developed for a graphite boundary [161]. However, similar results have already been shown for monolayer graphene [407, 408]. Interestingly, up to now variations in the D-band intensity have been observed, but never a complete absence of the D-band together with the presence of the D′-band. This result indicates that, up to now, no perfect zigzag structure has been measured by Raman spectroscopy. It should be emphasized that Raman spectroscopy provides one easy method for distinguishing between armchair and zigzag edges. Of course, high resolution transmission electron microscopy provides another experimental method.

13.5
Specificities of Disorder in the Raman Spectra of Carbon Nanotubes

When moving to carbon nanotubes, the quantum confinement of the electronic structure will constrain the double resonance processes, similar to what has been discussed for the G′-band. A multi-peak structure [409] and an oscillatory dispersive behavior [410] can be observed for the D-band in SWNT bundles. For isolated tubes, unusually sharp features can be seen (a D-band with a FWHM down to 7 cm^{-1} [242]) due to quantum confinement effects. Furthermore, the D-band frequency also depends on the nanotube diameter, with the following result for 514 nm (2.46 eV) laser excitation:

$$\omega_D = 1354.8 - 16.5/d_t \,, \tag{13.13}$$

which is in rough agreement with the D-band being a one-phonon process related to the G′-band (see Eq. (12.11)).

Some efforts have been made to quantify disorder in SWNTs, by studying irradiated samples [411], and SWNTs cut to different lengths [412]. Clear enhancement of the D-band is observed with increasing defect density or with reducing nanotube lengths. However, because of the lack of direct real-space characterization, the results are not as quantitative as the results obtained in graphene [194] and in nanographite [404], as discussed in Section 13.3.

Some aspects are still unclear such as why the D-band in metallic SWNTs is usually more intense than in semiconducting tubes. Although some theoretical efforts based on the double resonance process have addressed this problem [413], the predictions are still incomplete. It is also important to mention that the double resonance process could give rise to a multi-peak structure in the G-band of SWNTs. It was then proposed that the several peaks in the G-band spectra of defective SWNTs could originate from the double resonance process [237, 414], as discussed below.

Figure 13.14 shows the G-band Raman spectra obtained from a SWNT fiber at two different locations, as shown in the inset of Figure 13.14a [415]. The upper spectrum comes from location **1**, at the center of the fiber. Figure 13.14b shows the G-band spectrum from location **2**, which is at the edge of the same fiber, where misalignment and defects (structural and impurities) are expected. The G-band intensity is much lower in location **2**, about 35 times less intense than in location **1**. Many peaks are observed in the spectrum at location **2**, clearly different from the spectrum at location **1**. Eight Lorentzian peaks were used for fitting the spectra at location **2**, which can be related to first-order allowed Raman peaks and to several different double resonance defect-induced peaks. For example, the inset to the lower panel of Figure 13.14 shows the E_{laser} dependence of the two peaks indicated by arrows. The solid curves in this inset are predictions for the E_{laser} dependence of the G-band double resonance features [160, 237]. The dispersive peak seems to agree well with a double resonance mechanism, while a nondispersive behavior fits the lower frequency peak observed in location **2**, which can be assigned as the first-order Raman-allowed TO G-band. It is not clear at this time how these peaks are related to specific edge defects and, again, more effort is needed for a quantitative

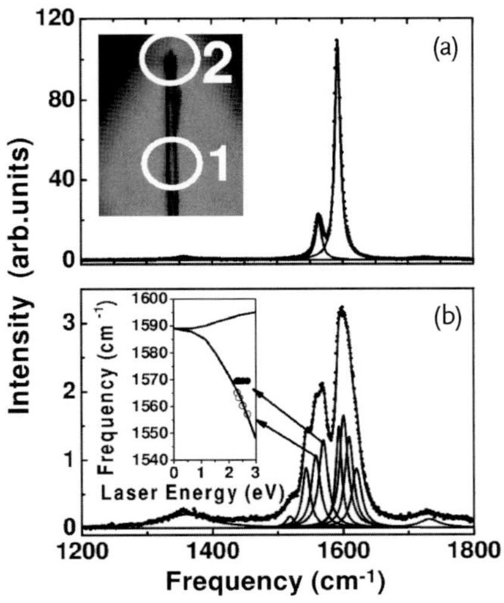

Figure 13.14 G-band resonance Raman spectrum from a fiber of aligned SWNT bundles ($E_{\text{laser}} = 2.71\,\text{eV}$). (a) The inset shows an optical image of the sample. The spectrum was taken from location **1** (see inset in (a)) with the light polarized along the fiber direction, and the spectrum in (b) was acquired from location **2**. The inset to (b) shows the E_{laser} dependence of the two peaks indicated by arrows. The solid curves in this inset are predictions for the E_{laser} dependence of the G-band double resonance features [415].

analysis of the disorder-induced features in carbon nanotubes, to make the RRS technique a more powerful tool for the characterization of disorder in sp^2 carbon materials.

13.6 Local Effects Revealed by Near-Field Measurements

Besides breaking momentum conservation, the presence of disorder is expected to change the local electron and phonon structure. The G′-band (see Chapter 12) is not a disorder-induced feature but it can nevertheless be used to probe changes in the electronic and vibrational structure related to disorder. The 2D vs. 3D stacking order of graphene layers is one example where the G′-band provides important information (see Section 12.2.4). Highly crystalline 3D graphite shows two G′ peaks (see Figure 12.8e). When the interlayer stacking order is lost, a one-peak feature starts to develop, identified with 2D graphite, and the peak is centered near the middle of the two peaks in the G′ lineshape from ordered graphite (see Figure 12.8f) [394]. More interesting, localized emission of a red-shifted G′-band was observed and is related to the local distortion of the nanotube lattice by a negatively charged defect. The defect position was initially located by local variations in the D-band intensity as described below [191].

Figure 13.15 shows near-field Raman and near-field photoluminescence spectra and their related spatial maps for an individual SWNT. The near-field technique can generate spectral information with spatial resolution Δx below the diffraction limit ($\Delta x \sim \lambda_{\text{laser}}/2$) [416]. In particular, Figure 13.15a,b show the photoluminescence and Raman spectra, respectively, with $\Delta x \sim 30\,\text{nm}$. The near-field microscopy measurements from the same SWNT are shown in Figure 13.15c–f. Figure 13.15c represents the near-field photoluminescence image of this SWNT, where the image contrast is provided by spectral integration over the photoluminescence peak centered at $\lambda_{\text{em}} = 900\,\text{nm}$ (see Figure 13.15a). The most striking feature in this image is the high degree of spatial localization of the photoluminescence emission along the SWNT. This is evident by inspection of the extended topography image of the nanotube shown in Figure 13.15f, and also of the near-field Raman image of

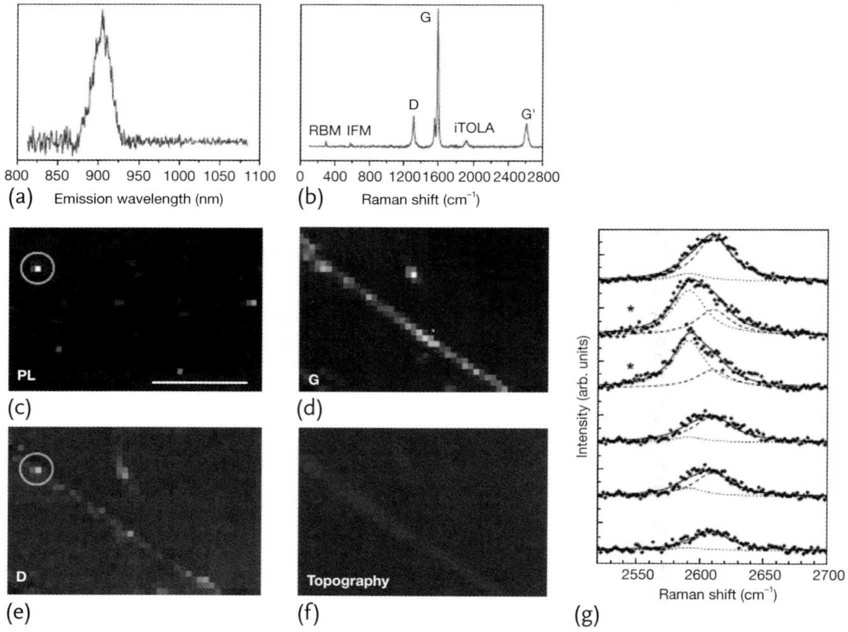

Figure 13.15 Localized excitonic emission in a semiconducting SWNT. (a) Photoluminescence emission at $\lambda_{\text{em}} = 900\,\text{nm}$. (b) Raman spectrum recorded from the same SWNT. The spectral position of the RBM, $\omega_{\text{RBM}} = 302\,\text{cm}^{-1}$, together with the $\lambda_{\text{em}} = 900\,\text{nm}$ information, leads to the (9,1) assignment for this tube. (c) Near-field photoluminescence image of the SWNT revealing localized excitonic emission. (d–e) Near-field Raman imaging of the same SWNT, where the image contrast is provided by spectral integration over the G and D-bands, respectively. (f) Corresponding topography image. The circles indicate localized photoluminescence (c) and defect-induced (D-band) Raman scattering (e). The scale bar in (c) denotes 250 nm. (g) Evolution of the G′-band spectra near the defective segment of the (9,1) SWNT. The spectra were taken in steps of 25 nm along the nanotube, showing the defect-induced G′ peak (dotted Lorentzian). The asterisks denote the spatial locations where localized photoluminescence and defect-induced D-band scattering were measured (see circles in (c) and (e), respectively) [191].

the G-band, with a peak intensity near 1590 cm^{-1} shown in Figure 13.15d. While from Figure 13.15d we observe that the G-band Raman scattering is present along the entire length of nanotube, from Figure 13.15e we observe an increased defect-induced (D-band, 1300 cm^{-1}) Raman scattering intensity localized in the *same* region where exciton emission was detected. Defects are known to act as trapping states for electron–hole recombination (i. e., exciton emission), thereby providing insights into the correlations observed between Figure 13.15c,e.

Interestingly, when measuring the Raman spectra across the defective spot, sudden changes in many Raman features are observed. Maciel *et al.* [191] have shown that substitutional doping in SWNTs causes changes in the G′-band spectra due to charge-induced renormalization of the electronic and vibrational energies. Figure 13.15g shows the G′-band measured on the same SWNT, moving along the position where the local D-band and photoluminescence emission is observed (circle in Figure 13.15e). The two spectra marked by "*" in Figure 13.15g were obtained at this defect location, and a new peak is observed at the G′-band. The frequency and intensity of this new peak depend on the type and level of doping, respectively [417–419]. This makes the G′-band a probe for studying and quantifying doping, which is more accurate than the D-band, since the D-band can also be related to amorphous carbon and any other defective sp^2 structure.

13.7 Summary

This chapter discusses how Raman spectroscopy can be used to probe defects in sp^2 nanocarbons. The break-down of momentum conservation together with the double resonance mechanisms make E_{laser}-dependent resonance Raman spectroscopy a powerful probe for electronic and phonon dispersion relations. Like the two-phonon double resonance peaks, the disorder-induced peaks represent a break-through in optical spectroscopy, because probing the phonon dispersion inside the Brillouin zone is usually done solely with neutron, electron or X-ray scattering, due to momentum conservation considerations. Besides, the possibility to probe particle size, layer stacking, defect concentration, edge structure and doping, and individual defects shows new power for optical spectroscopy techniques, that goes beyond the simple analysis of crystalline lattices. Here nanoscience is making possible the study of materials science from a completely new perspective.

The phenomenological model presented in Section 13.3.2 describes well the evolution of the D-band intensity with increasing defect density. However, the physics ruling the intensity of the double resonance features, in general, is an open question. It is clear that when moving into the use of near-field Raman spectroscopy, a new world opens for exploration. Although this research area still presents several technical difficulties, near-field techniques seem to provide major future opportunities for Raman spectroscopy studies of nanocarbons.

Problems

[13-1] In Eq. (13.3), if V is periodic in the crystal, show that the matrix elements vanish except for $k' = k$ in the first Brillouin zone. Here use the fact that $\langle \phi(R_{s'})|V|\phi(R_s)\rangle$ depends only on $R_{s'} - R_s$. Show that the matrix elements have a large value for $k' = k + G$, where G is a reciprocal lattice vector.

[13-2] Obtain the next order correction to the wavefunction of Eq. (13.5) by substituting $\Phi(k')$ iteratively for $\Psi^c(k')$ in the last term of Eq. (13.5).

[13-3] Using the previous results, obtain Eq. (13.6).

[13-4] Higher-order correction terms can be obtained similarly by substituting $\Phi(k')$ iteratively for $\Psi^c(k')$ in the last term of Eq. (13.5). Using this approach, get a general expression for $\Phi(k')$ and $M_{k'k}$.

[13-5] After consulting some textbook on scattering theory, explain that the T-matrix is expressed by $T = V + VG_0T$ or $T = V + VGV$, where $V \equiv H - H_0$ is an impurity potential operator and where H and H_0, respectively, denote the perturbed and nonperturbed Hamiltonians, while $G = (E - H)^{-1}$ and $G_0 = (E - H_0)^{-1}$ denote perturbed and nonperturbed Green's function. Using proper basis functions shows that we can make a matrix for each operator and get the T-matrix.

[13-6] In scattering theory, the S-matrix is also frequently used. Explain the difference between the T-matrix and the S-matrix. What are the advantages and disadvantages for using the T and S matrices?

[13-7] Considering the electron and phonon dispersions near the K point in graphene, derive a quantitative description which explains why $\omega_{G'} \neq 2\omega_D$ for one-phonon vs. two-phonon processes in the double resonance of the iTO phonon $\omega(q)$ branch near the K point, where q denotes the phonon wave vector.

[13-8] Sketch all the possible defect-induced double resonance one-phonon processes in monolayer graphene for Stokes and anti-Stokes, intravalley and intervalley processes.

[13-9] Consider the electronic dispersion of graphene in the linear approximation, that is, $E^{\pm}(k) = \pm\hbar v_F|k|$, where v_F is the Fermi velocity of the electrons given by $v_F = \sqrt{3}(\gamma_0 a/2\hbar)$, and $a = \sqrt{3}a_{C\text{–}C}$ is the lattice constant of graphene and $a_{C\text{–}C} = 1.42$ Å is the nearest neighbor carbon–carbon distance. Calculate the pair of (q, k) wave vectors which would fulfill the one-phonon defect-induced double resonance conditions for $E_{laser} = 1.98$ eV and 2.41 eV. Considering the phonon dispersion in Figure 13.5b, give the approximate frequency for all the defect-induced Raman peaks that should be observed for these two laser lines. Check your results against the values in Figure 13.5a.

[13-10] Calculate the number of Ar^+ ion-induced defects per C atom expected for a 10^{11} Ar^+/cm^2 and a 10^{15} Ar^+/cm^2 ion bombardment dose on a monolayer of graphene.

[13-11] In our phenomenological model for the D-band intensity dependence on Ar^+ ion impacts, consider S_S, S_A and S_T as the structurally damaged, activated and total areas of the graphene sheet, respectively, and N as the number of Ar^+ ion collisions. Derive Eq. (13.8) using rate equations (for dS_S/dN and dS_A/dN) and considering the initial conditions $f_S = 0$ and $f_A = 0$ for $\sigma = 0$.

[13-12] Derive Eq. (13.9). Suppose that the width δ of the boundary is responsible for the E_{laser}^4 dependence observed experimentally for I_D/I_G. How should δ vary with E_{laser} to obtain this relation? In this case, what would be the expected E_{laser} dependence of I_D/I_G for point defects inducing the D-band?

[13-13] Show that $(560/E_{\text{laser}}^4) = (2.4 \times 10^{-10}\lambda^4)$, as shown in Eq. (13.11).

[13-14] Explain why Eq. (13.12) is responsible for the polarization dependence observed for the D-band in armchair edges, and show that a $I_D \propto \cos\theta^4$ dependence is expected for the D-band intensity when both the incident and scattered light are analyzed along an angle θ with the edge direction.

[13-15] Calculate the intervalley matrix element $M_{k'k}$ for an armchair edged nanoribbon of width L_a. Here we assume that an electron is reflected at the armchair edge. Show that this assumption predicts a width dependence for the D/G intensity ratio given by $I_D/I_G \propto L_a^{-2}$. Such a relation has not been observed experimentally, showing that more work (both experimental and theoretical) is still needed.

[13-16] The G′-band was observed to shift to higher frequencies under acceptor (p-type) doping, and to lower frequencies under donor (n-type) doping. Explain qualitatively which doping-induced changes should happen to the electronic and/or phonon dispersion relations to generate such a shift.

14
Summary of Raman Spectroscopy on sp^2 Nanocarbons

This chapter provides a brief summary of the behavior of the various sp^2 nanocarbon Raman features, focusing on those properties which can be used for sample characterization. The physics behind each feature in the Raman spectrum has been discussed in detail in this book, and the main results are summarized in this chapter, ending with a short perspective for the future of Raman spectroscopy in sp^2 nanocarbons.

14.1
Mode Assignments, Electron, and Phonon Dispersions

The Raman spectra from sp^2 nanocarbons are very rich, composed of first-order and higher-order Raman modes, as well as disorder-induced features. The Raman features can all be related to phonons in graphene, at the Γ point, within the interior of the Brillouin zone and near the Brillouin zone boundary. The modes associated with interior points are, activated either as higher-order (combination modes and overtones) or as defects-induced processes. The Brillouin zone center modes can be dispersive and, therefore, they can be used for measuring the electron and phonon dispersion of sp^2 nanocarbons, using the double resonance (DR) model (see Figure 12.11 and Figure 13.5). The advantages, when comparing phonon dispersion relations obtained by Raman spectroscopy using the DR model with those obtained by neutron or electron inelastic scattering, is the simplicity and the high precision of the Raman measurements regarding phonon properties. The drawback is that the DR Raman scattering mechanism selects only the $|\boldsymbol{k}|$ and $|\boldsymbol{q}|$ wave vector modulus for electrons and phonons, respectively. Therefore, the Raman features are composed of averages of phonons around the high symmetry Γ or K points. Furthermore, the response depends both on the electron and phonon dispersions, and theory is needed to decouple these two dispersions. Table 14.1 provides a summary for the assignment of many of these features in the Raman spectra. Most notations have been introduced in the book, and the footnotes explain notations which are especially created for the table. The results give average values that usually exhibit small deviations depending on the sp^2 structure (single vs. multi-layer

Raman Spectroscopy in Graphene Related Systems. Ado Jorio, Riichiro Saito, Gene Dresselhaus, and Mildred S. Dresselhaus

ISBN: 978-3-527-40811-5

graphene, graphene nanoribbons, nanotubes with different diameters and chiral angles, etc.), and on the ambient conditions (temperature, pressure, environment, etc.). Summaries of the characteristics for the most intense Raman features are given in the following sections: the G-band in Section 14.2, the RBM in Section 14.3, the G′-band in Section 14.4 and the D-band in Section 14.5.

14.2
The G-band

The science behind the G-band is discussed in Chapters 7 and 8. Below we list a summary of the general properties of the G-band:

G1 The G-band is the Raman signature for sp^2 carbons, and is observed as a peak (or a multi-peak feature) at around 1585 cm^{-1} for all sp^2 carbons (see Figure 1.5).

G2 Hydrostatic pressure on graphene: shifts ω_G (see Section 7.1.1).

G3 Uniaxial stretching of graphene: splits the G peak into G^- and G^+, which are, respectively, related to atomic motion along and perpendicular to the stretching direction. Increasing the stretching, red-shifts both ω_G^+ and ω_G^- (see Figure 7.1).

G4 Doping graphene: blue-shifts ω_G (see Section 8.3.1 for 1-LG, and Section 8.3.2 for 2-LG) for weak doping (changes in the Fermi level near the K point). Higher doping levels can cause blue(red)-shift for $p(n)$ doping.

G5 Temperature: generally, increasing T, red-shifts ω_G. Different effects take place, such as changes in the electron–phonon renormalization, phonon–phonon coupling and ω_G shifts due to thermal expansion-induced volume changes (see Section 8.2.1).

G6 Polarization: when choosing light polarized in the graphene plane (propagation perpendicular to the sheet), then rotating the polarization is irrelevant for unstrained or homogeneously strained graphene. If graphene is non-homogeneously strained, then the relative intensity between the G^+ and G^- peaks I_{G^+}/I_{G^-} will give the strain direction (see [198, 230, 231]).

G7 Linewidth: usually in the range of 10–15 cm^{-1}, although it changes with strain, temperature and doping (see Section 8.3.1).

G8 Bending the graphene sheet: splits the G-band into ω_G^+ and ω_G^-, which have their atomic vibrations preferentially along and perpendicular to the folding axis, respectively (see Section 7.2).

G9 Rolling up the graphene sheet into a seamless tube (SWNT): (1) bending splits the G-band into ω_G^+ and ω_G^- (see Figure 4.12), which are preferentially along (LO) and perpendicular (TO) to the tube (folding) axis, respectively, for semiconducting SWNTs (see Section 7.2.1). For metallic tubes, electron–phonon coupling softens the LO modes, so that ω_G^+ and ω_G^- are actually associated with TO and LO modes, respectively. In the case of achiral tubes, ω_G^+/ω_G^- is strictly proportional to I_{LO}/I_{TO}. (2) Quantum confinement gener-

Table 14.1 Assignments and frequency behavior for the Raman modes from sp^2 carbon materials.

Name[a]	ω [cm^{-1}][b]	Res.[c]	$\partial\omega/\partial E$[d]	Notes[e]
iTA	288	DRd1	129	intraV ($q \sim 2k$ near Γ)
LA	453	DRd1	216	intraV ($q \sim 2k$ near Γ)
RBM[f]	$227/d_t$	SR	0	SWNT vibration of radius
IFM$^-$(oTO-LA)	750	DR2	–220	intraV+intraV ($q \sim 2k$ near Γ)
oTO	860	DRd1	0	intraV ($q \sim 0$ near Γ), IR active
IFM$^+$(oTO+LA)	960	DR2	180	intraV+intraV ($q \sim 2k$ near Γ)
D (iTO)	1350	DRd1	53	interV ($q \sim 2k$ near K)
G (iTO,LO)[g]	1585	SR	0	$q = 0$, i.e., at Γ
D′ (LO)	1620	DRd1	10	intraV ($q \sim 2k$ near Γ)
M$^-$ (2oTO)	1732	DR2	–26	intraV+intraV ($q \sim 2k$ near Γ)
M$^+$ (2oTO)	1755	DR2	0	intraV+intraV ($q \sim 0$ near Γ)
iTOLA (iTO+LA)	1950	DR2	230	intraV+intraV ($q \sim 2k$ near Γ)
G* (LA+iTO)	2450	DR2	–10	interV+interV ($q \sim 2k$ near K)[h]
G′ (2iTO)[i]	2700	DR2	100	interV+interV ($q \sim 2k$ near K)
G + D	2935	DRd2	50	intraV+interV[j]
D′ + D	2970	DRd2	60	intraV+interV[j]
2G	3170	SR	0	overtone of G mode
G + D′	3205	DRd2	10	intraV+intraV
G″(2LO)[i]	3240	DR2	20	intraV+intraV ($q \sim 2k$ near Γ)

a Usually the respective graphene phonon branch labels the Raman peaks. When other names are given in the literature, the respective phonon branch appears between parenthesis.

b The frequencies quoted in the table are observed at $E_{\text{laser}} = 2.41$ eV.

c The notation for resonances is: SR: single resonance, Raman allowed; DR2: double resonance, 2-phonon Raman allowed; DRd1: double resonance Raman activated by disorder, 1-phonon; DRd2: double resonance Raman activated by disorder, 2-phonons.

d The change of phonon frequency in cm^{-1} obtained by changing the laser excitation energy by 1 eV defines the phonon dispersion.

e Terms intraV: intravalley scattering; interV: intervalley scattering.

f The radial breathing mode (RBM) only occurs for carbon nanotubes and is unique for carbon nanotubes.

g The iTO and LO phonons are degenerate at Γ for graphene. For nanoribbons and SWNTs, the G-band splits into several peaks due to symmetry, and differs for metallic and semiconducting nanotubes. The G-band frequency depends strongly on doping and strain.

h There is another assignment for G* of 2iTO ($q \sim 0$ near K) with $\partial\omega/\partial E \sim 0$.

i Notation G′ and G″ are frequently named 2D and 2D′, respectively. Strictly speaking, the overtone assignment is not fully correct, since one of the elastic scattering event which appears in the one-phonon emission spectra in the D(D′) mode does not exist in the two-phonon emission Raman spectra in the G′(G″) mode (see Section 12.2.1). Here we adopt the notation where the letter D only appears for defect-induced features.

j This combination mode consists of intra V + inter V scattering and thus the elastic scattering process also exists for some combination modes.

ates up to six Raman-allowed G-band peaks, three of each exhibiting LO- or TO-like vibrations, two totally symmetric A_1 modes, two E_1 modes and two E_2 symmetry modes (see Section 7.2.1). Due to the depolarization effect and special resonance conditions, the A_1 modes usually dominate the G-band spectra.

G10 Decreasing the SWNT diameter: increases the effect of bending and shifts mostly ω_G^-. The ω_G^- shift can be used to measure the SWNT diameter (see Section 7.2.2 and Eq. (7.15)).

G11 Changing the chiral angle: changes the intensity ratio between LO-like and TO-like modes [274].

G12 Hydrostatic pressure on SWNT bundles: shifts ω_G (see [236, 243]).

G13 Strain on isolated SWNTs: hydrostatic and uniaxial deformation, torsion, bending, etc., change G^+ and G^-, depending on (n, m) (see Section 7.4).

G14 Doping SWNTs: changes ω_G, mainly for metallic SWNTs. There is a rich doping dependence on (n, m), but a strong effect is felt mostly on the broad and down-shifted G^- peak in metallic SWNTs, with doping usually causing an upshift and sharpening of the G^- feature (see Section 8.4).

G15 Temperature change: similar effect in SWNTs and graphene. Increasing T softens and broadens the G-band peaks in SWNTs (see Section 8.2.1 and [242]).

G16 Polarization analysis in SWNTs: can be used to assign the G-band mode symmetries (see Sections 7.3.1 and 7.3.2).

14.3 The Radial Breathing Mode (RBM)

The science behind the RBMs is discussed in Chapters 9 and [282], 10 and 11. Below we list the general properties of the RBMs:

RBM1 RBM: the Raman signature for the presence of nanotubes, and the RBM is observed as a peak (or a multi-peak feature) in the 50–760 cm^{-1} range[1)] (see Figure 1.5).

RBM2 The ω_{RBM} dependence on diameter (d_t):

$$\omega_{RBM} = \frac{227}{d_t}\sqrt{1 + C_e * d_t^2}\,, \tag{14.1}$$

where C_e (nm^{-2}) probes the effect of the environment on ω_{RBM}. Table 9.1 gives the C_e values fitting the RBM results for several samples in the literature (see Section 9.1).

RBM3 The ω_{RBM} dependence on chiral angle θ: predicted to depend on SWNT diameter and E_{laser}, but the dependence of ω_{RBM} on θ is rather weak,

1) The highest RBM frequency (760 cm^{-1}) is that for the (2,2) nanotube [420].

even for SWNTs with $d_t < 1$ nm, where the dependence of ω_{RBM} reaches a few wave numbers (see Section 9.1.5).

RBM4 E_{laser} dependence: for a given SWNT, the RBM peak intensity $I(E_{laser})$ is a function of E_{laser} and can be evaluated by Eq. (9.9). The RBM is intense when the incident light (E_{laser}) or the scattered light ($E_{laser} \pm \hbar\omega_{RBM}$) is in resonance with the SWNT optical transition energies E_{ii}.

RBM5 The optical transition energies E_{ii}: The E_{ii} can be obtained using an empirical formula (see Eq. (9.10)), in which the various fitting parameters are discussed in Section 9.3.2. The theoretical description depends on an accurate analysis of the nanotube structure, exciton effects and dielectric screening, as discussed in detail in Chapter 10.

RBM6 Intensity dependence on (n, m): the electron–photon and electron–phonon matrix elements, as well as the resonance broadening factor γ_r strongly depend on (n, m). This dependence has been obtained experimentally (see Section 11.2) and described theoretically (see Chapter 11).

RBM7 Polarization: the RBM is a totally symmetric mode. The polarization dependence is dominated by the antenna effect, where a strong Raman signal is observed when both the incident and scattered light are chosen along the tube axis (see Section 9.2.3).

RBM8 Tube–tube interaction: the same inner (n, m) tube within a DWNT can exhibit different ω_{RBM} values if surrounded by different outer (n', m') tubes (see Section 9.1.3).

RBM9 Linewidth: usually in the range of $3\,\text{cm}^{-1}$, although it can reach much larger values (by one order of magnitude) due to environmental effects (see Section 9.1.4), or smaller (also by as much as one order of magnitude) when measured for the inner tube of a DWNT and at low temperature (see [292–294]).

RBM10 Due to the relatively low RBM frequency, changes in ω_{RBM} with temperature, doping, strain and other such effects are less pronounced in the RBM than in the G-band. However, the RBM becomes important when looking for the effects on one single (n, m) specie among many SWNTs, since the RBM feature is unique for each (n, m) (ω_{RBM} depends strongly on tube diameter), while the G-band appears within the same frequency range for most SWNTs (very weak d_t dependence).

RBM11 Changing E_{ii}: as discussed above, changes in temperature, pressure or the dielectric constant of the environment does not change ω_{RBM} significantly. However, these factors do change E_{ii}, and changing the resonance condition changes the RBM intensity. Therefore, the RBM can be used to probe resonance effects sensitively (see Chapter 9), and for understanding the importance of excitonic effects for a theoretical description of the observed Raman spectra (see Chapter 10). Increasing the temperature decreases E_{ii}, and the temperature-dependent change in E_{ii} also depends on (n, m). Increasing the pressure changes E_{ii}, and the pressure-dependent changes in E_{ii} also depend on (n, m). Here a change in E_{ii} can be positive or negative, depending on i and on the mod$(2n + m, 3)$ type.

Increasing the dielectric constant of a SWNT wrapping agent decreases E_{ii} (see Section 10.5.2).

RBM12 Stokes vs. anti-Stokes RBM intensities: the S/aS intensity ratio for the RBM features is strongly sensitive to the energy displacement of E_{laser} with respect to E_{ii} (see Section 9.2.2).

14.4 G′-band

The science of the G′-band is discussed in Chapter 12. Below we list the general properties of the G′-band:

G′1 G′-band: also an sp^2 Raman signature, observed for all sp^2 carbons as a peak (or a multi-peak feature) in the range 2500–2800 cm^{-1} (see Figure 1.5).

G′2 The G′ frequency: ω'_{G} appears at ~2700 cm^{-1} for $E_{\text{laser}} = 2.41$ eV, but its frequency changes by changing E_{laser} (see Section 12.2.2). Its dispersion is $(\partial\omega_{\text{G}'}/\partial E_{\text{laser}}) \simeq 90\,\text{cm}^{-1}/\text{eV}$ for monolayer graphene, and this dispersion changes slightly by changing the sp^2 nanocarbon structure (see Section 12.2.2). The sensitivity of $\omega_{\text{G}'}$ to the detailed sp^2 structure makes this band a powerful tool for quantifying the number of graphene layers and the stacking order in few layer graphenes and graphite, and for characterizing SWNTs by the diameter and chiral angle dependence of $\omega_{\text{G}'}$ and of the G′-band intensity.

G′3 The number of graphene layers: 1-LG exhibits a single Lorentzian peak in the G′-band, and the intensity of the G′-band is larger (2–4 times) than that of the G-band in 1-LG. In contrast, 2-LG with AB Bernal stacking exhibits four Lorentzian peaks in the G′-band, and the intensity of the G′-band with respect to the G-band is strongly reduced (same magnitude or smaller). For 3-LG with AB Bernal stacking, 15 scattering processes are possible for the G′-band, but the 15 peaks occur close in frequency and cannot all be distinguished from each other. Usually the G′-band from 3-LG is fitted with six peaks (see Section 12.2.3). Highly oriented pyrolytic graphite (HOPG) exhibits two peaks. Turbostratic graphite exhibits only a single G′ peak, and care should be taken when assigning the number of layers based on the G′ feature. The single G′ peak in turbostratic graphite is slightly blue-shifted ($\sim$ 8 cm^{-1}) from the G′ peak in 1-LG (see Section 12.2.4).

G′4 Stacking order: while HOPG (considered a three-dimensional structure) exhibits a two-peak G′ feature, turbostratic graphite (no AB Bernal stacking order, and considered a two-dimensional structure) exhibits a single Lorentzian line. Therefore, the single vs. double peak G′ structure can be used to assign the amount of stacking order present in actual graphite samples (see Section 12.2.4).

G′5 Probing electron and phonon dispersion: Eq. (12.8) gives the electron and phonon wave vectors selected by the double resonance process. By changing

E_{laser}, it is possible to probe different electrons and phonons in the interior of the Brillouin zone. The G′-band probes the iTO phonons near the K point, where the strongest electron–phonon coupling occurs (see Section 12.2).

G′6 Doping: the G′ feature can be used to assign p and n type doping in graphene and SWNTs. A blue-shift (red-shift) is observed for p (n) doping. The magnitude of the shift depends also on the specific type of doping atom, while the relative intensity between doped-shifted and undoped-pristine G′-band peaks can be used to obtain the dopant concentration (see Section 13.6).

G′7 SWNTs and (n, m) dependence: carbon nanotubes show a very special G′-band feature, where the number of peaks and their frequencies depend on (n, m) due to both curvature-induced strain and the quantum confinement of their electronic and vibrational structures. The resonance condition is restricted to $E_{\text{laser}} \approx E_{ii}$ and $E_{\text{laser}} \approx E_{ii} + E_{\text{G}'}$, and this fact gives rise to a $\omega_{\text{G}'}$ dependence on the SWNT diameter and chiral angle (see Section 12.4.2).

14.5 D-band

The science behind the D-band is discussed in Chapter 13. Below we list the properties of the D-band:

D1 D-band: The dominant sp^2 Raman signature of disorder. The D-band is observed as a peak in the range 1250–1400 cm^{-1} (see Figure 1.5).

D2 The D-band frequency: ω_{D} appears at ~1350 cm^{-1} for $E_{\text{laser}} = 2.41$ eV, but its frequency changes by changing E_{laser} (see Section 13.2). Its dispersion is $(\partial\omega_{\text{G}'}/\partial E_{\text{laser}}) \simeq 50\,\text{cm}^{-1}/\text{eV}$ for monolayer graphene, and it changes slightly by changing the sp^2 nanocarbon structure (see Chapter 13). For SWNTs, the frequency ω_{D} depends on the nanotube diameter, according to Eq. (13.13).

D3 The D-band intensity: can be used to quantify disorder. The effect of nanocrystallite size and Ar^+ bombardment dose has been used to characterize the disorder in both SWNTs and graphene and a quantitative phenomenological model has been developed for explaining the D-band intensity evolution with the amount of disorder (see Section 13.3).

D4 The D-band linewidth: Being disorder related, the D-band linewidth can change from 7 cm^{-1} (observed for isolated SWNTs [242]) to a hundred wavenumbers (for very defective carbon materials see Figure 13.6).

D5 Type of graphene edge: The D-band scattering is forbidden at edges with zigzag structure. This property can be used to analyze the edge structure and to distinguish zigzag from armchair edges (see Section 13.4).

D6 The $I_{\text{D}}/I_{\text{G}}$ ratio: Because absolute intensity measurement is a difficult task in Raman spectroscopy, the normalized intensity $I_{\text{D}}/I_{\text{G}}$ ratio is largely used to measure the amount of disorder. This ratio depends not only on the amount of disorder, but also on the excitation laser energy, since $I_{\text{G}} \propto E_{\text{laser}}^4$, while I_{D} is E_{laser}-independent for graphite/graphene nanocrystallites (when measured in the 1.9–2.7 eV range) (see Sections 13.3.3 and 13.3.4). The E_{laser} dependence for I_D/I_G in ion bombarded graphene has not been established.

D7 Coherence length for Raman scattering: Since the D-band is activated by defects, it can only be observed near the defect within a coherence length ℓ. In [194] the D-band was used to obtain $\ell = 2$ nm for ion bombarded graphene measured with $E_{\text{laser}} = 2.41$ eV (see Section 13.3.2).

14.6 Perspectives

Over nearly one century Raman spectroscopy has been used to study the science of sp^2 materials, revealing more and more fundamental aspects of their electronic and vibrational properties. This non-stop development is due to improvements in experimental techniques, theoretical calculations, and to advances in the nanosciences generally. For the future perspectives, from the experimental side, near-field optics can now unravel Raman spectra with spatial resolution below the diffraction limit [191], a former limitation for Raman spectroscopy. Time-dependent Raman and coherent-phonon spectroscopy provide new frontiers for vibrational spectroscopy [43]. From the theoretical side, the simplicity of sp^2 carbon materials (only one atom species on a hexagonal structure) is making possible the development of fancy tight binding and first-principles calculations, reaching unprecedented levels of accuracy for the description of electronic and vibrational levels. New theoretical insights, such as electron–electron correlation, excitonic effects and electron–phonon interactions were successfully applied to sp^2 nanocarbons, but now can perhaps also be applied to other systems. In the field of carbon, the transition from molecular to crystalline behavior, and the transition from low levels to high levels of disorder might be addressable with detailed experimental inputs. Measuring the Raman signal from one single sheet of atoms, or one single rolled up tube, together with the ability of applying controlled perturbations to these nanomaterials (strain, doping, etc.), will keep generating unprecedented levels of detail in describing the physics of sp^2 carbons. Such knowledge has raised more and more fundamental questions. The experience gained from the studies discussed in this book indicates that Raman spectroscopy provides a powerful tool for finding and addressing the new physics that is needed.

References

1 Bassani, F. and Pastori-Parravicini, G. (1975) *Electronic States and Optical Transitions in Solids*, Pergamon Press, Oxford.

2 Kelly, B.T. (1981) in *Physics of Graphite*, Applied Science, London.

3 Zhao, Y., Ando, Y., Liu, Y., Jisino, M., and Suzuki, T. (2003) *Phys. Rev. Lett.*, **90**, 187401.

4 Fantini, C., Cruz, E., Jorio, A., Terrones, M., Terrones, H., Van Lier, G., Charlier, J.-C., Dresselhaus, M.S., Saito, R., Kim, Y.A., Hayashi, T., Muramatsu, H., Endo, M., and Pimenta, M.A. (2006) *Phys. Rev. B*, **73**, 193408-1–4.

5 Charlier, J.-C., Eklund, P.C., Zhu, J., and Ferrari, A.C. (2008) Electron and phonon properties of graphene: their relationship with carbon nanotubes, in *Springer Series on Topics in Appl. Phys.*, vol. 111 (eds A. Jorio, M.S. Dresselhaus, and G. Dresselhaus), Springer-Verlag, Berlin, pp. 673–708.

6 Dresselhaus, M.S., Dresselhaus, G., and Eklund, P.C. (1996) *Science of Fullerenes and Carbon Nanotubes*, Academic Press, New York, NY, San Diego, CA.

7 Castro Neto, A.H., Guinea, F., Peres, N.M.R., Novoselov, K.S., and Geim, A.K. (2009) The Electronic Properties of Graphene. *Rev. Mod. Phys.*, **81**, 109–162.

8 Castro Neto, A.H. and Guinea, F. (2007) *Phys. Rev. B*, **75**, 045404.

9 Heremans, J., Olk, C.H., Eesley, G.L., Steinbeck, J., and Dresselhaus, G. (1988) *Phys. Rev. Lett.*, **60**, 452.

10 Moore, A.W. (1973) in *Chemistry and Physics of Carbon* (eds P.L. Walker, Jr. and P.A. Thrower), Marcel Dekker, Inc., New York, **11**, 69.

11 Moore, A.W. (1969) *Nature*, **221**, 1133.

12 Moore, A.W. (1981) in *Chemistry and Physics of Carbon*, vol. 17 (eds P.L. Walker, Jr. and P.A. Thrower), Marcel Dekker, Inc., New York, p. 233.

13 Moore, A.W., Ubbelohde, A.R., and Young, D.A. (1962) *Brit. J. Appl. Phys.*, **13**, 393.

14 Ubbelohde, A.R. (1969) *Carbon*, **7**, 523.

15 Spain, I.L., Ubbelohde, A.R., and Young, D.A. (1967) *Phil. Trans. Roy. Soc. (London) A*, **262**, 345.

16 Dresselhaus, M.S., Dresselhaus, G., Sugihara, K., Spain, I.L., and Goldberg, H.A. (1988) Graphite Fibers and Filaments, in *Springer Series in Materials Science*, vol. 5, Springer-Verlag, Berlin.

17 Bacon, R. (1960) *J. Appl. Phys.*, **31**, 283–290.

18 Endo, M., Koyama, T., and Hishiyama, Y. (1976) *Japan. J. Appl. Phys.*, **15**, 2073–2076.

19 Endo, M., Strano, M.S., and Ajayan, P.M. (2008) Potential applications of carbon nanotubes, in *Springer Series on Topics in Appl. Phys.*, vol. 111 (eds A. Jorio, M.S. Dresselhaus, and G. Dresselhaus), Springer-Verlag, Berlin, pp. 13–61.

20 Jorio, A., Dresselhaus, M.S., and Dresselhaus, G. (2008) Carbon nanotubes: Advanced topics in the synthesis, structure, properties and applications, in *Springer Series on Topics in Appl. Phys.*, vol. 111, Springer-Verlag, Berlin.

Raman Spectroscopy in Graphene Related Systems. Ado Jorio, Riichiro Saito, Gene Dresselhaus, and Mildred S. Dresselhaus

ISBN: 978-3-527-40811-5

21 Kroto, H.W., Heath, J.R., O'Brien, S.C., Curl, R.F., and Smalley, R.E. (1985) *Nature (London)*, **318**, 162–163.

22 Monthioux, M. and Kuznetsov, V.L. (2006) *Carbon*, **44**, 1621–1623, Guest Editorial.

23 Radushkevich, L.V. and Luk'Yanovich, V.M. (1952) *Zh. Fiz. Khim.*, **26**, 88.

24 Oberlin, A., Endo, M., and Koyama, T. (1976) *J. Crystal Growth*, **32**, 335–349.

25 Iijima, S. and Ichihashi, T. (1993) *Nature (London)*, **363**, 603.

26 Bethune, D.S., Kiang, C.H., de Vries, M.S., Gorman, G., Savoy, R., Vazquez, J., and Beyers, R. (1993) *Nature (London)*, **363**, 605.

27 Iijima, S. (1991) *Nature (London)*, **354**, 56–58.

28 Oberlin, A., Endo, M., and Koyama, T. (1976) *Carbon*, **14**, 133.

29 Oberlin, A., Endo, M., and Koyama, T. (1976) *J. Crystal Growth*, **32**, 335–349.

30 Oberlin, A. (1984) *Carbon*, **22**, 521.

31 Saito, R., Dresselhaus, G., and Dresselhaus, M.S. (1998) *Physical Properties of Carbon Nanotubes*, Imperial College Press, London.

32 Reich, S., Thomsen, C., and Maultzsch, J. (2004) *Carbon Nanotubes: Physical Concepts and Physical Properties*, Wiley-VCH Verlag GmbH, Weinheim.

33 Joselevich, E., Dai, H., Liu, J., Hata, K., and Windle, A.H. (2008) Carbon nanotube synthesis and organization, in *Springer Series on Topics in Appl. Phys.*, vol. 111 (eds A. Jorio, M.S. Dresselhaus, and G. Dresselhaus), Springer-Verlag, Berlin, pp. 101–163.

34 Terrones, M., Souza Filho, A.G., and Rao, A.M. (2008) Doped carbon nanotubes: Synthesis, characterization and applications, in *Springer Series on Topics in Appl. Phys.*, vol. 111 (eds A. Jorio, M.S. Dresselhaus, and G. Dresselhaus), Springer-Verlag, Berlin, pp. 531–566.

35 Yamamoto, T., Watanabe, K., and Hernandez, E.R. (2008) Mechanical properties, thermal stability and heat transport in carbon nanotubes, in *Springer Series on Topics in Appl. Phys.*, vol. 111 (eds A. Jorio, M.S. Dresselhaus, and G. Dresselhaus), Springer-Verlag, Berlin, pp. 165–194.

36 Yakobson, B.I. (1998) *Appl. Phys. Lett.*, **72**, 918.

37 Spataru, C.D., Ismail-Beigi, S., Capaz, R.B., and Louie, S. (2008) Quasiparticle and excitonic effects: Topics in the synthesis, structure, properties and applications, in *Springer Series on Topics in Appl. Phys.*, vol. 111 (eds A. Jorio, M.S. Dresselhaus, and G. Dresselhaus), Springer-Verlag, Berlin, pp. 195–228.

38 Ando, T. (2008) Role of the Aharonov–Bohm phase in the optical properties of carbon nanotube, in *Springer Series on Topics in Appl. Phys.*, vol. 111 (eds A. Jorio, M.S. Dresselhaus, and G. Dresselhaus), Springer-Verlag, Berlin, pp. 229–249

39 Saito, R., Fantini, C., and Jiang, J. (2008) Excitonic states and resonance Raman Spectroscopy on single-wall carbon nanotube, in *Springer Series on Topics in Appl. Phys.*, vol. 111 (eds A. Jorio, M.S. Dresselhaus, and G. Dresselhaus), Springer-Verlag, Berlin, pp. 251–285

40 Lefebvre, J., Maruyama, S., and Finnie, P. (2008) Photoluminescence: Science and applications: Topics in the synthesis, structure, properties and applications, in *Springer Series on Topics in Appl. Phys.*, vol. 111 (eds A. Jorio, M.S. Dresselhaus, and G. Dresselhaus), Springer-Verlag, Berlin, pp. 287–320.

41 Ma, Y.-Z., Hertel, T., Vardeny, Z.V., Fleming, G.R., and Valkunas, L. (2008) Ultrafast spectroscopy of carbon nanotubes: Science and applications: Topics in the synthesis, structure, properties and applications, in *Springer Series on Topics in Appl. Phys.*, vol. 111 (eds A. Jorio, M.S. Dresselhaus, and G. Dresselhaus), Springer-Verlag, Berlin, pp. 321–352.

42 Heinz, T.F. (2008) Rayleigh scattering spectroscopy, in *Springer Series on Topics in Appl. Phys.*, vol. 111 (eds A. Jorio, M.S. Dresselhaus, and G. Dresselhaus), Springer-Verlag, Berlin, pp. 353–368.

43 Hartschuh, A. (2008) New techniques for carbon-nanotube study and characterization, in *Springer Series on Topics in Appl. Phys.*, vol. 111 (eds A. Jorio, M.S. Dresselhaus, and G. Dresselhaus), Springer-Verlag, Berlin, pp. 371–392.

44 Kono, J., Nicholas, R.J., and Roche, S. (2008) High magnetic field phenomena in carbon nanotubes, in *Springer Series on Topics in Appl. Phys.*, vol. 111 (eds A. Jorio, M.S. Dresselhaus, and G. Dresselhaus), Springer-Verlag, Berlin, pp. 393–421.

45 Avouris, P., Freitag, M., and Perebeinos, V. (2008) Carbon nanotube optoelectronics: Advanced topics in the synthesis, structure, properties and applications, in *Springer Series on Topics in Appl. Phys.*, vol. 111 (eds A. Jorio, M.S. Dresselhaus, and G. Dresselhaus), Springer-Verlag, Berlin, pp. 423–454.

46 Wu, J., Walukiewicz, W., Shan, W., Bourret-Courchesne, E., Ager, J.W., Yu, K.M., Haller, E.E., Kissell, K., Bachilo, S.M., Weisman, R.B., and Smalley, R.E. (2004) *Phys. Rev. Lett.*, **93**(1), 017404.

47 Biercut, M.J., Ilani, S., Marcus, C.M., and McEuen, P.L. (2008) Electrical transport in single wall carbon nanotubes, in *Springer Series on Topics in Appl. Phys.*, vol. 111 (eds A. Jorio, M.S. Dresselhaus, and G. Dresselhaus), Springer-Verlag, Berlin, pp. 455–492.

48 Kavan, L. and Dunsch, L. (2008) Electrochemistru of carbon nanotubes: Advanced topics in the synthesis, structure, properties and applications, in *Springer Series on Topics in Appl. Phys.*, vol. 111, (eds A. Jorio, M.S. Dresselhaus, and G. Dresselhaus), Springer-Verlag, Berlin, pp. 567–604.

49 Kalbac, M., Farhat, H., Kavan, L., Kong, J., and Dresselhaus, M.S. (2008) *Nanoletters*, **8**, 2532–2537.

50 Affoune, A.M., Prasad, B.L.V., Sato, H., Enok, T., Kaburagi, Y., and Hishiyama, Y. (2001) Experimental evidence of a single nano-graphene. *Chem. Phys. Lett.*, **348**(1–2), 17–20.

51 Novoselov, K.S., Geim, A.K., Morozov, S.V., Jiang, D., Katsnelson, M.I., Grigorieva, I.V., Dubonos, S.V., and Firsov, A.A. (2004) *Science*, **306**, 666–669.

52 Geim, A.K. and Novoselov, K.S. (2007) *Nat. Mater.*, **6** 183–191.

53 Wallace, P.R. (1947) *Phys. Rev.*, **71**, 622.

54 Lee, C., Wei, X., Kysar, J.W., and Hone, J. (2008) *Science*, **321**(5887), 385.

55 Bolotin, K.I., Sikes, K.J., Jiang, Z., Klima, M., Fudenberg, G., Hone, J., Kim, P., and Stormer, H.L. (2008) *Solid State Commun.*, **146**, 351–355.

56 Morozov, S.V., Novoselov, K.S., Katsnelson, M.I., Schedin, F., Elias, D.C., Jaszczak, J.A., and Geim, A.K. (2008) *Phys. Rev. Lett.*, **100**, 016602.

57 Novoselov, K.S., Geim, A.K., Morozov, S.V., Jiang, D., Katsnelson, M.I., Grigorieva, I.V., Dubonus, S.V., and Firsov, A.A. (2005) *Nature*, **438**, 197–200.

58 Beenakker, C.W.J. (2008) *Rev. Mod. Phys.*, **80**, 1337–1354.

59 Katsnelson, M.I., Novoselov, K.S., and Geim, A.K. (2006) *Nat. Phys.*, **2**, 620–625.

60 Cheianov, V.V. and Falko, V.I. (2006) *Phys. Rev. B*, **74**, 041403(R).

61 Pereira, J.M. Jr., Mlinar, V., Peeters, F.M., and Vasilopoulos, P. (2006) *Phys. Rev. B*, **74**, 045424.

62 Cheianov, V.V., Falko, V.I., and Altshuler, B.L. (2007) *Science*, **315**, 1252–1255.

63 Beenakker, C.W.J. (2006) *Phys. Rev. Lett.*, **97**, 067007.

64 Miao, F., Wijeratne, S., Zhang, Y., Coskun, U.C., Bao, W., and Lau, C.N. (2007) *Science*, **317**, 1530–1533.

65 Ossipov, A., Titov, M., and Beenakker, C.W.J. (2007) *Phys. Rev. B*, **75**, 241401(R).

66 Beenakker, C., Akhmerov, A., Recher, P., and Tworzydo, J. (2007) *Phys. Rev. B*, **77**, 075409.

67 Pederson, T.G., Flindt, C., Pedersen, J., Mortensen, N.A., Jauho, A.-P., and Pedersen, K. (2008) *Phys. Rev. Lett.*, **100**, 136804.

68 Park, C.-H., Yang, L., Son, Y.-W., Cohen, M. L., and Louie, S.G. (2008) *Nat. Phys.*, **4**, 213–217.

69 Elias, D.C., Nair, R.R., Mohiuddin, T.M.G., Morozov, S.V., Blake, P., Halsall, M.P., Ferrari, A.C., Boukhvalov, D.W., Katsnelson, M.A.K., and Novoselov, K.S. (2009) *Science*, **323**(5914), 610–613.

70 Geim, A.K. (2009) *Science*, **324**, 1530.

71 Berger, C., Song, Z., Li, X., Wu, X., Brown, N., Naud, C., Mayou, D., Li, T., Hass, J., Marchenkov, A.N., Conrad Conrad, E.H., First, W.A., and de Heer, P.N. (2006) *Science*, **312**, 1191–1196.

72 Kumazaki, H. and Hirashima, D.S. (2009) *J. Phys. Soc. Jpn.*, **78**, 094701–094706.

73 Li, X., Wang, X., Zhang, L., Lee, S., and Dai, H. (2008) *Science*, **319**, 1229.

74 Yang, X., Dou, X., Rouhanipour, A., Zhi, L., Rader, H.J., and Mullen, K. (2008) *J. Am. Chem. Soc.*, **130**(13), 4216–4217.

75 Kosynkin, D.V., Higginbotham, A.L., Sinitskii, A., Lomeda, J.R., Dimiev, A., Price, B.K., and Tour, J.M. (2009) *Nature*, **458**(7240), 872–876.

76 Jiao, L., Zhang, L., Wang, X., Diankov, G., and Dai, H. (2009) *Nature*, **458**(7240), 877–880.

77 Dresselhaus, M.S. and Spencer, W. (2007) *NAS Publication* National Research Council, Condensed-Matter and Materials Physics: The Science of the World Around Us, Washington, D.C., The National Academies Press, 2007.

78 Krishnan, R.S., Chandrasekharan, V., and Rajagopal, E.S. (1958) *Nature*, **182**, 518–520.

79 McSkimin, H.J. and Andreatch, J.P. (1972) *J. Appl. Phys.*, **43**, 2944–2948.

80 Dresselhaus, M.S., Dresselhaus, G., Saito, R., and Jorio, A. (2005) *Phys. Rep.*, **409**, 47–99.

81 Klett, J., Hardy, R., Romine, E., Walls, C., and Burchell, T. (2000) *Carbon*, **38**, 953.

82 Barros, E.B., Demir, N.S., Souza Filho, A.G., Mendes Filho, J., Jorio, A., Dresselhaus, G., and Dresselhaus, M.S. (2005) *Phys. Rev. B*, **71**, 165422.

83 Cançado, L.G., Pimenta, M.A., Neves, R.A., Medeiros-Ribeiro, G., Enoki, T., Kobayashi, Y., Takai, K., Fukui, K., Dresselhaus, M.S., Saito, R., and Jorio, A. (2004) *Phys. Rev. Lett.*, **93**, 047403.

84 Malard, L.M., Pimenta, M.A., Dresselhaus, G., and Dresselhaus, M.S. (2009) *Phys. Rep.*, **473**, 51–87.

85 Dresselhaus, M.S., Jorio, A., Hofmann, M., Dresselhaus, G., and Saito, R. (2010) *Nano Lett.*, **10**(3), 751–758.

86 Ferrari, A.C., Meyer, J.C., Scardaci, V., Casiraghi, C., Lazzeri, M., Mauri, F., Piscanec, S., Jiang, D., Novoselov, K.S., Roth, S., and Geim, A.K. (2006) *Phys. Rev. Lett.*, **97**, 187401.

87 Yudasaka, M., Iijima, S., and Crespi, V.H. (2008) Single-wall carbon nanohorns and nanocones: Advanced topics in the synthesis, structure, properties and applications, in *Springer Series on Topics in Appl. Phys.*, vol. 111 (eds M.S. Dresselhaus, G. Dresselhaus and A. Jorio), Springer-Verlag, Berlin.

88 Ferrari, A.C. and Robertson, J. (2000) *Phys. Rev. B*, **61**, 14095.

89 http://nobelprize.org/nobel_prizes/physics/laureates/1930/raman-lecture.html

90 Dresselhaus, M.S., Dresselhaus, G., Rao, A.M., Jorio, A., Souza Filho, A.G., Samsonidze, G.G., and Saito, R. (2003) *Ind. J. Phys.*, **77B**, 75–99.

91 Martin, R.M. and Falicov, L.M. (1975) *Light-Scattering in Solids* (ed. M. Cardona), Springer-Verlag, Berlin, p. 80.

92 Eisberg, R. and Resnick, R. (1974) Quantum physics of atoms, molecules, solids, nuclei and particles, in *Multielectron Atoms*, chap. 9, John Wiley & Sons, Inc., New York, N.Y.

93 Slater, J. (1951) *Phys. Rev.*, **81**, 385.

94 Dresselhaus, M.S., Dresselhaus, G., and Jorio, A. (2008) *Group Theory: Application to the Physics of Condensed Matter*, Springer-Verlag, Berlin.

95 Kittel, C. (1986) *Introduction to Solid State Physics*, 6th edn, John Wiley & Sons, New York.

96 Painter, G.S. and Ellis, D.E. (1970) *Phys. Rev. B*, **1**, 4747.

97 Slonczewski, J.C. and Weiss, P.R. (1958) *Phys. Rev.*, **109**, 272.

98 Malard, L.M., Guimarães, M.H.D., Mafra, D.L., Mazzoni, M.S.C., and Jorio, A. (2009) *Phys. Rev. B*, **79**, 125426.

99 McClure, J.W. (1957) *Phys. Rev.*, **108**, 612.

100 Slonczewski, J.C. and Weiss, P.R. (1958) *Phys. Rev.*, **109**, 272.

101 Koshino, M. and Ando, T. (2008) *Phys. Rev. B*, **77**, 115313.

102 Nakada, K., Fujita, M., Dresselhaus, G., and Dresselhaus, M.S. (1996) *Phys. Rev. B*, **54**, 17954–17961.

103 Jia, X., Hofmann, M., Meunier, V., Sumpter, B.G., Campos-Delgado, J., Romo-Herrera, J.M., Son, H., Hsieh, Y.-P., Reina, A., Kong, J., Terrones, M.,

and Dresselhaus, M.S. (2009) *Science*, **323**, 1701–1705.

104 Son, Y.W., Cohen, M.L., and Louie, S.G. (2006) *Phys. Rev. Lett.*, **97**(21), 216803.

105 Han, M.Y., Oezyilmaz, B., Zhang, Y., and Kim, P. (2007) *Phys. Rev. Lett.*, **98**(20), 206805.

106 Pisani, L., Chan, J.A., Montanari, B., and Harrison, N.M. (2007) *Phys. Rev. B*, **75**(6), 64418.

107 Wang, N., Tang, Z.K., Li, G.D., and Chen, J.S. (2000) *Nature*, **408**, 50.

108 Cabria, I., Mintmire, J.W., and White, C.T. (2003) *Phys. Rev. B*, **67**, 121406(R).

109 Zare, A. (2008) *Raman Spectroscopy of Single Wall Carbon Nanotubes of different lengths*. PhD thesis, Massachusetts Institute of Technology, Department of Electrical Engineering and Computer Science, Doctor of Philosophy.

110 Samsonidze, G.G., Saito, R., Jorio, A., Pimenta, M.A., Souza Filho, A.G., Grüneis, A., Dresselhaus, G., and Dresselhaus, M.S. (2003) *J. Nanosci. Nanotechnol.*, **3**, 431–458.

111 Jishi, R.A., Inomata, D., Nakao, K., Dresselhaus, M.S., and Dresselhaus, G. (1994) *J. Phys. Soc. Jpn.*, **63**, 2252–2260.

112 Dresselhaus, M.S. and Eklund, P.C. (2000) *Adv. Phys.*, **49**, 705–814.

113 Kataura, H., Kumazawa, Y., Kojima, N., Maniwa, Y., Umezu, I., Masubuchi, S., Kazama, S., Zhao, X., Ando, Y., Ohtsuka, Y., Suzuki, S., and Achiba, Y. (1999) Proc. of the Int. Winter School on Electronic Properties of Novel Materials (IWEPNM'99), in *AIP Conference Proceedings*, vol. 486 (eds H. Kuzmany, M. Mehring, and J. Fink), American Institute of Physics, Woodbury, N.Y., pp. 328–332.

114 Blase, X., Benedict, L.X., Shirley, E.L., and Louie, S.G. (1994) *Phys. Rev. Lett.*, **72**, 1878.

115 Reich, S., Maultzsch, J., Thomsen, C., and Ordejón, P. (2002) *Phys. Rev. B*, **66**, 035412(1–5).

116 Dubay, O. and Kresse, G. (2003) *Phys. Rev. B*, **67**, 035401.

117 Samsonidze, G.G., Barros, E.B., Saito, R., Jiang, J., Dresselhaus, G., and Dresselhaus, M.S. (2007) *Phys. Rev. B*, **75**, 155420.

118 Ashcroft, N.W. and Merman, N.D. (1976) *Solid State Physics*, Holt, Rinehart and Winston, New York, NY, p. 141.

119 Ferraro, J.R., Nakamoto, K., and Brown, C.W. (2003) *Introductory Raman Spectroscopy*, Academic Press, San Diego.

120 Cohen-Tannoudji, C., Diu, B., and Laloe, F. (2009) *Quantum Mechanics*, (2 VOL SET), Amazon.com.

121 Sala, O. (1996) *Fundamentos de Espectroscopia Raman e no Infravermelho*, Editora Unesp.

122 Kuzmany, H. (2009) *Solid-State Spectroscopy: An Introduction*, Springer-Verlag, Berlin.

123 Jishi, R.A., Venkataraman, L., Dresselhaus, M.S., and Dresselhaus, G. (1993) *Chem. Phys. Lett.*, **209**, 77–82.

124 Dubay, O., Kresse, G., and Kuzmany, H. (2002) *Phys. Rev. Lett.*, **88**, 235506.

125 Dresselhaus, G. and Dresselhaus, M.S. (1968) *Int. J. Quant. Chem.*, **IIS**, 333–345.

126 Johnson, L.G. and Dresselhaus, G. (1973) *Phys. Rev. B*, **7**, 2275–2285.

127 Aizawa, T., Souda, R., Otani, S., Ishizawa, Y., and Oshima, C. (1990) *Phys. Rev. B*, **42**, 11469.

128 Oshima, C., Aizawa, T., Souda, R., Ishizawa, Y., and Sumiyoshi, Y. (1988) *Solid State Commun.*, **65**, 1601.

129 Maultzsch, J., Reich, S., Thomsen, C., Requardt, H., and Ordejón, P. (2004) *Phys. Rev. Lett.*, **92**, 075501.

130 Grüneis, A., Serrano, J., Bosak, A., Lazzeri, M., Molodtsov, S.L., Wirtz, L., Attaccalite, C., Krisch, M., Rubio, A., Mauri, F., and Pichler, T. (2009) *Phys. Rev. B*, **80**, 085423.

131 Tanaka, T., Tajima, A., Moriizumi, R., Hosoda, M., Ohno, R., Rokuta, E., Oshimaa, C., and Otani, S. (2002) *Solid State Commun.*, **123**, 33–36.

132 Ren, W., Saito, R., Gao, L., Zheng, F., Wu, Z., Liu, B., Furukawa, M., Zhao, J., Chen, Z., and Cheng, H.M. (2010) *Phys. Rev. B*, **81**, 035412-1–7.

133 Grüneis, A., Saito, R., Kimura, T., Cançado, L.G., Pimenta, M.A., Jorio, A., Souza Filho, A.G., Dresselhaus, G., and Dresselhaus, M.S. (2002) *Phys. Rev. B*, **65**, 155405–1–7.

134 Alon, O.E. (2001) *Phys. Rev. B*, **63**, 201403(R).

135 Barros, E.B., Jorio, A., Samsonidze, G.G., Capaz, R.B., Souza Filho, A.G., Mendez Filho, J., Dresselhaus, G., and Dresselhaus, M.S. (2006) *Phys. Rep.*, **431**, 261–302.

136 Rao, A.M., Richter, E., Bandow, S., Chase, B., Eklund, P.C., Williams, K.W., Fang, S., Subbaswamy, K.R., Menon, M., Thess, A., Smalley, R.E., Dresselhaus, G., and Dresselhaus, M.S. (1997) *Science*, **275**, 187–191.

137 Kürti, J., Kresse, G., and Kuzmany, H. (1998) *Phys. Rev. B*, **58**, R8869.

138 Sanchez-Portal, D., Artacho, E., Solar, J.M., Rubio, A., and Ordejon, P. (1999) *Phys. Rev. B*, **59**, 12678–12688.

139 Menon, M., Richter, E., and Subbaswamy, K.R. (1996) *J. Chem. Phys.*, **104**, 5875.

140 Samsonidze, G. (2006) Photophysics of Carbon Nanotubes. PhD thesis, Massachusetts Institute of Technology, Department of Electrical Engineering and Computer Science, October.

141 Piscanec, S., Lazzeri, M., Mauri, M., Ferrari, A. C., and Robertson, J. (2004) *Phys. Rev. Lett.*, **93**, 185503.

142 Cardona, M. (1982) *Light-Scattering in Solids* (eds M. Cardona and G. Güntherodt), Springer-Verlag, Berlin, p. 19.

143 Chou, S.G., Plentz Filho, F., Jiang, J., Saito, R., Nezich, D., Ribeiro, H.B., Jorio, A., Pimenta, M.A., Samsonidze, G.G., Santos, A.P., Zheng, M., Onoa, G.B., Semke, E.D., Dresselhaus, G., and Dresselhaus, M.S. (2005) *Phys. Rev. Lett.*, **94**, 127402.

144 Leite, R.C.C. and Porto, S.P.S. (1966) *Phys. Rev. Lett.*, **17**, 1012.

145 Alfano, R.R. and Shapiro, S.L. (1971) *Phys. Rev. Lett.*, **26**, 1247–1251.

146 Song, D., Wang, F., Dukovic, G., Zheng, M., Semke, E.D., Brus, L.E., and Heinz, T.F. (2008) *Phys. Rev. Lett.*, **100**, 225503.

147 Brown, S.D.M., Jorio, A., Corio, P., Dresselhaus, M.S., Dresselhaus, G., Saito, R., and Kneipp, K. (2001) *Phys. Rev. B*, **63**, 155414.

148 Tuinstra, F. and Koenig, J.L. (1970) *J. Chem. Phys.*, **53**, 1126.

149 Tuinstra, F. and Koenig, J.L. (1970) *J. Comp. Mater.*, **4**, 492.

150 Nemanich, R.J. and Solin, S.A. (1977) *Solid State Commun.*, **23**, 417.

151 Nemanich, R.J. and Solin, S.A. (1979) *Phys. Rev. B*, **20**, 392.

152 Mernagh, T.P., Cooney, R.P., and Johnson, R.A. (1984) *Carbon*, **22**, 39–42.

153 Tsu, R., Gonzalez, J.H., and Hernandez, I.C. (1978) *Solid State Commun.*, **27**, 507.

154 Vidano, R.P., Fishbach, D.B., Willis, L.J., and Loehr, T.M. (1981) *Solid State Commun.*, **39**, 341.

155 Lespade, P., Marchand, A., Couzi, M., and Cruege, F. (1984) *Carbon*, **22**, 375.

156 Lespade, P., Al-Jishi, R., and Dresselhaus, M.S. (1982) *Carbon*, **20**, 427–431.

157 Wilhelm, H., Lelausian, M., McRae, E., and Humbert, B. (1998) *J. Appl. Phys.*, **84**, 6552–6558.

158 Baranov, A.V., Bekhterev, A.N., Bobovich, Y.S., and Petrov, V.I. (1987) *Opt. Spectrosk.*, **62**, 1036.

159 Thomsen, C. and Reich, S. (2000) *Phys. Rev. Lett.*, **85**, 5214.

160 Saito, R., Jorio, A., Souza Filho, A.G., Dresselhaus, G., Dresselhaus, M.S., and Pimenta, M.A. (2002) *Phys. Rev. Lett.*, **88**, 027401.

161 Cançado, L.G., Pimenta, M.A., Neves, B.R., Dantas, M.S., and Jorio, A. (2004) *Phys. Rev. Lett.*, **93**, 247401.

162 Mapelli, C.M., Castiglioni, C., Zerbi, G., and Müllen, K. (1999) *Phys. Rev. B*, **60**, 12710–12725.

163 Castiglioni, C., Mapelli, C.M., Negri, F., and Zerbi, G. (2001) *J. Chem. Phys.*, **114**, 963.

164 Negri, F., Emanuele, E., and Calabretta, R.A. (2002) *J. Comput. Meth. Sci. Engin.*, **2**, 133–141.

165 Watson, M.D., Fechtenkoetter, A., and Muellen, K. (2001) *Chem. Rev.*, **101**(5), 1267–1300.

166 Kötter, K. *et al.* (2008) *J. Phys. Chem. C*, **112**, 10637–10640.

167 Tommasini, M., Di Donato, E., Castiglioni, C., and Zerbi, G. (2005) *Chem. Phys. Lett.*, **414**, 166–173.

168 Ferrari, A.C. and Robertson, J. (eds) (2004) Raman spectroscopy in carbons: from nanotubes to diamond. *Phyl. Trans. of the Royal Soc. A* **362**(1824), 2267–2565.

169 Holden, J.M., Zhou, P., Bi, X.-X., Eklund, P.C., Bandow, S., Jishi, R.A., Das Chowdhury, K., Dresselhaus, G., and Dresselhaus, M.S. (1994) *Chem. Phys. Lett.*, **220**, 186–191.

170 Thess, A., Lee, R., Nikolaev, P., Dai, H., Petit, P., Robert, J., Xu, C., Lee, Y.H., Kim, S.G., Rinzler, A.G., Colbert, D.T., Scuseria, G.E., Tománek, D., Fischer, J.E., and Smalley, R.E. (1996) *Science*, **273**, 483–487.

171 Pimenta, M.A., Marucci, A., Empedocles, S., Bawendi, M., Hanlon, E.B., Rao, A.M., Eklund, P.C., Smalley, R.E., Dresselhaus, G., and Dresselhaus, M.S. (1998) *Phys. Rev. B Rapid*, **58**, R16016–R16019.

172 Milnera, M., Kürti, J., Hulman, M., and Kuzmany, H. (2000) *Phys. Rev. Lett.*, **84**, 1324–1327.

173 Pimenta, M.A., Hanlon, E.B., Marucci, A., Corio, P., Brown, S.D.M., Empedocles, S.A., Bawendi, M.G., Dresselhaus, G., and Dresselhaus, M.S. (2000) *Braz. J. Phys.*, **30**, 423–427.

174 Kataura, H., Kumazawa, Y., Maniwa, Y., Umezu, I., Suzuki, S., Ohtsuka, Y., and Achiba, Y. (1999) *Synth. Met.*, **103**, 2555–2558.

175 Liu, J., Fan, S., and Dai, H. (2004) *Bull. Mater. Res. Soc.*, **29**, 244–250.

176 Jorio, A., Saito, R., Hafner, J.H., Lieber, C.M., Hunter, M., McClure, T., Dresselhaus, G., and Dresselhaus, M.S. (2001) *Phys. Rev. Lett.*, **86**, 1118–1121.

177 Jorio, A., Pimenta, M.A., Souza Filho, A.G., Saito, R., Dresselhaus, G., and Dresselhaus, M.S. (2003) *New J. Phys.*, **5**, 1.1–1.17.

178 Saito, R., Grüneis, A., Samsonidze, G.G., Brar, V.W., Dresselhaus, G., Dresselhaus, M.S., Jorio, A., Cançado, L.G., Fantini, C., Pimenta, M.A., and Souza Filho, A.G. (2003) *New J. Phys.*, **5**, 157.1–157.15.

179 Jorio, A., Souza Filho, A.G., Dresselhaus, G., Dresselhaus, M.S., Swan, A.K., Ünlü, M.S., Goldberg, B., Pimenta, M.A., Hafner, J.H., Lieber, C.M., and Saito, R. (2002) *Phys. Rev. B*, **65**, 155412.

180 Bachilo, S.M., Strano, M.S., Kittrell, C., Hauge, R.H., Smalley, R.E., and Weisman, R.B. (2002) *Science*, **298**, 2361–2366.

181 Kane, C.L. and Mele, E.J. (2003) *Phys. Rev. Lett.*, **90**, 207401.

182 Samsonidze, G.G., Saito, R., Kobayashi, N., Grüneis, A., Jiang, J., Jorio, A., Chou, S.G., Dresselhaus, G., and Dresselhaus, M.S. (2004) *Appl. Phys. Lett.*, **85**, 5703–5705.

183 Fantini, C., Jorio, A., Souza, M., Strano, M.S., Dresselhaus, M.S., and Pimenta, M.A. (2004) *Phys. Rev. Lett.*, **93**, 147406.

184 Telg, H., Maultzsch, J., Reich, S., Hennrich, F., and Thomsen, C. (2004) *Phys. Rev. Lett.*, **93**, 177401.

185 O'Connell, M.J. Bachilo, S.M., Huffman, X.B., Moore, V.C., Strano, M.S., Haroz, E.H., Rialon, K.L., Boul, P.J., Noon, W.H., Kittrell, C., Ma, J., Hauge, R.H., Weisman, R.B., and Smalley, R.E. (2002) *Science*, **297**, 593–596.

186 Jiang, J., Saito, R., Samsonidze, G.G., Jorio, A., Chou, S.G., Dresselhaus, G., and Dresselhaus, M.S. (2007) *Phys. Rev. B*, **75**, 035407-1–13.

187 Dresselhaus, M.S., Dresselhaus, G., Saito, R., and Jorio, A. (2007) *Annual Reviews of Physical Chemistry Chemical Physics* (eds S.R. Leone, J.T. Groves, R.F. Ismagilov, and G. Richmond), Annual Reviews, Palo Alto, CA, pp. 719–747.

188 Jiang, J., Saito, R., Sato, K., Park, J.S., Ge. Samsonidze, G., Jorio, A., Dresselhaus, G., and Dresselhaus, M.S. (2007) *Phys. Rev. B*, **75**, 035405.

189 Araujo, P.T., Maciel, I.O., Pesce, P.B.C., Pimenta, M.A., Doorn, S.K., Qian, H., Hartschuh, A., Steiner, M., Grigorian, L., Hata, K., and Jorio, A. (2008) *Phys. Rev. B*, **77**, 241403.

190 Araujo, P.T., Jorio, A., Dresselhaus, M.S., Sato, K., and Saito, R. (2009) *Phys. Rev. Lett.*, **103**, 146802.

191 Maciel, I.O., Anderson, N., Pimenta, M.A., Hartschuh, A., Qian, H., Terrones, M., Terrones, H., Campos-Delgado, J., Rao, A.M., Novotny, L., and Jorio, A. (2008) *Nat. Mater.*, **7**, 878.

192 Malard, L.M., Nilsson, J., Elias, D.C., Brant, J.C., Plentz, F., Alves, E.S., Castro Neto, A.H., and Pimenta, M.A. (2007) *Phys. Rev. B*, *76*, 201401.

193 Ni, Z.H., Wang, H.M., Kasim, J., Fan, H.M., Yu, T., Wu, Y.H., Feng, Y.P., Shen, Z.X. (2008) *J. Phys. Chem. C*, **112**, 10637–10640.

194 Lucchese, M.M., Stavale, F., Ferreira, E.H., Vilane, C., Moutinho, M.V.O., Capaz, R.B., Achete, C.A., Jorio, A. (2010) *Carbon*, **48**(5), 1592–1597.

195 Jorio, A., Lucchese, M.M., Stavale, F., Martins Ferreira, E.H., Moutinho, M.V.O., Capaz, R.B., and Achete, C.A. (2010) *J. Phys. Cond. Matt.*, **22**, 334204.

196 Das, A., Pisana, S., Chakraborty, B., Piscanec, S., Saha, S.K., Waghmare, U.V., Novoselov, K.S., Krishnamurthy, H.R., Geim, A.K., Ferrari, A.C., and Sood, A.K. (2008) *Nat. Nanotech.*, **3**, 210.

197 Gupta, A., Chen, G., Joshi, P., Tadigadapa, S., and Eklund, P.C. (2006) *Nano Lett.*, **6**, 2667.

198 Ni, Z.H., Yu, T., Lu, Y.H., Wang, Y.Y., Feng, Y.P., and Shen, Z.X. (2008) *ACS Nano*, **2**(11), 2301–2305.

199 Ni, Z., Wang, Y., Yu, T., You, Y. and Shen, Z. (2008) *Phys. Rev. B*, **77**, 235403.

200 Schiff, L.I. (1968) *Quantum Mechanics*, 3rd edn, McGraw-Hill, New York.

201 Morse, P.M. and Feshbach, H. (1953) *Methods of Theoretical Physics*, chap. 5, McGraw-Hill, New York.

202 Yu, P.Y. and Cardona, M. (1995) Light scattering in solids II, in *Fundamentals of Semiconductors: Physics and Materials Properties*, vol. 52 (eds M. Cardona and G. Güntherodt), Springer.

203 Jiang, J., Saito, R., Grüneis, A., Chou, S.G., Samsonidze, G.G., Jorio, A., Dresselhaus, G., and Dresselhaus, M.S. (2005) *Phys. Rev. B*, **71**, 205420-1–13.

204 Khan, K. and Allen, P. (1984) *Phys. Rev. B*, **29**, 3341.

205 Machón, M., Reich, S., Telg, H., Maultzsch, J., Ordejón, P., and Thomsen, C. (2005) *Phys. Rev. B*, **71**, 035416.

206 Trulson, M.O. and Mathies, R.A. (1986) *J. Chem. Phys.*, **84**, 2068.

207 Loudon, R. (2001) *Adv. Phys.*, **50**, 813.

208 Ganguly, A.K. and Birman, J.L. (1967) *Phys. Rev.*, **162**, 806.

209 Menendéz, J. and Cardona, M. (1985) *Phys. Rev. B*, **31**, 3696.

210 Cantarero, A., Trallero-Giner, C., and Cardona, M. (1989) *Phys. Rev. B*, **39**, 8388.

211 Trallero-Giner, C., Cantarero, A., and Cardona, M. (1989) *Phys. Rev. B*, **40**, 4030.

212 Trallero-Giner, C., Cantarero, A., and Cardona, M. (1989) *Phys. Rev. B*, **40**, 12290.

213 Alexandrou, A., Trallero-Giner, C., Cantarero, A., and Cardona, M. (1989) *Phys. Rev. B*, **40**, 1603.

214 Gavrilenko, V.I., Martínez, D., Cantarero, A., Cardona, M., and Trallero-Giner, C. (1990) *Phys. Rev. B*, **42**, 11718.

215 Trallero-Giner, C., Cantarero, A., and Cardona, M. (1992) *Phys. Rev. B*, **42**, 6601.

216 Landau, L.D. and Lifshitz, E.M. (1973) *The Classical Theory of Fields*, vol. 2, 4th revised english edn, Butterworth Heinemann, Amsterdam.

217 Malard Moreira, L. (2009) Raman spectroscopy of graphene: probing phonons, electrons and electron–phonon interactions. PhD Thesis, Universidade Federal de Minas Gerais, Departamento de Fisica.

218 Barros, E.B., Capaz, R.B., Jorio, A., Samsonidze, G.G., Souza Filho, A.G., Ismail-Beigi, S., Spataru, C.D., Louie, S.G., Dresselhaus, G., and Dresselhaus, M.S. (2006) *Phys. Rev. B Rapid*, **73**, 241406(R).

219 Nugraha, A.R.T., Saito, R., Sato, K., Araujo, P.T., Jorio, A., Dresselhaus, M.S. (2010) *Appl. Phys. Lett.*, **97**, 091905.

220 Grüneis, A., Saito, R., Samsonidze, G.G., Kimura, T., Pimenta, M.A., Jorio, A., Souza Filho, A.G., Dresselhaus, G., and Dresselhaus, M.S. (2003) *Phys. Rev. B*, **67**, 165402-1-1–65402-7.

221 Partoens, B. and Peeters, F.M. (2006) *Phys. Rev. B*, **74**, 075404.

222 Jiang, J., Saito, R., Samsonidze, G.G., Chou, S.G., Jorio, A., Dresselhaus, G., and Dresselhaus, M.S. (2005) *Phys. Rev. B*, **72**, 235408-1–11.

223 Saha, S.K., Waghmare, U.V., Krishnamurthy, H.R. and Sood, A.K. (2008) *Phys. Rev. B*, **78**, 165421.

224 Jiang, J., Tang, H., Wang, B., and Su, Z. (2008) *Phys. Rev. B*, **77**, 235421.

225 White, C.T., Roberston, D.H., and Mintmire, J.W. (1993) *Phys. Rev. B*, **47**, 5485.

226 Jorio, A., Pimenta, M.A., Souza Filho, A.G., Samsonidze, G.G., Swan, A.K., Ünlü, M.S., Goldberg, B.B., Saito, R., Dresselhaus, G., and Dresselhaus, M.S. (2003) *Phys. Rev. Lett.*, **90**, 107403.

227 Jorio, A., Dresselhaus, G., Dresselhaus, M.S., Souza, M., Dantas, M.S.S., Pimenta, M.A., Rao, A.M., Saito, R., Liu, C., and Cheng, H.M. (2000) *Phys. Rev. Lett.*, **85**, 2617–2620.

228 Jorio, A., Souza Filho, A.G., Brar, V.W., Swan, A.K., Ünlü, M.S., Goldberg, B.B., Righi, A., Hafner, J.H., Lieber, C.M., Saito, R., Dresselhaus, G., and Dresselhaus, M.S. (2002) *Phys. Rev. B Rapid*, **65**, R121402.

229 Saito, R., Dresselhaus, G., and Dresselhaus, M.S. (2000) *Phys. Rev. B*, **61**, 2981–2990.

230 Huang, N., Yan, H., Chen, C., Song, D., Heinz, T.F., and Hone, J. (2009) *PNAS*, **106**(18), 7304–7308.

231 Mohiuddin, T.M.G., Lombardo, A., Nair, R.R., Bonetti, A., Savini, G., Jalil, R., Bonini, N., Basko, D.M., Galiotis, C., Marzari, N., Novoselov, K.S., Geim, A.K., and Ferrari, A.C. (2009) *Phys. Rev. B*, **79**, 205433.

232 Reich, S., Jantoljak, H., and Thomsen, C. (2000) *Phys. Rev. B*, **61**, R13389–R13392.

233 Dresselhaus, M.S., Dresselhaus, G., Jorio, A., Souza Filho, A.G., and Saito, R. (2002) *Carbon*, **40**, 2043–2061.

234 Reich, S., Thomsen, C., and Ordejón, P. (2001) *Phys. Rev. B*, **64**, 195416.

235 Duesberg, G.S., Loa, I., Burghard, M., Syassen, K., and Roth, S. (2000) *Phys. Rev. Lett.*, **85**, 5436–5439.

236 Thomsen, C., Reich, S., Goni, A.R., Jantoliak, H., Rafailov, P.M., Loa, I., Syassen, K., Journet, C., and Bernier, P. (1999) *Phys. Status Solidi (b)*, **215**, 435–441.

237 Maultzsch, J., Reich, S., and Thomsen, C. (2002) *Phys. Rev. B*, **65**, 233402.

238 Ajiki, H. and Ando, T. (1994) *Phys. B Condens. Matter*, **201**, 349.

239 Marinopoulos, A.G., Reining, L., Rubio, A., and Vast, N. (2003) *Phys. Rev. Lett.*, **91**, 046402.

240 Hwang, J., Gommans, H.H., Ugawa, A., Tashiro, H., Haggenmueller, R., Winey, K.I., Fischer, J.E., Tanner, D.B., and Rinzler, A.G. (2000) *Phys. Rev. B*, **62**, R13310–R13313.

241 Rao, A.M., Jorio, A., Pimenta, M.A., Dantas, M.S.S., Saito, R., Dresselhaus, G., and Dresselhaus, M.S. (2000) *Phys. Rev. Lett.*, **84**, 1820–1823.

242 Jorio, A., Fantini, C., Dantas, M.S.S., Pimenta, M.A., Souza Filho, A.G., Samsonidze, G.G., Brar, V.W., Dresselhaus, G., Dresselhaus, M.S., Swan, A.K., Ünlü, M.S., Goldberg, B.B., and Saito, R. (2002) *Phys. Rev. B*, **66**, 115411.

243 Venkateswaran, U.D., Rao, A.M., Richter, E., Menon, M., Rinzler, A., Smalley, R.E., and Eklund, P.C. (1999) *Phys. Rev. B*, **59**, 10928–10934.

244 Cronin, S.B., Swan, A.K., Ünlü, M.S., Goldberg, B.B., Dresselhaus, M.S., and Tinkham, M. (2004) *Phys. Rev. Lett.*, **93**, 167401.

245 Cronin, S.B., Swan, A.K., Ünlü, M.S., Goldberg, B.B., Dresselhaus, M.S., and Tinkham, M. (2005) *Phys. Rev. B*, **72**, 035425.

246 Souza Filho, A.G., Kobayasi, N., Jiang, J., Grüneis, A., Saito, R., Cronin, S.B., Mendes Filho, J., Samsonidze, G.G., Dresselhaus, G., and Dresselhaus, M.S. (2005) *Phys. Rev. Lett.*, **95**, 217403.

247 Duan, X., Son, H., Gao, B., Zhang, J., Gao, B., Wu, T., Samsonidze, G.G., Dresselhaus, M.S., Liu, Z., and Kong, J. (2007) *Nano Lett.*, **7**, 2116–2121.

248 Pisana, S., Lazzeri, M., Casiraghi, C., Novoselov, K.S., Geim, A.K., Ferrari, A.C. and Mauri, F. (2007) *Nat. Mater.*, **6**, 198.

249 Zhang, Y., Zhang, J., Son, H., Kong, J., and Liu, Z. (2006) *JACS*, **127**, 17156–17157.

250 Moos, G., Gahl, C., Fasel, R., Wolf, M., and Hertel, T. (2001) *Phys. Rev. Lett.*, **87**, 267402–1.

251 Kampfrath, T., Perfetti, L., Schapper, F., Frischkorn, C., and Wolf, M. (2005) *Phys. Rev. Lett.*, **95**(18), 187403.

252 Postmus, C., Ferraro, J.F., Mitra, S.S. (1968) *Phys. Rev.*, 174, 983.

253 Calizo, I., Balandin, A.A., Bao, W., Miao, F. and Lau, C.N. (2007) *Nano Lett.*, **7**, 2645–2649.

254 Raravikar, N.R., Keblinski, P., Rao, A.M., Dresselhaus, M.S., Schadler, L.S., and Ajayan, P.M. (2003) *Phys. Rev. B*, **66**, 235424(1–9).

255 Bassil, A., Puech, P., Tubery, L., Bacsa, W., and Flahaut, E. (2006) *Appl. Phys. Lett.*, **88**, 173113.

256 Tan, P.H., Deng, Y., Zhao, Q., and Cheng, W. (1999) *Appl. Phys. Lett.*, **74**, 1818.

257 Ghosh, S., Calizo, I., Teweldebrhan, D., Pokatilov, E.P., Nika, D.L., Balandin, A.A., Bao, W., Miao, F., and Lau, C.N. (2008) *Appl. Phys. Lett.*, **92**, 151911.

258 Lazzeri, M. and Mauri, F. (2006) *Phys. Rev. Lett.*, **97**, 266407.

259 Popov, V.N. and Lambin, P. (2006) *Phys. Rev. B*, **73**, 085407.

260 Sasaki, K., Saito, R., Dresselhaus, G., Dresselhaus, M.S., Farhat, H., and Kong, J. (2008) *Phys. Rev. B*, **77**, 245441.

261 Farhat, H., Son, H., Samsonidze, G.G., Reich, S., Dresselhaus, M.S., and Kong, J. (2007) *Phys. Rev. Lett.*, **99**, 145506.

262 Yan, J., Henriksen, E.A., Kim, P., and Pinczuk, A. (2008) *Phys. Rev. Lett.*, **101**, 136804.

263 Porezag, D., Frauenheim, T., Köhler, T., Seifert, G., and Kaschner, R. (1995) *Phys. Rev. B*, **51**, 12947.

264 Porezag, D. and Mark Pederson, R. (1996) *Phys. Rev. B*, **54**(11), 7830–7836.

265 Jiang, A. *et al.* (unpublished).

266 Malard, L.M., Nishide, D., Dias, L.G., Capaz, R.B., Gomes, A.P., Jorio, A., Axete, C.A., Saito, R., Achiba, Y., Shinohara, H., and Pimenta, M.A. (2007) *Phys. Rev. B*, **76**, 233412.

267 Samsonidze, G.G., Saito, R., Jiang, J., Grüneis, A., Kobayashi, N., Jorio, A., Chou, S.G., Dresselhaus, G., and Dresselhaus, M.S. (2005) *Functional Carbon Nanotubes: MRS Symposium Proceedings, Boston, December 2004* (eds D.L. Carroll, B. Weisman, S. Roth, and A. Rubio), Materials Research Society Press, Warrendale, PA p. HH7.2.

268 Saito, R., Fujita, M., Dresselhaus, G., and Dresselhaus, M.S. (1992) *Phys. Rev. B*, **46**, 1804–1811.

269 Ouyang, M., Huan, J.L., Cheung, C.L., and Lieber, C.M. (2001) *Science*, **292**, 702.

270 Kavan, L., Dunsch, L., Kataura, H., Oshiyama, A., Otani, M., Okada, S. (2003) *J. Phys. Chem. B*, **107**, 7666–7675.

271 Nguyen, K.T., Gaur, A., and Shim, M. (2007) *Phys. Rev. Lett.*, **98**, 145504.

272 Wu, Y., Maultzsch, J., Knoesel, E., Chandra, B., Huang, M.Y., Sfeir, M.Y., Brus, L.E., Hone, J., and Heinz, T.F. (2007) *Phys. Rev. Lett.*, **99**, 027402.

273 Bushmaker, A.W., Deshpande, V.V., Hsieh, S., Bockrath, M.W., and Cronin, S.B. (2009) *Nano Lett.*, **9**(2), 607–611.

274 Saito, R., Jorio, A., Hafner, J.H., Lieber, C.M., Hunter, M., McClure, T., Dresselhaus, G., Dresselhaus, M.S. (2001) *Phys. Rev. B*, **64**, 085312–085319.

275 Sasaki, K., Saito, R., Dresselhaus, G., Dresselhaus, M.S., Farhat, H., and Kong, J. (2008) *Phys. Rev. B*, **78**, 235405-1–11.

276 Tsang, J.C., Freitag, M., Perebeinos, V., Liu, J., and Avouris, P. (2007) *Nat. Nanotech.*, **2**, 725.

277 Souza Filho, A.G., Jorio, A., Samsonidze, G.G., Dresselhaus, G., Saito, R., and Dresselhaus, M.S. (2003) *Nanotechnology*, **14**, 1130–1139.

278 Saito, R., Takeya, T., Kimura, T., Dresselhaus, G., and Dresselhaus, M.S. (1998) *Phys. Rev. B*, **57**, 4145–4153.

279 Mahan, G.D. (2002) *Phys. Rev. B*, **65**, 235402.

280 Venkateswaran, U.D., Masica, D.L., Sumanasekera, G.U., Furtado, C.A., Kim, U.J., and Eklund, P.C. (2003) *Phys. Rev. B*, **68**, 241406(R).

281 Hata, K., Futaba, D.N., Mizuno, K., Namai, T., Yumura, M., and Iijima, S. (2004) *Science*, **306**, 1362–1365.

282 Araujo, P.T., Pesce, P.B.C., Dresselhaus, M.S., Sato, K., Saito, R., Jorio, A. (2010) *Physica E*, **42**(5), 1251–1261.

283 Hartschuh, A., Pedrosa, H.N., Novotny, L., and Krauss, T.D. (2003) *Science*, **301**, 1354–1356.

284 Strano, M.S. (2003) *J. Am. Chem. Soc.*, **125**, 16148–16153.

285 Doorn, S.K., Heller, D.A., Barone, P.W., Usrey, M.L., and Strano, M.S. (2004) *Appl. Phys. A*, **78**, 1147.

286 Paillet, M., Ponchara, P., and Zahab, A. (2006) *Phys. Rev. Lett.*, **96**, 039704.

287 Araujo, P.T., Doorn, S.K., Kilina, S., Tretiak, S., Einarsson, E., Maruyama, S., Chacham, H., Pimenta, M.A., and Jorio, A. (2007) Third and fourth optical transitions in semiconducting carbon nanotubes. *Phys. Rev. Lett.*, **98**, 067401.

288 Bachilo, S.M., Balzano, L., Herrera, J.E., Pompeo, F., Resasco, D.E., and Weisman, R.B. (2003) *J. Am. Chem. Soc.*, **125**, 11186–11187.

289 Villalpando-Paez,F., Son, H., Nezich, D., Hsieh, Y.P., Kong, J., Kim, Y.A., Shimamoto, D., Muramatsu, H., Hayashi, T., Endo, M., Terrones, M., and Dresselhaus, M.S. (2008) *Nano Lett.*, **8**, 3879–3886.

290 Villalpando-Paez, F., Muramatsu, H., Kim, Y.A., Farhat, H., Endo, M.M., Terrones, M., and Dresselhaus, M.S. (2009) *Nanoscale*, **2**, 406–411.

291 Pfeiffer, R., Pichler, T., Kim, Y.A., and Kuzmany, H. (2008) Double-wall carbon nanotubes, in *Springer Series on Topics in Appl. Phys.*, vol. 111, (eds A. Jorio, M.S. Dresselhaus, and G. Dresselhaus), Springer-Verlag, Berlin, pp. 495–530.

292 Pfeiffer, R., Kramberger, C., Simon, F., Kuzmany, H., Popov, V.N., and Kataura, H. (2004) *Eur. Phys. J. B*, **42**(3), 345–350.

293 Pfeiffer, R., Simon, F., Kuzmany, H., and Popov, V.N. (2005) *Phys. Rev. B*, **72**(16), 161404.

294 Pfeiffer, R., Simon, F., Kuzmany, H., Popov, V.N., Zolyomi, V., and Kurti, J. (2006) *Phys. Status Solidi (b)*, **243**(13), 3268–3272.

295 Pfeiffer, R., Peterlik, H., Kuzmany, H., Simon, F., Pressi, K., Knoll, P., Rummeli, M.H., Shiozawa, H., Muramatsu, H., Kim, Y.A., Hayashi, T., and Endo, M. (2008) *Phys. Status Solidi (b)*, **245**(10), 1943–1946.

296 Kuzmany, H., Plank, W., Pfeiffer, R., and Simon, F. (2008) *J. Raman Spectrosc.*, **39**(2), 134–140.

297 Zhao, X., Ando, Y., Qin, L.-C., Kataura, H., Maniwa, Y., and Saito, R. (2002) *Chem. Phys. Lett.*, **361**, 169–174.

298 Soares, J.S., Barboza, A.P.M., Araujo, P.T., Barbosa Neto, N.M., Nakabayashi, D., Shadmi, N., Yarden, T.S., Ismach, A., Geblinger, N., Joselevich, E., Vilani, C., Cançado, L.G., Novotny, L., Dresselhaus, G., Dresselhaus, M.S., Neves, B.R.A., Mazzoni, M.S.C., Jorio, A. (2010) submitted.

299 Jorio, A., Fantini, C., Pimenta, M.A., Capaz, R.B., Samsonidze, G.G., Dresselhaus, G., Dresselhaus, M.S., Jiang, J., Kobayashi, N., Grüneis, A., and Saito, R. (2005) *Phys. Rev. B*, **71**, 075401.

300 Kürti, J., Zólyomi, V., Kertesz, M., and Sun, G.Y. (2003) *New J. Phys.*, **5**, 125.

301 Farhat, H., Sasaki, K., Kalbac, M., Hofmann, M., Saito, R., Dresselhaus, M.S., and Kong, J. (2009) *Phys. Rev. Lett.*, **102**, 126804(1–4).

302 Jorio, A., Souza Filho, A.G., Dresselhaus, G., Dresselhaus, M.S., Saito, R., Hafner, J.H., Lieber, C.M., Matinaga, F.M., Dantas, M.S.S., and Pimenta, M.A. (2001) *Phys. Rev. B*, **63**, 245416–(1–4).

303 Brown, S. D. M., Corio, P., Marucci, A., Dresselhaus, M. S., Pimenta, M. A., and Kneipp, K. (2000) *Phys. Rev. B Rapid*, **61**, R5137–R5140.

304 Duesberg, G.S., Blau, W.J., Byrne, H.J., Muster, J., Burghard, M., and Roth, S. (1999) *Chem. Phys. Lett.*, **310**, 8–14.

305 Azoulay, J., Debarre, A., Richard, A., and Tchenio, P. (2000) *J. Phys. IV France*, **10**, Pr8–223.

306 Kneipp, K., Kneipp, H., Corio, P., Brown, S.D.M., Shafer, K., Motz, J., Perelman, L.T., Hanlon, E.B., Marucci, A., Dresselhaus, G., and Dresselhaus, M.S. (2000) *Phys. Rev. Lett.*, **84**, 3470–3473.

307 Hafner, J.H., Cheung, C.L., Oosterkamp, T.H., and Lieber, C.M. (2001) *J. Phys. Chem. B*, **105**, 743.

308 Park, J.S., Oyama, Y., Saito, R., Izumida, W., Jiang, J., Sato, K., Fantini, C., Jorio, A., Dresselhaus, G., and Dresselhaus, M.S. (2006) *Phys. Rev. B*, **74**, 165414.

309 Rafailov, P.M., Jantoliak, H., and Thomsen, C. (2000) *Phys. Rev. B*, **61**, 16179–16182.

310 Souza Filho, A.G., Jorio, A., Hafner, J.H., Lieber, C.M., Saito, R., Pimenta, M.A., Dresselhaus, G., and Dresselhaus, M.S. (2001) *Phys. Rev. B*, **63**, 241404R.

311 Ando, T. (2005) *J. Phys. Soc. Jpn.*, **74**, 777–817.

312 Miyauchi, Y., Oba, M., and Maruyama, S. (2006) *Phys. Rev. B*, **74**, 205440.

313 Jorio, A., Pimenta, M.A., Fantini, C., Souza, M., Souza Filho, A.G., Samsonidze, G.G., Dresselhaus, G., Dresselhaus, M.S., and Saito, R. (2004) *Carbon*, **42**, 1067–1069.

314 Wang, Y., Kempa, K., Kimball, B., Carlson, J.B., Benham, G., Li, W.Z., Kempa, T., Rybcznski, J., Herczynski, A., and Ren, Z.F. (2004) *Appl. Phys. Lett.*, **85**, 2607.

315 Uryu, S. and Ando, T. (2006) *J. Phys. Soc. Jpn.*, **75**, 024707.

316 Grüneis, A., Saito, R., Jiang, J., Samsonidze, G.G., Pimenta, M.A., Jorio, A., Souza Filho, A.G., Dresselhaus, G. and Dresselhaus, M.S. (2004) *Chem. Phys. Lett.*, **387**, 301–306.

317 Araujo, P.T. and Jorio, A. (2008) *Phys. Status Solidi (b)*, **245**, 2201–2204.

318 Doorn, S.K., Araujo, P.T., Hata, K., and Jorio, A. (2008) *Phys. Rev. B*, **78**, 165408.

319 Ando, T. (1997) *J. Phys. Soc. Jpn.*, **66**, 1066–1073.

320 Spataru, C.D., Ismail-Beigi, S., Capaz, R.B., and Louie, S.G. (2005) *Phys. Rev. Lett.*, **95**, 247402.

321 Spataru, C.D., Ismail-Beigi, S., Benedict, L.X., and Louie, S.G. (2004) *Phys. Rev. Lett.*, **92**, 077402.

322 Spataru, C.D., Ismail-Beigi, S., Benedict, L.X., and Louie, S.G. (2004) *Appl. Phys. A*, **78**, 1129–1136.

323 Chang, E., Bussi, G., Ruini, A., and Molinari, E. (2004) *Phys. Rev. Lett.*, **92**, 113410.

324 Perebeinos, V., Tersoff, J., and Avouris, P. (2004) *Phys. Rev. Lett.*, **92**, 257402.

325 Perebeinos, V., Tersoff, J., and Avouris, P. (2005) *Phys. Rev. Lett.*, **94**, 027402.

326 Pedersen, T.G. (2003) *Phys. Rev. B*, **67**, 073401.

327 Zhao, H. and Mazumdar, S. (2005) *Synth. Met.*, **155**, 250.

328 Capaz, R.B., Spataru, C.D., Ismail-Beigi, S., and Louie, S.G. (2006) *Phys. Rev. B*, **74**, 121401(R).

329 Popov, V.N. (2004) *New J. Phys.*, **6**, 17.

330 Reich, S., Thomsen, C., and Ordejón, P. (2002) *Phys. Rev. B*, **65**, 155411.

331 Koster, G.F. (1957) *Solid State Physics*, Academic Press, New York, NY.

332 Rohlfing, M. and Louie, S.G. (2000) *Phys. Rev. B*, **62**, 4927.

333 Knupfer, M., Schwieger, T., Fink, J., Leo, K., and Hoffmann, M. (2002) *Phys. Rev. B*, **66**, 035208.

334 Pichler, T. (2007) *Nat. Mater.*, **6**, 332–333.

335 Gunnarsson, O. (2004) *Alkali-doped Fullerides*, World Scientific, Singapore, p. 282.

336 Saito, R., Sato, K., Oyama, Y., Jiang, J., Samsonidze, G.G., Dresselhaus, G., and Dresselhaus, M.S. (2005) *Phys. Rev. B*, **71**, 153413.

337 Knox, R.S. (1963) Theory of excitons, in *Solid State Physics*, suppl. 5 (eds F. Seitz, D. Turnbull, and H. Ehrenreich), Academic Press, New York.

338 Saito, R., Fujita, M., Dresselhaus, G., and Dresselhaus, M.S. (1992) *Appl. Phys. Lett.*, **60**, 2204–2206.

339 Wang, F., Dukovic, G., Brus, L.E., and Heinz, T.F. (2005) *Science*, **308**, 838–841.

340 Maultzsch, J., Pomraenke, R., Reich, S., Chang, E., Prezzi, D., Ruini, A., Molinari, E., Strano, M.S., Thomsen, C., and Lienau, C. (2005) *Phys. Rev. B*, **72**, 241402(R).

341 Qiu, X., Freitag, M., Perebeinos, V., and Avouris, P. (2005) *Nano Lett.*, **5**, 749–752.

342 Ando, T. (2009) *J. Phys. Soc. Jpn.*, **78**, 104703.

343 Sato, K., Saito, R., Jiang, J., Dresselhaus, G., and Dresselhaus, M.S. (2007) *Phys. Rev. B*, **76**, 195446.

344 Miyauchi, Y., Saito, R., Sato, K., Ohno, Y., Iwasaki, S., Mizutani, T., Jiang, J., and Maruyama, S. (2007) *Chem. Phys. Lett.*, **442**, 394.

345 Miyauchi, Y. and Maruyama, S. (2006) *Phys. Rev. B*, **74**, 035415.

346 Murakami, Y., Einarsson, E., Edamura, T., and Maruyama, S. (2005) *Carbon*, **43**, 2664.

347 Iwasaki, S., Ohno, Y., Murakami, Y., Kishimoto, S., Maruyama, S., and Mizutani, T. (2006) unpublished. paper at APS March Meeting, Baltimore, Maryland, 13–17, March, 2006.

348 Ohno, Y., Iwasaki, S., Murakami, Y., Kishimoto, S., Maruyama, S., and Mizutan, T. (2006) *Phys. Rev. B*, **73**, 235427.

349 Pesce, P.B.C., Araujo, P.T., Nikolaev, P., Doorn, S.K., Hata, K., Saito, R., Dres-

selhaus, M.S., and Jorio, A. (2010) *Appl. Phys. Lett.*, **96**, 051910.

350 Grüneis, A., Attaccalite, C., Wirtz, L., Shiozawa, H., Saito, R., Pichler, T., and Rubio, A. (2008) *Phys. Rev. B*, **78**, 205425-1–16.

351 Bostwick, A., Ohta, T., McChesney, J.L. (2007) *Solid State Commun.*, **143**, 63–71.

352 Guha, S., Menéndez, J., Page, J.B., and Adams, G.B. (1996) *Phys. Rev. B*, **53**, 13106.

353 Chantry, G.W. (1971) *The Raman Effect*, Dekker, New York, NY.

354 Snoke, D.W., Cardona, M., Sanguinetti, S., and Benedek, G. (1996) *Phys. Rev. B*, **53**, 12641.

355 Zimmermann, J., Pavone, P., and Cuniberti, G. (2008) *Phys. Rev. B*, **78**, 045410.

356 Madelung, O. (1978) *Solid State Theory*, Springer-Verlag, Berlin.

357 Oyama, Y., Saito, R., Sato, K., Jiang, J., Samsonidze, G.G., Grüneis, A., Miyauchi, Y., Maruyama, S., Jorio, A., Dresselhaus, G., and Dresselhaus, M.S. (2006) *Carbon*, **44**, 873–879.

358 Saito, R., Grüneis, A., Samsonidze, G.G., Dresselhaus, G., Dresselhaus, M.S., Jorio, A., Cançado, L.G., Pimenta, M.A., and Souza, A.G. (2004) *Appl. Phys. A*, **78**, 1099–1105.

359 Jiang, J., Saito, R., Grüneis, A., Dresselhaus, G., and Dresselhaus, M.S. (2004) *Carbon*, **42**, 3169–3176.

360 Jiang, J., Saito, R., Grüneis, A., Dresselhaus, G., and Dresselhaus, M.S. (2004) *Chem. Phys. Lett.*, **392**, 383–389.

361 Jiang, J., Saito, R., Grüneis, A., Chou, S.G., Samsonidze, G.G. Jorio, A., Dresselhaus, G., and Dresselhaus, M.S. (2005) *Phys. Rev. B*, **71**, 045417-1–9.

362 Grüneis, A. (2004) Resonance Raman spectroscopy of single wall carbon nanotubes. PhD thesis, Tohoku University, Sendai, Japan, Department of Physics.

363 Saito, R. and Kamimura, H. (1983) *J. Phys. Soc. Jpn.*, **52**, 407.

364 Fantini, C., Jorio, A., Souza, M., Saito, R., Samsonidze, G.G., Dresselhaus, M.S., and Pimenta, M.A. (2005) *Phys. Rev. B*, **72**, 085446.

365 Sanders, G.D. (2009) *Phys. Rev. B*, **79**, 205434.

366 Lazzeri, M., Attaccalite, C., and Mauri, F. (2008) *Phys. Rev. B*, **78**, 081406(R).

367 Cançado, L.G., Pimenta, M.A., Saito, R., Jorio, A., Ladeira, L.O., Grüneis, A., Souza Filho, A.G., Dresselhaus, G., and Dresselhaus, M.S. (2002) *Phys. Rev. B*, **66**, 035415.

368 Saito, R., Grüneis, A., Cançado, L.G., Pimenta, M.A., Jorio, A., Souza Filho, A.G., Dresselhaus, M.S., and Dresselhaus, G. (2002) *Mol. Cryst. Liq. Cryst.*, **387**, 287–296.

369 Basko, D.M.B. (2007) *Phys. Rev. B*, **76**, 081405(R).

370 Basko, D.M.B. (2008) *Phys. Rev. B*, **78**, 125418.

371 Maultzsch, J., Reich, S., and Thomsen, C. (2004) *Phys. Rev. B*, **70**, 155403.

372 Mafra, D.L., Samsonidze, G., Malard, L.M., Elias, D.C., Brant, J.C., Plentz, F., Alves, E.S., and Pimenta, M.A. (2007) *Phys. Rev. B*, **76**, 233407.

373 Shimada, T., Sugai, T., Fantini, C., Souza, M., Cançado, L.G., Jorio, A., Pimenta, M.A., Saito, R., Grüneis, A., Dresselhaus, G., Dresselhaus, M.S., Ohno, Y., Mizutani, T., and Shinohara, H. (2005) *Carbon*, **43**, 1049–1054.

374 Park, J.S., Cecco, A.R., Saito, R., Jiang, J., Dresselhaus, G., and Dresselhaus, M.S. (2009) *Carbon*, **47**, 1303–1310.

375 Cançado, L.G., Cecco, A.R., Kong, J., and Dresselhaus, M.S. (2008) *Phys. Rev. B*, **77**, 245408.

376 Cançado, L.G., Takai, K., Enoki, T., Endo, M., Kim, Y.A., Mizusaki, H., Jorio, A., Coelho, L.N., Magalhaes-Paniago, R., and Pimenta, M.A. (2006) *Appl. Phys. Lett.*, **88**, 3106.

377 Reina, A., Son, H., Liying Jiao, Fan, B., Dresselhaus, M.S., Liu, Z.F., and Kong, J. (2008) *J. Phys. Chem. C Lett.*, **112**, 17741–17749.

378 Reina, A., Thiele, S., Jia, X., Bhaviripudi, S., Dresselhaus, M.S., Schaefer, J.A., and Kong, J. (2009) *Nano Res.*, **2**, 509–516.

379 Tan, P., Hu, C.Y., Dong, J., Shen, W.C., and Zhang, B.F. (2001) *Phys. Rev. B*, **64**, 214301.

380 Brar, V.W., Samsonidze, G.G. Dresselhaus, G., Dresselhaus, M.S., Saito, R., Swan, A.K., Ünlü, M.S., Goldberg, B.B.,

Souza Filho, A.G., and Jorio, A. (2002) *Phys. Rev. B*, **66**, 155418.
381 Fantini, C., Jorio, A., Souza, M., Ladeira, L.O., Pimenta, M.A., Souza Filho, A.G., Saito, R., Samsonidze, G.G., Dresselhaus, G., and Dresselhaus, M.S. (2004) *Phys. Rev. Lett.*, **93**, 087401.
382 Fantini, C., Jorio, A., Souza, M., Saito, R., Samsonidze, G.G., Dresselhaus, M.S., and Pimenta, M.A. (2005) *Phys. Rev. B*, **72**, 5446.
383 Saito, R., Takeya, T., Kimura, T., Dresselhaus, G., and Dresselhaus, M.S. (1999) *Phys. Rev. B*, **59**, 2388–2392.
384 Chou, S.G., Son, H., Zheng, M., Saito, R., Jorio, A., Kong, J., Dresselhaus, G., and Dresselhaus, M.S. (2007) *Chem. Phys. Lett.*, **443**, 328–332.
385 Villalpando-Paez, F., Zamudio, A., Elias, A.L., Son, H., Barros, E.B., Chou, S.G., Kim, Y.A., Muramatsu, H., Hayashi, T., Kong, J., Terrones, H., Dresselhaus, G., Endo, M., Terrones, M., and Dresselhaus, M.S. (2006) *Chem. Phys. Lett.*, **424**, 345–352.
386 Villalpando-Paez, F., Son, H., Nezich, D., Hsieh, Y.P, Kong, J., Kim, Y.A., Shimamoto, D., Muramatsu, H., Hayashi, T., Endo, M., Terrones, M., and Dresselhaus, M.S. (2009) *Nano Lett.*, **8**, 3879–3886.
387 Tan, P.H., An, L., Liu, L.Q., Guo, Z.X., Czerw, R., Carroll, D.L., Ajayan, P.M., Zhang, N., and Guo, H.L. (2002) *Phys. Rev. B*, **66**, 245410.
388 Souza Filho, A.G., Jorio, A., Dresselhaus, G., Dresselhaus, M.S., Saito, R., Swan, A.K., Ünlü, M.S., Goldberg, B.B., Hafner, J.H., Lieber, C.M., and Pimenta, M.A. (2002) *Phys. Rev. B*, **65**, 035404-(1–6).
389 Souza Filho, A.G., Jorio, A., Samsonidze, G.G., Dresselhaus, G., Pimenta, M.A., Dresselhaus, M.S., Swan, A.K., Ünlü, M.S., Goldberg, B.B., and Saito, R. (2003) *Phys. Rev. B*, **67**, 035427(1–7).
390 Samsonidze, G.G., Saito, R., Jorio, A., Souza Filho, A.G., Grüneis, A., Pimenta, M.A., Dresselhaus, G., and Dresselhaus, M.S. (2003) *Phys. Rev. Lett.*, **90**, 027403.
391 Souza Filho, A.G., Jorio, A., Swan, A.K., Ünlü, M.S., Goldberg, B.B., Saito, R., Hafner, J.H., Lieber, C.M., Pimenta, M.A., Dresselhaus, G., and Dresselhaus, M.S. (2002) *Phys. Rev. B*, **65**, 085417.
392 Souza Filho, A.G., Jorio, A., Samsonidze, G.G., Dresselhaus, G., Dresselhaus, M.S., Swan, A.K., Ünlü, M.S., Goldberg, B.B., Saito, R., Hafner, J.H., Lieber, C.M., and Pimenta, M.A. (2002) *Chem. Phys. Lett.*, **354**, 62–68.
393 Saito, R., Jorio, A., Souza Filho, A.G., Dresselhaus, G., Dresselhaus, M.S., Grüneis, A., Cançado, L.G., and Pimenta, M.A. (2002) *Japan. J. Appl. Phys.*, **41**, 4878–4882.
394 Pimenta, M.A., Dresselhaus, G., Dresselhaus, M.S., Cançado, L.G., Jorio, A., and Saito, R. (2007) *Phys. Chem. Chem. Phys.*, **9**, 1276–1291.
395 Sato, K., Saito, R., Oyama, Y., Jiang, J., Cançado, L.G., Pimenta, M.A., Jorio, A., Samsonidze, G.G., Dresselhaus, G., and Dresselhaus, M.S. (2006) *Chem. Phys. Lett.*, **427**, 117–121.
396 Ando, T. and Nakkanishi, T. (1998) *J. Phys. Soc. Jpn.*, **67**, 1704.
397 Ando, T., Nakanishi, T., and Saito, R. (1998) *J. Phys. Soc. Jpn.*, **67**, 2857–2862.
398 Pócsik, I., Hundhausen, M., Koós, M., and Ley, L. (1998) *J. Non-Crystall. Solids*, **227–230**, 1083–1086.
399 Wang, Y., Aolsmeyer, D.C., and McCreery, R.L. (1990) *Chem. Mater.*, **2**, 557.
400 Matthews, M.J., Pimenta, M.A., Dresselhaus, G., Dresselhaus, M.S., and Endo, M. (1999) *Phys. Rev. B*, **59**, R6585.
401 Dresselhaus, M.S. and Kalish, R. (1992) Ion implantation in diamond, graphite and related materials, in *Series in Materials Science*, vol. 22, Springer-Verlag, Berlin.
402 Jorio, A., Lucchese, M.M., Stavale, F., Achete, C.A. (2009) *Phys. Status Solidi (b)*, **246**, 2689–2692.
403 Knight, D.S. and White, W.B. (1989) *J. Mater. Res.*, **4**, 385.
404 Cançado, L.G., Takai, K., Enoki, T., Endo, M., Kim, Y.A., Mizusaki, H., Speziali, N.L., Jorio, A., and Pimenta, M.A. (2008) *Carbon*, **46**, 272–275.
405 Cancado, L.G., Jorio, A., Pimenta, M.A. (2007) *Phys. Rev. B*, **76**, 064304.
406 Takai, K., Oga, M., Sato, H., Enoki, T., Ohki, Y., Taomoto, A., Suenaga, K., and Iijima, S. (2003) *Phys. Rev. B*, **67**, 214202.

407 Casiraghi, C., Hartschuh, A., Qian, H., Piscanec, S., Georgi, C., Fasoli, A., Novoselov, K.S., Basko, D.M., and Ferrari, A.C. (2009) Raman Spectroscopy of Graphene Edges. *Nano Lett.*, 9(4), 1433–1441.

408 Gupta, A.K., Russin, T.J., Gutierrez, H.R., and Eklund, P.C. (2009) Probing graphene edges via Raman scattering. *ACS Nano.*, **3**, 45–52.

409 Zolyomi, V., Kurti, J., Grüneis, A., Kuzmany, H. (2003) Origin of the Fine Structure of the Raman D Band in Single-Wall Carbon Nanotubes. *Phys. Rev. Lett.*, **90**, 157401.

410 Brown, S.D.M., Jorio, A., Dresselhaus, M.S., and Dresselhaus, G. (2001) *Phys. Rev. B*, **64**, 073403–(1–4).

411 Hulman, M., Skakalova, V., Roth, S., and Kuzmany, H. (2005) *J. Appl. Phys.*, **98**, 024311.

412 Chou, S.G., Son, H., Kong, J., Jorio, A., Saito, R., Zheng, M., Dresselhaus, G., and Dresselhaus, M.S. (2007) *Appl. Phys. Lett.*, **90**, 131109.

413 Maultzsch, J., Reich, S., and Thomsen, C. (2001) *Phys. Rev. B*, **64**, 121407(R).

414 Maultzsch, J., Reich, S., Schlecht, U., and Thomsen, C. (2003) *Phys. Rev. Lett.*, **91**, 087402.

415 Souza, M., Jorio, A., Fantini, C., Neves, B.R.A., Pimenta, M.A., Saito, R., Ismach, A., Joselevich, E., Brar, V.W., Samsonidze, G.G., Dresselhaus, G., and Dresselhaus, M.S. (2004) *Phys. Rev. B*, **69**, R15424-1–4.

416 Novotny, L. and Hecht, B. (2006) *Principles of Nano-Optics*, Cambridge University Press, Cambridge, UK.

417 Maciel, I.O., Pimenta, M.A., Terrones, M., Terrones, H., Campos-Delgado, J., and Jorio, A. (2008) *Phys. Status Solidi (b)*, **254**, 2197–2200.

418 Maciel, I.O., Campos-Delgado, J., Cruz-Silva, E., Pimenta, M., Sumpter, B.G., Meunier, V., Lopez-Uriaz, F., Munoz-Sandoval, E., Terrones, H., Terrones, M., and Jorio, A. (2009) *Nano Lett.*, **9**, 2267–2272.

419 Maciel, I.O., Campos-Delgado, J., Pimenta, M.A., Terrones, M., Terrones, H., Rao, A.M., Jorio, A. (2009) *Phys. Status Solidi (b)*, **246**(11–12), 2432–2435.

420 Tang, Z.K., Zhai, J.P., Tong, Y.Y., Hu, X.J., Saito, R., Feng, Y.J., and Sheng, P. (2008) *Phys. Rev. Lett.*, **101**, 047402.

Index

Raman Spectroscopy in Graphene Related Systems. Ado Jorio, Riichiro Saito,
Gene Dresselhaus, and Mildred S. Dresselhaus

ISBN: 978-3-527-40811-5